AF279175

How did an Austrian-born misfit who had never risen higher in military service than the rank of lance-corporal attain mastery over Germany and most of Europe? Much of that dubious credit can be attributed to the actions of his earliest paramilitary army, the *Sturmabteilungen* (SA, Storm Troops), and the men chosen by the *Führer* to lead it. This series analyses the lives and careers of those men, the first volume covering 49 officers, 35 of whom were, like their leader, veterans of the First World War who had found themselves stunned, bitterly disillusioned, and in many cases unemployed and destitute in the aftermath of that four-year struggle. They eagerly sought the opportunity to return to uniform, battled the enemies of the Nazi Party in the streets of interwar Germany, and saw their efforts rewarded by their own leader's betrayal, as he essentially decapitated his SA in favour of its own subordinate formation, Heinrich Himmler's SS, in the 'Night of the Long Knives' (30 June–1 July 1934). But the SA did not end with that devastating blow, and despite its loss of prestige and power it was to play an important role in military training and internal security within and outside the borders of the Reich. During World War II, many of its leaders were tasked with administering occupied territories and representing Germany as ambassadors to other Axis nations. Still others, men of all SA ranks, served individually as members of the German armed forces, tens of thousands of them losing their lives on all fronts and many of them receiving the highest awards for bravery and leadership. Relying primarily on contemporary documentation, including the official personnel files of these men, Michael Miller and Andreas Schulz have compiled the first in-depth study yet produced on the SA leadership corps, a series designed to provide as comprehensive a picture as possible of the *hauptamtlicher* (full-time, actively serving) and *ehrenamtlicher* (honorary) *SA-Führer*.

Michael Miller was born in Torrance, California on 28 May 1971. He served in the U.S. Navy from 1989 to 1993, achieving the rank of Yeoman Petty Officer 3rd Class with assignments to USS *Orion* (AS-18) and the Port Services Department of Commander, Submarine Squadron 22 in La Maddalena, Sardinia as well as Fleet Air Reconnaissance Squadron TWO (VQ-2) in Rota, Spain. His interest in the Second World War dates back to 1978. Founder of the website *Axis Biographical Research* in 1999, he is the author or coauthor of the following:

Leaders of the SS & German Police, Volume I: Reichsführer-SS – SS-Gruppenführer (Georg Ahrens to Karl Gutenberger). R. James Bender Publishing, 2006.

Knight's Cross Holders of the SS & German Police, 1940-1945. Self-published CD-Rom, 2009.

With Andreas Schulz:
The SS-Brigadeführer, 1933-1945. Self-published CD-Rom, 2004.
Gauleiter: The Regional Leaders of the Nazi Party and Their Deputies, Volume I (Herbert Albrecht – H. Wilhelm Hüttmann. R. James Bender Publishing, 2012.
Leaders of the SS & German Police, Volume II: Reichsführer-SS – SS-Gruppenführer (Hans Haltermann to Walter Krüger). R. James Bender Publishing, 2015.

Andreas Schulz was born in Stendal (Altmark) on 12 April 1965. After completing his studies in Magdeburg, he worked as a hydraulic engineer. For over a decade, he has served in the Bundespresseamt (Federal Press Office) of the German Government. A diligent and passionate research historian, he is the coauthor of the following works:

With Joachim Lilla and Martin Döring:
Statisten in Uniform. Die Mitglieder des Reichstags 1933-1945. Droste Verlag, 2004.

With Dr. med. Dieter Zinke:
Die Generale der Waffen-SS und der Polizei 1933-1945, Band 1 (Abraham-Gutenberger). Biblio-Verlag, 2003.
Die Generale der Waffen-SS und der Polizei 1933-1945, Band 2 (Hachtel-Kutschera). Biblio-Verlag, 2005.
Die Generale der Waffen-SS und der Polizei 1933-1945, Band 3 (Lammerding-Plesch). Biblio-Verlag, 2008.
Die Generale der Waffen-SS und der Polizei 1933-1945, Band 4 (Podzun-Schimana). Biblio-Verlag, 2009.
Die Generale der Waffen-SS und der Polizei 1933-1945, Band 5 (Schlake-Turner). Biblio-Verlag, 2011.
Die Generale der Waffen-SS und der Polizei 1933-1945, Band 6 (Ullmann-Zottmann). Biblio-Verlag, 2012.

With Michael D. Miller:
The SS-Brigadeführer, 1933-1945. Self-published CD-Rom, 2004.
Gauleiter: The Regional Leaders of the Nazi Party and Their Deputies, Volume I (Herbert Albrecht – H. Wilhelm Hüttmann. R. James Bender Publishing, 2012.

LEADERS OF THE STORM TROOPS VOLUME 1

OBERSTER SA-FÜHRER and *SA-STABSCHEF*

and

SA-OBERGRUPPENFÜHRER (B – J)

Michael D Miller and Andreas Schulz

Helion & Company

Helion & Company Limited
26 Willow Road
Solihull
West Midlands
B91 1UE
England
Tel. 0121 705 3393
Fax 0121 711 4075
Email: info@helion.co.uk
Website: www.helion.co.uk
Twitter: @helionbooks
Visit our blog http://blog.helion.co.uk/

Published by Helion & Company 2015

Designed and typeset by Farr out Publications, Wokingham, Berkshire
Cover designed by Paul Hewitt, Battlefield Design (www.battlefield-design.co.uk)
Printed by Lightning Source Ltd, Milton Keynes, Buckinghamshire

Text © Michael D. Miller & Andreas Schulz 2015
Photographs © as individually credited

Front cover: *SA-Stabschef* Viktor Lutze, ca. 1938, in a formal portrait taken ca. 1938. (Heinrich Hoffmann collection, NARA). Rear cover: *SA-Obergruppenführer* Franz Ritter von Epp and *SA-Gruppenführer* Friedrich Haselmayr in Nürnberg, September 1933. (Victor Gross Militärische Antiquitäten)

ISBN 978-1-909982-87-1

British Library Cataloguing-in-Publication Data.
A catalogue record for this book is available from the British Library.

For details of other military history titles published by Helion & Company Limited contact the above address, or visit our website: http://www.helion.co.uk.

We always welcome receiving book proposals from prospective authors.

Contents

Foreword

The reference works by Michael Miller and Andreas Schulz, painstakingly compiled on the basis of meticulous research, are a great help to historians of the Third Reich and to anyone wishing to learn precise biographical and career details of the Nazi leadership. This latest volume, providing a plethora of personal data on the *SA* leadership, is no exception. I know of no comparable compilation. It would have saved me much endeavour had it been available when I was writing my biography of Hitler. I'm sure it will now be warmly welcomed, as it deserves to be, by scholars and students of Nazi Germany, and I'm very happy to record my own pleasure at its publication.

Sir Ian Kershaw

Acknowledgements

The following individuals provided varying degrees of invaluable assistance to us in the creation of this and future volumes.

Stefan Barber
Roger Bender
Patrick Butcher
Prof. Randall Bytwerk
Prof. Bruce Campbell
Jeff Clark
Mike Constandy
Mark Costa
Howard Davies
Ann Farr
Frederick Froberg
Victor Gross Militärische Antiquitäten
Richard Hargreaves
Laurens Hessels
Prof. Peter Hoffmann
Georges Jerome
Glenn Jewison.
Igor Karpov
Sten Lundø
Rick Lundström
Ken McCanliss
Prof. Donald M McKale
Gary Merlie
Mary Miller
Hugo Morsink
Phil Nix (1938-2013)
Duncan Rogers
Ian Sayer
Michal Sika
Thomas Rief & Hermann-Historica Auctioneers, München
Piotr Szarafinski
Andrew Wackerfuss, PhD
James Webb
Marcus Wendel
Max Williams
Mark C.Yerger
Dr. med. Dieter Zinke

Commonly used abbreviations

Abbreviation	German Meaning	English Meaning
a. D.	*ausser Dienst*	Retired; on the inactive list
d. R.	*der Reserve*	Of the reserves [term following *Wehrmacht* or Waffen-SS rank, e.g. "*Oberst d. R.*"]
Dr. h. c.	*doctor honoris causa*	Honorary doctor's degree
Dr. Ing.	*Doktor Ingenieur*	Doctor of Engineering
Dr. jur.	*doctor juris*	Doctor of Jurisprudence / Law
Dr. med.	*doctor medicinae*	Doctor of Medicine
Dr. phil.	*doctor philosophiae*	Doctotr of Philosophy
Dr. rer. Pol.	*doctor rerum politicarum*	Doctor of Political Science
e. h.	*ehrenhalber*	Honorary
e. V.	*eingetragener Verein*	A registered society (e.g. "Lebensborn e.V.")
Gestapo	*Geheime Staatspolizei*	Secret State Police
HSSPF	*Höhere SS-und Polizeiführer*	Higher SS and Police Leader
i. G.	*im Generalstab*	In the General Staff [term following *Wehrmacht* rank, e.g. "*Major* i. G."]
Kripo	*Kriminalpolizei*	Criminal Police
k.u.k.	*kaiserlich und königlich (k.u.k.)*	Of or pertaining to the Imperial and Royal armed forces of the Austro-Hungarian Empire
m.d.F.b.	*mit der Führung beauftragt*	Temporarily charged with leadership or command
m.d.F.d.G.b.	*mit der Führung der Geschäfte beauftragt*	Temporarily charged with the leadership of affairs
m.d.W.d.G.b.	*mit der Wahrnehmung des Geschäfte beauftragt*	Temporarily charged with the conduct of affairs
NSDAP	*Nationalsozialistische Deutsche Arbeiter Partei*	National-Socialist German Workers Party
NSKK	*Nationalsozialistische-Kraftfahrkorps*	National-Socialist Motor Corps
NSV	*Nationalsozialistische Volkswohlfahrt*	National-Socialist People's Welfare Organization
SA	*Sturmabteilungen*	Storm Troops
SD	*Sicherheitsdienst*	Security Service
Sipo	*Sicherheitspolizei*	Security Police
SS	*Schutzstaffel*	Lit. Protection or Guard Detachment
ß	*Eszett (ss)*	Not an abbreviation per se, but a character which represents the "scharfes S" (sharp S) in certain German words, such as "preußisch" (Prussian) and "Stoßtrupp" (Shock Troop)

Introduction

How did Adolf Hitler, who had never risen higher in military service than the rank of lance-corporal, attain mastery over Germany and most of Europe? Much of that dubious credit can be attributed to the actions of his earliest paramilitary army, the *Sturmabteilungen* (*SA*, Storm Troops), and the men chosen by the *Führer* to lead it. This series analyzes the lives and careers of those men, the first volume covering 49 officers, 35 of whom were, like their leader, veterans of the First World War who had found themselves stunned, bitterly disillusioned, and in many cases unemployed and destitute in the aftermath of that four-year struggle. They eagerly sought the opportunity to return to uniform, battled the enemies of the Nazi Party in the streets of interwar Germany, and saw their efforts rewarded by their own leader's betrayal, as he essentially decapitated his *SA* in favor of its own subordinate formation, Heinrich Himmlers' SS, in the so-called "Night of the Long Knives" (30 June-1 July 1934). But the *SA* did not end with that devastating blow, and despite its loss of prestige and power was to play an important role in military training and internal security within and outside the borders of the Reich. During World War II, a number of its leaders were tasked with administering occupied territories and representing Germany's interests as ambassadors to other Axis nations. Still others, men of all *SA* ranks, served individually as members of the German *Wehrmacht*, tens of thousands of them losing their lives on all fronts and many of them receiving the highest awards for bravery and leadership.

This volume, based largely on official personnel files and other contemporary documentation, analyzes the lives and careers of the highest-ranking *SA* leaders. It is not a sociological study of the kinds of men who became *SA* leaders; that was already done 15 years ago (see Bruce Campbell's *The SA Generals and Hitler's Rise to Power*). It is, as are my previous books on leaders of the *SS* and *NSDAP*, simply a compendium of biographical data and photographs, designed to provide as comprehensive a picture as possible of the *hauptamtlicher* (full-time, actively serving) and *ehrenamtlicher* (honorary) *SA-Führer*.

Michael Miller, 2014

Part I

Oberster SA-Führer and *Stabschef der SA*

Adolf Hitler
Oberster SA-Führer
(01.01.1931–30.04.1945)

The supreme leader of the *SA* consecrates the banners of *SA* regiments by touching the *Blutfahne* – the blood-spattered flag carried during the march to the Feldherrnhalle during the "München-Putsch" of 09.11.1923 – to them. The official bearer of the *Blutfahne*, Jakob Grimminger, stands at far left. (NARA photo, from the Heinrich Hoffmann collection)

Viktor Lutze and the top leadership of the *SA*, photographed at the *Reich Propaganda Ministry* in Berlin, 01.11.1934. Pictured from left to right, beginning with the 1st row, are: Otto Marxer, Viktor Lutze, and Karl-Siegmund Litzmann. 2nd row: Max Luyken, Wilhelm Schepmann, Siegfried Kasche, and Hermann Reschny; 3rd row: Otto Schramme, Dietrich von Jagow, Adolf-Heinz Beckerle, Arthur Böckenhauer, and Gustav Zunkel; 4th row: Hans Friedrich, Herbert Fust, Hans-Günther von Obernitz, Heinrich August Knickmann, Otto Herzog, and Arno Manthey; 5th row: Johann Heinrich Böhmcker, Joachim Meyer-Quade, Wilhelm Helfer, Heinrich Schoene, Dr. Emil Ketterer, Max Juttner, and Emil Steinhoff; 6th row: Farthest up on the staircase is Hanns-Elard Ludin, with Arthur Rakobrandt below him. The officer behind Rakobrandt has been identified in another work – perhaps erroneously – as "*SA-Obersturmführer* Morenga" (there was, however, a Jakobus Morenga who led the 1904-1908 Herero uprising against the German colonial authorities in German Southwest Africa).

SA-Stabschef Schepmann presides over an *SA* leadership
conference in 1943. (Igor Karpov photo)

Berlin, 19.11.1943: *SA-Stabschef* Schepmann addresses a conference
of active and honorary *SA* leaders. (NARA photo)

Bevergern / Westfalen, 04.05.1944: *SA-Stabschef* Schepmann and Hans Jüttner (2nd from right) at a memorial ceremony marking the first anniversary of Viktor Lutze's death. At extreme right is the deputy Gauleiter of Westfalen-Nord, Peter Stangier. Behind Frau Lutze is Johann Heinrich Böhmcker. (NARA photo)

Emil ("Moritzl") Maurice

12.11.1920–00.11.1921(?)	*Organisator und Führer der "Turn-und Sportabteilung" der NSDAP / Organisator der SA*

Born:	19.01.1897 in Westermoor / Kreis Eckernförde / Harburg / Schleswig-Holstein.
Died:	06.02.1972 in Starnberg bei München / Regierungsbezirk Oberbayern / Bayern. Gravesite: Nordfriedhof in München.

NSDAP-Nr:	39 (Joined the *DAP* with Nr. 594, 01.12.1919; Party banned following the *München-Putsch* of 09.11.1923; Officially reenrolled in the *NSDAP* with Nr. 39 on 07.01.1926)
SS-Nr.:	2 (Joined 21.09.1925)

Promotions

01.12.1917	*Kanonier*
12.11.1920–00.08.1921	*Organisator und Führer der "Turn- und Sportabteilung" der NSDAP* (forerunner of the SA)
00.08.1921–00.11.1921(?)	*Organisator der SA*
21.09.1925	*SS-Anwärter*
00.00.1926	*SS-Mann*
04.05.1933	*SS-Sturmbannführer*
09.11.1933	*SS-Obersturmbannführer*
01.06.1934	*SS-Standartenführer* (mit Wirkung vom 01.07.1934)
30.01.1939	*SS-Oberführer*
01.01.1940 (?)	*Oberleutnant d. R. (Luftwaffe)*

Career

00.00.1904–00.00.1907	Attended *Volksschule* in Owschlag.
00.00.1907–00.00.1914	Attended *Realschule* in Eckernförde.
14.01.1914–01.10.1917	Served an apprenticeship with the master watchmaker Adolf Christen in Gettorf / Kreis Eckernförde.
00.04.1916	Attempted to join the *Landsturm* (militia) in Eckernförde, but deemed unfit due to his generally debilitated physical condition. He remained thus categorized for one year, again attempting to enter service in March 1917; he was again classified unfit for service for a further six months.
00.09.1917	Passed his *Gehilfenprüfung* (assistant's examination) as a watchmaker's assistant.
02.10.1917	Resettled in München.
02.10.1917–30.11.1917	Worked as a watchmaker's assistant in a shop on the Sendlinger Straße in München.
01.12.1917–25.01.1919	Finally succeeded in entering military service as a *Kanonier* with *4. Ersatz-Batterie / 1. Kgl. Bayerisches Feldartillerie-Regiment "Prinzregent Luitpold"*.

SS-Oberführer Emil Maurice in 1939. (Bundesarchiv Bild 146-1980-073-19A)

03.01.1918–23.01.1918	Hospitalized in the *Reserve-Lazarett München* due to influenza.
24.01.1918–23.02.1918	Hospitalized for unknown reasons.
01.02.1919–01.10.1919	Resumed work as an assistant watchmaker on the Sendlinger Straße, München.
01.05.1919–01.09.1919	Member of *Freikorps Oberland* in München.
00.00.1919–00.00.1920	Service as a *Freiwilliger* (volunteer) with *Infanterie-Regiment 5* of the *Einwohnerwehr*.
13.11.1919	Attended a meeting of the *Deutsche Arbeiterpartei* (*DAP*, German Workers' Party) in the Münchener Hofbräuhaus, with Hitler as the featured speaker.
01.12.1919	Joined the *DAP*.
01.12.1919	Appointed as an *Ordnungsmann der NSDAP* (Nr. 8). He became one of Hitler's few "Duz-Freunden" during this period.
12.11.1920–00.11.1921	*Organisator und Führer der "Turn- und Sportabteilung" der NSDAP* (Organizer and Leader of the Gymnastics and Sports Section of the Nazi Party). When these units were redesignated *Sturmabteilung* (Storm Troops) in August 1921, Maurice was granted the new title *Organisator der SA*. He was succeeded by Hans Ulrich Klintzsch, who was given the new title *Führer der SA*. Konrad Heiden writes of Maurice's contributions as "the spiritual father of the movement's tactics": "He was the first to resort to direct action, in the sense that he roved through the

Landsberg, 1924. Hitler poses with Hermann Kriebel (left) and a smiling Emil Maurice.

Landsberg, 1924. From left to right: Hitler, Maurice, Hermann
Kriebel, Rudolf Hess, and Dr. Friedrich Weber.

Emil Maurice marches beside Jakob Grimminger, bearer of the *"Blutfahne" (Blood Banner)*, during a commemoration of the *"München-Putsch"*, ca. 1934. (NARA photo, Heinrich Hoffmann collection)

	streets at night seeking opponenents- especially Jews- that he might beat them up." (Heiden, *Der Fuehrer: Hitler's Rise to Power*, p. 413)
00.00.1921–00.00.1921	Attached, as a member of the *Bund Oberland*, to the *Grenzschutz* (border defense formations) in Oberschlesien.
00.07.1921–00.08.1923	Personal chauffeur to Adolf Hitler.
00.08.1921	Entered *NSDAP* service as a clerical employee, receiving a monthly salary of 250 RM.
00.08.1921	Joined the *SA*.
29.10.1921	Arrested for involvement in a 25.10.1921 assassination attempt on the Social Democratic delegate Erhard Auer. Released after just 9 days due to lack of evidence.
04.11.1921	Seriously injured in a "Saalschlacht" (meeting hall brawl) at the Münchener Hofbräuhaus.
00.05.1922	Arrested for burning the national flag of the *Reich* and for acts of battery against political opponents on the Münchener Bahnhofsplatz.
00.06.1922	Rearrested for involvement in a brawl.
14.10.1922–15.10.1922	Participated in the annual *Coburger Tag*.
00.10.1922–17.01.1923	Arrested and interned in the *Landesgefängnis* (state prison) of Mannheim / Baden for his role in a bomb attack on the *Mannheimer Börse* (Mannheim stock exchange).

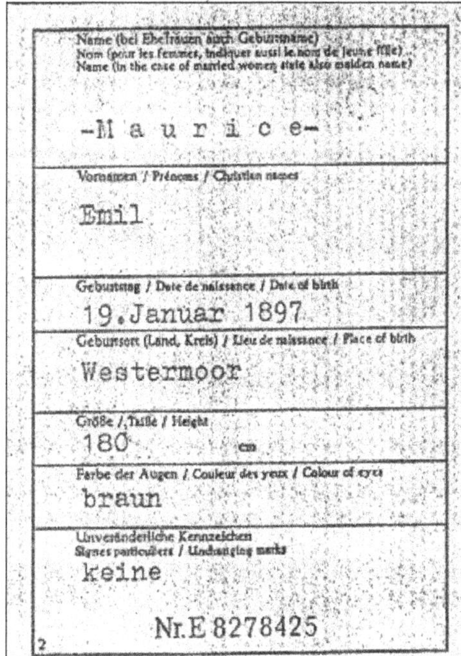

A postwar identification card belonging to Emil Maurice.

00.03.1923–09.11.1923	Member of the *Stabswache Hitler*, an *SA* company tasked with maintaining Hitler's personal security. The forerunner of the later SS, it was redesignated as the *Stoßtrupp Adolf Hitler* in May 1923.
00.05.1923–09.11.1923	Adjutant of the *Stoßtrupp Adolf Hitler*, under Joseph Berchtold.
00.08.1923	Resumed work as a watchmaker in München.
08./09.11.1923	Participated in the *München-Putsch* as Berchtold's *Adjutant* in the *Stoßtrupp Adolf Hitler*.
29.04.1924	Sentenced by the *Volksgericht München I* to 18 years' fortress arrest for being an accessory to high treason (i.e., participating in the *München-Putsch*) and 3 months' imprisonment for illegal possession of weapons. He then fled to Pommern where he remained under a false name until turning himself in to police.
00.00.1924–27.01.1925	Imprisoned with Hitler and other Nazis at the *Staatliche Gefangenenanstalt* in Landsberg am Lech, approximately 65 km west of München.
27.01.1925	Released from Landsberg Prison.
00.02.1925–00.01.1928	Permanent *Begleiter* (escort) to and chauffeur of Adolf Hitler. Succeeded by Julius Schreck.
00.00.1925–00.00.1925	Member of the *Stabswache Hitler*.
21.09.1925	Joined the *SS*.
21.09.1925–00.01.1928	*1. Inspekteur der SS* (1st Inspector of the *SS*).
07.01.1926	Officially reentered the *NSDAP*.

00.12.1927	Break in the friendship between Maurice and Hitler. He had by this time become a member of the innermost circle of the *NSDAP*, and had earned the special affection of his *Führer* (who referred to him as "Moritzl"). In the 1920's, Maurice boasted of being Hitler's best friend. Not even the rumor- constantly flaring up- that Maurice had Jewish ancestors could spoil the friendship. But in the autumn of 1927, Maurice made the mistake of falling in love with Hitler's niece, Angela ("Geli") Raubal (born in Linz on 04.01.1908), who had just moved in with her "Onkel Adolf" in München; the two had first met in 1926 at an *NSDAP* function in Weimar. They secretly became engaged, and when Hitler learned of the affair, he flew into a rage that caused Maurice to fear he'd be shot. Hitler ultimately prevailed upon Geli to break the engagement, and ordered her to separate from Maurice for two years. The matter surely played a role in Hitler's decision to remove Maurice, in January 1928, from all his offices and banish him from the inner circle of the Party. The *Führer* did not fire him outright, but gradually excluded and lapsed in paying him. Maurice ultimately resigned. In April 1928, he brought a lawsuit against Hitler before the *Arbeitsgericht* (labor court) in München, suing his former friend and leader for back pay of 3,000 *Reichsmarks*; the *Führer* was ordered to pay the sum of 500 RM. After this trial, Rudolf Hess formally removed Maurice from the *SS*. Only in April 1933, during a meeting in Hitler's private apartment in München, did the two finally reconcile.
00.00.1928–00.00.1945	Self-employed as a watchmaker and salesman of electrical equipment in München.
00.04.1933–01.10.1935	*Stadtrat* in München.
04.05.1933–01.07.1934	Readmitted to the *SS* with the status of a *Rangführer* in *SS-Sturmbann I/1* (München).
01.07.1934–01.10.1934	Assigned as a *Rangführer* to *1.SS-Standarte* (München).
30.06.1934	Participated in the "suppression" of the so-called "Röhm Revolte" in München. As a member of a *Sonderkommando* of the *Bayerischen Politischen Polizei* (*BPP*, Bavarian Political Police), he took part in the arrest of the *SA* leaders Hans-Peter von Heydebreck and Karl Graf von Spreti. As a reward for this, Hitler promoted him to *SS-Standartenführer*.
01.10.1934–01.04.1936	Assigned as an *SS-Führer z.b.V.* to *1.SS-Standarte* (München).
Spring 1935	In accordance with *SS* marriage laws, in order to marry on 11.05.1935, Maurice was required to submit proof of his bride's and his own "pure German-blood descent" to the *SS-Rasse- und Siedlungshauptamt*. In the course of the subsequent genealogical investigation, the groom-to-be was found to be of partly Jewish blood. His great-grandfather was the Jewish founder of Hamburg's Thalia Theater, Charles Maurice Schwarzenberger (1805–1896), and this fact prompted the *Reichsführer-SS* to demand that Hitler order his removal from the *SS*. Hitler refused to do so, as "Moritzl" was one of his earliest assistants and a founding member of the *SS*. The *Führer* authorized a *Gnadeerlaß* (amnesty) declaring Maurice and his brothers as "Ehrenarier" (honorary Aryans). Hitler's remarkable

decision was promulgated via the following official file notice, dated 31.08.1935, of Heinrich Himmler:

> 1. *SS-Standartenführer* Emil Maurice is, without doubt, according to his family tree, not of Aryan descent.
>
> 2. On the occasion of *SS-Standartenführer* Maurice's marriage when he had to submit the family tree, I reported to the *Führer* my position to the effect that Maurice must be removed from the ranks of the *SS*.
>
> 3. The *Führer* has decided that in this one and only exceptional case Maurice as well as his brothers could remain in the SS, because Maurice had been his very first companion, and because his brothers and the entire family Maurice had served the Movement with rare bravery and loyalty in the first and most difficult months and years.
>
> 4. I decree that Maurice must not be entered in the *SS-Sippenbuch* [Clan Book], and that none of the descendants of the Maurice family may be admitted into the *SS*.
>
> 5. The Chief of the *Rasse- und Siedlungshauptamt* receives a copy of this Minute with the request for its most strictly confidential treatment; only the Chief of the *Sippenamt* [Clan Office] is to be informed of the decision.
>
> 6. For myself and for all successors as *Reichsführer-SS*, I state that only Adolf Hitler himself had and has the right to decree such an exception with regard to blood. No *Reichsführer-SS* has, or will have for all future time, the right to allow exceptions from the requirements of the *SS* regarding blood.
>
> 7. I oblige all my successors to maintain most strictly the position laid down in point 6.
>
> > 1. Two copies to the Chief of the Race and Settlement Head Office
> >
> > A. one closed and sealed for Maurice's marriage file.
> >
> > B. one for the information of the Chief of the *Sippenamt*.
> > (Peter Hoffmann, *Hitler's Personal Security*, pp. 50–51)

00.09.1935	During the *7. Reichsparteitag der NSDAP* in Nürnberg, a 71-year-old Party member had the misfortune of getting in the way of Maurice's car, and not moving aside quickly enough. Maurice then battered him with a rubber truncheon. Charges were brought against him for this assault, but were quashed on Hitler's insistence.
01.10.1935–08.05.1945	*Ratsherr der Hauptstadt der Bewegung München* (Councilman of the Capital City of the [National-Socialist] Movement, München).
01.10.1935–08.05.1945	*Bayerischer Staatsrat.*
01.01.1936–08.05.1945	*Stellvertreter des Reichshandwerkermeisters in der Hauptstadt der Bewegung München* (Deputy to the *Reich* Craftsmans' Leader in the Capital of the [National-Socialist] Movement, München).

29.03.1936–08.05.1945	Member of the *Reichstag (Wahlkreis 29, Leipzig)*.
01.04.1936–00.05.1945	Attached to *SS-Abschnitt I* (München).
20.09.1936	Appointed as *Landeshandwerksmeister von Bayern* (kommissarischer [acting] until 01.01.1937, then permanent).
01.04.1937	Appointed as *Präsident* of the *Handwerkskammer für München und Oberbayern*.
00.00.193_	Member of the *Vorstand* (Board of Directors), *Landes-Gewerbebank* München.
00.00.193_-00.00.19__	Member of the *Aufsichtsrat* (Supervisory Board) of the firm *Lenz*.
04.03.1938–16.03.1938	While participating in reserve training exercises with *3.Scheinwerfer-Abteilung / Flak-Regiment 5*, took part in military operations in connection with the "Anschluß" (annexation of Austria).
04.08.1938–18.08.1938	Participated in military training exercises with *Flak-Regiment 5*.
26.09.1938–12.10.1938	While participating in training exercises with *Flak-Regiment 5*, took part in the occupation of the Sudetenland.
01.01.1940–01.10.1940	*Luftwaffe* service with the rank of *Oberleutnant d. R.*
00.00.1943–00.05.1945	*Vizepräsident* of the *Gauwirtschaftskammer München-Oberbayern*.
00.00.1943–00.05.1945	*Leiter* of the *Handwerksabteilung* in the *Gauwirtschaftskammer München-Oberbayern*.
00.05.1943–00.05.1945	*Gauhandwerksmeister für den Gau München-Oberbayern*.

Postwar Prosecution

Arrested by U.S. troops in Starnberg, 25.05.1945 and interned, Maurice was transferred in September 1946 to the Lager Cham bei Regensburg. On 18.12.1947, he was indicted by the *Spruchkammer* in Regensburg, and on 13.05.1948, convicted in a de-Nazification proceeding that resulted in his classification as a "Belasteter" (Incriminated person, in de-Nazification Gruppe II); among the charges against him was his role in the murders of *SA* leaders on 30.06.1934. He was sentenced to 4 years' labor camp and confiscation of 30% of his assets. As of 26.03.1951, he owned a watch shop at Schumannstr. 5 in München.

Decorations and Awards

00.00.194_	*Kriegsverdienstkreuz II. Klasse ohne Schwerter*
20.08.1934	*Ehrenkreuz des Weltkrieges 1914–1918 mit Schwertern*
ca. 1939	*Medaille zur Erinnerung an den 1 Oktober 1938*
ca. 1938	*Medaille zur Erinnerung an den 13 März 1938*
00.00.1933	*Goldenes Ehrenzeichen der NSDAP*
00.00.1934	*Ehrenzeichen des 9 November 1923 (Blutorden*; Nr. 495, with effect from 09.11.1933)
00.10.1932	*Coburger Abzeichen 1922*
00.00.194_	*Dienstauszeichnung der NSDAP in Gold*
00.00.194_	*Dienstauszeichnung der NSDAP in Silber*
00.00.194_	*Dienstauszeichnung der NSDAP in Bronze*
00.00.193_	*Goldene Ehrennadel des Handwerks*
00.00.193_	*Ehrendolch der SS*
ca. 1933	*Ehrendegen des Reichsführers-SS*
00.00.1933	*Totenkopfring der SS*

00.00.193_	*Ärmelband der "Stroßtrupp Hitler"*
16.12.1935	*Julleuchter der SS*
00.02.1934	*Ehrenwinkel für alte Kämpfer*

Notes

- Son of the former factory owner *Karl* (aka Charles) Emil Amandus Maurice (born 00.00.1864 in Hamburg) and his wife Amanda, née Hennings. In addition to Emil, at least two other sons, both of whom later served in the SS, were born to this marriage.
- Religion: Protestant until 00.00.1936, then left the church and declared himself "gottgläubig".
- Became engaged on 31.03.1935 to the medical student [later Dr. med.] Hedwig Maria Anna Ploetz (born 26.08.1911 in Aschaffenburg, died 00.00.2003; daughter of *Oberst* Rudolf Ploetz [1865–1942]). They married in München on 11.05.1935, with Christian Weber and Prof. Heinrich Hoffmann appearing as witnesses.

Sources:

Bayerisches Hauptstaatsarchiv, München, Abteilung IV· Kriegsarchiv: Excerpts from various Kriegsranglisten containing data on the Royal Bavarian Army service of Emil Maurice.

Dornberg, John: *Munich 1923. The Story of Hitler's First Grab for Power.* Harper & Row, 1982.

Flood, Charles Bracelen: *Hitler: The Path to Power.* Houghton Mifflin Company, 1989.

Gordon, Harold J., Jr.: *Hitler and the Beer Hall Putsch.* Princeton University Press, 1972.

Hamilton, Charles: *Leaders and Personalities of the Third Reich, Volume I.* R. James Bender Publishing, 1984.

Hanfstaengl, Ernst Franz Sedgwick ("Putzi"): *Hitler: The Missing Years.* Arcade Publishing, 1957.

Hayman, Ronald: *Hitler and Geli.* Bloomsbury Publishing USA, 1999.

Heiden, Konrad: *Der Fuehrer: Hitler's Rise to Power.* Houghton Mifflin, 1944.

Höhne, Heinz: *The Order of the Death's Head.* Martin Secker and Warburg, 1969.

Hoffmann, Peter: *Hitler's Personal Security.* Macmillan, 1979.

Lilla, Joachim; Döring, Martin; & Schulz, Andreas: *Statisten in Uniform. Die Mitglieder des Reichstags 1933–1945.* Droste Verlag, 2004.

Mitchell, Otis C.: *Hitler's Stormtroopers And The Attack On The German Republic, 1919–1933.* McFarland, 2008.

National Archives and Records Administration, College Park, Maryland: *SS-Personalakte* of Emil Maurice. Microfilm document collection A3343SS.

Ryback, Timothy W.: *Hitler's Private Library: The Books that Shaped His Life.* Random House Digital, 2009.

Johann ("Hans") Ulrich Klintzsch

00.11.1921–00.02.1923   *Führer der SA*

Born:	04.11.1898 in Lübbenau (Spreewald).
Died:	17.08.1959 while attending the wedding of his son, Fridthjof. Gravesite: Judischer Friedhof Ohlsdorfer, Hamburg.
NSDAP-Nr.:	3,603 (Joined 20.07.1921)

Promotions

17.09.1917	*Fähnrich zur See (ohne Patent)*
28.09.1919	*Leutnant zur See d. R.*
00.00.191_	*Leutnant zur See*
00.00.191_	*Oberleutnant zur See*
00.11.1921–00.02.1923	*Führer der SA*
00.00.1936	*Major*
00.00.194_	*Oberstleutnant*
01.02.1943	*Oberst*

Career

00.10.1916	Entered the *Kaiserliche Marine* (Imperial German Navy).
00.10.1916–00.03.1917	Underwent training at the *Marineschule* and aboard the training vessel SMS *Freya* (formerly a protected cruiser) in Flensburg.
00.03.1917–00.05.1917	Assigned to the torpedo boat *G 95*.
00.05.1917–00.07.1917	Attached to *1.Marine-Artillerie-Regiment*.
00.07.1917–00.06.1918	Participated in various training courses.
00.06.1918–00.11.1918	Assigned as *Wachoffizier* (watch officer) to the *Zerstörerflotte Flandern* (destroyer flotilla in Flanders).
00.12.1918	Transferred to the naval reserves.
00.00.1919–00.00.1921	Service with the *Marinebrigade Ehrhardt* (in the rank of *Oberleutnant*).
10.01.1920–02.06.1921	Returned to service as a *Leutnant* zur See in the *Reichsmarine*, then discharged 02.06.1921.
[00.00.1921]	Student in München.
26.08.1921	Involved, as a member of *Organisation Consul*, in the shooting death of Matthias Erzberger (20.09.1875–26.08.1921). A prominent figure in the Catholic *Zentrumspartei*, Erzberger was appointed by President Friedrich Ebert as head of the German Armistice Commission and later served as Finance Minister until March 1920. The fact that he had signed the Armistice of 11.11.1918 led to his murder, which occurred in Bad Griesbach (Schwarzwald) / Baden.
20.07.1921	Joined the *NSDAP*
20.07.1921–00.11.1921	*Ausbilder* (training coordinator) of the *Turn- und Sportabteilung der NSDAP* (later redesignated *Sturmabteilung*).
03.08.1921–00.05.1923	Employed as an *Organisationsleiter* (organization coordinator) of the *NSDAP*.

Hans Ulrich Klintzsch as *Führer der SA*, 1923.

00.09.1921–00.09.1921	Arrested and briefly detained in connection with the murder of Matthias Erzberger.
00.11.1921–00.02.1923	*Führer der SA*. Succeeded Emil Maurice. Resigned and succeeded by Hermann Göring.
00.03.1923–11.05.1923	Member of the *Oberkommando der SA (SA* High Command).
00.04.1923–00.05.1923	*1. Adjutant* to the *Kommandeur der SA,* Hermann Göring.
11.05.1923	Returned to the *Brigade Ehrhardt* as a *Bataillonsführer,* and later served as a leader of its successor organization, the *Bund-Wiking.*
00.00.1927	Proceedings initiated against Klintzsch by the *Landgericht I* in Berlin for his continued work with the government-banned *Bund-Wiking.* The case was dismissed, 00.00.1929, in accordance with the *Reichsamnestiegesetz* (*Reich* Amnesty Law).
00.00.1936–31.10.1939	Joined the *Luftwaffe,* assigned as *Kommandeur* of the *Blindflugschule (BFS) Brandis* (at Brandis-Waldpolenz). Succeeded by *Oberst* Paul Aue.
31.10.1939–00.08.1940	Assigned to the *Reichsluftfahrtministerium (Luftwaffeninspektion für Flugnavigation [L.In. 12].*
00.08.1940–00.06.1941	*Kommandeur* of the *Seenotgruppe der Luftwaffe* (air-sea rescue group of the *Luftwaffe*). On 16.09.1940, he was injured while serving as *Beobachter* (observer) of *Heinkel He 59c DA+AG* of *Seenotflugkommando 2* when it crashed on takeoff near Ostende.

05.06.1941–00.11.1942	*Seenotdienstführer West* (redesignated 01.06.1942 as *Seenotdienstführer 3*), based in Cherbourg.
01.12.1942–29.08.1944	*Inspekteur des Seenotdienstes (L.In.16) / Generalstab der Luftwaffe.*
29.08.1944–01.03.1945	Assignment[s] unknown.
01.03.1945–08.05.1945	*Fliegerführer 6 (Luftflotte 6)*, comprised of *Seeaufklärungsgruppe 126, Bordfliegergruppe 196*, and *Seenotgruppe 60.* Succeeded *Oberst* Karl Stockmann. Final holder of this post, which was formed at Bad Ahlbeck on 26.11.1944 and remained headquartered there until 00.04.1945, when HQ transferred to Värlöse (Denmark). During Klintzsch's tenure, it was subordinated to *Luftwaffenkommando Ostpreußen* (24.01.1945–21.04.1945), then to *Luftwaffenkommando Nordost* (21.04.1945–08.05.1945).

Decorations and Awards

00.00.191_	*1914 Eisernes Kreuz I. Klasse*
00.00.191_	*1914 Eisernes Kreuz II. Klasse*
ca. 1921	*Schlesisches Bewährungsabzeichen (Schesischer Adlerorden) 2.Stufe*
ca. 1934	*Ehrenkreuz des Weltkrieges 1914–1918 mit Schwertern*

Notes

- Son of a pastor.
- Married to Hildegard _______ (born 05.01.1906, died 24.07.1992). At least two sons resulted from this marriage (Fridtjof and Heidjer; the latter is known to be deceased, and is buried under the same gravestone as his parents).

Sources:

Campbell, Bruce: *The SA Generals and the Rise of Nazism.* University Press of Kentucky, 1998.

Hermann Wilhelm Göring
01.03.1923–00.11.1923   Oberster SA-*Führer*

Born: 12.01.1893 at the Sanatorium Marienbad bei Rosenheim / Regierungsbezirk Oberbayern / Bayern.

Suicide: 15.10.1946 in Cell #5, Nürnberg Prison at approximately 2245 hours, swallowing a cyanide capsule he had successfully concealed since his arrest. In a statement concerning Göring's final act, Supreme Court Justice Robert H. Jackson, chief United States prosecutor at the Nürnberg Trials, declared:

> ... When he took his own life he killed the myth of Nazi bravery and stoicism and deep conviction ... The real significance of this event is its effect on Germany ... If Goering had been made of the stuff that could walk to thegallows voicing some patriotic sentiment such as our Nathan Hale's regret that he had but one life to give to his country, he might well have become a German martyr-hero. The end comes, from the viewpoint of his hero worshipers as anti-climactic as a burlesque after a Wagnerian overture. Goering could not keep up the part when it began to hurt. The founder of the concentration camps, where death was handed out to millions, could not face the gallows himself. His end betrayed the weakness of his whole life - cunning and crafty, always outwitting somebody, bullying and cowardly. (Associated Press, "Goering Ends 'Martyr Myth,' Jackson Says", 17.10.1946)

NSDAP-Nr.:   23 (Joined 00.12.1922 with unknown Party Nr.; later stricken from the membership rolls during his exile in Italy and Sweden, then readmitted with Nr. 23 on 01.04.1928)

Promotions

13.05.1911	*Fähnrich*
20.01.1914	*Leutnant* (mit Patent vom 22.06.1912)
18.08.1916	*Oberleutnant*
08.06.1920	*Charakter als Hauptmann* (mit Wirkung vom 11.03.1920)
01.03.1923–00.11.1923	*Oberster SA-Führer*
18.12.1931	*SA-Gruppenführer (NSFK)*
01.01.1933	*SA-Obergruppenführer*
25.03.1933	*DLV-Minister*
05.05.1933–28.04.1945	*Reichsminister der Luftfahrt*
30.08.1933	*Charakter als General der Infanterie* (mit RDA vom 01.10.1931)

The following is excerpted from the 31.08.1933 official notice of appointment by *Reichspräsident* von Hindenburg:

Kadett Göring at the *Kadettenanstalt*
in Karlsruhe, 1907. (*Bundesarchiv
Bild* 183-R25668)

Leutnant Göring (left, in the background) as a pilot of *Feldflieger-Abteilung 25*, ca. 1916.
Second from left is Lothar von Richthofen (40 aerial victories by war's end), younger brother
of Manfred von Richthofen (the "Red Baron"). (Hermann-Historica, Auctioneers, München)

Leutnant Göring, ca. 1916.

Oberleutnant Göring wearing
his Pour le Mérite in 1918.

Oberleutnant Göring poses beside his Fokker D VII.

> With effect as per yesterday's date, the *Reich* President has
> ... conferred upon the Prussian *Ministerpräsident*, former
> *Hauptmann* Göring, Knight of the *Pour le mérite*, the rank of
> *General der Infanterie* in recognition of his extraordinary merits
> both in war and peace, by virtue of which he is entitled to wear
> the uniform of the *Reich* Army. (Dr. Max Domarus, ed, *Hitler.*
> *Speeches and Proclamations, 1932–1945: The Years 1932–1934*, p.
> 352)

14.09.1933	*General der Landespolizei*
21.05.1935	*General der Flieger*
20.04.1936	*Generaloberst* (mit RDA vom 01.04.1936)
04.02.1938	*Generalfeldmarschall*
19.07.1940–28.04.1945	*Reichsmarschall des Großdeutschen Reiches*

Career

00.00.1893–00.00.1896	Lived in Fürth with his mother's friend, Frau Erna Graf. His parents were in the Caribbean during this period as Herr Göring was serving as German *Generalkonsul* in Haiti.
00.00.1896–00.00.1900	Lived in Berlin-Friedenau with his parents.
00.00.1900–00.00.1904	Attended *Volksschule* in Fürth.
00.00.1904–00.00.1905	Attended *Gymnasium* in Ansbach.
00.00.1905–00.00.1909	Attended the *Kadettenanstalt* in Karlsruhe.
00.00.1909–20.01.1914	Attended *Haupt-Kadettenanstalt Groß-Lichterfelde*, where, in the Spring of 1911, he passed his *Fähnrich* (ensign) examination with the grade *summa cum laude*.
13.10.1910–26.10.1910	Hospitalized with tonsillitis in the infirmary of the *Haupt-Kadettenanstalt Groß-Lichterfelde*.
20.01.1914–19.08.1914	Entered service, assigned as *Zugführer* and *Kompanie-Offizier* to 8. *Kompanie / 4. Badische Infanterie-Regiment "Prinz Wilhelm" Nr. 112* (Mühlhausen [*Fr.*: Mulhouse] / Elsaß, approximately one mile from the French border).
20.08.1914–23.09.1914	*Bataillons-Adjutant* of II. *Bataillon / 4. Badische Infanterie-Regiment "Prinz Wilhelm" Nr. 112.*
23.09.1914	Fell ill with rheumatic fever in Thiancourt, then hospitalized in the following medical facilities: 23.09.1914 – *Feldlazarett Thiancourt* 23.09.1914–24.09.1914 – *Festungslazarett Haushaltungsschule Metz-Montigny* 24.09.1914–12.10.1914 – *Klinischen Krankenhaus Freiburg in Baden* While hospitalized, he was visited by a comrade from his regiment, Bruno Loerzer, who had recently completed flight training. Loerzer convinced him to volunteer for the flying service.
13.10.1914–27.10.1914	Training as *Beobachter* (observer) with *Fliegerersatz-Abteilung 3 (FEA 3).*
28.10.1914–29.06.1915	Assigned as *Flugzeugbeobachter* (aerial observer) of Bruno Loerzer in *Feldflieger-Abteilung 25 (FFA 25)* at Stenay Airfield.

01.07.1915–14.09.1915	Pilot training at *Fliegerschule Freiburg.*
15.09.1915–00.00.191_	Assigned to the *Armeeflugpark* of *5.Armee.*
15.09.1915–08.07.1916	Assigned as a pilot to *Feldflieger-Abteilung 25.* He flew his first mission as a fighter pilot on 03.10.1915.
09.07.1916–04.08.1916	Attached as a pilot to *Artillerieflieger-Abteilung 203* (under *3.Armee*).
04.08.1916–04.09.1916	Returned to duty with *Feldflieger-Abteilung 25.*
06.09.1916–27.09.1916	Attached to *Kampfstaffel Metz* (*Kampfeinsitzer Kommando* [Kek, single-seat fighter unit] *Metz*).
28.09.1916–27.09.1916	Assigned as pilot to *Jagdstaffel 7 (Jasta 7).*
20.10.1916–21.12.1916	Pilot with *Jagdstaffel 5 (Jasta 5)* on the Somme.
02.11.1916	While attacking a British Handley-Page bomber, severely wounded in the right hip and grazed on the right knuckles by machine gun fire from a flight of Sopwith Camels over Combles. He was subsequently hospitalized in the following:

02.11.1916–12.11.1916   *Feldlazarett 3 / XV.Armee-Korps*
15.11.1916–30.11.1916   *Etappen-Lazarett 6* in Valenciennes
03.12.1916–11.12.1916   *Reservelazarett I* in Bochum
19.12.1916–15.01.1917   *Reservelazarett G* in München

22.12.1916–14.02.1917	Assigned for training to *Fliegerersatz-Abteilung 10* in Böblingen / Württemberg.
15.02.1917–16.05.1917	Pilot with *Jagdstaffel 26 (Jasta 26),* based in Mühlhausen and commanded by Bruno Loerzer.

The following is a list of Göring's aerial victories:

Number	Date & Time	Enemy Aircraft Type & Unit	Location
Unconfirmed	03.06.1915	Unknown	Stenay
1	16.11.1915	Maurice Farman	Tahure
2	14.03.1916	Caudron G.IV bomber	Forced down and captured intact southeast of Haumont Forest
3	30.07.1916	Caudron G.IV bomber of Escadrille C 10 (Pilot: Sergent Girard-Varet)	Mamey, 30 km southwest of Metz
4	23.04.1917–1720 hrs.	F.E.2b of 18 Sqdn (Pilot: 2nd Lt. George B. Bate)	Northeast of Arras
5	28.04.1917–1830 hrs	Sopwith 1 ½ Strutter of Sop 27 Esc	South of St. Quentin
6	29.04.1917–1945 hrs.	Nieuport XVII of 6(N) Sqdn (Pilot: Sub-Lt. Albert H.V. Fletcher)	Ramicourt
7	10.05.1917–1505 hrs.	Airco D.H.4 of 55 Sqdn	Northeast of Le Pavé
8	08.06.1917–0730 hrs.	Nieuport 23 of 1 Sqdn (Pilot: 2nd Lt. Frank D. Slee)	Near Moorslede
9	16.07.1917–2005 hrs.	SE5 of 56 Sqdn (Pilot: 2nd Lt. Robert G. Jardine)	Northeast of Ypres

10	24.07.1917–2045 hrs.	Sopwith Triplane of the Royal Naval Air Service (Pilot: Flight Sub-Lt. Theodore C. May)	South of Passchendaele
11	05.08.1917–2015 hrs.	Sopwith F.1 Camel of 70 Sqdn (Pilot: Lt. Gilbert Budden)	Northeast of Ypres
12	25.08.1917–0845 hrs.	Sopwith Camel of 70 Sqdn (Pilot: 2nd Lt. Orlando C. Brideman)	Southwest of Ypres
13	03.09.1917–1955 hrs.	Airco D.H. 4 (probably of 57 Sqdn)	North of Lampemisse
14	21.09.1917–0905 hrs.	Bristol BF2b (2-seater, flown by Lt. R. L. Curtis [15 victories] and Lt. D.P Fitzgerald-Uniacke [13 victories] of 48 Sqdn	
15	21.10.1917–1545 hrs.	S.E.5 of 84 Sqdn (Pilot: 2nd Lt. Arthur E. Hempel)	Linselles
16	07.11.1917–0915 hrs.	Airco D.H.5 of 32 Sqdn	Northwest of Poelcapelle
17	21.02.1918–1000 hrs.	S.E.5a of 60 Sqdn (Pilot: 2nd Lt. George B. Craig)	Ledeghem
18	07.04.1918–1250 hrs.	R.E.8 of 42 Sqdn (Pilot: 2nd Lt. Harry W. Collier)	Southwest of Merville
19	05.06.1918–1000 hrs.	A.R.1? (No French loss report for that type reported; Possibly a Bréguet 14 B2)	North of Villers Cotterets
20	09.06.1918–0845 hrs.	Spad S.VII of Escadrille Spa 94 (Pilot: Caporal Pierre Chan)	South of Coroy
21	17.06.1918–0830 hrs.	Spad of Escadrille Spa 93	Ambleny
22	18.07.1918–0815 hrs.	Spad E	Bandry Wood

17.05.1917–06.07.1918   *Führer* of *Jagdstaffel 27 (Jasta 27)* in Ypres, flying two *Fokker D VII*'s (Nr. 278/18 and 324/18). Succeeded by Lothar von Richthofen.

07.07.1918–30.07.1919   *Kommandeur* of *Jagdgeschwader "Freiherr von Richthofen Nr. 1" (JG 1)*. Succeeded *Hauptmann* Wilhelm Reinhard, who had been killed in the crash of his aircraft. In his book *Richthofen Flieger* (1930), one of Göring's former pilots, Richard Wenzl, wrote:

> Our new commander had arrived during our last days in Beugneux, Oblt Göring, the former leader of Jasta 27. If there had been some friction now and then during Reinhard's time, Göring understood how to make himself at home in the Geschwader in short order. He was a good comrade to us to the last and a good commander, who upheld the traditions of the Geschwader. (Translation in Norman Franks & Greg VanWyngarden, *Fokker D VII Aces of World War I, Part 1*, p. 15)

Göring flew two *D VII*'s during this period: *D VII (F) 294/18* and *D VII (F) 5128/18*.

Göring and Bruno Loerzer pose
with Friedrich Seekatz, right-hand
man to the Dutch aircraft designer,
Anthony Fokker, in 1918.

Heinrich Hoffmann's portrait of Göring
as *Oberster SA-Führer*, taken shortly
before the "*München-Putsch*" in 1923.

Oberleutnant Göring posing in the cockpit of his *Fokker D VII*.

26.07.1918–22.08.1918	On leave.
22.08.1918–00.00.1918	Hospitalized at the *Kriegslazarett Maresolm Deinse* due to tonsillitis.
00.11.1918	In violation of Armistice terms, led *Jagdgeschwader 1* to Darmstadt / Hessen, then formally demobilized the unit at a paper factory in Aschaffenburg. He was later to write of the fall of the German *Reich* and the country's descent into revolutionary violence:

> ... [O]ne will ... be amazed to see what a high percentage of the Social Democratic leaders and agitators were Jews. But now in the days of the post-War rising these Jewish leaders sprouted from the ground like poisonous fungi. Wherever soldiers' councils were formed Jews were the leaders, those very same Jews who had not been seen out at the front, but had been employed in the supply departments at the base or had filled indispensable official and military posts at home. In the streets the mob raged. Soldiers had their badges and shoulder-straps torn off. The flag, which for decades had symbolized the greatness of the *Reich*, was trampled in the mud. On all buildings fluttered the red flag of rebellion; everywhere there was disorder and dissolution. (Göring, *Germany Reborn* [translation of *Aufbau einer Nation*, 1934], pp. 27–28)

In his opening testimony before the International Military Tribunal at Nürnberg on 13.03.1946, Göring stated:

> After the collapse in the first World War I had to demobilize my squadron. I rejected the invitation to enter the *Reichswehr* because from the very beginning I was opposed in every way to the republic which had come to power through the revolution; I could not bring it into harmony with my convictions. (*Trial of the Major War Criminals Before the International Military Tribunal Nuremberg, 14 November 1945 to 1 October 1946, Volume IX*)

00.12.1918–00.00.1919	Lived in Berlin with fellow air ace Ernst Udet, then with his widowed mother in München.
00.04.1919–00.00.1919	Invited by the *Fokker-Flugzeugwerke* to Denmark to demonstrate their latest aircraft (agreed on the condition that he be allowed to keep the plane as payment). He was then employed by Fokker as a test pilot and aviation advisor, also staging aerobatic displays together with four other veterans of the *"Richthofen" Geschwader*.
Summer 1919–	Flew from Denmark to Sweden, selling his *Fokker* at Malmstät. After receiving his license to fly passenger aircraft in Sweden on 02.08.1919, he served as a commercial pilot and flight chief for the fledgling Swedish airline *Firma Svenska Lufttravik AB*, co-owned by Karl Lignell, in Stockholm. He then worked as an aerial chauffeur to wealthy Swedes during this period and received his Swedish license as a passenger aircraft pilot.

13.02.1920	In a letter from Stockholm to Germany, requested demobilization from the Army at the rank of *Hauptmann*, thus forfeiting his pension and disability rights. His application was approved on 08.06.1920, with retroactive effect from 11.03.1920.
Summer 1921	Returned to Germany, eventually buying a villa at Nr. 30 Döbereinerstraße in München-Obermenzing which was later seized by Bavarian authorities in the wake of the abortive *München-Putsch*.
00.00.1922–00.11.1923	Sporadic studies in economics and history at the University of München. There, together with Rudolf Hess, he attended lectures by the nationalist historian Alexander von Müller, brother-in-law of *NSDAP*-cofounder Gottfried Feder. He was also employed as salesman of aircraft equipment during this period.
00.12.1922	Joined the *NSDAP*. In his courtroom testimony of 13.03.1946, he spoke of how he had come to make this decision:

> Then one day, on a Sunday in November or October of 1922, the demand having been made again by the Entente for the extradition of our military leaders, at a protest demonstration in Munich – I went to this protest demonstration as a spectator, without having any connection with it. Various speakers from parties and organizations spoke there. At the end Hitler, too, was called for. I had heard his name once before briefly and wanted to hear what he had to say. He declined to speak and it was pure coincidence that I stood nearby and heard the reasons for his refusal. He did not want to disturb the unanimity of the demonstration; he could not see himself speaking, as he put it, to these tame, bourgeois pirates. He considered it senseless to launch protests with no weight behind them. This made a deep impression on me; I was of, the same opinion.
>
> I inquired and found that on the following Monday evening I could hear Hitler speak, as he held a meeting every Monday evening. I went there, and there Hitler spoke in connection with that demonstration, about Versailles, the treaty of Versailles, and the repudiation of Versailles.
>
> He said that such empty protests as that of Sunday had no sense at all – one would just pass on from it to the agenda – that a protest is successful only if backed by power to give it weight. Until Germany had become strong, this kind of thing was of no purpose.
>
> This conviction was spoken word for word as if from my own soul. On one of the following days I went to the office of the *NSDAP*. At that time I knew nothing of the program of the *NSDAP*, and nothing further than that it was a small party. I had also investigated other parties. When the National Assembly was elected, with a then completely unpolitical attitude I had even voted democratic. Then, when I saw whom I had elected, I avoided

SA-Führer Göring in full regalia, 1923.

Another photo of Göring as commander of the *SA*, walking beside *General* Erich Ludendorff in the autumn of 1923. At far right is the leader of the *Bund Oberland*, Dr. Friedrich Weber.

A Heinrich Hoffmann postcard of 'Hauptmann Göring' from 1932. (Roger Bender photo)

Two views of Hermann Göring attending the *4. Reichsparteitag der NSDAP* in Nürnberg, August 1929. The bespectacled man behind him in the photo at left is Hinrich Lohse, *Gauleiter* of Schleswig-Holstein.

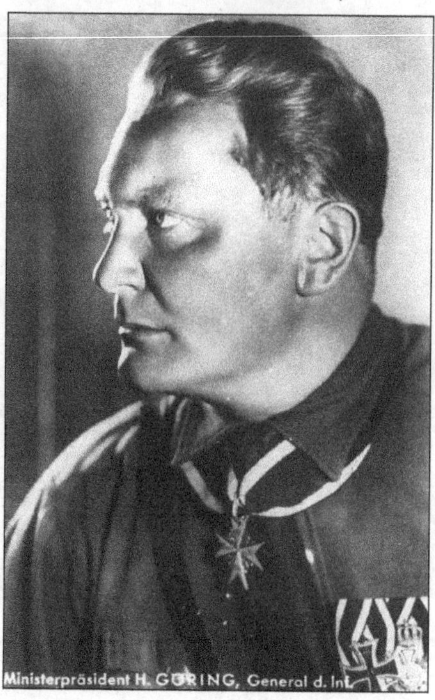

Heinrich Hoffmann's 1933 postcard of Hermann Göring. (NARA, Hoffmann collection).

politics for some time. Now, finally I saw a man here who had a clear and definite aim. I just wanted to speak to him at first to see if I could assist him in any way. He received me at once and after I had introduced myself he said it was an extraordinary turn of fate that we should meet. We spoke at once about the things which were close to our hearts – the defeat of the fatherland, and that one could not let it rest with that.

The chief theme of this conversation was again Versailles. I told him that I myself to the fullest extent, and all I was, and all I possessed, were completely at his disposal for this, in my opinion, most essential and decisive matter: The fight against the Treaty of Versailles.

The second point which impressed me very strongly at the time and which I felt very deeply and really considered to be a basic condition, was the fact that he explained to me at length that it was not possible under the conditions then prevailing to bring about, in co-operation with only that element which at that time considered itself national – whether it be the political so-called nationalist parties or those which still called themselves national, or the then existing clubs, fighter organizations, the Free Corps, et cetera – with these people alone it was not possible to bring about a reconstruction with the aim of creating a strong national will among the German people, as long as the masses of German labor opposed this idea. One could only rebuild Germany again if one could enlist the masses of German labor. This could be achieved only if the will to become free from the unbearable shackles of the Treaty of Versailles were really felt by the broad masses of the people, and that would be possible only by combining the national conception with a social goal.

He gave me on that occasion for the first time a very wonderful and profound explanation of the concept of National Socialism; the unity of the two concepts of nationalism on the me hand and socialism on the other, which should prove themselves the absolute supporters of nationalism as well as of socialism – the nationalism, if I may say so, of the bourgeois world and the socialism of the Marxist world. We must clarify these concepts again and through this union of the two ideas create a new vehicle for these new thoughts.

Then we proceeded to the practical side, in regard to which he asked me above all to support him in one point. Within the Party, as small as it was, he had made a special selection of these people who were convinced followers, and who were ready at any moment to devote themselves completely and unreservedly to the dissemination of our idea.

He said that I knew myself how strong Marxism and communism were everywhere at the time, and that actually he had been able to make himself heard at meetings only after he had opposed one physical force disturbing the meeting with another physical force protecting the meeting; for this purpose he had created the *SA*. The leaders at that time ware too young, and he had long been on the lookout for a leader who had distinguished himself in some way in the last war, which was only a few years ago, so that there would be the necessary authority. He had always tried to find a 'Pour le Merite' aviator or a 'Pour le Merite' submarine man for this purpose, and now it seemed to him especially fortunate that I in particular, the last commander of the 'Richthofen Squadron', should place myself at his disposal.

I told him that in itself it would not be very pleasant for me to have a leading part from the very beginning, since it might appear that I had come merely because of this position. We finally reached an, agreement that for 1 to 2 months I was to remain officially in the background and take over leadership only after that, but actually I was to make my influence felt immediately. I agreed to this, and in that way I came together with Adolf Hitler." (ibid)

01.03.1923–00.11.1923	*Oberster SA-Führer* (on extended leave following the *München-Putsch* of 09.11.1923). Succeeded Hans Ulrich Klintzsch. In his IMT testimony on 13.03.1946, he described his primary tasks in leading the early SA:

At first it was important to weld the *SA* into a stable organization, to discipline it, and to make of it a completely reliable unit which had to carry out the orders which I or Adolf Hitler should give it. Up to that point it had been just a club which had been very active, but which still lacked the necessary construction and discipline. I strove from the beginning to bring into the *SA* those members of the Party who were young and idealistic enough to devote their free time and their entire energies to it. For at that time things were very difficult for these good men. We were very small in number and our opponents were far more numerous. Even in those days these men were exposed to very considerable annoyances and had to suffer all sorts of things.

In the second place I tried to find recruits among workmen; for I knew that among workmen particularly I should enroll many members for the *SA*. At the same time we had naturally to see to it that the meetings of the Party, which generally were limited at that time to Munich, Bavaria and Franconia, could actually be carried through in a satisfactory manner, and disturbances prevented. In most cases we succeeded. But sometimes we had a strong party of our opponents present. One side or the other still had weapons from the war and sometimes critical situations arose, and in some cases we had to send the *SA* as reinforcements to other localities. (ibid)

08./09.11.1923 As *SA* commander, played a leading role in the abortive *München-Putsch*, during which he was seriously wounded in the groin (where a bullet had missed an artery by millimeters) and thigh by police bullets and concrete fragments as he advanced with Hitler and General Ludendorff toward the Feldherrnhalle. In the aftermath of the revolt, he was taken by his *SA* men to the home of a furniture dealer, Herr Ballin, whose wife, Ilse, had been a nurse during the war. Frau Ballin was able to stop the bloodflow and cared for him until nightfall. Although the Ballins were Jewish and knew the Party to which Göring belonged, they did not call the police. From the Ballins' home, Göring was carried to the clinic of Professor Alwin Ritter von Ach, who was sympathetic to the Nazi cause. From von Ach's clinic, Göring was driven by his wife to Partenkirchen, 70 miles south of München, where he was briefly hidden at the villa of a Dutch Nazi sympathizer, *Major* (retired) Schüler van Krieken.

10.11.1923 Arrest warrant for Göring issued by the München police.

10.11.1923 Driven by *SA* man Franz Thanner from Partenkirchen to the Austrian border. An officer of the *Schutzpolizei* station in Garmisch-Partenkirchen, *Leutnant* Nikolaus Maier (a future *Luftwaffe Generalmajor*), had issued instructions to the frontier post at Mittenwald that Göring was to be arrested if he arrived there. When the car carrying Göring arrived at the border, he and the other occupants were escorted under police guard back to Garmisch. Briefly held at the Wiggers Sanatorium in that city, he

	soon fled together with his wife. *SA* man Thanner once again drove him to the Mittenwald frontier post, this time breaking through the barrier into Austria. Once there, Göring was given a false passport, then taken to the Golden Lamb Inn in Seefeld.
12.11.1923	Checked into the Tirolerhof hotel in Innsbruck, owned by a Nazi sympathizer, then given medical treatment by a Dr. Sopelsa.
13.11.1923–24.12.1923	Hospitalized in Innsbruck. His severe blood loss and pain from infection led to his being treated heavily with opiates, and as a result, he became severely addicted to morphine.
24.12.1923	Released from hospital and returned to the Tirolerhof.
05.03.1924	Obtained an Austrian passport in Innsbruck.
24.05.1924	Issuance of an order for Göring's arrest due to his leading role in the *München-Putsch*.
Early May 1924-Spring 1924	Sent to Italy by order of the imprisoned Adolf Hitler, for the purpose of meeting with Benito Mussolini and obtaining from the Italian leader a loan of 2 million lire to rebuild the shattered Nazi Party. After several days of sightseeing in Venice and Florence, the Görings traveled to Rome, staying at the elegant Hotel Eden from 11.05.1924. Refused an audience with Mussolini, he managed only to meet with the diplomat Giuseppe Bastianini, to whom the journalist and member of Mussolini's staff, Dr. Leo Negrelli, had introduced him. In a letter dated 19.09.1924, Göring had written: "Italy must seek to find strong allies. One such ally would be a National-Socialist Germany under Hitler's leadership." (Charles R. Hamilton, *Leaders and Personalities of the Third Reich, Volume 1*) For financial reasons, the Görings eventually had to move to the less expensive Hotel de Russie. His negotiations with the Fascists failed dismally, with Bastianini ultimately asking him to leave Rome. The Görings left Rome for Venice on 11.08.1924, staying at the Grand Hotel Britannia.
Spring 1925– Summer 1926	Traveled with Frau Göring to her home country of Sweden and lived in a modest apartment at No. 23 Odengatan in Stockholm. He was employed for several weeks as a pilot for *Nordiska Flygrederiet*.
06.08.1925–30.08.1925	Voluntarily registered himself at the Aspuddens Nursing Home in Stockholm due to his morphine addiction. On 30.08.1925, he broke open a medicine cabinet to obtain Eukodal, a synthetic derivative of morphine, giving himself two injections. On 01.08.1925, he leaped from his bed, demanding more medication. When this was refused, he dressed himself and attempted to leave the building, but found the doors locked. He ran to his room, retrieving a sword cane that he brandished at nursing staff until they gave him the requested injections. When police and firemen came to the nursing home that evening, he vainly attempted to resist but was placed in a straitjacket and transported to the Katarina Hospital.
02.09.1925–07.10.1925	Committed by his wife to the Långbro Asylum near Stockholm. The doctor who attended to Göring described him as depressed, suicidal, egocentric, and a Jew-hater. His condition was seen to improve, resulting

SA-Gruppenführer Göring in 1932. (Hermann-Historica, Auctioneers, München)

Hitler is flanked by the newly appointed *Reich* Interior Minister Dr. Wilhelm Frick and Göring shortly after the *"Machtergreifung"* (Seizure of Power), 1933.

Göring attends the 02.10.1932 Gauparteitag in Wien. Behind
him are Julius Streicher and Paul Schulz.

	in his discharge from Långbro, with a clean bill of mental health, by Prof. Olof Kinberg on 07.10.1925.
12.11.1925	Order for Göring's arrest in Germany officially lifted.
14.05.1926	Indictment against Göring for high treason officially nullified due to an amnesty issued by *Reich* President Paul von Hindenburg.
22.05.1925–Early June 1926	At his own request, returned to the Långbro Asylum to complete their program of detoxification from Eukodal. He was ultimately certified as completely cured by the assistant medical superintendent, Dr. C. Franke.
Summer 1926–00.01.1927	Employed by *Bayerische Motorenwerke (BMW)* as a sales representative for aircraft engines in Sweden.
00.01.1927	Returned to Germany with a commission to sell parachutes for the Swedish firm of *Tornblad*. He soon met with Hitler who dismissively advised him to settle in Berlin and establish contacts with members of society in that city, then rented a hotel room off the Kurfurstendamm. He befriended his later adjutant, the wealthy Paul "Pilli" Körner, during this period. With Körner as chauffeur, Göring attempted to sell parachutes in Germany.
00.00.1927	Visited Turkey (possibly for the purpose of obtaining opium) and Great Britain.
07.09.1927–26.09.1927	Returned to Sweden and readmitted – due to continued morphine abuse – to the Långbro Asylum.
00.01.1928	Returned to Berlin and settled in an office on the Geisberg-Straße where he was again engaged in parachute sales. He shared this office with another aviation businessman, Fritz Siebel.
01.04.1928	Rejoined the *NSDAP* (having been deleted from the membership files by Alfred Rosenberg).
20.05.1928–23.04.1945	Member of the *Reichstag* (*NSDAP* Faction; representing *Wahlkreis 4, Potsdam I* from 31.07.1932). He received his parliamentary seat by means of extortion, threatening a lawsuit against the *NSDAP* for all monies owed him by the Party back to 1922. He thus forced the Party to include him on its list of candidates for the *Reichstag* in the 20.05.1928

SA-Obergruppenführer Göring chatting with Dr. Goebbels during the *5.Reichsparteitag der NSDAP* (30.08.1933-03.09.1933) in Nürnberg. Visible in the background are *Reichsleiters* Max Amann, Franz Xaver Schwarz, and Dr. Wilhelm Frick. (NARA photo)

	elections. Hitler agreed, guaranteeing him a seat provided at least 7 Nazi delegates were selected – the Party won 12 seats in the Social Democratic and Communist-dominated German Parliament, this number rising to 107 in the elections of 14.09.1930.
00.00.1928	Hired as a "consultant" to *BMW* and *Heinkel* and as a paid advisor to the German airline *Lufthansa*.
00.11.1928	Moved with Frau Göring into a luxurious apartment on the Badensche-Straße, Berlin.
00.10.1930–00.00.19__	*Reichsschlichter der NSDAP* (*Reich* Arbitrator of the Nazi Party).
14.09.1930–00.01.1933	Personal and political representative of the *Führer* in Berlin. During this period, he was active in raising funds for the Party through his contacts with *Krupp, Thyssen, Lufthansa, Heinkel, Messerschmitt*, the *Deutsche Bank, BMW*, etc.
13.10.1930–04.06.1932	*2. Stellvertreter* (2nd Deputy) to the Chairman of the *NSDAP* Faction and *Schriftführer* (secretary) of the *Reichstag*.
00.11.1930–00.00.1931	*Leiter* of the *Fliegerstaffel* of *NSDAP-Kreisgruppe Frankfurt-am-Main*.
00.04.1931–00.00.1933	*Politischer Kommissar Oberost der NSDAP* in Berlin.
00.05.1931	Sent by Hitler to Rome for three weeks, tasked with meeting Vatican leaders and allaying their fears that the *NSDAP* was a pagan organization (also met with Mussolini, King Vittorio Emmanuel III, and Italo Balbo).

00.10.1931–00.00.193_	*Führer* of the *Nationalen Deutscher Luftverband* (National German Air Association).
00.10.1931	Traveled to Sweden for the funeral of his wife, Carin (who had fallen to tuberculosis on 17.10.1931), then returned to Berlin and moved into the Kaiserhof Hotel.
Mid-November 1932	Sent by Hitler to Rome to meet with Benito Mussolini (possibly to raise funds for the *NSDAP*, which had lost 2 million votes in the *Reichstag* elections of 06.11.1932 and was therefore reduced from 230 seats to 196).
00.12.1931	Appointed to the following posts:

- *Inspekteur* of the *Inspektion für Flugwesen der NSDAP* (Aviation Inspectorate of the Nazi Party)
- 1.*Vorsitzender* (1st Chairman) of the *NS-Fliegerkorps*
- *Vizepräsident* of the *Nationale Deutsche Automobilclub* (*NDAC*)
- *1. Vorsitzender* of the *Abteilung Luftfahrt* in the *NDAC*

18.12.1931–13.04.1932	Assigned as an *SA-Führer z. V.* to the *Obersten SA-Führung.*
Spring 1932	Vacation on the Isle of Capri.
01.07.1932–26.01.1934	Assigned to the *Stab der Obersten SA-Führer.*
30.08.1932–23.04.1945	*Präsident des Deutscher Reichstag* (President of the German National Parliament).
22.01.1933–30.01.1933	*Bevollmächtigter für weitere Verhandlungen* (Plenipotentiary for Further Negotiations [toward the establishment of a Hitler-led coalition government]) of the *NSDAP.*
30.01.1933–05.05.1933	*Reichsminister ohne Geschäftsbereich* (*Reich* Minister without Portfolio).
30.01.1933–05.05.1933	*Reichskommissar für den Luftverkehr* (*Reich* Commissioner for Air Traffic; redesignated as *Reichskommissar für die Luftfahrt* [*Reich* Commissioner for Aviation] on 02.02.1933).
00.02.1933	Issued an order which contained the following:

Each bullet which now leaves the barrel of a policeman's revolver is my bullet. If you call that a murder, it is I who am the murderer. These are my orders and I give them my full support. I assume the responsibility and I am not afraid. (Jacques Delarue, *Gestapo: A History of Horror*, p. 34)

17.02.1933	Delivered a speech to Prussian Police officials from which the following are excerpts:

No hesitation to shoot in case of need. Every policeman must grasp the fact that inaction is a more serious crime than an error committed in the execution of orders received.

You shall act; that is always better than inaction. In this case, if you follow my principles, you shall be backed up by me fully and entirely. And have no doubt about this: As long as possible we intend to apply ordinary means, but if conditions require it, we shall use all means and we shall not hesitate to use the heaviest weaons. (Translations from Erich Gritzbach, ed, *Hermann Göring, Reden-und Aufsätze*, in ibid)

22.02.1933   Established the *Hilfspolizei (Hipo)*, an auxiliary police force composed of approximately 25,000 *SA* and 15,000 *SS* men. Among other duties, the Hilfspolizei rounded up political opponents and sent them to the Oranienburg and Papenburg concentration camps which Göring had also set up. By October 1933, some 700 political prisoners had been murdered in these and other "wild camps" of the *SA*.

25.02.1933   Established the *Polizeiabteilung z.b.V. "Wecke"*, an elite 414-man police unit designed to seek out and destroy "enemies of the State" (principally communists). It was named for its commander, *Polizei-Major* (promoted to *Polizei-Oberstleutnant* on 21.03.1933) Walther Wecke, who led the unit until 05.06.1934 (he was succeeded on 06.06.1934 by *Oberstleutnant der Landespolizei* Friedrich Wilhelm Jakoby). On 17.07.1933, the unit was redesignated *Landespolizeigruppe Wecke z.b.V.*, and on 22.12.1933 it became *Landespolizeigruppe General Göring*. Further redesignation, as *Regiment General Göring*, took place on 01.04.1935.

27.02.1933   The "Reichstagsbrand" (*Reichstag* Fire) in Berlin. The building was largely gutted, however Göring's office remained intact. A mentally troubled ex-communist Dutchman, Marinus van der Lubbe, was discovered by firefighters as they attempted to douse the flames, and he thus became the chief suspect. Göring was advised of the fires outbreak at 2130 hours, and he raced to the scene. On the 28th, at Göring's command, some 4,000 communists were arrested and all Social Democratic and Communist newspapers banned. Hitler issued a decree, signed by *Reichspräsident* von Hindenburg, concerning a "Law for the Protection of the People and State" (also known as the "*Reichstag* Fire Decree"). This law stated:

> Sections 114, 115, 117, 118, 123, 124, and 153 of the Constitution of the German *Reich* are suspended until further notice. Accordingly, restrictions of the freedom of the person, of speech, of press, of assembly and association, interference with the privacy of letters, of mails, of telegraph and telephone, also authorized search of domicile and confiscation and restraint of property are admissible also outside the limits normally set by the law … " (*Trials of War Criminals Before the Nuernberg Military Tribunals Under Control Council Law No. 10, Volume V: "The RuSHA Case" / "The Pohl Case"*)

Thus were basic freedoms, until then guaranteed by the Weimar Constitution, removed from the German people. As a result of the decree, Göring's *Hilfspolizei* was free to terrorize all real and perceived enemies of the new regime, and numerous "wild" concentration camps were set up for their detention and torture. Within a week of the fire, accusations arose, in Germany and abroad, that it had been a clever ploy devised by the Hitler, Göring, and other Nazis to permit their unchallenged dictatorship over the *Reich*. In postwar years, the rather disreputable ex-police officials Rudolf Diels and Hans Bernd Gisevius each placed the blame for the conflagration squarely on the Nazi leadership, in Diels' case Göring ("Göring knew exactly how the fire was to be started [and ordered me] to prepare, prior to the fire, a list of people who were to be arrested immediately after it.") and according to Gisevius, Goebbels (" … it was Goebbels who first thought of setting the *Reichstag* on fire"). It is now generally believed that van der Lubbe acted, as per his own claim, alone. On 23.12.1933, van der Lubbe was sentenced to death by the *Reichsgericht* in Leipzig, and beheaded on 10.01.1934.

03.03.1933 Delivered a speech ("We Bear the Responsibility") in Frankfurt-am-Main in which he declared:

> Fellow Germans, my measures will not be crippled by any judicial thinking. My measures will not be crippled by any bureaucracy. Here I don't have to give Justice, my mission is only to destroy and exterminate, nothing more! This struggle, fellow Germans, will be a struggle against chaos and such a struggle I shall not conduct with the power of any police. A bourgeois state might have done that. Certainly, I shall use the power of the state and the police to the utmost my dear Communists, so that you won't draw any false conclusions but the struggle to the death, in which my fist will grasp your necks, I shall lead with those down there,- those are the brown shirts. (Partial Translation of Document 1856-PS, in *Nazi Conspiracy and Aggression, Volume III*)

11.03.1933 Stated the following in a speech at Essen:

> I have only just begun my purge; it is far from finished. For us the people are divided into two parts: One which professes faith in the nation, the other which wants to poison and destroy. I thank my Maker that I do not know what objective is. I am subjective. I repudiate the idea that the police are a defense force for Jewish department stores. We must put an end to the absurdity of every rogue shouting for the police. The police are not there to protect rogues, vagabonds, usurers, and traitors. If people say that here and there someone has been taken away and maltreated, I can only reply: You can't make omelets without breaking eggs. Don't shout for justice so much, otherwise there might be a justice that is to be found in the stars and not in your paragraphs! Even if we make a lot of mistakes, we shall at least act and keep our nerve. I'd rather shoot a few times too short or too wide, but at least I shoot. (Translation from Johannes Hohlfeld, ed, *Deutsche Reichsgeschichte in Dokumenten, Vol. IV*, in Joachim C. Fest, *The*

Göring in his uniform as a
General der Infanterie, ca. 1935.

Göring, as *Reichsjägermeister,* delivers
a speech at Carinhall, ca. 1938.

Face of the Third Reich, p. 77)

10.04.1933	Established the *Forschungsamt* (*FA*, Research Office), a powerful and super secret intelligence and wiretapping agency. Initially a component of Göring's Prussian State Government under the general management of Paul Körner, it had been transferred to the *Reichsluftfahrtministerium* by 1937.
10.04.1933–00.04.1933	Mission to Italy to strengthen relations between Germany and Italy, during which he and Franz von Papen met three times with Mussolini and once, on 12.04.1933, with Pope Pius XII.
11.04.1933–23.04.1945	*Preußischer Ministerpräsident* (Prussian Prime Minister). He was appointed to this post via the following telegram from Hitler, dated 11.04.1933:

> With effect as per today's date I hereby appoint you *Ministerpräsident* of Prussia. I may request that you be so kind and assume the duties of this office on 20 April in Berlin.
>
> I am happy for this opportunity to demonstrate to you my trust and my gratitude for the services which you have rendered in support of the new uprising of the German Volk for over ten years as a fighter in our Movement; for the victorious execution of the National Revolution as temporary Minister of the Interior in Prussia; and, last but not least, for the unique loyalty with which

A formal portrait of *Reichsmarschall* Göring, 1940. (NARA photo)

A postcard image of *Reichsmarschall* Göring in white summer tunic, ca. 1941.

you have bound your destiny to mine.
Reich Chancellor Adolf Hitler

11.04.1933–30.04.1934	*Preußischer Minister des Innern* (Prussian Minister of Interior) and *Chef der Preußischen Polizei* (Chief of the Prussian Police).
26.04.1933	Issued a law creating the *Geheime-Staatspolizeiamt* (*Gestapa*, Secret State Police Office) under the *Preußische Ministerium des Innern* (i.e., under his personal control), and appointed as its director Rudolf Diels, head of *Abteilung Ia* (*Politische-Polizei*, Political Police) of the Prussian Interior Ministry. Diels was dismissed at the end of September 1933 and succeeded as acting head of the *Gestapa* by the former *Gauleiter* Paul Hinkler. During the 15 days that Hinkler led the office, rumors spread in *SA* circles that he was mentally weak and dependent on alcohol. At the end of November or start of December 1933, Diels returned from his exile in Czechoslovakia at Göring's request and reassumed the post.
27.04.1933–28.02.1935	*Mit dem geheimen Aufbau der Luftwaffe beauftragt* (charged with the secret building of the *Luftwaffe*).
29.04.1933	Appoined as *Präsident der Reichsluftschutzverband* (President of the *Reich* Air Defense Association).
05.05.1933–23.04.1945	*Reichsminister der Luftfahrt* (*Reich* Minister of Aviation).
00.07.1933	Excerpt from a statement made by Göring at a press conference:

... Whoever in the future lifts his hand against a representative of the National Socialistic movement or of the State, must know that he will lose his life in a very short while. Furthermore, it will be entirely sufficient, if he is proven to have intended the act, or, if the act results not in a death, but only in an injury. (Partial Translation of Document 2494-PS [excerpt from the *Völkischer Beobachter*, Berlin Edition, 23./24.07.1933], in *Nazi Conspiracy and Aggression, Volume V*)

08.07.1933–23.04.1933	*Präsident der Preußische Staatsrat* (President of the Prussian State Council) and *Protektor der Preußischen Akademie der Künste* (Guardian of the Prussian Academy of the Arts) in Berlin.
20.10.1933–24.10.1933	To Sweden to visit the grave of his wife.
06.11.1933–07.11.1933	Meeting with Benito Mussolini in Rome to reassure the Italian dictator that Germany did not intend to annex Austria.
30.11.1933	Issued a decree that removed the *Gestapo* from *Reich* Interior Ministry Control.
02.12.1933–26.06.1936	*Chef der Preußische Geheime Staatspolizei* (Chief of the Prussian Secret State Police [*Gestapo*]).
31.12.1933	Letter from Hitler to Göring:

My dear Göring:

When in November 1923 the Party tried for the first time to conquer the power of the State, you as Commander of the *SA* created within an extraordinarily short time that instrument with which I could bear that struggle. Highest necessity had forced us to act, but a wise providence at that time denied the success. After receiving a grave wound you again entered the ranks as soon as circumstances permitted as my most loyal comrade in the battle for power. You contributed essentially to creating the basis for, the 30th of January. Therefore, at the end of a year of the National Socialist Revolution, I desire to thank you wholeheartedly, my dear Party Comrade Göring, for the great values which you have for the National Socialist Revolution and consequently for the German people.

In cordial friendship and grateful appreciation.

Yours,

(signed) Adolf Hitler! (Translation of Document 3259-PS, in *Nazi Conspiracy and Aggression, Volume II*)

27.01.1934	Appointed as *Ehrenführer* of *SA-Standarte 1*, Berlin.
00.03.1934	Visited the Polish State Hunting Ground at Bialowiéza, at the invitation of Józef Lipski (Polish Ambassador in Berlin).
09.04.1934	Appointed as *Präsident* of the *Kameradschaft vom 30 Januar 1933*.
20.04.1934	Appointed *Reichsführer-SS* Heinrich Himmler as Inspector and Deputy Head of the *Geheime-Staatspolizei* (*Gestapo*), in succession

to Rudolf Diels, with Reinhard Heydrich as Director of the *Geheime-Staatspolizeiamt*. Göring relinquished directive functions over the *Gestapo* to Himmler in November 1934. He retained titular command until 26.06.1936, when the *Gestapo* was placed under Heydrich in his capacity as *Chef* of the *Sicherheitspolizei* (composed of the *Gestapo* and *Kriminalpolizei*).

02.05.1934	Appointed as *Oberster Leiter der Landespolizei in Preußen* (Supreme Head of the State Police in Prussia).

15.05.1934–25.05.1934(?) "Vacation" in southeastern Europe (accompanied by Paul Körner, Erhard Milch, Prince Philipp of Hessen, Hanns Kerrl, and his future wife, Emmy Sonnemann). The actual purpose of this tour, during which he made stops in Budapest, Belgrade, and Athens, was to convince these countries of Germany's intention to safeguard them from Italian territorial ambitions.

00.00.1934–00.00.1936 Co-chairman (together with General Werner von Blomberg) of the *Deutschen Olympia-Komittees für Reiterei* (German Olympic Committee for Equestrian Events).

26.06.1934 Delivered a speech at an *NSDAP* gathering in Hamburg in which he railed against "interest cliques" and "unproductive critics [i.e. Kurt von Schleicher and Gregor Strasser, both of whom were snuffed out in Göring's and Himmler's organized bloodbath of 30.06–01.07.1934]:

> When one day the cup runneth over, I will strike! We have worked as no one has ever worked before, because we have behind us a Volk which trusts us. Anyone who gnaws away at this trust is committing a crime against the Volk, he is committing treachery and high treason. He who designs to destroy this trust, destroys Germany. He who sins against this trust has put his own head in the noose." (Dr. Max Domarus, ed, *Hitler. Speeches and Proclamations, 1932–1945: The Years 1932–1934*, p. 466)

28.06.1934 Together with Hitler, served as a witness to the wedding of his friend *Gauleiter* Josef Terboven in Essen. He returned that afternoon to Berlin.

30.06.1934–01.07.1934 Presided over the purge of the *SA* leadership and others from his villa in Berlin. Together with Himmler, Heydrich, *Generalmajor* Walter von Reichenau, and others, Göring decided who would live and who would be shot for their part in an alleged conspiracy to make Kurt von Schleicher (murdered in his home together his wife) Chancellor and *SA-Stabschef* Ernst Röhm (shot by Theodor Eicke at Stadelheim Prison near München) Defense Minister. Göring used this "Night of the Long Knives" as an opportunity to settle old scores. A witness to Göring's role in the purge was the *Kripo* official (and later member of the conspiracy to assassinate Hitler), Hans Bernd von Gisevius, who recalled:

> Suddenly loud shouting reached us from the adjoining room. This room was Göring's study, and here the execution committee was

meeting. Now and then couriers from the Gestapo rushed into and out of this room, slips of white paper in their hands. Through the door we could see Göring, Himmler, Heydrich, and little Pilli Koerner, under-secretary to Göring in his capacity as minister-president. We could see them conferring, but naturally we could not hear what was being said. Occasionally, however, we could catch a muffled sound: 'Away!' or 'Aha!' or 'Shoot him!' For the most part we heard nothing but raucous laughter. The whole crew of them seemed to be in the best humor.

Göring radiated cheerful complacency. It was easy to see that he was in his element. He swaggered abou the room, his long hair waving, in a white military tunic and blue-gray military trousers, with high black boots that reached over his fat knees. There goes Puss-in-Boots, I thought suddenly. The epithet seemed appropriate, not so much because of the spiteful, catlike attitude of the man as because there was something puffed-up, counterfeit, and ridiculous about him.

But to continue – we suddenly heard loud shouting. A police major, his face flaming, rushed out of the room, and behind him came Göring's hoarse, booming voice: 'Shoot them ... take a whole company ... shoot them ... shoot them at once!' The written word cannot reproduce the undisguised blood lust, fury, vicious vengefulness, and, at the same time, the fear, the pure funk, that the scene revealed ... (Gisevius, *To the Bitter End*, p. 152)

On 02.07.1934, *Reich* President von Hindenburg sent the following telegram to Göring:

I may express to you my thanks and appreciation for your vigorous and successful action in crushing the attempt to commit high treason.

With comradely regards, von Hindenburg (Dr. Max Domarus, ed, *Hitler. Speeches and Proclamations, 1932–1945: The Years 1932–1934*, p. 480)

An interview with Göring published in the 03.07.1934 issue of the Swedish newspaper *Nya Dagligt Allehand* quoted him as follows:

... the action became necessary not only due to the plans for a putsch. Röhm's private life as well as that of the other persons who have now been arrested was such that it meant a scandal for the entire *SA*. They were a moral cancer which had to be cut out. (ibid, p. 482)

According to Hitler's *Reichstag* speech of 13.07.1934, 74 deaths had resulted from the purge (postwar research indicates the minimum figure was probably 85, including 12 members of the *Reichstag*; Göring cynically conceded that "in the general excitement some mistakes were made." According to official reports, 1,124 people had been arrested on Göring's orders). In his delivery to the *Reichstag*, Hitler gave special recognition to Göring:

> Meanwhile Minister President Goering had previously received my instructions that in case of a purge, he was to take analogous measures at once in Berlin and in Prussia. With an iron fist he beat down the attack on the National Socialist State before it could develop. (Excerpt from Document 3442-PS, in *Nazi Conspiracy and Aggression, Volume II*)

The Prussian *Ministerpräsident* celebrated the successful conclusion of the operation with a crab feast to which were invited his colleagues in murder – Werner Blomberg, Himmler, Erhard Milch, and Paul Körner. A second celebration took place at Carinhall on 07.03.1934, with Göring playing host to Reinhard Heydrich and his *Gestapo* staff. He later justified the murders with: "Their treason was as clear as day … . After all, there had been a plot against the *Führer*'s life. The whole point was to act fast, as a deterrent." On 12.07.1934, Göring delivered a speech to the Public Prosecutors of Prussia from which the following is an excerpt:

> The action of the Government in the days of the Roehm revolt was the highest realization of the legal consciousness of the people. Later the action which itself was justified, now has been made legal by the passage of a law. (Translation of Document 2496-PS, in *Nazi Conspiracy and Aggression, Volume V*)

In his testimony of 14.03.1946, Göring gave his version of the events of June/July 1934:

> The main controversy between Röhm and us was that unlike his predecessor Pfeffer, wanted a stronger revolutionary way to be adopted, whereas the *Führer*, as I said earlier, had ordered a legal development, the final victory of which could be expected.
>
> After the seizure of power Röhm desired, under all circumstances, to get hold of the *Reich* Defense Ministry. The *Führer* refused that point-blank, as he did not wish the Armed Forces to be conducted politically in any way, or to have any political influence brought to bear on the Armed Forces.
>
> The contrast between the Armed Forces and the Röhm group am intentionally not speaking of a contrast between the Armed Forces and the *SA,* since there was none, but solely of this leadership group, which called itself at that time the *SA* Leadership and it actually was-was that Röhm wanted to remove the greater number

of the generals and higher officers who had been members of the *Reichswehr* all this time, since it was his view that these officers did not offer a guarantee for the new State, because, as he expressed it, their backbone had been broken in the course of the years and they were no longer capable of being active elements of the new National Socialist State.

The *Führer*, and I also, had exactly the opposite point of view in this connection.

Secondly, the aims of the Röhm-minded people, as I should like to call them, were directed in a different direction, towards a revolutionary act; and they were opposed to what they called reaction. They definitely desired to adopt a more Leftist attitude. They were also sharply opposed to the Church and also very strongly opposed to the Jews. Altogether, and I refer only to the clique consisting of certain persons, they wished to carry out a revolutionary act.

That Röhm placed all his people in leading positions in the *SA* and removed the decent elements, and misguided the decent *SA* people without their knowledge, is a well-known fact. If encroachments did occur at that time, they always involved the same persons, first of all the Berlin *SA* leader, Ernst, secondly the Breslau leader, Heines, the Munich and Stettin leaders, et cetera.

A few weeks before the Röhm Putsch a low-ranking *SA* leader confided in me that he had heard that an action against the *Führer* and his corps was being planned to replace the Third *Reich* as expeditiously as possible by a final Fourth *Reich*, an expression which these people used.

I myself was urged and begged to place outside my house not only guards from a police regiment but also to appoint an *SA* guard of honor. I had agreed, and later on I heard from the commander of these troops that the purpose of that guard of honor was to arrest me at a given moment. I knew Röhm very well. I had him brought to me. I put to him openly the things which I had heard. I reminded him of our mutual fight and I asked him to keep unconditional faith with the *Führer*.

I brought forward the same arguments which I have just mentioned, but he assured me that he naturally was not thinking of undertaking anything against the *Führer*. Shortly afterward I received further news to the effect that he had close connections with those circles who also were strongly opposed to us. There was, for instance, the group around the former *Reich* Chancellor Schleicher. There was the group around Gregor Straßer, the former member of the *Reichstag* and organizational leader of the Party, who had been excluded from the Party. These were groups who had belonged to the former trade unions and were rather inclined to the Left. I felt it my duty to consult the *Führer* now on this

An autographed Röhr postcard photo of Göring. (Hermann-Historica, Auctioneers, München).

Reichsmarschall Göring, ca. 1941. (Roger Bender photo)

subject. I was astonished when he told me that he, too, already knew about these things and considered them a great threat. He said that he wished, however, to await further developments and observe them carefully.

The next event occurred just about as the witness Körner described it here, and therefore I can skip it. I was given the order to proceed immediately against the implicated men of the Röhm group in northern Germany. It was decided that some of them were to be arrested. In the course of the day the *Führer* ordered the execution of the *SA* leader of Pomerania, Ernst, and two or three others. He himself went to Bavaria where the last meeting of a number of Röhm leaders was taking place and personally arrested Röhm and these people in Wiessee.

At that time this matter presented a real danger, as a few *SA* units, through the use of false passwords, had been armed and called up. At one spot only a very short fight ensued and two *SA* leaders were shot. I deputized the police, which in Prussia was then already under Himmler and Heydrich, to make the arrests. Only the headquarters of Röhm, who himself was not present, I had occupied by a regiment of the uniformed police subordinated to me. When the headquarters of the *SA* leader Ernst in Berlin were searched, we found in the cellars of those headquarters more submachine guns than the whole Prussian police had in its possession.

After the *Führer*, on the strength of the events which had been met with at Wiessee, had ordered who should be shot in view of the state of national emergency, the order for the execution of Ernst, Heydebreck, and some of the other Röhm collaborators was issued. There was no order to shoot the other people who had been arrested. In the course of the arrest of the former *Reich* Chancellor Schleicher, it happened that both he and his wife were killed. An investigation of this event took place and it was found that when Schleicher was arrested, according to the statements of the two witnesses, he reached for a pistol, possibly in order to kill himself, whereupon the two men raised their pistols and Frau Schleicher threw herself upon one of them to hold him, causing his revolver to go off. We deeply regretted that event.

In the course of that evening I heard that other people had been shot as well, even some people who had nothing at all to do with this Röhm Putsch. The *Führer* came to Berlin that same evening.

After I learned this, later that evening or night, I went to him at noon the next day and asked him to issue an order immediately, that any further execution was under any circumstances forbidden by him, the *Führer*, although two other people who were deeply involved and who had been ordered by the *Führer* to be executed, were still alive. These people were consequently left alive. I asked him to do that because I was worried lest the matter should get out of hand – as, in fact, it had already done to some extent – and I told the *Führer* that under no circumstances should there be any further bloodshed. This order was then given by the *Führer* in my presence, and it was communicated at once to all offices. The action was then announced in the *Reichstag*, and it was approved by the *Reichstag* and the *Reich* President as an action called for by the state of national emergency. It was regretted that, as in all such incidents, there were a number of blunders. The number of victims has been greatly exaggerated. As far as I can remember exactly today, there were 72 or 76 people, the majority of whom were executed in southern Germany. (*Trial of the Major War Criminals Before the International Military Tribunal, Nuremberg, 14 November 1945 – 1 October 1946, Volume IX*)

03.07.1934–23.04.1945	*Reichsforstmeister* (*Reich* Master of Forests) and *Preußischer Landesforstmeister* (Prussian State Master of Forests).
03.07.1934–23.04.1945	*Reichsjägermeister* (*Reich* Master of the Hunt).
00.10.1934	Attended the funeral of King Alexander of Yugoslavia in Belgrade.
07.12.1934–23.04.1945	*Stellvertreter des Reichskanzlers in allen Angelegenheiten der Staatsführung* (Deputy to the *Reich* Chancellor in All Aspects of National Government), appointed via *Führer* decree.

Reichsmarschall Göring, ca. 1941.
(Roger Bender photo)

Reichsmarschall Göring smokes an
ornate pipe, probably during a visit to
Luftwaffe units in France in the Autumn
of 1940. (Roger Bender photo)

27.01.1935–31.01.1935	Official visit to Warsaw where he was tasked by Hitler to investigate Poland's readiness to work with Germany in the event of conflict with the Soviet Union.
30.01.1935–23.04.1945	*Reichsstatthalter in Preußen.*
01.03.1935–23.04.1945	*Chef der Reichsluftwaffe* (Chief of the *Reich* Air Force; redesignated *Oberbefehlshaber der Luftwaffe* on 01.06.1935). The Treaty of Versailles had forbidden Germany an air force. Göring countered that prohibition with pronouncements such as the following:

> As long as I am Minister for Air, I shall not cease to tell the world again and again, that so long as Germany is debarred from development in the air, so long she will remain defenceless. We demand equality of rights and nothing less. When the other nations are prepared to disband their military air fleets, then Germany will also agree to do the same. But as the other nations are arming for war in the air, we also reserve to ourselves the right to make similar technical preparations, so that the necessary measure of safety may be assured to the German people ... (Erich Gritzbach, *Hermann Goering: The Man and His Work*, p. 93)

In Berlin on 20.05.1936, he swore in 1,000 newly commissioned *Luftwaffe* officers and delivered a speech in which he declared:

> I repeat: I intend to create a *Luftwaffe* which, if the hour should strike, shall burst upon the foe like a chorus of revenge [wie ein Chor der Rache]. The enemy must have a feeling of being lost already before even having fought … . (Translation from Erich Gritzbach, ed, *Hermann Göring, Reden-und Aufsätze*)

01.04.1935	Appointed as *Oberbefehlshaber der Flakartillerie* (Commander-in-Chief of Anti-Aircraft Artillery).
18.08.1935–21.08.1935	*Schirmherr der 23.Deutschen Ostmesse* (Patron of the 23rd Eastern Exhibition) in Königsberg.
01.03.1936–00.00.1936	Member of the *Ehrenausschuß* (Committee of Honor) of the *Deutschen Kulturausstellung in Finnland* (German Cultural Exposition in Finland).
00.00.1936	Appointed as *Generalbevollmächtigter für die Kraftstoff-Verteilung* (General Plenipotentiary for the Allocation of Fuel).
04.04.1936–23.04.1945	*Reichskommissar für Rohstoffe und Devisen* (*Reich* Commissioner for Raw Materials and Foreign Exchange; later redesignated *Reichsbeauftragter für Rohstoff-und Devisen-Fragen* [*Reich* Representative for Raw Materials and Foreign Exchange Questions]).
01.05.1936–06.07.1936	*Leiter* of the *Dienststelle "Ministerpräsident Generaloberst Göring-Rohstoff-und Devisenstab"* (Raw Materials and Foreign Exchange Staff).
15.05.1936	Appointed as *Vorsitzender* of the *Gutachterausschuß für Exportfragen* (Committee of Experts on Export Questions).
29.07.1936–23.04.1945	*Präsident* of the *Deutsche Akademie für Luftfahrtforschung* (German Academy for Aviation Research).
29.07.1936	Appointed as a member of the Committee for the *Braune Band von Deutschland* (an organization set up to maintain the quality of horses for sporting activities and to control the quality of such stock for mounted units of the *SS* and other organizations).
18.10.1936–23.04.1945	*Beauftragter des Führers für den Vierjahresplan* (Representative of the *Führer* for the Four-Year Plan) and *Vorsitzender* of the *Generalrate der Vierjahresplan* (General Council of the Four-Year Plan).
18.10.1936–23.04.1945	*Vorsitzender der Kleine Ministerrat* (Chairman of the Small Ministerial Council).
00.01.1937	In Rome for meetings with Mussolini and the Italian Government.
12.01.1937	Appointed as *Chef* of the *SA-Standarte "Feldherrnhalle"* (München).
14.07.1937–00.04.1945	*Protektor der Preußischen Akademie der Künste* (Guardian of the Prussian Academy of Arts), Berlin.
15.07.1937–23.04.1945	Founder and Hauptleiter of the *Reichswerke AG für Erzbergbau und Eisenhütten "Hermann Göring"* in Salzgitter.
27.11.1937–15.01.1938	*Reichs-und Preußische Wirtschaftsminister* (m.d.F.d.G.b.). Given responsibility for the business of the *Reich* – and Prussian Ministries of Economics in the interim between the dismissal of Hjalmar Schacht and the appointment of Walter Funk.

04.02.1938–23.04.1945	Member of the *Geheime Kabinettsrat* (Secret Cabinet Council).
17.03.1938–18.03.1938	Chairman of the *Ehrengericht* (Court of Honor) against *Generaloberst* Werner Freiherr von Fritsch, who had been charged with alleged homosexual conduct. These charges were later dismissed, but von Fritsch never recovered from them and was to be killed in action during the siege of Warsaw, on 22.09.1939, as honorary colonel of *Artillerie-Regiment 12.*
26.03.1938	In a speech delivered in Vienna on that date, shortly after Austria's incorporation into the *Reich*, made the following statements:

> May everybody, friend or foe, know that in Germany a man is put to death only after the court has sentenced him to death and the *Führer* has confirmed the sentence. In Germany only one man decides over life and death: The *Führer*. Nobody else has such a right. Anybody, any agency, whether of state or Party, intrudes upon the most sacred right of the *Führer* if it would intrude here.
>
> Where there are 300,000 Jews, one cannot speak of [Vienna as] a German city ... [Vienna must again become a German city by means of] the elimination of the Jews ... As Commissioner of the Four Year Plan, I ... commission the *Reich* Governor in Austria [Dr. Seyss-Inquart] together with the *Reich* Plenipotentiary to take quietly all measures necessary for the expert transformation of the Jewish economy, i.e., the Aryanization of business and economic life, and to execute this program according to the laws but without mercy. (Translation from Erich Gritzbach, ed, *Hermann Göring, Reden-und Aufsätze*)

11.04.1938	Received the following telegram from the *Führer*:

> Dear *Generalfeldmarschall* Göring,
> Five years ago today you entered office as *Ministerpräsident* of Prussia. With heartfelt gratitude, I honor your loyal cooperation in the rebuilding of Germany. The feats you have accomplished within the past five years in the service of strengthening Germany belong to history. It is my sincere desire that I may continue to count upon your loyal assistance for many years to come.
>
> Yours in old friendship, Adolf Hitler. (Dr. Max Domarus, ed, *Hitler. Speeches and Proclamations, 1932–1945: The Years 1935–1938*, p. 1091)

19.04.1938	Appointed as *Ehrenmeister des deutschen Handwerks* (Honorary Master of German Craftsmanship).
12.11.1938	In the wake of "Kristallnacht", convened an interministerial meeting (from 1100 to 1440 hours) on "the Jewish Question" at the *Reichsluftfahrtministerium* in Berlin. Göring, outraged by the extent of the damage caused by Goebbels' "spontaneous" pogrom, opened the meeting as follows: "Today's meeting is of a decisive nature. I have received

Two formal portraits of Goring taken by Prof. Heinrich Hoffmann. (NARA photos)

a letter written on the *Führer*'s orders by the Stabsleiter of the *Führer*'s deputy Bormann, requesting that the Jewish question be now, once and for all, coordinated and solved one way or another. And yesterday once again did the *Führer* request by phone for me to take coordinated action in the matter." Other participants included *Reichsminister* Dr. Walter Funk (Ministry of Economics); *Reichsminister* Dr. Wilhelm Frick (Ministry of Interior); *Reichsminister* Lutz Graf Schwerin von Krosigk (Ministry of Finance); *Reichsminister* Franz Gürtner (Ministry of Justice); *Staatssekretär* Ernst Woermann (Foreign Office); *Staatssekretär* Fritz Reinhardt (Ministry of Finance); *Staatssekretär* Dr. Wilhelm Stuckart (Ministry of Interior); *SS-Obergruppenführer und General der Polizei* Kurt Daluege (*Chef* of *Hauptamt Ordnungspolizei*); *Reichskommissar* Josef Bürckel; Eduard Hilgard (Leader of the *Reich* Group for Insurance); *Staatssekretär* Hans Fischböck; *Ministerialdirektor* Rudolf Schmeer; *Reichsminister* Hanns Kerrl (Ministry of Church Affairs); *Reichsbankdirektor* Karl Blessing; and *Staatssekretär* Rudolf Brinkmann (Ministry of Economics). During the conference, the following exchange concerning passenger train accommodations for Jews (which is rather chilling in light of the wartime use of the German railway system with regard to them) took place between Göring and Goebbels:

Goebbels: ... Furthermore, I advocate that the Jews be eliminated

A formal portrait of Göring
in his *Reichsmarschall*'s cloak.
(Roger Bender photo)

Walter E. Herbst's woodcut
portrait of Hermann Göring, ca.
1942. (Roger Bender photo)

from all positions in public life in which they may prove to be provocative. It is still possible today that a Jew shares a compartment in a sleeping car with a German. Therefore, we need a decree by the *Reich* Ministry for Communications stating that separate compartments for Jews shall be available; in case where compartments are filled up, Jews cannot claim a seat. They shall be given a separate compartment only after all Germans have secured seats. They shall not mix with Germans, and if there is no more room, they shall have to stand in the corridor.

Goering: In that case, I think it would make more sense to give them separate compartments.

Goebbels: Not if the train is overcrowded!

Goering: Just a moment. There'll be only one Jewish coach. If that is filled up, the other Jews will have to stay at home.

Goebbels: Suppose, though, there won't be many Jews going on the express train to Munich, suppose there would be two Jews in the train and the other compartments would be overcrowded. These two Jews would then have a compartment all themselves. Therefore, Jews may 'claim a seat only after all Germans have secured a seat.

Goering: I'd give the Jews one coach or one compartment. And should a case like you mention arise and the train be

Two portraits of *Reichsmarschall* Göring by Heinrich Hoffmann, ca. 1942. (NARA photos)

overcrowded, believe me, we won't need a law. We'll kick him out and he'll have to sit all alone in the toilet all the way!

Goebbels: I don't agree. I don't believe in this. There ought to be a law... (Translation of Document 1816-PS, in *Nazi Conspiracy and Aggression, Volume IV*)

On the same date, he signed the following decrees (published in *Reichsgesetzblatt, 1938*, I, p. 1580):

Decree relating to the payment of a fine by the Jews of German nationality of 12 November 1938:

The hostile attitude of Jewry towards the German people and *Reich*, an attitude which does not even shrink from committing cowardly murder, makes decisive defensive action and harsh atonement necessary. I order, therefore, by virtue of the decree concerningthe execution of the Four Year Plan of 18 October 1936 as follows:

1 On the community of Jews in Germany the payment of a contribution of 1,000,000,000 *Reichsmark* to the German *Reich* is imposed.

2 Provisions for the implementation will be issued by the *Reich* Minister of Finance in agreement with the *Reich* Ministers

concerned.

Berlin, 12 November 1938

The Commissioner for the Four Year Plan

Göring, *Generalfeldmarschall*

(Translation of Document 1412-PS, in *Nazi Conspiracy and Aggression, Volume IV*)

Order eliminating Jews from German economic life of 12 November 1938:

On the basis of the Decree of 18 October 1936 for the execution of the Four Year Plan, the following is ordered:

ARTICLE 1

1 From 1 January 1939 operation of retail shops or mail order houses as well as independent handicrafts businesses is forbidden to Jews ...

2 Moreover from the same date it is forbidden to Jews to offer goods or services in markets of any kind, fairs, or exhibitions, or to advertise such or accept orders therefor.

3 Jewish shops operated in violation of this order will be closed by police ...

ARTICLE 2

1 No Jew can manage a firm according to the interpretation of the term 'manager' under the Law for National Labor of 20 January 1934.

2 If a Jew is an executive in a business concern he may be dismissed with notice of six weeks. At expiration of this period all claims resulting from the employee's contract, especially claims for severance pay or pensions, become null and void.

ARTICLE 3

1 No Jew can be a member of a cooperative society.

2 Jewish members of cooperatives lose membership from 21 December 1938. No notice is necessary.

ARTICLE 4

The *Reich* Economic Minister in consultation with other *Reich* ministers whose competencies are involved are empowered to issue regulations required by this decree. They may permit exceptions insofar as this is necessary for the transfer of Jewish firms into non-Jewish hands, the liquidation of Jewish businesses, or in special cases to insure the availability of supplies.

Berlin, 12 November 1938

The Commissioner for the Four Year Plan

Göring, *Generalfeldmarschall*"

(Translation of Document 1662-PS, in ibid)

00.01.1939   Appointed as *Beauftragter für die Regelung der Judenfrage* (Representative for the Control of the Jewish Question). In this capacity, appointed Reinhard Heydrich as *Leiter der Reichszentrale für jüdische Auswanderung* (Head of the *Reich* Central Office for Jewish Emigration) within the *Reichsministerium des Innern* on 24.01.1939 (see entry of 31.07.1941, below). In 1934, he had summed up his attitude toward the Jews:

> Altogether the Jew had for long taken the lead in the fight against us. It was he who pulled the strings behind all our various opponents. At times he would appear as a reactionary, as a supporter of the German Nationalists, at times he was to be found as the soft and hypocritical and, on that account, craftier member of the Centre Party; and then again he would be the peaceful bourgeois of the People's Party. At other times he would look at us with the satiated middle-class face of a Marxist politician; and then again he would stare at us with the hate-distorted features of a Communist from the underworld. However different the masks might be, the face behind was always the same – Ahasuerus, the Wandering Jew, always burrowing and agitating and considering every means legitimate. (Göring, *Germany Reborn* [English translation of *Aufbau einer Nation*], pp. 73–74)

> By spreading these stories [of Nazi atrocities], the Jews of Germany have proved more conclusively than we could do in our speeches and attacks how right we were in our defensive action against them. Here the Jew is in his element, lying and concocting atrocity stories, from a safe distance throwing buckets full of mud at the people and country whose hospitality he had enjoyed for decades. The decent Jews have only the members of their own race to thank that they are now treated all alike. They can send their protests to the Jewish organizations abroad which play the chief part in the atrocity campaign. Our case against the Jews is not merely that the part they played in every profession was out of all proportion to their total numbers; it is not merely that they had made themselves masters of finance, capital; it is not merely that they carried on usury and corruption on a vast scale and that they exploited Gemrany and sucked the blood from her veins; it is not merely that they were primarily to blame for the crime of the inflation, that they pitilessly strangled their economically weaker German hosts. Our chief accusation against the Jews is that it was they who provided the Marxists and Communists with their leaders, and it was they who occupied the editorial offices of those subversive and defamatory newspapers which besmirched with their venom and hatred all that to us Germans was sacred; they it was who cynically distorted and ridiculed the words "German" and "National," and the ideas of honour and freedom, marriage

and loyalty. No wonder, then, that the German people was at last seized with a righteous anger and was at last unwilling to allow these parasites and oppressors to play the part of master any longer. Only he who has observed the activities of the Jews in Gemrany, only he who knows the Jew from his behavior in Germany, can fully understand the necessity of what has now been done. The Jewish question has not yet been completely solved. All that has happened up to the present has simply been defence of the people, a reaction against the ruin and corruption produced by the Jewish race ... (ibid, pp. 130–132)

30.01.1939	In response to Hitler's *Reichstag* speech of 30.01.1939, declared:

My *Führer*!
Your comrades of the first hour are seated before you willing to follow your lead loyally as one united whole; to stride forth at your side in the future also, suffused by the single desire to follow you blindly toward the attainment of the greatest of victories: the victory of our great German Volk. You have led us onward to victories unfathomable. You have restored to us a life worth living, a life splendid and magnificent. It was you who created Greater Germany. How feeble are our expressions of gratitude; words to express our gratitude to you simply defy us! The cries with which we jubilantly hail you presently, my *Führer*, these shouts of Heil sum up everything we feel within ourselves in respect to inspiration, dedication, love, and loyalty.

Comrades! To our dearly beloved *Führer*, the creator of Greater Germany: Sieg Heil! Sieg Heil! Sieg Heil! (Dr. Max Domarus, ed, *Hitler. Speeches and Proclamations, 1932–1945: The Years 1935–1938*, p. 1459)

07.04.1939–12.04.1939	Departed Naples aboard the Hamburg-Amerika liner *Monserrat* for a visit to the Italian colony of Libya. On arrival in Tripoli, he met with the Governor-General, Italo Balbo (himself a distinguished aviator and a personal friend of Göring), and spent several days sightseeing before returning to Italy.
15.04.1939–18.04.1939	Meeting with Mussolini in Rome, then returned to Berlin.
00.00.1939	Appointed as *Schirmherr* of the *Lilienthal-Gesellschaft für Luftfahrtforschung* ([Otto] Lilienthal Society for Aviation Research).
00.00.1939	Appointed as an *Ehrensenator* (honorary senator) of the *Technischen Hochschule* Braunschweig.
30.08.1939–23.04.1945	*Vorsitzender des Ministerrates für die Reichsverteidigung* (Chairman of the Ministerial Council for *Reich* Defense).
01.09.1939–06.10.1939	"Fall Weiß" (Case White), the German invasion of Poland. Two *Luftwaffe* air fleets – *Luftflotte 1* (807 aircraft under *General der Flieger* Albert Kesselring, with assistance from an additional 92 seaplanes under

the *Fliegerführer der Seeluftstreitkräfte*) and *Luftflotte 4* (627 aircraft under *General der Flieger* Alexander Löhr, as well as 30 machines of the Slovak Air Force) were deployed over Poland as key components of this Blitzkrieg campaign. An additional 333 reconnaissance planes were allocated to the Army. Within the first two weeks of the campaign, the Polish Air Force was reduced from 397 to 54 aircraft. The *Luftwaffe*, and most notably its *Ju 87* (*Stuka*) dive bombers, were thus able to destroy Polish ground targets at will. German bombers dropped some 3,000 tons of explosives on Poland. According to Cajus Bekker's postwar research of *Luftwaffe* documents and interviews with pilots, German aerial losses in Poland were 285 planes destroyed, 379 planes damaged, and 734 airmen killed (Bekker, *Angriffshohe 4000. Ein Kriegstagebuch der deutschen Luftwaffe*).

01.09.1939	Officially named as Hitler's First Deputy and designated successor as *Reichskanzler*.
03.01.1940–23.04.1945	*Leiter der gesamten Kriegswirtschaft* (Coordinator of the Entire War Economy).
10.05.1940–00.06.1940	"Fell Gelb" (Case Yellow), the German invasion of France and the Low Countries (Belgium, The Netherlands, and Luxembourg). The numbers of *Luftwaffe* aircraft deployed in the campaign – in *Luftflotten 2* (Kesselring) and 3 (Hugo Sperrle)- were as follows: 1,482 bombers, 1,016 single-engined fighters, and 248 twin-engined fighters (*Messerschmitt Bf 110s*). On 09.05.1940, Hitler left Berlin in Göring's hands and entrained for the newly-opened Western Front. Göring remained for the next six days on his new Sonderzug (special train) *Asia* at his headquarters in Kurfürst near Berlin. On the 15th, he departed for the Western Front, arriving in the west of Germany, in the Eiffel mountains, the following day. He remained in the West until 30.05.1940, then returned to the *Reich*.
19.07.1940–23.04.1945	*Reichsmarschall des Großdeutschen Reiches* (*Reich* Marshal of the Greater German *Reich*). In his lengthy speech to the *Reichstag* on 19.07.1940, Hitler acknowledged the contributions to victory of Göring and his *Luftwaffe*:

> At dawn on the morning of 10 May, thousands of fighter planes and dive bombers, under the cover of fighters and destroyers, descended on enemy airfields. Within a few days uncontested air superiority wasassured. And not for one minute in the further course of the battle was it allowed to slip.
>
> Only where temporarily no German airplanes were sighted, could enemy fighters and bombers make short appearances. Besides this, their activities were restricted to night action. The Field Marshal had the *Luftwaffe* under his orders during this mission in the war.
>
> Its tasks were:
>
> 1. to destroy the enemy air forces, i.e. to remove these from

Hermann Göring wearing his double-breasted *Reichsmarschall* tunic in 1943. (NARA photo)

Reichsmarschall Göring delivers a speech in 1943. (NARA photo)

the skies;

2. to support directly or indirectly the fighting troops by uninterrupted attacks;

3. to destroy the enemy's means of command and movement;

4. to wear down and break the enemy's morale and will to resist;

5. to land parachute troops as advance units.

The manner of their deployment in the operation in general, as well as their adjustment to the tactical demands of the moment, was exceptional. Without the valor of the Army, the successes attained should never have been possible. Equally true is it that, without the heroic mission of the *Luftwaffe*, the valor of the Army should have been for naught. Both Army and *Luftwaffe* are deserving of the greatest glory!

The deployment of the *Luftwaffe* in the West took place under the personal command of Field *Marshal* Göring. His Chief of Staff: *Generalleutnant* Jeschonnek ...

Generalfeldmarschall Göring as creator of the German *Luftwaffe*, and as an individual man, has made the greatest contribution to the rebuilding of the German *Wehrmacht*. As the leader of the German *Luftwaffe* he has, in the course of the war up to date, created the prerequisites for victory. His merits are unequaled!

I name him *Reichsmarschall* of the Greater German *Reich* and award him the Grand Cross of the Iron Cross.(Dr. Max Domarus, ed, *Hitler. Speeches and Proclamations, 1932–1945: The Year's 1939–1940*, pp. 2053–2054 and p. 2056)

08.08.1940–15.09.1940   The Battle of Britain, which took part in three phases, as follows.

First Phase (08.08.1940–18.08.1940)

2,800 aircraft, including 1,300 bombers and 900 fighters distributed among three *Luftflotten* (2 [*Generalfeldmarschall* Albert Kesselring, based in northern France; 3 [*Generalfeldmarschall* Hugo Sperrle, with bases in Belgium and The Netherlands; and 5 [*Generaloberst* Hans-Jürgen Stumpff, in Norway). *Luftwaffe* forces were opposed by just 650 fighters of Air Marshal Sir Hugh "Stuffy" Dowding's RAF Fighter Command. The *Luftwaffe* made 1,485 sorties on the first day (08.08.1940), and most attacks consisted of strafing runs against fighter bases and seaports, the aim being to lure British fighters into combat and shoot them down. With the aid of radar, a recent British innovation, Dowding was able to determine the areas most in need of protection and concentrate his units accordingly. As a result, the RAF maintained dominance of the skies over Britain.

Second Phase (24.08.1940–05.09.1940)

The focus of *Luftwaffe* bombing changed to main inland bases of the RAF, and substantial damage was done by large bomber formations abundantly protected by fighters. The *Luftwaffe* enjoyed noteworthy success during this period, downing 450 RAF fighters and bringing Fighter Command to the brink of collapse.

Third Phase (07.09.1940–30.09.1940)

Target: London. A non-stop wave of attacks was launched against the British capital, culminating with a great and extremely destructive raid by over 1,000 bombers and approximately 700 fighters on the 15th of the month. The British refused to give in, however, and the high losses of *Luftwaffe* aircraft resulted in discontinuance of daytime bombing raids (the last of these took place 30.09.1940). In a series of attacks conducted 14./15.09.1940, RAF Bomber Command destroyed some 200 barges in the ports of France, Belgium, and the Netherlands, leading Hitler to suspend his plans for "*Unternehmen Seelöwe*" (Operation Sea Lion), the invasion of Great Britain planned for 27.09.1940.

Final Phase (01.10.1940–30.10.1940)

Luftwaffe attacks became sporadic, but London was hit hard by a bombing rad on 10.10.1940. The *Führer* cancelled *Seelöwe* on the 12th. German losses were horrendous: 1,733 aircraft destroyed, with enemy losses numbering 915. The following month, however, a renewed air offensive – known as the "Blitz" – began, and continued through May 1941, by which time over 43,000 British cvilians had been killed and 51,000 severely wounded. Among the most devastating attacks were those against Coventry (almost completely destroyed by approximately 500 bombers on 14./15.11.1940) and London (set aflame in a raid on 29.12.1940).

In a letter of 21.11.1940, sent from his hunting estate Reichsjägerhof Rominten in East Prussia to his brother-in-law, Eric von Rosen, Göring gave the following defense of his air force's performance in the Battle of Britain and the "Blitz":

Dear Eric,

I thank you for your letters and the attached items. I am sorry to learn from it that insidious and untrue reporting of the great Swedish newspapers gave you the impression too that the air warfare and the destruction caused by it are about the same in England as in Germany. I should like to point out that England has suffered terrible blows from my *Luftwaffe* and that a city such as the important armament town of Coventry, was actually completely levelled, while London shows immense destruction and earthquakelike annihilation of entire districts. On the other hand, Mary will be able to tell you that with the exception of a few burnt attics and a few destroyed houses, nothing at all has been destroyed in Berlin. The exaggerations of the British are simply grotesque. A comparison will be able to show how different the achieved results are. Until 1 November 1940 the British have dropped 31 tons on Berlin, while we have dropped during the same period 15,872 tons of bombs on London. Up to that date (1 November 1940) the British have dropped a total of 130 tons of bombs over Germany while we have dropped 21,500 tons of bombs over England. It is also to be considered that the heaviest type of bombs dropped by the British was a bomb of 250 kg which until now has been dropped in a very few instances only, while the majority of the high explosive bombs weighed 100 kg. The greater part of bombs used were incendiary bombs of the 25 kg type. On the other hand, the heaviest German bomb weighs 1800 kg, while the bulk of the bombs which we have dropped, consisted of bombs weighing 250 to 500 kg.

While the British fly with 60 to 90 aircraft per night over Germany, and do not dare to come at all during daytime with a single plane with the exception of three instances when they attacked with one plane, the German *Luftwaffe* sends on the average four to five hundred planes to England per night and attacks without interruption during the day.

A further comparison may be derived from the fact that until now London had almost 350 air raid alarms while Berlin had not even one tenth this number. During many nights when the British did not send any planes at all because of the weather, such as just tonight, the German planes were over England in all sorts of weather.

Since the attack on London on 7 September, London was not bombed for one night only and was spared during daytime on six days. Otherwise bombs are thundering day and night in London without interruption, while during the war Berlin was never attacked in daytime and had one to two nightly air raid alarms per week, on the average.

Alone from the amount of bombs dropped by either side you can see the tremendous difference between the German and British Air Forces. Until now in no part of the entire German *Reich* were damages caused which could not have been repaired within a few hours, while in England photographs prove that important armament factories, etc. were completely levelled. Besides, I have not even mentioned the innumerable ships which have been sunk by the *Luftwaffe*.

It is especially easy for Sweden to investigate the ridiculously small damage which the British Air Force has caused in Germany. Time and again opportunity was offered to Swedish reporters for this purpose, the more so as we here in Germany are outraged at the impossible attitude of especially the large Swedish bourgeois newspapers. I do not make any single exception; even though the Stockholm Tidningen may be slightly better, even this paper is far away from an attitude which could be called friendly or even strict neutrality.

I have done everything to point to this fact time and again and to warn. I did not leave any ways unexplored. If Sweden believes that its freedom of press, i.e., its lack of discipline is more important than its future, then we have to accept such an attitude. But Sweden should not be surprised later on if Germany's attitude will take the consequences from these facts one day. However, I would like to tell something more pleasant, namely, the fact that your Finnish friends can be completely reassured as to their future, even after the visit of Molotov. In this matter I have sent my confidential agent to Mannerheim months ago and am doing it right now again. The Finns were clever enough to realize their completely wrong policy towards Germany, which nearly cost them their existence and to change their attitude radically in the sense of a pro-German attitude. A friendly Finland can and will never be deserted by Germany.

Furthermore, I am quite prepared to give to a small Swedish group to which you could belong too, an opportunity to personally inspect the so-called damage caused by the British and, on the other hand to show to this commission how the German *Luftwaffe* dealt with Holland, Belgium and France, as we are not in the position right now to prove the damages caused in England by our side. But I am completely convinced that this will become possible one day.

Unfortunately, the British government has a completely different attitude, because it exercises a rigorous censorship in order to prevent that anything should become known about German destructions in England. Nevertheless, we have received sufficient reports from neutral side, particularly from Americans, partly with photographic evidence, and we ourselves'have made enough air photos in order to prove how terribly severe the destructions are in England.

    With kindest regards I remain

    Your faithful brother-in-law

    (signed) Hermann Göring (Translation of Document 3775-PS, in *Nazi Conspiracy and Aggression, Volume VI*)

00.11.1940–10.01.1941   Convalescent leave at his hunting lodge in Röminten, during which he consulted with the heart specialist Prof. Siebert. He left Erhard Milch in command of the *Luftwaffe* during this period.

22.06.1941   Opening of "Unternehmen Barbarossa", the German invasion of the Soviet Union on a 2,780-mile front stretching from the Barents Sea in the north to the Black Sea in the south. As in the previous Blitzkrieg campaigns, Göring's *Luftwaffe* – including *Luftflotten 1, 2, 4,* and *5*- played a key role in supporting the advancing ground units and decimating the Red Air Force. 2,598 frontline aircraft were deployed to the new Eastern Front. 1,489 Soviet aircraft were reported destroyed on the ground and 322 in the air during the first day of Barbarossa, compared to German losses of just 35 aircraft. By 29.06.1941, the *Luftwaffe* reported the destruction of 4,990 enemy aircraft in the air and on the ground, while Germany's well-trained and equipped squadrons lost only 179 in the first week of the campaign. These happy and successful days were not to last long, as the Russian winter would cause considerable setbacks to the invasion, both on the ground and in the air.

29.06.1941   *Führer* Decree confirming Göring as his designated successor:

In the event that I am impeded in the discharge of my duties by sickness or other circumstance, even temporarily ... I denote as my deputy in all my offices the *Reichsmarschall* of the Greater German *Reich*, Hermann Göring.

31.07.1941   Signed an administrative directive which authorized Reinhard Heydrich to implement a "Solution of the Jewish Question" in Europe (the document was drafted by *SS-Sturmbannführer* Adolf Eichmann at Heydrich's instruction, and submitted to Göring for signature). The text of this order was as follows:

Complementing the task already assigned to you in [my] directive of 24.01.1939, to undertake, by emigration or evacuation, a solution of the Jewish question as advantageous as possible under the conditions at the time, I hereby charge you with making all necessary organizational, functional, and material preparations for a complete solution of the Jewish question in the German sphere of influence in Europe.

Insofar as the jurisdiction of other central agencies may be touched thereby, they are to be involved.

I charge you furthermore with submitting to me in the near future an overall plan of the organizational, functional, and material measures to be taken in preparing for the implementation of the aspired final solution [Endlosung] of the Jewish question. (Translation of Document 710-PS, *in Nazi Conspiracy and Aggression, Volume II*)

23.01.1942	Assigned by Hitler as chairman of a *Sondersenat* (special senate) of the *Reichskriegsgericht* (*Reich* Court-Martial) at *Führerhauptquartier* in the trial of *Generalleutnant* Hans Graf von Sponeck. Charged with disloyalty for making an unauthorized withdrawal of his forces from the Kerch Peninsula on the Russian Front, von Sponeck was reluctantly sentenced to death, at Göring's urging, by this court. Hitler later commuted the sentence to fortress arrest, but he was executed on 23.07.1944.
00.01.1942–00.02.1942	Visit to Rome to meet with Mussolini.
21.03.1942	Diary entry of *Reichsminister* Dr. Goebbels:

In the afternoon I had a more than three-hour talk with Göring, which came off in an atmosphere of the greatest friendliness and cordiality. I was happy we could let our hair down. We surveyed the overall situation, and I was gratified to note that we agree 100 per cent on all important problems. Without having consulted each other we have arrived at almost exactly the same appraisal of the situation.

Göring is in exceptionally good condition physically. He works hard, achieves enormous successes, tackles problems with a healthy common sense, without much theorizing, and for this reason is pretty skeptical about certain trends in the Party. I can't blame him for this. He has the rare good fortune of not being dependent on the Party in his work, so that he can risk being more independent. In many respects he is to be envied ...

We are in complete agreement about the *Wehrmacht*. Göring has nothing but abysmal contempt for the cowardly generals ... . *Generalfeldmarschall* Keitel, he said, was not tough enough. He was probably responsible for the fact that the plan of campaign in the East did not function properly. He carried the *Führer*'s commands to the OKH with trembling knees. While he bore a great part of the responsibility he was not told sufficiently plainly [by Keitel] that he must obey the *Führer*, and that if he did not do so, he would soon feel the consequences ... .

> Göring spoke in terms of highest praise about our [the Propaganda Ministry's] work. We resolve to meet more frequently and have frank talks about everything … The result of my talk with Göring is exceptionally satisfactory and favorable … (Louis P. Lochner, ed, *The Goebbels Diaries 1942–1943*)

00.04.1942–23.04.1945	*Vorsitzender der Zentrale Planungsrat* (Chairman of the Central Plannng Council). The other members were *Reich* Armaments Minister Albert Speer and Göring's *Staatssekretär, Generalfeldmarschall* Erhard Milch. The council first convened on 27.04.1942.
09.06.1942–23.04.1945	*Präsident der (neue) Reichsforschungsrat* (RFR, President of the [new] *Reich* Research Council).
00.10.1942–00.10.1942	One-week visit to Rome to meet with Mussolini.
23.11.1942	In a phone conversation with Hitler, agreed that a successful airlift of fuel, ammunition, food, and other supplies to the 300,000 men of Friedrich Paulus's encircled *6.Armee* at Stalingrad seemed feasible. He gave the *Führer* his promise that he would mobilize all the *Luftwaffe*'s transport aircraft, and the following day, the airlift began. It was a failure, largely due to the fact that it gave Hitler the false impression that *6.Armee* could survive in the Stalingrad cauldron, when in fact the best course of action at this early stage would have been for Paulus to lead a breakout from the Soviet encirclement. Wolfram von Richthofen's *Luftflotte 4*, despite being very shorthanded and using every type of aircraft that could be scavenged, dropped approximately 5,300 tons of cargo into the city. The planes were subjected to attack by Soviet fighters, anti-aircraft fire, a shortage of fuel, and extreme challenges to their effective ranges, and the operation – like *6.Armee*'s attempt to take Stalingrad – was a miserable failure. Hitler was infuriated with the *Luftwaffe* and its leader, and his relationship with Göring would never recover. Regarding losses, David Irving writes:

> Göring's air force had airlifted into Stalingrad 8,350 tons of supplies, a daily average of 116 tons. But he had lost 266 Junkers 52 transports, 165 Heinkel 111 bombers, and 42 Junkers 86 bombers – virtually an entire air corps. (Irving, *Göring: A Biography*, p. 558)

30.11.1942	Visit to Rome, together with *Generalfeldmarschall* Erwin Rommel, to meet with Mussolini concerning the dismal status of the North African campaign.
01.03.1943	Lengthy meeting (nearly four hours) with *Reichsminister* Dr. Goebbels at Carinhall, during which the two discussed the war situation and their mutual displeasure with other senior members of the *Reich* leadership (notably *Reichsministers* Rosenberg and von Ribbentrop and the so-called *Dreimännerkollegium*, or "Committee of Three" [Martin Bormann, Dr. Hans Heinrich Lammers, and Wilhelm Keitel]). The following is an excerpt from Goebbels' diary entry of 02.03.1943:

The little dissensions that have crept into our work in the course of time were not even mentioned. They seem quite unimportant compared with the historic tasks that we have to discuss ...

Göring evidenced the greatest concern about the *Führer*. To him too, the *Führer* seems to have aged fifteen years during three-and-a-half years of war. It is a tragic thing that the *Führer* has become such a recluse and leads so unhealthy a life. He doesn't get out into the fresh air. He does not relax. He sits in his bunker, fusses and broods. If one could only transfer him to other surroundings! But he has made up his mind to conduct this war in his own Spartan manner, and I suppose nothing can be done about it.

But it is equally essential that we succeed somehow in making up for the lack of leadership in our domestic and foreign policy. One must not bother the *Führer* with everything. The *Führer* must be kept free for the military leadership.... As was always the case during crises of the Party, the duty of the *Führer*'s closest friends in time of need consists in gathering about him and forming a solid phalanx around his person....

Göring realizes perfectly what is in store for all of us if we show any weakness in this war. He has no illusions about that. On the Jewish question, especially, we have taken a position from which there is no escape. That is a good thing. Experience teaches that a movement and a people who have burned their bridges fight with much greater determination than those who are still able to retreat.

... I introduce my proposals. I express the opinion that we'd be 'over the hump' if we succeeded in transferring the political leadership tasks of the *Reich* from the Committee of Three to the Ministerial Council for the Defense of the *Reich*. This Ministerial Council would then have to be composed of the strong men who assisted the *Führer* in the Revolution. These will certainly also muster the strength to bring this war to a victorious conclusion....

While talking I gained the spontaneous impression that my presentation visibly pepped up Göring. He became very enthusiastic about my proposals and immediately asked how we were to proceed specifically...

I am very happy that a clear basis of mutual trust was established with Göring. I believe that the *Führer*, too, will be very happy about this. I hope we shall render him the very greatest service possible. (Louis P. Lochner, ed, *The Goebbels Diaries 1942–1943*)

Goebbels' grand plans of heightening his power and realizing his dreams of totalizing the war with Göring's help faded quickly. The *Reichsmarschall*'s stock had plummeted in all quarters, owing mainly to the many failures of his *Luftwaffe* to adequately defend the *Reich* from Allied bombing and to provide sufficient aerial supply and support to the doomed 6th Army at Stalingrad. In testimony before the International Military Tribunal at Nürnberg on 08.03.1946, Göring's former liaison to *Führer* HQ, *General der Flieger* Karl Bodenschatz stated:

> According to my personal opinion and conviction, Hermann Göring began to lose influence with Hitler in the spring of 1943 ... That was the beginning of large-scale air attacks by night by the R.A.F. on German towns, and from that moment there were differences of opinion between Hitler and Göring which became more serious as time went on. Even though Göring made tremendous efforts, he could not recapture his influence with the *Führer* to the same extent as before. The outward symptoms of this waning influence were the following: First, the *Führer* criticized Göring most severely. Secondly, the eternal conversations between Adolf Hitler and Hermann Göring became shorter, less frequent, and finally ceased altogether. Thirdly, as far as important conferences were, concerned, the *Reich Marshal* was not called in. Fourthly, during the last months and weeks the tension between Adolf Hitler and Hermann Göring increased to such a degree that he was finally arrested ... (*Trial of the Major War Criminals Before the International Military Tribunal Nuremberg, 14 November 1945 to 1 October 1946, Volume IX*)

24.07.1944–00.08.1944	At Rominten, ostensibly due to illness.
20.04.1945	Visited Adolf Hitler for the last time in the *Führerbunker*, on the occasion of the *Führer's* 56th birthday. After departing the Bunker, he made one last visit to "Carinhall", then fled by air to Berchtesgaden.
21.04.1945	Arrived at the Obersalzberg, Berchtesgaden (approximately 2300 hours). Joined his wife and daughter, who had fled there from Carinhall in January 1945 due to the advance of Soviet forces; after removing several trainloads of innumerable art treasures and other belongings, Göring had Carinhall destroyed.
23.04.1945–06.05.1945	In reaction to reports that Hitler had suffered a nervous collapse in Berlin, Göring (at the Berghof) consulted with Dr. Hans Heinrich Lammers and Philipp Bouhler and drafted a telegram to Hitler. The message read as follows:

> My *Führer*!
> In view of your decision to remain in the fortress of Berlin, do you agree that I should take over the leadership of the whole *Reich* as your deputy with complete liberty of action at home and abroad, in accordance with your decree 29 April 1941? If I receive no reply by 22[:00 hours] I shall assume that your freedom of action has been curtailed. I shall then regard the conditions of the decree as

fulfilled. I shall act in the best interest of Volk and Fatherland. You are aware of my feelings for you in this darkest of hours of my life, as words fail me. May the Lord protect you and allow you, despite everything, to come here as soon as possible.

Your loyal Hermann Göring. (Dr. Max Domarus, ed, *Hitler. Speeches and Proclamations, 1932–1945: The Years 1941–1945*, p. 1459)

Hitler's reply, sent by radio to the Berghof, read:

The *Führer* decree of 29 June 1941, is herewith declared null and void. Your behavior and your measures constitute a betrayal of my person and the National Socialist cause. I am in complete possession of my freedom of action and forbid any further measures.

Adolf Hitler (ibid)

Shortly afterward, the following announcement was issued on the *Führer*'s order:

Reichsmarschal Hermann Göring has become acutely ill due to a chronic heart problem which has long troubled him. He himself has requested to be relieved of the command of the *Luftwaffe* and the connected tasks in view of the current situation, which demands the deployment of all forces. The *Führer* has granted this request. He has appointed *Generaloberst* [Robert] Ritter von Greim as the new commander in chief of the *Luftwaffe* and promoted him to *Generalfeldmarschall*. (ibid, p. 3047)

The same day, he was arrested on Hitler's orders by *SS-Obersturmbannführer* Dr. Bernhard Frank and detained at the Berghof until its destruction in an Allied bombing raid forced him and his captors to flee. The following is "Annex III: Goering's Arrest" (22.10.1945), from the interrogation report of *SS-Brigadeführer und Generalmajor der Waffen-SS und der Polizei* Ernst August Rode:

On the morning of 23 Apr 45 Rode was visited by one of Goering's adjutants, with the request that he transmit the following message to Himmler: "*Mein Fuehrer, nach Vortrag des Generals Koller befurchte ich, dass Sie nicht mehr die volle Handlungsfreiheit besetzen, sondern in Berlin eingeschlossen sind. Ich nehme deshalb an, dass meine Vollmacht in Kraft tritt, falls ich bis 22 Uhr keine Antwort habe.*

GOERING."

["My Fuehrer, according to the report of Gen Koller, I am afraid

09.05.1945: Hermann Göring on arrival at the 7th Army Interrogation Center in Augsburg. At right is his adjutant, Major Bernd von Brauchitsch. (U.S. Army Photos)

that you no longer have a free hand, but are encircled. Therefore, unless I hear from you by 2200 hrs, I shall assume full power. GOERING."]

On the following night Rode received a phone call from O/Stubaf Frank, summoning him immediately. Frank informed Rode that Hitler, through ... Bormann (under Landes Gericht *Praesident* Mueller) had ordered Goering's arrest. Additional messages expelled Goering from the *NSDAP* and ordered him to relinquish all his duties. Goering complied by wire. Goering was put under house arrest by guards of the Waffen *SS*. Dr. Lammers was also arrested in Bischofswiesen. Later an order arrived from Bormann to shoot Goering and his associates in the event that Berlin surrendered. Rode ordered Frank to shoot no one under any circumstances. Rode visited Gen Koller in Berchtesgaden, 28 Apr 45, in order to ask if any further instructions regarding Goering had been received from Gen/FM von Greim, Goering's successor. [On 28.04.1945,] Goering was transferred to [Schloss] Mauterndorf [in Taunach-Tal] by order of Kaltenbrunner because of air attacks on Obersalzberg. ("Interrogation Records Prepared

Göring shortly after his capture. (Roger Bender photos)

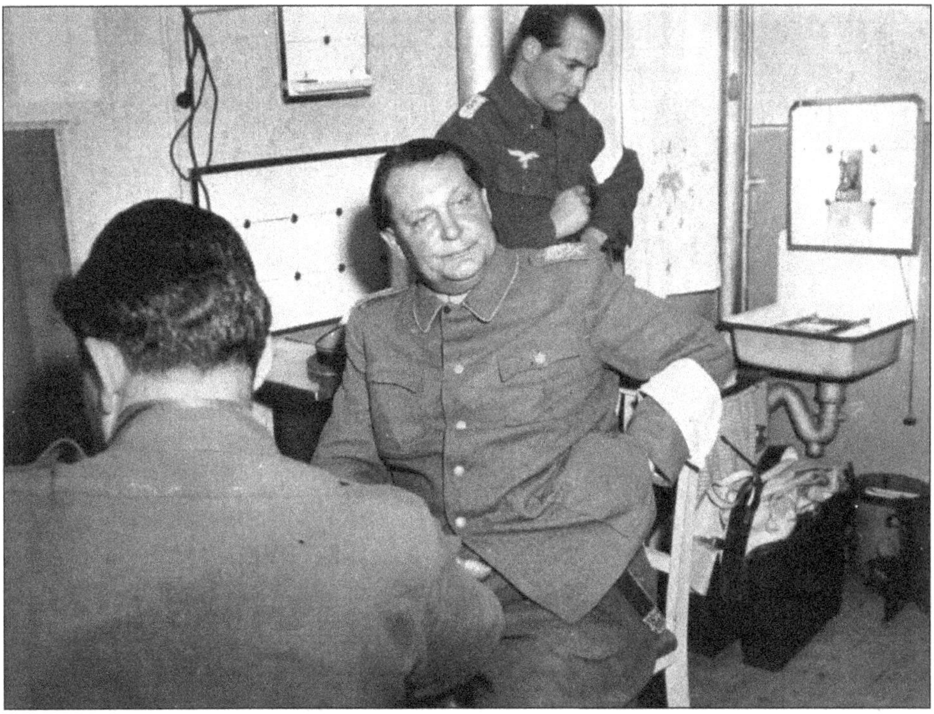

Göring shortly after his capture. (Roger Bender photos)

Detention report with mug shots for Prisoner of War "Goering, Hermann".

for War Crimes Proceedings at Nuernberg, 1945–1947/OCCPAC Interrogation Transcripts And Related Records: Rode, Ernst August; Publication Number M1270, Record Group RG238)

On 06.05.1945, on orders of *Generalfeldmarschall* Albert Kesselring (*Oberbefehlshaber Süd*), Göring was released by his final captor, the *SS* legal officer *SS-Standartenführer* Ernst Brausse. On 29.04.1945, he was formally expelled from all posts by Hitler, who wrote in his final testament:

> Before my death, I expel the former *Reichsmarschall* Hermann Göring from the Party and strip him of all his rights that might be derived from the decree of 29 June 1941, and my *Reichstag* declaration of 1 September 1939. In his place, I appoint *Großadmiral* Dönitz as *Reich* President and supreme commander of the *Wehrmacht* ...
>
> Göring and Himmler, by their secret negotiations with the enemy, which took place without my knowledge and contrary to my will, as well as by attempting to usurp power in the state in violation of the law, have done immeasurable damage to the country and the entire Volk, not to mention their disloyalty to my person. (Excerpt from translation of Document 3569-PS, in *Nazi Conspiracy and Aggression, Volume VI*)

Defendant Göring enters the courtroom in Nürnberg.

Postwar Prosecution

On 07.05.1945, Göring, together with his wife, daughter, and members of his staff, surrendered approximately 30 miles from Salzburg to Brigadier General Robert J. Stack, Deputy Commander of the 36th (Texas) Division. Interned at Schloß Fischhorn at Bruck near Zell-am-See, he was driven on 08.05.1945 to the U.S. 7th Army Headquarters at Kitzbühel where he met with USAAF General Carl Spaatz and was treated to a party by his captors (upon hearing of this affair, General Eisenhower ordered that Göring was to be treated as an ordinary prisoner of war). Flown to the 7th Army Interrogation Center in Augsburg on 09.05.1945, he remained there until 21.05.1945. He was then transported to the Allied internment and interrogation center "Ashcan" at the Palace Hotel in Mondorf-les-Bains, Luxembourg. From there he was sent to Nürnberg Prison to stand trial as lead defendant before the International Military Tribunal (IMT), Nürnberg (Defense Counsel: Dr. Otto Stahmer). On trial from 20.11.1945 to 01.10.1946, he was found guilty on all four counts of the indictment: Count 1, Conspiracy to Commit Crimes Alleged in Other Counts; Count 2, Crimes Against Peace; Count 3, War Crimes; and Count 4, Crimes Against Humanity. Sentenced to death by hanging on 01.10.1946, Göring indignantly requested that, as a soldier, he be permitted to die before a firing squad. Refusing to allow himself to be hanged, he committed suicide in his prison cell. The question of how he was able to conceal the cyanide capsule has never been conclusively answered. It is possible he received it from a U.S. Army lieutenant, Jack G. "Tex" Wheelis (died 1954) whom he had befriended at Nürnberg. The following is the judgment of the International Military Tribunal against Defendant Göring:

Göring dines and reads
in his cell. (U.S. Army
Signal Corps photos)

Göring is indicted on all four counts. The evidence shows that after Hitler he was the most prominent man in the Nazi Régime. He was Commander-in-Chief of the *Luftwaffe*, Plenipotentiary for the Four Year Plan, and had tremendous influence with Hitler, at least until 1943 when their relationship deteriorated, ending in his arrest in 1945. He testified that Hitler kept him informed of all important military and political problems.

Crimes against Peace

From the moment he joined the Party in 1922 and took command of the street-fighting organisation, the *SA,* Göring was the advisor, the active agent of Hitler and one of the prime leaders of the Nazi movement. As Hitler's political deputy he was largely instrumental in bringing the National Socialists to power in 1933, and was charged with consolidating this power and expanding German armed might. He developed the *Gestapo*, and created the first concentration camps, relinquishing them to Himmler in 1934, conducted the Röhm purge in that year, and engineered the sordid proceedings which resulted in the removal of von Blomberg and von Fritsch from the Army. In 1936 he became Plenipotentiary for the Four Year Plan, and in theory and in practice was the economic dictator of the *Reich*. Shortly after the Pact of Munich, he announced that he would embark on a five-fold expansion of the *Luftwaffe*, and speed rearmament with emphasis on offensive weapons.

Göring was one of the five important leaders present at the Hossbach Conference of 5th November, 1937, and he attended the other Important conferences already discussed in this Judgment. In the Austrian Anschluss, he was indeed the central figure, the ringleader. He said in Court: "I must take 100 per cent responsibility. I even overruled objections by the *Führer* and brought everything to its final develonment." In the seizure of the Sudetenland he played his role as *Luftwaffe* chief by planning an air offensive which proved unnecessary and his role as a politician by lulling the Czechs with false promises of friendship. The night before the invasion of Czechoslovakia and the absorption of Bohemia and Moravia, at a conference with Hitler and President Hacha he threatened to bomb Prague if Hacha did not submit. This threat he admitted in his testimony.

Göring attended the *Reich* Chancellery meeting of 23rd May, 1939, when Hitler told his military leaders "there is, therefore, no question of sparing Poland," and was present at the Obersalzburg briefing of 22nd August, 1939. And the evidence shows he was active in the diplomatic maneuvers which followed. With Hitler's connivance, he used the Swedish businessman, Dahlerus, as a go-between to the British, as described by Dahlerus to this Tribunal, to try to prevent the British Government from keeping its guarantee to the Poles.

He commanded the *Luftwaffe* in the attack on Poland and throughout the aggressive wars which followed. Even if he opposed Hitler's plans against Norway and the Soviet Union, as he alleged, it is clear that he did so only for strategic reasons once Hitler had decided the issue, he followed him without hesitation. He made it clear in his testimony that these differences were never ideological or legal. He was "in a rage" about the invasion of Norway, but only because he had not received sufficient warning to prepare the *Luftwaffe* offensive. He admitted he approved of the attack: "My attitude was perfectly positive." He was active in preparing and executing the Yugoslavian and Greek campaigns, and testified that "Plan Marita," the attack on Greece had been prepared long beforehand. The Soviet Union he regarded as the "most threatening menace to Germany," but said there was no immediate

Goring testifies before the IMT, Nürnberg, 1946.

military necessity for the attack. Indeed, his only objection to the war of aggression against the U.S.S.R. was its timing; he wished for strategic reasons to delay until Britain was conquered. He testified: "My point of view was decided by political and military reasons only."

After his own admissions to this Tribunal, from the positions which he held, the conferences he attended, and the public words he uttered, there can remain no doubt that Göring was the moving force for aggressive war second only to Hitler. He was the planner and prime mover in the military and diplomatic preparation for war which Germany pursued.

War Crimes and Crimes against Humanity

The record is filled with Göring's admissions of his complicity in the use of slave labor. "We did use this labor for security reasons so that they would not be active in their own country and would not work against us. On the other hand, they served to help in the economic war." And again: "Workers were forced to come to the *Reich*. That is something I have not denied." The man who spoke these words was Plenipotentiary for the Four Year Plan charged with the recruitment and allocation of manpower. As *Luftwaffe* Commander-in-Chief he demanded from Himmler more slave laborers for his underground aircraft factories: "That I requested inmates of concentration camps for the armament of the *Luftwaffe* is correct and it is to be taken as a matter of course."

As Plenipotentiary, Göring signed a directive concerning the treatment of Polish workers in Germany and implemented it by regulations of the *SD* including "special

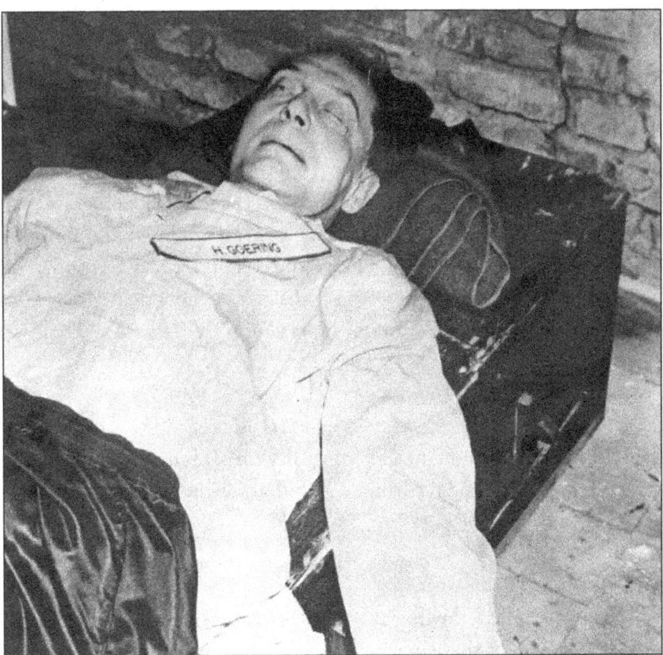

The end of *Reichsmarschall* Hermann Wilhelm Göring, 16.10.1946.

treatment." He issued directives to use Soviet and French prisoners of war in the armament industry; he spoke of seizing Poles and Dutch and making them prisoners of war if necessary, and using work. He agrees Russian prisoners of war were used to man anti-aircraft batteries.

As Plenipotentiary, Göring was the active authority in the spoliation of conquered territory. He made plans for the spoliation of Soviet territory long before the war on the Soviet Union. Two months prior to the invasion of the Soviet Union, Hitler gave Göring the over-all direction for the economic administration in the territory. Göring set up an economic staff for this function. As *Reichsmarschall* of the Greater German *Reich*, "the orders of the Reichmarschall cover all economic fields, including nutrition and agriculture." His so-called "Green" folder, printed by the *Wehrmacht*, set up an "Economic Executive Staff, East." This directive contemplated plundering and abandonment of all industry in the food deficit regions and, from the food surplus regions, a diversion of food to German needs. Göring claims its purposes have been misunderstood but admits "that as a matter of course and a matter of duty we would have used Russia for our purposes," when conquered.

And he participated in the conference of 16th July, 1941, when Hitler said the National Socialists had no intention of ever leaving the occupied countries, and that "all necessary measures – shooting, desettling, etc." should be taken.

Göring persecuted the Jews, particularly after the November, 1938 riots, and not only in Germany where he raised the billion mark fine as stated elsewhere, but in the conquered territories as well. His own utterances then and his testimony now show this interest was primarily economic – how to get their property and how to force them out of the economic life of Europe. As these countries fell before the German army, he extended the *Reich*'s anti-Jewish laws to them; the Reichsgesetzblatt for 1939, 1940, and 1941 contains several anti-Jewish decrees signed by Göring. Although their extermination was in Himmler's hands,

Göring was far from disinterested or inactive, despite his protestations in the witness box. By decree of 31st July, 1941, he directed Himmler and Heydrich to bring "about a complete solution of the Jewish question in the German sphere of influence in Europe."

There is nothing to be said in mitigation. For Göring was often, indeed almost always, the moving force, second only to his leader. He was the leading war aggressor, both as political and as military leader; he was the director of the slave labor program and the creator of the oppressive program against the Jews and other races, at home and abroad. All of these crimes he has frankly admitted. On some specific cases there may be conflict of testimony, but in terms of the broad outline, his own admissions are more than sufficiently wide to be conclusive of his guilt. His guilt is unique in its enormity. The record discloses no excuses for this man.

Conclusion

The Tribunal finds the defendant Göring guilty on all four counts of the Indictment. (Judgment of the International Military Tribunal for the Trial of German *Major* War Criminals)

Published Works

Aufbau einer Nation (1934). The following are noteworthy excerpts:

[Pages 86–87] I declared at that time before thousands of fellow Germans, each bullet which leaves the barrel of a police pistol now, is my bullet. If one calls this murder, then I have murdered; I ordered all this, I back it up; I assume the responsibility, and I am not afraid to do so. Through a network of outer offices converging into the headquarters in Berlin, I am daily, one could almost say hourly, informed about everything that happens in widespread Prussia. [Page 891] Against the enemies of the State, we must proceed ruthlessly. It cannot be forgotten that at the moment of our rise to power, according to the official election figures of March 1933, six million people still confess their sympathy for Communism and eight million for Marxism ... .

Therefore, the concentration camps have been created, where we have first confined thousands of Communists and Social Democrat functionaries.

[Page 921] The *Gestapo* deserves a great deal of credit for the success of the revolution and for the consolidation of its achievements. Right in the middle of this constructive work occurred the blaze that destroyed the high cupola and the auditorium of the *Reichstag*. Criminal hands had set this fire, had put the German *Reichstag* in flames, in order to give a last beacon to dying Communism, so that it could make one last desperate thrust before the Hitler government was consolidated. The blaze was to be the signal for the Communist party for general terror, for a general uprising and for civil war. That it did not have these consequences, Germany and the world owe not to the noble motives of Communism, but solely to the iron resolution and the hard fist of Adolf Hitler and his closest collaborators, who struck more quickly than the enemy had expected, and harder than he could imagine, and with the first blow suppressed Communism once and for all. That night, when I ordered the arrest of 4,000 Communist functionaries, I knew that by dawn Communism had lost a great battle. (Partial Translation of Document 2324-PS, in *Nazi Conspiracy and Aggression. Volume III*)

Der Geist des neuen Staates (1934)

Reden und Aufsätze (1938, edited by Erich Gritzbach)

Decorations and Awards
German

19.08.1940	*Großkreuz des Eisernes Kreuzes* (In honor of the *Luftwaffe*'s achievements in the Western Campaign of May / June 1940 (first and last Third *Reich* recipient of this award)
30.09.1939	*Ritterkreuz des Eisernes Kreuzes* (for his command of the *Luftwaffe* in the Polish Campaign)
02.06.1918	*Orden Pour le mérite* as *Oberleutnant* and *Führer* of *Jagdstaffel* 27, Western Front
20.10.1917	*Ritterkreuz mit Schwertern des königlich Preußischen Hausordens von Hohenzollern* as *Oberleutnant* and *Führer* of *Jagdstaffel* 27
02.06.1918	*Ritterkreuz des Großherzoglich Badischen Militär Karl-Friedrich-Verdienstordens* as *Oberleutnant* and *Führer* of *Jagdstaffel* 27
15.04.1916	*Ehrenbecher für den Sieger im Luftkampf* (awarded for second aerial victory, a Maurice Farman MF.11 Shorthorn [reconaissance and light bomber] of the French Air Force, shot down over Tahure on 16.11.1915)
18.07.1915	*Orden vom Großherzoglich Badischen Ordens vom Zähringer Löwen, Ritterkreuz II. Klasse mit Schwertern*
00.09.1939	*1939 Spange zum 1914 Eisernes Kreuz I. Klasse*
00.09.1939	*1939 Spange zum 1914 Eisernes Kreuz II. Klasse*
22.03.1915	*1914 Eisernes Kreuz I. Klasse* as *Leutnant* and *Flugzeugbeobachter* in *Feldflieger-Abteilung 25*
15.09.1914	*1914 Eisernes Kreuz II. Klasse* as *Leutnant* and *Bataillonsadjutant* of II. *Bataillon / 4. Badische Infanterie-Regiment "Prinz Wilhelm" Nr. 112*
00.00.193_	*Großkreuz des Herzoglich Sachsen-Ernestinischen Hausordens*
00.00.1918	*Verwundetenabzeichen, 1918 in Schwarz*
12.10.1915	*Abzeichen für Militär-Flugzeugführer*
15.11.1914	*Abzeichen für Beobachtungsoffiziere*
00.00.1935	*Gemeinsames Flugzeugführer-und Beobachter-Abzeichen in Gold mit Brillanten* (Combined Pilot's and Oberver's Badge in Gold with Diamonds)
00.00.194_	*U-Boot-Kriegsabzeichen Modell 1939 in Gold mit Brillianten* (presented as a gift by *Großadmiral* Karl Dönitz)
19.01.1935	*Fliegerschaftsabzeichen*
ca. 1939	*Spange "Prager Burg" zur Medaille zur Erinnerung an den 1 Oktober 1938*
ca. 1939	*Medaille zur Erinnerung an den 1 Oktober 1938*
ca. 1938	*Medaille zur Erinnerung an den 13 März 1938*
00.00.19__	*Deutsche Ehrendenkmünze des Weltkrieges mit dem Kampfabzeichen*
ca. 1934	*Ehrenkreuz des Weltkrieges 1914–1918 mit Schwertern*
00.00.193_	*Wehrmacht-Dienstauszeichnung IV.Klasse*
00.10.1935	*Luftwaffen-Ärmelband "Jagdgeschwader Frhr. v.Richthofen Nr. 1 1917/1918"* (Commemorative cuff title for his service with that unit in World War I)
08.09.1940	Mentioned in the *Wehrmachtbericht*

21.01.1938	*Amtskette des Präsidenten und der Präsidiumsmitglieder der Deutsche Akademie der Luftfahrtforschung*
01.12.1933	*Goldenes Ehrenzeichen der NSDAP*
00.00.1934	*Ehrenzeichen des 9 November 1923 (Blutorden)* (with effect from 09.11.1933)
00.00.1929	*Nürnberger Parteitagsabzeichen 1929*
00.00.1936	*Goldenes Gauehrenzeichen des Gaues Groß-Berlin der NSDAP*
20.04.1937	*Gauehrenzeichen "Silberner Gauadler" des Gaues Thüringen der NSDAP*
00.00.194_	*Dienstauszeichnung der NSDAP in Gold*
00.00.194_	*Dienstauszeichnung der NSDAP in Silber*
00.00.194_	*Dienstauszeichnung der NSDAP in Bronze*
00.10.1939	*Kreuz von Danzig I. Klasse*
00.10.1939	*Kreuz von Danzig II. Klasse*
00.00.1939	*Deutsches Schutzwall-Ehrenzeichen*
20.04.1938	*Luftschutz-Ehrenzeichen 1. Stufe*
00.00.1936	*Deutsches Olympia-Ehrenzeichen I. Klasse*
00.00.1934	*Großkreuz des Ehrenzeichens des Deutschen Roten Kreuzes*
00.00.1934	*Ehrenzeichen des Deutschen Roten Kreuzes I. Klasse*
14.01.1939	*Ehrenring des deutschen Handwerks*
00.00.193_	*Ehrenzeichen der Ring der Nationalen Kraftfahrt und Luftfahrtbewegung in Gold*
05.05.1934	*Ehrenurkunde des Internationalen Jagdrates* (Honor Certificate of the International Hunting Council)
06.05.1939	*Ehrenbürgerrecht der Universität Göttingen*
01.03.1937	*Ehrenbürgerrecht der Stadt Saarbrücken*
30.01.1937	*Ehrenbürgerrecht der Stadt Hamburg*
00.00.1936	*Ehrenbürgerrecht der Stadt Wesermünde* (status revoked, 00.00.1949)
26.10.1933	*Ehrenbürgerrecht der Stadt Stettin*
13.09.1933	*Ehrenbürgerrecht der Kreisstadt Ottweiler im Saar*
28.08.1933	*Ehrenbürgerrecht der Stadt Allenstein*

Foreign

Kingdom of Bulgaria

00.00.19__	Order of Saints Cyril and Methodius, the Small Collar (*Малко огърлие на ордена "Св. св. равноапостоли Кирил и Методий"*)
00.00.19__	Order of Saint Alexander, Grand Cross with the Breast Star with Diamonds (*Велик кръст на ордена "Свети Александръ", звезда с брилянти*)
00.00.19__	Order of Saint Alexander, the small collar (*Малко огърлие на ордена "Свети Александръ"*)
00.00.19__	Order of Bravery, I Grade, 1st Class (*Орден "За храброст" I степен 1 клас*)

Independent State of Croatia (*Nezavisna Države Hrvatska, NDH*)

06.04.1942	Order of the Crown of King Zvonimir, Grand Cross with the Breast Badge with Oak Wreath (*Red krune kralja Zvonimira 1.stupnja SA zvijezdom i hrastove grančice*)

00.00.194_	Military Pilot Badge (*Oznaka Zrakoplovstva Nezavisne Države Hrvatske*)

Democratic Republic of Georgia (refugee government, reestablished in the castle of Leuville-sur-Orge in France)

00.00.1933 (?)	Order of Saint Tamara

Kingdom of Denmark

06.08.1938	Order of the Dannebrog, Grand Cross with Diamonds (*Storkorset af Dannebrogordenen med bryststjeme i diamanter*)

Republic of Finland

00.00.1941	Order of the White Rose of Finland, Grand Cross with Swords (*Suomen Valkoisen Ruusun suurristi miekkojen kera*)
00.00.19__	Order of the Lion of Finland, Grand Cross with Swords (*Suomen Leijonan suurristi miekkojen kera*)
25.03.1942	Order of the Cross of Liberty, Grand Cross with Swords (*Vapaudenristin suurristi miekkojen kera*)
27.08.1941	Pilot Badge with diamonds (*Suomen lentomerkki briljantein koristeltuna*)

Kingdom of Hungary

00.00.19__	Royal Hungarian Order of Saint Stephen, Grand Cross (*Nagykeresztje Magyar Királyi Szent István-rend*)
00.00.19__	Order of Merit, Grand Cross with the Crown of Saint Stephen (*Nagykeresztje Magyar Érdemrend a koronát a Szent István*)

Kingdom of Italy

26.05.1933	Order of Saints Maurice and Lazarus, Grand Cross (*Gran Croce dell'Ordine dei Santi Maurizio e Lazzaro*)
27.11.1941	Military Order of Savoy, Grand Cross (*Gran Croce dell'Ordine militare di Savoia*)
22.05.1940	Supreme Order of the Most Holy Annunciation, the small collar (*Piccola Collare dell'Ordine Supremo della Santissima Annunziata*) (or 06.05.1940)
10.01.1943	Military Order of the Roman Eagle, Grand Cross (*Gran Croce dell'Ordine Militare dell'Aquila Romana*)

Empire of Japan

29.09.1943	Order of the Rising Sun, Grand Cross with Paulownia Flowers
00.00.194_	Order of the Rising Sun, Grand Cross with the Grand Cordon
00.00.194_	Honorable Samurai Sword of the Japanese nationalist organization "*Shochoku Seishin Shinkokai*" (詔勅精神振興会), awarded in recognition of Göring's "loyalty to the Great Germany and exceptional military merit"

Ottoman Empire

ca. 1916	War Medal (aka Eiserner Halbmond, or "Gallipoli Star" (*Harb Madalyasi*)

Kingdom of Romania

21.10.1941	Order of Michael the Brave, Model 1941, 1st, 2nd and 3rd Classes (*Ordinul Mihai Viteazul, Clasele I-III*)
11.10.1941	Order of Aviation Merit, Commander Cross with Military Decoration and Swords (*Ordinul Virtutea Aeronautică în grad de comandor, cu însemne de război și cu spade*)

00.00.19__	Order of Carol I, Grand Collar (*Ordinul Carol I, Colan*)
00.00.19__	Order of Carol I, Grand Cross (*Ordinul Carol I, Mare Cruce*)
00.00.194_	Military Pilot Badge (*Insignă de pilot brevetat*, Model 1941)

Slovak Republic

| 00.00.194_ | Slovak War Victory Cross, Grand Cross (*Velký kříž Slovenského válečného vítězného kříže*) |
| ca. 1943 | Slovak War Victory Cross, 1st Class without Swords (personally presented by President Jozef Tiso) |

Republic of Spain

| ca. 1939 | Imperial Order of the Yoke and Arrows, the Large Collar (*Gran collar de la Orden Imperial del Yugo y las Flechas*) |

Kingdom of Sweden

| 02.02.1939 | Royal Order of the Sword, Grand Cross (*Kungliga Svärdsorden, Kommendör med stora korset*) |

Kingdom of Yugoslavia

| 00.00.19__ | Order of the White Eagle, Grand Cross (*Велики крст Ордена Белог Орла*) |

Notes

- Son of Dr. jur. Ernst *Heinrich* Göring (born 31.10.1838 in Emmerich, died 06.12.1913 in München) and his second wife, Franziska (Fanny) Tiefenbrunn (born 21.04.1859 in München, died 15.07.1923 in München; daughter of Peter Paul Tiefenbrunn from Reit / Tirol and his Elisabeth, née März).

- Dr. Heinrich Göring was the son of the *Landesgerichtsdirektor* Wilhelm Göring [1791–1874] and his wife Caroline Maria, née de Nerée [1815–1886]). A Prussian cavalry officer during the Austro-Prussian War of 1866 and the Franco-Prussian War of 1870–1871, he later became the *Kaiserlicher Kommissar* (Imperial Commissioner) of Deutsche Südwest-Afrika (German Southwest Africa, today Namibia), serving as *Reichskommissar* and official representative resident in that German colony. He arrived there in May 1885, and in August, together with his secretary Dr. Nel and a police superintendent named von Goldammer, established his headquarters in the mission school building at Otjimbingwe (which became the capital of Namibia). Concerning his role in Namibia, the following is excerpted from the *House of Commons publication Papers by Command, Volume 17* (1918):

> Goering was a kind of commercial agent, with certain limited powers and jurisdiction over German settlers in the country, but with not the slightest authority to promise the protection of Germany to the natives. Aided by the missionary, Carl Büttner, Dr. Goering immediately proceeded, however, to make "Protection Agreements" with such native chiefs as he had persuaded to ask for the protection and good will of the Emperor. In return for such protection the chiefs were required to give Germans favoured-nation treatment, and they undertook to give no facilities or rights to others that Germans, without the Emperor's consent. Amongst these, Kamaherero, the chief of the Okahandja Hereros, styled "Chief Captain of the Hereros in Damaraland," entered into such an agreement on 21st October 1885. Writing of these agreements ... Governor [Theodor] Leutwein remarks, "those persons who promised the protection in the name

Hermann's father, Heinrich Göring.

Göring with his mother and siblings in 1899.

of the German Emperor had not the slightest authority to do so." (*Elf Jahre Gouverneur*, page 13)

• In 1888, leaders of the Ovaherero tribe succeeded in expelling Dr. Göring from the colony. David Killingray writes:

> In October 1888-only three years after it had been concluded – the Herero in South-West Africa unilaterally withdrew fromt he protection agreement. Their reason was that the Deutsche Kolonialgesellschaft für Südwestafrika (DKGfSWA, German Colonial Society for Southwest Africa, of which Göring was director) was unable to effectively honour their guarantee them against attack from their neighbours. Heinrich Goering, the local director of the company and representative of the *Reich*, immediately withdrew to [Walvis Bay on] the coast. He then asked the Chancellor [Otto von Bismarck] to send 400–500 troops, the number he thought were necessary to re-occupy the lost territories ... (Killingray, *Guardians of Empire: The Armed Forces of the Colonial Powers, c. 1700–1964*, p. 92)

• Dr. Göring departed Africa in August 1890, and became German consul in Haiti and San Domingo, a post he held until his retirement in 1895. He was twice married, first at Neuwied-Heddesdorf on 24.07.1869 to *Ida* Friederike Remy (born 26.09.1847 in Rasselstein, died of "maternity fever", 23.04.1879 in Devant les Ponts, near Metz). From his first marriage were born the following older half-siblings of Hermann Göring:

Friedrich Wilhelm Göring (born 29.10.1870 in Heddesdorf bei Neuwied, died 14.12.1959 in Wiesbaden). Joined the *Schutztruppe für Deutsch-Ostafrika* (1901–1905, Stations-und Bezirkschef; 1906, *Kompaniechef* in Deutsch-Ostafrika; 1908: Promoted *Hauptmann*; 1910: Appointed as *Resident für Urundi*; 1911: Retired; At the start of World War I, assigned to *Ersatz-Bataillon / Infanterie-Regiment* Nr. 87; Start of October 1914: Attached to the *Kommando der Schutztruppen* in the *Reichskolonialamt*; 22.03.1915, promoted to *Major*.

Ida Göring (born 13.09.1872, died 30.09.1872)

Ernst Albert Göring (born 05.10.1873 in Metz, died 21.04.1909 in Berlin), a lawyer.

Friederike (Frieda) Wilhelmine Clara Göring (born 16.07.1875 in Metz, died 09.10.1929 in Kiel). Married to *Korvettenkapitän* Otto Burchardt (born 06.09.1865 in Rostock, died 10.01.1904 in Kiel), that marriage producing one daughter (Ilse Burchard, born 28.04.1898 in Kiel).

Dr. med. Heinrich Carl Göring (born 17.04.1879 in Devant les Ponts). Director of the Augenklinik (eye clinic) in Wiesbaden. He was married twice, first to Dora Barth (born 15.02.1881 in Berlin) and later to Edith Zambona (born 17.08.1898 in Hildesheim). His son, and Hermann Göring's eldest nephew, was Dr. jur. Heinz Göring (born 04.09.1907 in Wiesbaden, killed in action 29.07.1944 near Pogorzel, east of Warsaw as *Hauptmann* and *Chef* of *10.Batterie / III.(Stug)Abteilung / Panzer-Regiment "Hermann Göring"*; he completed his law studies in Wiesbaden on 20.07.1936).

• In London on 26.05.1884, Dr. Heinrich Göring remarried, to Franziska (Fanny) Tiefenbrunn. To this marriage were born the following:

◊ **Karl-Ernst Göring** (born 03.08.1885 in Rosenheim, died 04.10.1932 in Hannover). He also served in the *Deutsch-Ostafrika Schutztruppe*, and was Adjutant in Daressalaam in 1914. In November of that year, he took command of *4.Feldkompanie* of the *DOA-Schutztruppe*, and was promoted *Hauptmann* by Lettow-Vorbeck on 25.02.1915. He was appointed an *Abteilungsführer* in 1916, and was severely wounded on 30.08.1918. He was captured by British troops near Lioma on 06.09.1918. In the postwar years, he served as a police officer, rising to the rank of *Polizei-Oberstleutnant*. Married to Ilse Burchard (born 28.04.1898 in Kiel), he had one son, Peter Göring (born 14.04.1922 in Weissenfels, killed in action as a *Leutnant* with *Stab / Jagdgeschwader 26 "Schlageter"* over Hubersent, France, 13.10.1941).

◊ **Olga Therese Sophie Göring** (born 16.01.1889 in Walvis Bay, German Southwest Africa, died 00.00.1970). Married on 27.05.1912 to Dr. jur. Friedrich (Fritz) Rigele (born 12.08.1878 in Wolkersdorf, died from injuries sustained in a mountain climbing accident in Bad Reichenhall near Berchtesgaden, 10.10.1937), a notary public in Saalfelden (Salzburg) and later in Linz. From 1931 to 1935, Dr. Rigele administered Göring's personal budget.

◊ **Paula Elisabeth Rosa Göring** (born 08.05.1890 in Rosenheim, died 00.00.1960). Married 06.09.1920 to Dr. jur. Franz Hueber (born 10.01.1894 in Untergrünberg / Oberösterreich, died 10.07.1981 in Salzburg), a notary public in Mattsee (Salzburg) and later in Wels (Oberösterreich). He served as Austrian Minister of Justice [1930 and 1938–1939] and rose to the rank of *SA-Brigadeführer*.

◊ **Albert Günter Göring** (born 09.03.1895 in Berlin-Friedenau, died 12.12.1966 in München), a businessman. A signals *Oberleutnant* in the Bavarian Army during World War I, he subsequently worked as an engineer in Wien. He ultimately rose to become director of the Skoda armaments works, a post he held from 1938 to 1945. On 10.05.1945, he voluntarily reported to the U.S. CIC in Salzburg and was arrested on 13.05.1945. Eventually released, he was, att the request of Czech authorities, rearrested by the Americans in Germany on 15.11.1945, and extradited to Czechoslovakia. Held at Pankrac prison, Prague, to await prosecution for war crimes, he was released due to lack of evidence in March 1947. Married 16.03.1921 to Maria von Ammon; later divorced and remarried 02.10.1923 to Ernestine Mathilde Emma Buchlmeyer (born 01.11.1886 in München). One daughter was born to this marriage (ca. January 1945). Disgusted by the Nazi regime and its crimes, he used his family connection to the *Reichsmarschall* and his position at Škoda to save numerous Jews, members of the Czech resistance, and others from Nazi persecution.

• Godfather to all the Göring children was Dr. med. Hermann Ritter von Epenstein (1849–1934), a wealthy Austrian landowner, former Army physician, and friend of Hermann's father from his years in Africa. Although Roman Catholic, his father was a Jew. Von Epenstein provided his castle, Veldenstein, near Neuhaus an der Pegnitz / Franken (which he'd purchased on 29.11.1897) to the Göring family, and Hermann spent his boyhood there beginning in 1901. Von Epenstein retained a room of the castle, and was to take Fanny Göring as his mistress for 15 years before marrying a young woman named Lilli in 1913.

• Religion: Protestant (confirmed in 1908).

• After flying Count Carl Gustaf Bloomfield *Eric* von Rosen (1879–1948) to his castle at Rockelstad, Sweden on 20.02.1920, Göring met the beautiful sister of the Swedish explorer's wife (Mary Fock [1886–1967]), Carin von Kantzow, née Freiin von Fock (born 21.10.1888 in Stockholm, died of tuberculosis 17.10.1931 in Stockholm; daughter of the Swedish Army Colonel Karl Alexander Freiherr Fock and his wife Huldine, née Beamish); she had been married to the Swedish Army officer Captain Nils von Kantzow on 07.07.1910, and by this first marriage had a son, Thomas (born 01.03.1913, died 27.05.1973). An affair soon developed, and Carin divorced Nils von

The first Frau Göring, Carin.

Berlin, 11.04.1935: The most extravagant
wedding in the history of the Third *Reich*.

Kantzow on 13.12.1922. Göring and Carin were married in a civil ceremony in Stockholm on
25.01.1923, and the marriage was solemnized at München-Obermenzing on 3.02.1923. A virulent
anti-Semite and anti-Communist who equated Judaism with Bolshevism, Carin soon became
an ardent National-Socialist. Fiercely loyal to her Nazi husband, she was his primary supporter
throughout his troubled years in the 1920's. Her death left him inconsolable. On 20.06.1934,
Göring had her remains brought from Stockholm to Germany, where he reinterred them at a
massive country lodge, named "Carinhall", which he'd built in her honor.

- In Weimar in 1932, Göring became romantically involved with a stage actress, Emma Johanna
Henny ("Emmy") Sonnemann (born 24.03.1893 in Hamburg, died 08.06.1973 in München;
daughter of the successful Hamburg chocolate factory owner Heinrich Sonnemann and his wife
Emmy, née Sagell), who had been separated from her husband, the actor Karl Köstlin (married
on 13.01.1916, their divorce was finalized early in 1935). He married her, in one of the most
elaborate ceremonies of any kind ever arranged in the Third *Reich*, in Berlin on 11.04.1935; Hitler
was present as Göring's best man. At the christening of their only child (a daughter, Edda, born
02.02.1938) on 04.11.1938, Hitler presented Frau Göring with a Golden Party Badge engraved
with *NSDAP*-Nr. 744 606. Emmy Sonnemann-Göring was arrested by Allied authorities on
25.10.1945, and Edda was cared for by farmers until eventually joining her mother in Straubing
Prison. Emmy and Edda were released from Straubing on 19.02.1946, and met with Hermann for
the last time on 07.10.1946. Rearrested by German authorities as the result of a warrant issued
29.05.1947, she was held at an internment camp in Göggingen and was charged at the internment
camp in Garmisch-Partenkirchen on 20.07.1948 with having having benefited from the Nazi
regime. Placed in de-Nazification Group II, she had 30% of her assets confiscated, was prohibited

An image from Göring's personal wedding album. (Hermann-Historica, Auctioneers, München)

An autographed portrait of Hermann and Emmy Göring. (Hermann-Historica, Auctioneers, München).

from practicing as an actress for five years, and was sentenced to 1 year in a labor camp, with credit for time served. She was immediately released. In 1967, she published her memoirs (and a shameless whitewash of her late husband), *An der Seite meines Mannes* (By My Husband's Side), which was ghostwritten by her lawyer, Dr. Erich Ebermeyer and Alfred Muhr, neither of whom were acknowledged in the book.

- At the time of his capture, Göring weighed 280 lbs., and his weight had been even higher in earlier years. After four months of Allied custody – drug-free, under constant interrogation, and eating U.S. Army food – his weight decreased dramatically; by September 1945, he weighed 200 lbs.

- In addition to his insatiable appetite for power, Göring's character was marked by a lust for extravagance that knew no bounds. His collection of palaces, precious jewels, uniforms, and medals was unrivalled in the Third *Reich*. The former British naval attaché in Berlin, Captain Hawes, once reported, "He can be described as a mountain of egotism and pomposity." A shrewd and voracious collector of art, Göring was to amass – through means both legal and illegal – a collection of thousands of works valued in the hundreds of millions of dollars. He dedicated more attention and care to maintaining his lavish lifestyle than he did to managing the *Luftwaffe*, and he gradually fell out of favor with Hitler and the German people (with whom he had once enjoyed great popularity).

- The following are excerpts concerning Göring from the "Final Interrogation Report-Five Years of Nazi Germany 1933–1938- A Report by Former *Reichswehrminister* (Minister of War) Marshal [Werner] von Blomberg" (Prepared by Seventh Army Interrogation Center, Augsburg, 13.09.1945):

Hermann Göring with his daughter, Edda, 1940.

... (3) Association With the Party ...

... The Fuehrer could use this man, who turned out to be adept at public speaking, had a stentorian voice, and knew how to become popular. In addition to that he was still a Pour le Merite flier and a gentleman. Hitler welcomed Generals Litzmann and Epp for similar reasons. He gave Goering preferential treatment, however. As soon as the National Socialist Party had grown sufficiently strong to nominate a President for the German *Reichstag*, Goering was the man chosen to fill this place. He was also chosen to act as Hitler's personal political negotiator on many occasions. At that time Goering was diligent and ambitious.

His sole connecting link with the Party was the Fuehrer himself. He told me once that he had never quite finished reading the Party program. For Goering, the whole National Socialist Party was embodied in that one person, in Hitler. As a result of this belief, he was an independent member of the Party, apart from all others.

(4) The Rise to Power

After the National Socialists gained control over Germany, Goering wanted all sorts of privileges and offices for himself. He wanted to be top man in the army first of all—his command over the GAF [German Air Force] was a mere consolation prize. He also wanted to become a statesman—thus his position of Ministerpraesident of Prussia. But above all, he wanted power, recognition, and a life of luxury and popularity.

Goering was not cut out to be a great soldier. He did not have the necessary ability of decision, and he lacked the background of knowledge and talent. His knowledge was limited to a few pertinent facts about the *Luftwaffe*, which he had picked up in his conversation

A John Heartfield collage depicts Göring as "The Hangman of the Third *Reich*".

A bronze bust of Hermann Göring.

with competent men. His statesmanship in Prussia was little more than an act. He created a council which had no functions except to spend money. His ministries should have been incorporated in those of the central government at Berlin.

As long as I have known Goering, he was absolutely subservient to his chief. He was actually afraid of him. I remember times when Hitler spoke quite sharply to him, even in my presence, reprimanding him for very minor misbehaviors. It always has been a puzzle to me why Hitler should put this man in any important positions. Immediately after my departure, Goering was promoted to Field Marshal without any reason at all. When, after the campaign in the west in 1940, many other Field Marshals were appointed, he was advanced to *Reichsmarschall*. This was a position which had never been heard of before in the history of Prussia or Germany. Goering immediately made it ridiculous by creating a uniform to go with it which looked as though it had been designed for an operatic fantasy.

(5) Observations on the Goering Mentality

Goering's behavior was most likely conditioned by emotions and appetites rather than by deliberate thought. He was a man who lived just for the moment, a man lacking education and maturity. His was a primitive nature. Reading and self education were foreign to him. If he had not teamed up with Hitler, his career would probably have turned out to be that of a small time adventurer, or a stumble-bum and parasite. But even that might not have phased him a lot, because he cared little for matters of respectability. He had but one important interest, the person of Hermann Goering.

> *FOREWORD*
>
> I welcome this opportunity of presenting to the English-speaking peoples a few of my ideas about the struggle of the German people for freedom and honour. I hope that these words will also be accepted by our opponents as a frank expression of my boundless love for my country, to whose service alone I have pledged my whole life.
>
> Berlin, February, 1934.

Author's foreword to *Germany Reborn*, the English language edition of Göring's *Aufbau einer Nation* (1934).

He loved to receive gifts, even to the extent of soliciting them. This custom began with his second marriage, getting worse from year to year at the celebration of his birthdays. It was a well disguised form of corruption. There was not reticence either on the part of the donor, nor on that of the recipient.

Goering liked to play the part of a 'good fellow', freely dispensing gifts and favors, whenever he found suitable company. He shared the 'live and let live' doctrine of most of his men, as long as he could get enough for himself first. He was an unscrupulous egotist, undaunted by the victims who had to fall on his way. Gossip has it that he was responsible for the suicides of Generals Udet and Jeschonnek. When necessary he would make a tearful show of sentimentality, which was coupled with inhuman brutality.

(6) Failure

I have no idea whether Goering contributed anything to our entry into the war. He would have found it difficult, however, to give sound advice on this subject, since he was ignorant of foreign affairs. Nevertheless he contributed greatly to Germany's defeat. He failed miserably in his position as head of the *Luftwaffe*. He also failed to dissuade Hitler from continuing to fight when it became clear that our defeat was inevitable. But this was not his greatest fault. No man on earth could have changed the Fuehrer's mind.

Hitler's final gesture towards his admirer of long standing, of sentencing him to death, was no more than an additional piece of evidence showing the degraded relationship between Germany's leading personalities. ("Interrogation Records Prepared for War Crimes Proceedings at Nuernberg, 1945–1947/OCCPAC Interrogation Transcripts And Related Records: Von Blomberg, Werner Eduard Fritz"; Publication Number M1270, Record Group RG238)

One of Göring's *Reichsmarschall* collar tabs. (Hermann-Historica, Auctioneers, München).

Sources

Angolia, LTC John (USAF, Ret.): - *For Führer and Fatherland – Military Awards of the Third Reich.* R. James Bender Publishing, 1976.

- *For Führer and Fatherland – Political and Civil Awards of the Third Reich.* R. James Bender Publishing, 1978.

- *On the Field of Honor – A History of the Knight's Cross Bearers* (2 volumes). R. James Bender Publishing, 1979 & 1980.

Barnett, Correlli (editor): *Hitler's Generals.* Grove-Weidenfeld, 1989.

Bekker, Cajus: *Angriffshöhe 4000. Ein Kriegstagebuch der deutschen Luftwaffe.* G. Stalling Verlag, 1964.

Bender, Roger James & Petersen, George A.: *Hermann Göring: From Regiment to Fallschirmpanzerkorps. R. James Bender Publishing, 1975.*

Bradley, Dermot / Hildebrand, Karl-Friedrich / Rövekamp, Markus: *Die Generale des Heeres 1921–1945.* Biblio Verlag, 1993–1999.

Browning, Christopher R.: *The Origins of the Final Solution.* University of Nebraska Press & Yad Vashem, 2004.

Campbell, Bruce: *The SA Generals and the Rise of Nazism.* University Press of Kentucky, 1998.

Delarue, Jacques: *The Gestapo: A History of Horror.* Macdonald & Co., 1964.

Deschner, Günther: *Reinhard Heydrich.* Stein and Day, 1981.

Dornberg, John: *Munich 1923. The Story of Hitler's First Grab for Power.* Harper & Row, 1982.

Fest, Joachim: *The Face of the Third Reich: Portraits of the Nazi Leadership.* Weidenfeld & Nicholson, 1970.

Franks, Norman L.R.; Bailey, Frank W.; & Guest, Russell: *Above the Lines: A Complete Record of the Fighter Aces of the German Air Service, Naval Air Service and Flanders Marine Corps, 1914–1918.* Grub Street, 1993.

Franks, Norman & VanWyngarden, Greg: *Fokker D VII Aces of World War I, Part 1 (Aircraft of the Aces 53).* Osprey Publishing, 2003.

Göring's shoulder boards as *Reichsmarschall* of Germany.
(Hermann-Historica, Auctioneers, München)

Gisevius, Hans Bernd: *To the Bitter End*. Houghton Mifflin Company, 1947.

Goebbels, Dr. Paul Joseph: *The Goebbels Diaries 1942–1943* (ed. Louis P. Lochner). Doubleday, 1948.

Great Britain. Parliament. House of Commons: *Papers by Command, Volume 17*. His Majesty's Stationery Office (HMSO), 1918.

Hamilton, Charles: - *Leaders and Personalities of the Third Reich, Volume I*. R. James Bender Publishing, 1984.

- *Leaders and Personalities of the Third Reich, Volume II*. R. James Bender Publishing, 1996.

Heiden, Konrad: *Der Fuehrer – Hitler's Rise to Power*. Houghton Mifflin, 1944.

Hilberg, Raul: *The Destruction of the European Jews*. (Revised and definitive edition – 3 Volumes). Holmes & Meier, 1985.

Höhne, Heinz: *The Order of the Death's Head*. Martin Secker & Warburg, 1969.

Irving, David: *Göring: A Biography*. Avon Books, 1990.

Jablonsky, David: *The Nazi Party in Dissolution: Hitler and the Verbotzeit, 1923–1925*. Psychology Press, 1989.

Kershaw, Ian: *Hitler: 1936–1945- Nemesis*. W.W. Norton, 2000.

Killingray, David: *Guardians of Empire: The Armed Forces of the Colonial Powers, c. 1700–1964*. Manchester University Press, 1999.

Knopp, Guido: *Göring: Eine Karriere*. C. Bertelsmann Verlag, 2009.

Kube, Alfred: *Pour le mérite und Hakenkreuz: Hermann Göring im Dritten Reich*. Oldenbourg Verlag, 1987.

Lilla, Joachim; Döring, Martin; & Schulz Andreas: *Statisten in Uniform. Die Mitglieder des Reichstags 1933–1945*. Droste Verlag, 2004.

Lyne-Gordon, David: *"And yet you have conquered"- Notable recipients of the Blood Order, Volume I*. Galago Publishing, 2000.

Miller, Michael D.: *Leaders of the SS & German Police, Volume I: Reichsführer-SS - SS-Gruppenführer (Georg Ahrens to Karl Gutenberger)*. R. James Bender Publishing, 2006.

Mosley, Leonard: *The Reich Marshal: A Biography of Hermann Göring*. Doubleday, 1974.

National Archives & Records Administration, College Park, Maryland: - *SA-Personalakte* of Hermann Göring (microfilm printout provided by Mike Constandy of Westmoreland Research).

- "Interrogation Records Prepared for War Crimes Proceedings at Nuernberg, 1945–1947/ OCCPAC Interrogation Transcripts and Related Records: Von Blomberg, Werner Eduard Fritz"; Publication Number M1270, Record Group RG238.

- "Interrogation Records Prepared for War Crimes Proceedings at Nuernberg, 1945–1947/ OCCPAC Interrogation Transcripts and Related Records: Rode, Ernst August"; Publication Number M1270, Record Group RG238.

Noakes, Jeremy & Pridham, Geoffrey (Editors): *Nazism 1919–1945: A History in Documents and Eyewitness Accounts, Volume II – Foreign Policy, War and Racial Extermination*. Schocken Books, 1990.

Office of United States Chief Counsel for Prosecution of Axis Criminality: *Nazi Conspiracy and Aggression (11 volumes)*. U.S. Government Printing Office, District of Columbia, 1946.

Padfield, Peter: *Himmler*. Henry Holt & Company, 1990.

Payne, Robert: *The Life and Death of Adolf Hitler*. Praeger Publishers, 1973.

Reitlinger, Gerald: *The SS. Alibi of a Nation 1922–1945*. Viking Press, 1968.

Schirach, Baldur von: *Die Pioniere des Dritten Reiches*. Zentralstelle fur der deutschen Freiheitskampf, 1933.

Schulz, Andreas & Zinke, Dr. Dieter: *Die Generale der Waffen-SS und Polizei 1933–1945, Band 1 (Abraham-Gutenberger)*. Biblio-Verlag, 2003.

Schuman, Frederick L.: *Hitler and the Nazi Dictatorship. A Study in Social Pathology and the Politics of Fascism*. Robert Hale & Company, 1934.

Sereny, Gitta: *Albert Speer: His Battle With with Truth*. Vintage Books, 1995.

Sigmund, Anna Maria: *Women of the Third Reich*. NDE Publishing, 2000.

Walter Hans Buch

13.11.1923–00.01.1924   *"mit der Führung der illegalen SA beauftragt"*
(charged with leadership of the illegal SA)

Born:	24.10.1883 in Bruchsal / Landeskommissariat Karlsruhe / Baden.
Suicide:	12.09.1949 in Schöndorf am Ammersee (slashed his wrists and drowned himself in the Ammersee, according to a report of the Bavarian State Police).

NSDAP-Nr.:	7,733 (First joined 09.12.1922; Party banned following the *München-Putsch* of 09.11.1923; Reenrolled 15.06.1925)
SS-Nr.:	81,353 (Joined 01.07.1933)

Promotions

30.09.1902	*Fahnenjunker*
27.01.1904	*Leutnant*
27.01.1913	*Oberleutnant*
24.12.1914	*Hauptmann*
20.11.1918	*Major a. D.*
13.11.1923–00.01.1924	*"mit der Führung der illegalen SA beauftragt"* (Charged with leadership of the illegal SA)
18.12.1931	*SA-Gruppenführer z. V.*
02.06.1933–08.05.1945	*Reichsleiter der NSDAP*
01.07.1933	*SS-Gruppenführer*
09.11.1934	*SS-Obergruppenführer*

Career:

ca. 1889–00.09.1902	Attended *Volksschule* in Bruchsal, then the humanistic *Gymnasien* in Karlsruhe and Konstanz (passed his *Abitur* in Konstanz).
30.09.1902–00.00.1915	Entered service as *Fahnenjunker*, assigned to *6.(badische) Infanterie-Regiment "Kaiser Friedrich III" Nr. 114* (Konstanz). Assigned as *Regiments-Adjutant*, 00.09.1914–00.00.1915.
27.01.1904–00.00.19__	Commissioned and assigned as a *Lehr-Offizier* (instructional officer).
[00.00.1912–00.00.1914]	Assigned to the *Unteroffizierschule* in Biebrich (wearing the uniform of *Infanterie-Regiment Nr. 114*).
00.00.1915–00.00.191_	Assigned to *8.(westfälisches) Infanterie-Regiment "Herzog Ferdinand von Braunschweig" Nr. 57*, then *47. Infanterie-Brigade*.
00.00.191_–00.00.191_	*Kompaniechef* in *4.(badische) Infanterie-Regiment "Prinz Wilhelm" Nr. 112*.
00.00.191_–00.12.1918	*Kommandeur* of *23.Maschine-Scharfschütze-Gewehrabteilung* (Machine-Gun Sharpshooter Detachment) in *6. (badische) Infanterie-Regiment "Kaiser Friedrich III" Nr. 114*.
00.03.1918–00.09.1918	*Kommandeur* of the *Offiziers-Aspiranten-Bataillon* and machine-gun instructor at the *Infanterie-Lehrschule* in Döberitz.

Buch as a young man. (Source: Walter Buch's personal
photo album, in collection of Gary Merlie)

00.09.1918–00.11.1918	Assigned to the *Kriegsministerium* in Berlin as a specialist in the employment of machine gun units.
20.11.1918	Retired from service as *Major a. D.*, having refused to swear the oath of loyalty to the new German republic. In a letter of 02.11.1931 to William Fanderl, editor of *Der Angriff*, Buch wrote that he "did not want to work in the new command under the flag of [Weimar Republic President Friedrich] Ebert" (Donald M. McKale, *The Nazi Party Courts: Hitler's Management of Conflict in His Movement, 1921–1945*, pp. 53–54). He was declared 50% "Kriegsbeschädigt" (war-disabled).
00.00.1918–00.00.1923	Owned a small chicken farm in Scheuern bei Gernsbach / Schwarzwald, near the estate of his father.
00.03.1920	First meeting with Adolf Hitler, as described in an interrogation report of 13.06.1945:

> Buch's father, an Amtsrichter, had written a book entitled "Vom Internationalen zum Nationalen Arbeitsstaat", published by Lehmann Verlag in Munich. Through correspondence with Lehmann, the older Buch became vitally interested in Hitler, and ordered his son to take a copy of the book he had written to Hitler personally. In this manner, in Mar 1920, Walter Buch met Hitler, and became one of his followers. He corresponded with Hitler on several occasions, and followed his efforts through hearsay and newspapers ... ("Special Report on Walter Buch, *Reichsleiter* and *Oberster Parteirichter*", Special Detention Center "Ashcan"-Detailed Interrogation Report, in "Interrogation Records Prepared for War Crimes Proceedings at Nuernberg, 1945–1947/

Buch on horseback, ca. 1914. (Source: Walter Buch's personal
photo album, in collection of Gary Merlie)

OCCPAC Interrogation Transcripts And Related Records:
Buch, Walter Hans"; Publication Number M1270, Record Group
RG238)

00.00.1919–00.00.1922	Member of the *Deutschvölkischer Schutz- und Trutzbund*, serving as its *Gaugeschäftsführer* (regional business manager) in Baden.
00.00.1919–00.00.1922	Member of the *Deutschnationale Volkspartei* (*DNVP*, German Nationalist People's Party). He rose to become Party Secretary in Karlsruhe by 1923.
00.00.19__-00.00.19__	Member of *Badische Verband ehemaliger Kriegsteilnehmer* (Baden Organization of War Veterans).
00.00.1920–00.00.192_	Member of the *Bund Oberland*.
00.03.1920	First meeting with Adolf Hitler. This resulted from his father asking him to personally deliver a copy of his own book, *Vom Internationalen zum Nationalen Arbeitsstaat*, to Hitler. The elder Buch had learned of the future *Führer* via correspondence with his publisher, *Lehmann Verlag* in München, and had become very enthusiastic about him.
00.00.1921–00.00.19__	Co-editor of the *Badische Wochenzeitung*, a small völkisch newspaper in Karlsruhe.
14./15.10.1922	Participated in the violent demonstration march of some 700 *SA* men to the third annual *Deutschen Tag* of the *Deutschvölkischer Schutz- und Trutzbund* in Coburg.
09.12.1922	Joined the *NSDAP*. Around this time, he wrote to Max Amann, business manager of the Party, regarding possible employment in the *NSDAP* or *SA*. Above all other reasons, Buch enrolled in the *NSDAP* because its platform mirrored his own virulent anti-Semitism (which had been instilled in him by his own father). Donald M. McKale writes:

Buch's anti-Semitic philosophy represented a synthesis of German
völkisch thought with the ideology of a new "revolutionary

Buch and other officers, ca. 1914.
(Source: Walter Buch's personal photo
album, in collection of Gary Merlie)

Leutnant Buch, ca. 1910. (Source:
Walter Buch's personal photo album,
in collection of Gary Merlie)

nationalism" that began to develop among far-right elements at the beginning of the 1920s. Thus, he despised Jews not so much for religious or cultural reasons but on biological and racial grounds. For Buch, whose anti-Semitism was typically Nazi, the Jew had been Germany's enemy throughout history... [H]e saw [the] "decay" brought about by Jews] most clearly in the decline of the German family and of the Germanic concepts of "honor" and "justice." Supposedly, the destructive characteristics of Jewish blood had been allowed to mix with pure Aryan blood during the nineteenth century, and this fateful development had led to Germany's collapse in World War I. The reason for this, Buch argued [in *Ehre und Recht*, 1932], had been the liberal Enlightenment and the French Revolution, whose sole aim was "the emancipation of Judaism." The result was the defeat of a superior people (that is, the Germans), whose destiny in the world was unlike that of other peoples. [In *Niedergang und Aufstieg*, he wrote:] There is no race with an equal or even similar depth of soul to the German ... The assertion that the French Revolution was contrived as a conspiracy by the Jewish enemy for the destruction of Germanic-Nordic blood can be proven with little difficulty today. (McKale, *The Nazi Party Courts: Hitler's Management of Conflict in His Movement, 1921–1945*, p. 55)

Buch wears a steel helmet of the reinforced
type designed for snipers, ca. 1916.
(Source: Walter Buch's personal photo
album, in collection of Gary Merlie)

Hauptmann Buch and a fellow officer
in the trenches on the Western Front.
(Source: Walter Buch's personal photo
album, in collection of Gary Merlie)

In an article that appeared in the 21.10.1938 issue of *Deutsche Justiz*,
Buch wrote:

> The Jew is not a human being. He is an appearance of putrescence.
> [*Der Jude ist kein Mensch. Er ist eine Fäulniserscheinung*] Just as
> the fission-fungus cannot permeate wood until it is rotting, so the
> Jew was able to creep into the German people, to bring on disaster
> only after the German nation... had begun to rot from within.

In the 02.01.1940 issue of *Der Parteirichter. Amtliches Mitteilungsblatt
des Obersten Parteigerichts der NSDAP* (München), he wrote:

> ... For what else did we struggle, take upon us want and deprivation,
> for what else did the courageous men of the *SA* and the SS, the
> boys of the Hitler Youth fall, if not for the possibility that one day
> the German people might start its struggle for liberation against
> its Jewish oppressors? In this struggle we are now involved ... .
> Neither Jew nor freemason will finish the war with treaties that
> bear in them new mischief. Victory will be attained by Adolf
> Hitler and he will bring Europe the peace that forever takes away
> from the Jewish sub-man the opportunity to decompose men and

Walter Buch in civilian attire, ca. 1927. (Source: Walter Buch's
personal photo album, in collection of Gary Merlie)

Hermann Buch (left) during the celebration of "*Deutscher Tag*" in Bayreuth, 29./30.09.1923.
Adolf Hitler appears in the background, toward the left in light-colored overcoat.
(Source: Walter Buch's personal photo album, in collection of Gary Merlie)

24.06.1927: Walter Buch, far left in second row, posing with other Nazis. (Source: Walter Buch's personal photo album, in collection of Gary Merlie)

peoples and play them off against each other. (S. Mendelsohn, *The Polish Jews behind the Nazi Ghetto Walls*)

00.00.1922–00.09.1923	*Ortsgruppenleiter* of the *Ortsgruppe Karlsruhe der NSDAP.*
01.01.1923	Joined the *SA.*
01.08.1923–09.11.1923	*Bezirksführer der SA* for Franken and *Führer* of *SA Sturmregiment 3* in Nürnberg (also known as *SA-Regiment "Frankenland"* and *SA-Kommando Franken*; HQ: Nürnberg). The *SA* force over which he assumed command from Walter Steinbeck (dismissed in July 1923 due to conflicts with Julius Streicher). numbered some 275 men as of July 1923, a number that grew to about 800 by October. Buch was succeeded by *Oberleutnant* Freund.
08./09.11.1923	Played a peripheral role in the *München-Putsch.* In the wake of the failed coup, he successfully eluded the police in Nürnberg and succeeded in transferring the files of the Nürnberg *SA* to Erlangen. Buch also called for a pogrom against the Jews, although there was no evidence that Jews had given any attention- positive or negative- to the Nazi bid for power in München. He issued the following order of the day for 11.11.1923:

The first period of the National Revolution is over. It has brought the desired clearing [of the air]. Our highly revered leader, Adolf Hitler, has again bled for the German people. The most shameful treachery that the world has ever seen has victimized him and the German people. Through Hitler's blood and the steel directed against our comrades in München by the hands of traitors the

Walter Buch wearing the short-lived collar tabs for senior officials of the *Reichsleitung der NSDAP* (not to be mistaken for those of an *SA-Gruppenführer*, which were very similar), 1933. (Source: Walter Buch's personal photo album, in collection of Gary Merlie)

patriotic Kampfverbände are welded together for better or for worse. The second phase of the National Revolution begins [now].... (Harold J. Gordon, Jr., *Hitler and the Beer Hall Putsch*, pp. 432–433)

00.11.1923–00.04.1924	Employed as a wine and cigar salesman in München.
00.00.1924–01.01.1928	*Gau-SA-Führer Oberbayern-Schwaben.*
13.11.1923–00.01.1924	*Mit der Führung der illegalen SA beauftragt* (charged with leadership of the illegal SA), leading the *SA* during the ban imposed in the wake of the München-Putsch. Appointed by Alfred Rosenberg, Max Amann, and Anton Drexler who jointly led the Party following Hitler's arrest. He succeeded Hermann Göring who had been severely wounded during the Putsch. During this period he and his adjutant, Emil Danzeisen, worked diligently to reorganize and camouflage the *SA,* setting up rifle clubs and choral societies as a cover for illegal *SA* activities.
00.02.1924	Arrested in München, charged with attempting to reestablish the *NSDAP* (released after submitting a statement to the Police).
00.02.1925–23.11.1927	*SA-Führer in München.*
15.06.1925	Officially reenrolled in the *NSDAP*.
ca. 1925–00.00.192_	*Vorsitzender* of the *Ehrengericht "Deutsche Vaterländische Orden"* (Court of Honor of the German Patriotic Order).
27.11.1927–01.01.1928	Acting *Vorsitzender* of the *Untersuchungs-und Schlichtungsausschuß* (USCHLA, Investigation and Arbitration Committee) in the *Reichsleitung der NSDAP*. Succeeded *Generalleutnant a. D.* Bruno Heinemann who had retired due to differences with Hitler, and possibly also due to age and frequent illness-related absences from office. Buch worked closely with Hitler, whom he worshipped, on every significant

Walter Buch wearing the short-lived collar tabs for senior officials of the *Reichsleitung der NSDAP* 1933. Also of note: The rare Coburg-Abzeichen, instituted in October 1932 to recognize participation in the 14./15.10.1922 "Battle of Coburg". (Source: Walter Buch's personal photo album, in collection of Gary Merlie)

matter dealt with by the Party courts. " ... [H]e did not always agree with Hitler on matters involving the *Reichs-Uschla*," writes Donald M. McKale,

> ... but he was willing to subordinate his own ideas to those of the *Führer* and to carry through the latter's decisions at any cost ... [T]he *Führer* apparently had an extraordinary talent for convincing Buch in private discussions. Yet at the same time, Buch recognized early a fateful trait in Hitler that was eventually to cost the world dearly[:] Hitler's "contempt for humanity" (*Menschenverachtung*) ... (McKale, *The Nazi Party Courts: Hitler's Management of Conflict in His Movement, 1921–1945*, pp. 61–62)

On 01.10.1928, Buch wrote a remarkable letter to Hitler, the handwritten original – in pencil – which is found in the *Bundesarchiv* (NS 26/Folder 1375), however it is uncertain if it was ever sent:

> I have recently acquired an impression about a number of things and feel it is my difficult duty to tell you, Herr Hitler, that you have a contempt for humanity that fills me with grave uneasiness. I

Heinrich Hoffmann's portrait of *Reichsleiter* Walter Buch, ca. 1934. (NARA photo).

do not believe, Herr Hitler, that a person who is filled with human contempt from recent experiences can continue to fulfill a task that has burdened a people's fate for centuries … . It is quite clear to me that for some time you have believed yourself capable of building on the bitter disappointments of people and that during the past months you have received such a vote. I hope, however, these votes will be permanent. (ibid, p. 62)

01.01.1928–01.01.1934	*Vorsitzender* of the USCHLA in the *Reichsleitung der NSDAP*.
20.05.1928–18.07.1930	Member of the *Reichstag (Wahlkreis 24, Oberbayern-Schwaben)*.
00.00.1930	Involved in a conspiracy to murder Ernst Röhm and certain members of his staff (including Graf Spreti, Graf du Moulin-Eckart, and Georg Bell). The originator of this plan, Emil Traugott Danzeisen, a former *SA-Standartenführer* and Adjutant to Buch, was tried and sentenced to six month's imprisonment. Charges were not brought against Buch and the other conspirators due to lack of evidence.
14.09.1930–14.10.1933	Member of the *Reichstag*.
11.06.1930–00.10.1931	*Leiter* of the *Jugendamt* (Youth Office) in the *Reichsleitung der NSDAP*.
00.00.193_–00.00.1933	Assigned as an editor of the *Völkischer Beobachter*.
17./18.10.1931	Participated in the *SA-Aufmarsch in Braunschweig*.
01.06.1932–10.10.1935	*Vorsitzender* of the *I.Kammer der Reichs-USCHLA* (redesignated *I. Kammer des Obersten Parteigerichts der NSDAP*, 01.01.1934). He resigned from this post and was succeeded by Johannes Schneider.

30.08.1932–12.09.1932	Member of the *2. Ausschuß (Auswärtige Angelegenheiten) der Reichstag* (2nd Committee [Foreign Affairs] of the *Reichstag*).
06.12.1932–01.02.1933	Member of the *2. Ausschuß (Auswärtige Angelegenheiten) der Reichstag*.
01.07.1933–01.10.1934	Joined the *SS*, assigned as *Ehrenführer* to *63. SS-Standarte*.
12.12.1933–28.03.1936	Member of the *Reichstag (Wahlkreis 15, Osthannover)*.
01.01.1934–08.05.1945	*Vorsitzender des Obersten Parteigerichts der NSDAP* (Chairman of the Supreme Party Court of the *NSDAP*, also known as *Oberster Parteirichter* [Chief Party Judge] *der NSDAP*), München. Buch fervently believed that National Socialism should represent a revolution in morality as well as politics and power. He naively expected Adolf Hitler to be a champion of law and decency, spearheading a crusade against vice and corruption. "For Buch," writes Dietrich Orlow, "morality consisted primarily of a series of negatives, chief among them anti-Semitism, antifreemasonry, and antipornography." (*History of the Nazi Party: 1933–1945*, p. 67) But Buch was given little power or direction by his *Führer* in overseeing the preservation of moral rectitude among the Party membership, and it did not help that neither Buch nor his judges were trained in law. "Moreover," continues Orlow, "Buch had little control over disputes within the *PO* [*politische Organisation* of the *NSDAP*]. In quarrels between *Reichsleiters* and *Gauleiters*, for example, his role was restricted to that of a mediator ..." (ibid) Under interrogation at Nürnberg by Lt. Col. Thomas S. Hinkel, Buch made the following statements concerning his role in this assignment:

> [05.09.1945]: I was charged with the administration of all party courts in the different states and I was in charge of higher court methods in cases of higher party members. In other words, when higher party members stood before the court, I took charge of the court procedure. The higher court I divided in different departments, and they have independently without my influence handled the different cases ... . I myself was only a judge in the case of higher party officials ...
>
> [31.10.1945]: ... I would like to ask you to consider the position of a man who for seventeen years was forced to do against his will, a job which he didn't like. I kept asking that I be relieved of my functions, and be permitted to return to my original profession as a regular army officer. This was particularly true during the war when my boys were out in the field, and my mind was not on my job as much as it was on the military situation. This is why the things that I may have done, which had to be done in the execution of my functions, are not as clear in my memory as they would have been if I had taken more interest in my job, which I had to do against my own will.

Another interrogation report, dated 18.07.1945, states:

Buch convenes a meeting in 1933. (Source: Walter Buch's personal photo album, in collection of Gary Merlie)

Studio portrait of Walter Buch in brown shirt, 1933. (Source: Walter Buch's personal photo album, in collection of Gary Merlie)

Walter Buch on a boating excursion with his daughter, Gerda, ca. 1935. (Source: Walter Buch's personal photo album, in collection of Gary Merlie)

An inaccurate illustration of Walter Buch wearing *SA-Obergruppenführer* insignia (his highest rank in the *SA* was *Gruppenführer*). (*Orientalischen Cigaretten-Compagnie "Rosma", Männer des Dritten Reich*, 1934)

Buch tried to organize the party courts along the lines of the Ehrengerichte of the old German Army. He was a member of the Army himself, and a passionate soldier. His father who was a judge of the Supreme Court (*Oberlandesgericht*) in Baden, had warned the boy to stay away from the study of law. Buch had followed this advice, and apparently kept a deep-rooted distrust of all legal procedures. Just the same, he was always fascinated by the problem of justice. He found the highest expression of justice in the army, and consciously organized his party justice so as to be similar to that of the army. Even the punishments of the party courts were the same as the punishments in the military court of honor, viz. Verweis, Verwarnung (after 1933 Verwarnung with Aberkennung der Fähigkeit, ins Amt zu bekleiden) (reprimand with the prohibition of holding an office), Entlassung [discharge], Ausschluss [dismissal], and Ausstosung [expulsion]. The latter punishment involved the prohibition for any party offices or organizations to employ the culprit in [the] future. He became an outcast of the party. Private firms or private parties, however, were not bound by this prohibition. The Ausstossung ... was generally pronounced only when a criminal act had been committed, and was followed, as a rule, by a trial before a regular civilian court.

It was, however, not the only task of the party courts to

Walter Buch poses with his family in 1933.

take disciplinary action against party members who had sinned against the spirit of the party. A vast sector of the party judiciary was concerned with arbitration between party members. In all cases, the party courts were entitled to conduct investigations. In no case could the party courts pronounce a sentence which had effect outside the party proper; it could not impose punishments other than disciplinary actions which only affected the defendant within the party.

If a case involved any criminal acts, the party court would notify the regular courts, and even furnish the material of its findings. Regularly, according to Buch, an action of the party courts would precede a trial before the regular courts ... A pronouncement of the party court did not bind the regular courts, e.g. if a man was expelled from the party because he had embezzled party funds, a regular court was not compelled to sentence this man; he might have refunded the embezzled money, or might be acquitted before the regular court for other reasons. By the same token, a man who was acquitted by a party court might yet be prosecuted before a regular court. However, Buch is not sure whether it happened in practice that cases that were dismissed by the party courts were prosecuted further by regular courts.

The jurisdiction of the party courts was not based upon any codified law. Hitler ordered Buch in 1933 to draft a codified criminal law for offenses of party members which would be prosecuted by party justice. For that purpose Buch was to get in contact with the Reichsjustizminister Gürtner. Buch stalled all during 1933, because he was opposed to the use of codified laws; finally, in December 1933 he found his way to Gürtner, with whom he had a long conversation. The result of this conversation was that both men agreed that such a law would be against the spirit of the

Studio portraits of *SS-Obergruppenführer* Buch, ca. 1934. (Source: Walter Buch's personal photo album, in collection of Gary Merlie)

Walter Buch chats with an *NSDAP* leader, ca. 1937. (Source: Walter Buch's personal photo album, in collection of Gary Merlie)

Walter Buch, his father (at far left), and Frau Buch (extreme right) dine with *SS* and *Luftwaffe* officers. Standing, 3rd from left, is Felix Steiner, and seated, 2nd from right, is Bernhard Voss. (Source: Walter Buch's personal photo album, in collection of Gary Merlie)

Reichsleiter Buch, ca. 1935. (Source: Walter Buch's personal
photo album, in collection of Gary Merlie)

institution of party courts. Buch reported this agreement to Hitler
who let himself be convinced ...

... Buch tried to avoid choosing professional lawyers as party
judges. He tried to fill these positions with old members of the
party who had partaken in the fight of the party for power, because
he trusted these men to maintain the spirit of the party. Most of
the judges worked on a honorary basis, only those of the Supreme
Court and the Gaugerichte were full time judges and paid as such.
They numbered, all in all, about 100 men.

When the war broke out, Buch insisted that his judges
become soldiers for a limited time. He wanted them to experience
the life of a front-line soldier, so as to be better fitted for their job
when they returned from the war. 150 full time judges were killed
in the war, which lead to a degeneration of the party court system.
More professional lawyers infiltrated into the courts, and the
best judges were absent or dead. At the same time, the expulsions
pronounced by party courts took a sharp increase (proportionally),
because the better elements of the party were at the front, and
the new party members who were accepted into the party, largely
under pressure from the party treasurer, were no longer carefully

Reichsleiter Buch shares a meal with Joachim Albrecht Eggeling (deputy *Gauleiter* of Magdeburg-Anhalt), ca. 1936. (Source: Walter Buch's personal photo album, in collection of Gary Merlie)

Reichsleiter Buch visiting a foreign country, possibly Turkey, ca. 1935. (Source: Walter Buch's personal photo album, in collection of Gary Merlie)

selected to represent the elite of the nation ...

... The *Oberste Parteigericht* needed the confirmation of Hitler himself to make its decisions effective. Since 1941, the Chief of the *Parteikanzlei*, Bormann, achieved the power of confirming the decisions of the Supreme Party Court. He became, by Hitler's delegation, the party's Supreme Justice (*Oberster Gerichtsherr*). As such, he not only was able to nullify sentences pronounced by the Supreme Court, but also inform the court that a certain decision was expected in an individual case. Buch tried to main his independence of decision, but since he was in practice unable to judge according to his personal convictions, he refused to preside in court sessions, and more or less retired from his position. This interference of Bormann, however, only applied when the court was sitting in first instance, e.g. proceedings against *Gauleiter*, but not in cases where the Supreme Court was judging an appeal from a lower court ... (Institut für Zeitgeschichte, München, Akte 4637/71)

In his postwar memoirs, the former Hamburg *Gauleiter* Dr. Albert Kreb described Buch as follows:

I had frequent dealings with Buch and in the process always found him to be an honorable man of the best intentions. Without a doubt he had the goodwill to operate his office in such a way as to keep out or remove all corrupt elements. But it is equally without doubt that he did not succeed in doing so. His concept of what constituted behavior injurious to the party differed from Hitler's so frequently that it was impossible for Buch to effectuate it. As a dry, somewhat pedantic man in his ideas and needs, Buch possessed neither the wit nor the charm that was needed to resist or even convince Hitler. That is, Buch presented opinions and proposals but then simply obeyed when Hitler held contrary opinions and ordered decisions in accordance with them.... On [one] occasion... Buch openly admitted to me his attitude and his method, saying it was caused by his military training in obedience. "You just click your heels together and say 'yes, sir!'" (Dr. Krebs, *The Infancy of Nazism*)

Buch's relevance in the *NSDAP* disappeared as a result of a crusade against immorality within the Party that backfired. In his official publication, *Der Parteirichter* (issues of 10.04.1935; 10.06.1935; 10.08.1935; 10.10.1935; 10.01.1936; and 10.06.1936) he stated that scrupulous marital fidelity and the maintenance of family stability were National-Socialist cornerstones. His greatest mistake was to demand the punishment by party courts of moral offenses by senior Party leaders. In a Party where marital infidelity was rather commonplace at all levels of the chain of command, this won Buch no friends (notable examples of those who would theoretically be affected by Buch's morality campaign were the powerful *Reichsminister* Dr. Goebbels, as well as two philandering *Gauleiters* (Wilhelm Kube and Julius Streicher). Hitler himself disapproved of Buch's statements and writings, as it was loyalty to *Führer* and Party that concerned him- not the amorous adventures of his otherwise stalwart satraps. It was likely only after being called to meet with an angry Hitler on 14.11.1935 that Buch learned that he was acting against the will of his *Führer*. His significance as a Party leader all but disappeared following that meeting.

01.01.1934–10.10.1935   *Vorsitzender der I. Kammer des Obersten Parteigerichts der NSDAP* (Chariman of the 1st Chamber of the Supreme Party Court of the *NSDAP*).

00.06.1934–00.07.1934   Played an important role in the purge of the *SA* leadership. He had amassed a large quantity of incriminating material against the *SA* from *Gauleiters* and other *NSDAP* officials, and collected hundreds of complaints from *SA* men and parents of Hitler Youth members. Also in his possession were "love letters" to male friends sent by Röhm to Bolivia. Buch demonstrated great zeal and brutality during the purge of 30.06./ 01.07.1934; unconfirmed accounts indicate that he was personally involved in the murder of *SA* leaders, directing *SS* executioners at

A gathering of old fighters in München, signed by Prince August Wilhelm ("AuWi"). Left to right: Buch, "AuWi", and Adolf Wagner. (Source: Walter Buch's personal photo album, in collection Gry Merlie)

Stadelheim Prison near München. The following is excerpted from a report of Buch's postwar interrogation:

The Purge of 1934

Sometime in 1933, PW Walter Buch received numerous letters and official complaints about the homosexual activities of the Leader of the *SA*, Roehm. Because Buch was the Party Judge, it was only right that he should be the one to whom such complaints from the mothers, wives, etc. of Roehm's subordinates, would be addressed. Buch immediately sought audience with Hitler and informed him of the complaints, also showing the written proof. But Hitler, who was always frantically loyal to his friends, refused to listen to Buch's advice and did not believe the charges brought against Roehm ...

On the 30th of Jun 1934, Buch was working in his office in Munich when Hess called on him and told him that Roehm had been arrested. Together with Hess, Buch went to a gathering of Party leaders whom Hitler addressed at that time. In his speech, Hitler told the assembly that Roehm was guilty of conspiracy, and that he had ordered his arrest. He then went on to say that Buch had warned him about Roehm a year before, but that he, (Hitler)

An avid horseman, Buch sits in the saddle in 1937. (Source: Walter Buch's personal photo album, in collection of Gary Merlie)

had not listened to Buch's advice. (PW says that since this day, Hitler and he were estranged, because Hitler bore a grudge against him for making him admit his mistake).

Although the case of Roehm should have come before the Supreme Party Judge, this was not done, and the entire Purge was handled by the *SS* in a very secretive fashion. Buch was told several days later that Roehm had been killed. ("Special Report on Walter Buch, *Reichsleiter* and *Oberst*er Parteirichter" [13.06.1945], Special Detention Center "Ashcan"-Detailed Interrogation Report, in "Interrogation Records Prepared for War Crimes Proceedings at Nuernberg, 1945–1947/OCCPAC Interrogation Transcripts And Related Records: Buch, Walter Hans"; Publication Number M1270, Record Group RG238)

01.10.1934–01.04.1936	Assigned as an *SS-Führer z.b.V.* to the *Reichsführer-SS*.
00.09.1934–30.04.1945	Member of the *Sachverständigenbeirat für Bevölkerungs- und Rassepolitik* (Council of Experts for Population and Racial Policy) attached to the *Reichsministerium des Innern*.
03.10.1934–00.00.1944	Member of the *Akademie für Deutsches Recht*, München.
05.02.1936	At a press conference in München on 05.02.1936, declared:

Right is what benefits the German people, and wrong is what would be harmful. To establish the limits between right and wrong is the task of the highest party court of justice. (Rolf Tell, *Sound and Fuehrer*, p. 128)

01.04.1936–30.04.1945	Assigned to the staff of the *Reichsführer-SS*.
30.01.1937–08.05.1945	Member of the *Reichstag (Wahlkreis 29, Leipzig)*.
00.00.193_-00.00.19__	Member of the Committee for the *Braune Band von Deutschland* (an organization set up to maintain the quality of horses for sporting

At a reception for the Japanese Ambassador to Germany, General Hiroshi Oshima, in München, 1942. Above: left to right, Bavarian *Ministerpräsident* Ludwig Siebert, Buch, and Oshima . Below: left to right, *SS-Obergruppenführer* Werner Lorenz, Oshima, and Buch. (Source for both photos: Walter Buch's personal photo album, in collection of Gary Merlie)

activities and to control the quality of such stock for mounted units of the *SS* and other Party organizations).

13.02.1939   Submitted the following secret report on his investigation into the *"Reichskristallnacht"* pogrom to Hermann Göring:

> Dear Party Comrade Göring!
> I enclose the report of my special senate about the procedure hitherto concluded concerning the excesses on the occasion of the anti-Jewish operations of 9. and 10 November 1938. Heil Hitler !
> [signed] Walter Buch ...

> Report about the events and judicial proceedings in connection with the antisemitic demonstrations of 9 November 1938.

> On the evening of 9 November 1938, *Reich* Propaganda Minister Party Comrade Dr. Goebbels told the Party leaders assembled at a social evening in the old town hall in München, that in the *Gaue* of Kurhessen and Magdeburg-Anhalt it had come to demonstrations against Jews, during which Jewish shops were demolished and synagogues were set on fire. The *Führer*, at Goebbels' suggestion, had decided that such demonstrations were not to be prepared or organized by the Party, but so far as they originated spontaneously, they were not to be discouraged either. In other respects, Party Comrade Dr. Goebbels carried out the purport of what was prescribed in the teletype of the *Reich* Propaganda Administration of 10 November 1938 (12:30 to 1 o'clock) (Enclosure 2). It was probably understood by all Party leaders present, from the oral instructions of the *Reich* Propaganda Minister, that the Party should not appear outwardly as the originator of the demonstrations but in reality should organize and execute them. Instructions in this sense were telephoned immediately (thus a considerable time before transmission of the first teletype) to the bureaus of their *Gaue* by a large part of the Party members present ... . At the end of November 1938, the *Oberstes Parteigericht*, through reports from several *Gau* courts, heard that these demonstrations of 9 November 1938 had to a considerable extent gone as far as plundering and killing of Jews and that they had already been the object of investigation by the police and the state prosecutor. The deputy *Führer* [Rudolf Hess] agreed with the interpretation of the chief Party Court, that known transgressions in any case should be investigated under the jurisdiction of the party. Because of the obvious connection between the events to be judged and the instructions which *Reich* Propaganda Minister Party Comrade Dr. Goebbels gave in the town hall at the social evening. Without investigation and evaluation of this connection, a just judgment did not appear

Walter Buch and son, 1941. (Source: Walter Buch's personal photo album, in collection of Gary Merlie)

possible. This investigation, however, could not be left to innumerable state courts, especially as the demonstrations had meanwhile been presented to the public as the spontaneous expression of the people's sentiments. According to the conception of the *Oberstes Parteigericht* it must, as a matter of principle, be impossible for political offences to be determined and judged by the state court without the Party having the possibility of first obtaining clarification about the happenings, so that, if occasion arises, the *Führer* could be asked in good time to cancel the proceedings at the state court. This concerns matters, which primarily concern the interests of the Party and which even though this be only from the viewpoint of the perpetrator, are desired by the Party. Due to such considerations, *Generalfeldmarschall* Party Comrade Göring ... has entrusted the Secret State Police and the Party jurisdiction with the investigation of excesses. The *Oberstes Parteigericht* has reserved for itself the investigation of killings, severe mistreatment and moral transgressions. On the basis of state police inquiries, the judges of the *Oberstes Parteigericht*, who were present with their alternates held quick trials of those cases about which facts were ascertained up to 17 January 1939. *Gau* leaders and Group leaders of the branches served as jurors at the trials and decisions. The decisions, which, for reasons to be discussed later, contain only portions of the statements of the facts, are attached. Party Member Frey, Heinrich, Party Member since 1932, residing in Rheinhausen, Horst Wessel Street 23, was ejected from the Party because of a moral crime and race violation perpetrated upon the thirteen-year old-school girl Ruth Kalter. Frey is in custody and has been handed over to the criminal court (Enclosure 5). Party Member Gerstner, Gustav, Party membership number 3,135,242, *SA-Oberscharführer*, residing at Niederwern, at present district court prison Würzburg, was expelled from the *NSDAP* and *SA* because of theft. Gerstner is in custody and has been handed over to the public court because suspected of race violation. (Enclosure 6) ... . 4. Party Member Norgall, Franz, Party membership number 342,751, *SA-Sturmführer* , residing at 58 Neuhoefer street, Heilsberg (East Prussia), was given a warning and sentenced to three years deprivation of the right to hold public office because of disciplinary violation, namely the killing of the Jewish couple Seelig in Heilsberg contrary to orders. (Enclosure 8) ... 16. Proceedings against Party Members Aichinger, Hans, *SS-Hauptsturmführer*, residing at 9 Seilergasse, Innsbruck, and Hofgartner, Walter, *SS-UnterSturmführer* residing at 21 Gavelsberger Street, Innsbruck, for killing the Jews Graubart, Dr. Bauer, and Berger, have already been quashed on the basis of inquiries on the part of the State Police and individual interrogations of the *Oberstes Parteigericht* (Enclosure 20). With regard to cases 3–16 the *Oberstes Parteigericht* asks the *Führer* to quash the proceedings in the State Criminal Courts. The *Reich* Minister of Justice has been informed of this petition and the decisions on which it was based handed down by the *Oberstes Parteigericht*. Cases 4–16 are killings committed by order, committed on the basis of a vague or presumed order, committed without orders but motivated by hatred against Jews or in the opinion that vengeance ought to be taken for the death of Party Comrade [Ernst] vom Rath upon the wish of the leaders, or killings motivated by a resolution suddenly formed in the excitement of the situation. The professed object of the entire action was the innermost reason, as well as the thought that reprisals had to be made in some form or other, on behalf of Party Comrade vom Rath. If a clearly defined order is at hand... the request to quash the proceedings against the immediate perpetrator needs no further argument. The order must shift the responsibility from the person who acted to the person who gave the order. Furthermore, the men often had to fight down strongest inner restraints in order to carry out the order. As was repeatedly expressed by the culprits, it is not our *SA* and *SS* men's affair to force their way into bedrooms by night, dressed in civilian clothes in order personally to do away with a despised political foe by his wife's side or together with his wife. Investigation into the circumstances under which the orders were given has shown that in all these cases a misunderstanding arose in some link or other of the chain of command, especially due to the fact that it was a

Walter Buch and family, ca. 1943. (Source: Walter Buch's
personal photo album, in collection of Gary Merlie)

matter of course to the National Socialist who was active in the days of the Party struggle that in drives in which the Party does not wish to appear as the organizer, orders are not given with final clarity and with full details. He is therefore used to deducing more from what he reads in such an order than is said literally, just as it had frequently become the practice on the part of the person issuing the order in the interest of the Party to refrain from saying anything to hint what he meant to achieve with the specific order especially when it concerned illegal political demonstrations. Therefore Party Comrade Dr. Goebbels' instruction that the Party was not to organize this demonstration was most likely interpreted by each Party leader present in the town-hall to mean that the Party should not appear as the organizer. Party Comrade Dr. Goebbels probably meant it in that way for politically interested and active circles, who might participate in such demonstrations are members of the Party and its branches. Naturally, they could be mobilised only through offices of the Party and its branches. Thus, a series of subordinate leaders understood some unfortunately phrased orders which reached them orally or by phone to mean that Jewish blood would now have to flow for the blood of Party Comrade vom Rath, that at any rate the leadership did not attach importance to the life of a

Rastenburg / Ostpreußen, 07.02.1943: Conference of *Reichs-* and *Gauleiter* at *Führer* HQ *"Wolfsschanze"*. Left to right: Heinrich Himmler, Friedrich Hildebrandt, Walter Buch, Emil Stürtz, Dr. Otto Dietrich, and Franx Xaver Schwarz. (NARA photo)

Jew, for example, not the Jew Gruenspan but all Jewry was guilty of the death of Party Comrade vom Rath. The German people were therefore taking revenge on all Jewry, the synagogues were burning in the entire *Reich*, Jewish residences and businesses were to be laid waste, life and property of Aryans had to be protected, foreign Jews were not to be harassed. The drive was being carried out by order of the *Führer*; the police were withdrawn; pistols were to be brought, and at the least resistance, the weapon was to be used without consideration, as each *SA* man would certainly know what he had to do etc ...

It is another question, whether an intentionally ambiguous order, given with the expectation that the order's recipient would recognise the intention and would act accordingly, is not an example of the discipline of the past. In times of struggle, such an order may, in individual cases, be necessary, in order to achieve political success without giving the government any possibility of discovering the origin of the Party. This viewpoint is now obsolete. The public, down to the last man, realizes that political drives like those of 9 November were organized and directed by the Party,

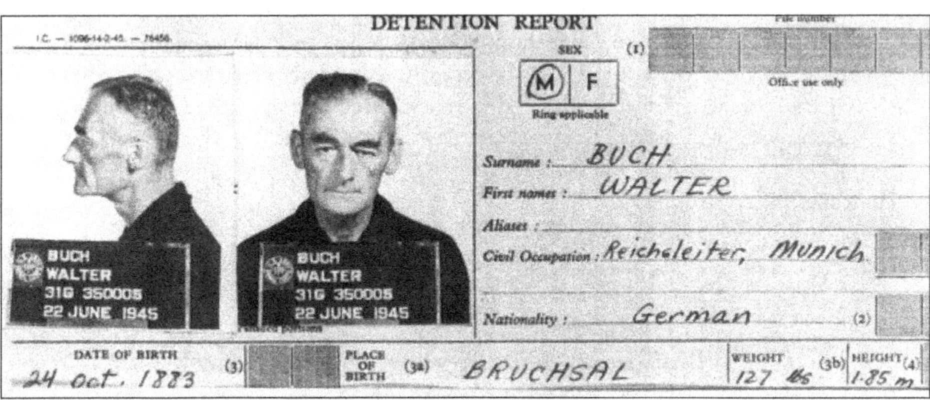

Detention report for Prisoner of War Walter Buch.

whether this is admitted or not ... (Translation of Document 3063-PS, in *Nazi Conspiracy and Aggression, Vol. V*)

00.07.1944   Together with several other Party judges, visited *Reichsgau Sudetenland*. During a banquet arranged in his honor by *Gauleiter* Henlein, Buch collapsed from a sudden attack of pleurisy, and was bedridden for some time thereafter.

Postwar Prosecution

Arrested by U.S. Army troops in München, 30.04.1945. From May through 12.08.1945, he was interned at POW Camp No. 32 (also known as "Ashcan", the Allied special interrogation and internment facility for senior Nazi leaders at the Palace Hotel in Mondorf-les-Bains, Luxembourg.), where he was interrogated numerous times concerning his activities during the Third *Reich* and on the subject of his missing (in later years, confirmed dead) son-in-law, Martin Bormann, who was being tried in absentia by the Internal Military Tribunal, Nürnberg. He was then transferred to Nürnberg, where he was held for a time before being transferred to an unknown facility. On 01.01.1947, he was returned to Nürnberg and underwent further interrogation. At his first de-Nazification trial, before the *Lagerspruchkammer* in Garmisch-Partenkirchen on 03.07.1948, he was classified as a "Hauptschuldiger" (Major Offender, i.e.- placed in de-Nazification Kategorie I) and sentenced to five years' hard labor, as well as confiscation of his assets up to 2000 DM. On 16.02.1949, he had an appeals hearing before the Weilheim Senate of the *Oberbayern Spruchkammer*, and was granted a second de-Nazification hearing. In this second hearing, held in München on 29.07.1949, he was again classified as a "Hauptschuldiger" but re-sentenced to 3½ years in a labor camp. Deemed to have served more than this amount of time in postwar internment, Buch was released from custody the same day.

Published Works

Nationalsozialismus: Volk und Familie (1932)
Niedergang und Aufstieg der Deutscher Familie (1932)
Fünfzig Jahre antisemitische Bewegung (1937)
Des nationalsozialistischen Menschen Ehre und Ehrenschutz (1939). The following description of this book is excerpted from the report of a 1945 interrogation of Buch:

[Buch wrote it] in order to define the position of the party members, and to establish a code of honor for the party... Hitler was very satisfied with the book, and ordered Buch to come to some agreement with the army to settle arguments between Army members and members of the party... (Institut für Zeitgeschichte, München, Akte 4637/71)

Decorations and Awards

00.00.191_	*1914 Eisernes Kreuz I. Klasse*
00.00.191_	*1914 Eisernes Kreuz II. Klasse*
00.00.194_	*Kriegsverdienstkreuz I. Klasse ohne Schwerter*
00.00.194_	*Kriegsverdienstkreuz II. Klasse ohne Schwerter*
ca. 1934	*Ehrenkreuz des Weltkrieges 1914–1918 mit Schwertern*
00.00.1934	*Goldenes Ehrenzeichen der NSDAP*
06.07.1936	*Ehrenzeichen des 9 November 1923 (Blutorden)* (Nr. 1 496)
00.00.1932	*Coburger-Abzeichen 1922*
00.00.19___	*Goldenes Hitler-Jugend Ehrenzeichen mit Eichenlaub*
00.00.194_	*Dienstauszeichnung der NSDAP in Gold*
00.00.194_	*Dienstauszeichnung der NSDAP in Silber*
00.00.194_	*Dienstauszeichnung der NSDAP in Bronze*
00.00.1929	*Nürnberger Parteitagsabzeichen 1929*
ca. 1931	*Abzeichen des SA-Treffens Braunschweig 1931*
00.00.1936	*Deutsche Olympia-Ehrenzeichen I. Klasse*
[01.12.1937]	*Ehrendegen des Reichsführers-SS*
25.12.1934	*Totenkopfring der SS*
00.00.193_	*SS-Zivilabzeichen*
00.02.1934	*Ehrenwinkel für alte Kämpfer*

Notes

- Son of Dr. jur. h. c. Hermann Buch, Senate President of the *Baden Oberlandesgericht* (born 28.07.1854, died 29.07.1921 in Karlsruhe, the son of Hermann Buch [00.00.1816–00.00.1882] and his wife Karoline Elisabeth, née Fischer [00.00.1823–00.00.1895]), and his wife, Hedwig Gertrud, née Heidlauff. Walter Buch had one sister, Hedwig Paula Buch (born 18.04.1886, died 00.00.1955 in Loerrach), who had three children by her husband Franz Josef Grimm (born 23.09.1881, also died in Loerrach in 1955).
- Religion: Protestant.
- Married on 23.09.1908 to Else Pleusser (born 11.05.1887 in Barmen, died 29.10.1944 of a heart attack; *NSDAP*-Nr. 7 732, recipient of the *Goldenes Ehrenzeichen der NSDAP*; *NS-Frauenschaft*-Nr. 5 721; *F.M.* [*Fördernder Mitglied*, sponsoring member of the *SS*] -Nr. 310 542). Two sons (____, born 11.01.1912 and Hermann, born 16.02.1920 in Gernsbach; *SS*-Nr. 357 263- Rose to *SS-Hauptsturmführer* and *IIa* [Divisional Adjutant] of *2.SS-Panzer-Division "Das Reich"*) and two daughters (Gerda, born 23.10.1909, died of cancer in the Merano civil hospital, 23.04.1946; and ____, born 10.03.1913). On 02.09.1929, Gerda Buch married the later *Reichsleiter* Martin Bormann. Walter Buch grew to detest his son-in-law and fellow *Reichsleiter*. Under interrogation at Nürnberg by Lt. Col. Thomas S. Hinkel, on 31.10.1945, Buch stated:

> ... I didn't get along with him, and when the Fuehrer put Bormann in charge of jurisprudence in the Party [Autumn 1943], my job was practically finished ... . The

Mug shot of a bearded Walter
Buch, taken at Nürnberg
in 1946. (Ian Sayer photo)

result of it was that the Party Chancellery actually acquired the power to decide on our judgments. In other words, the Party Chancellery would decide that a person should be expelled from the Party, for instance, and then it was up to the Party court to sanction this decision which had been taken beforehand. The procedure, in other words, had been reversed, and, of course, I opposed this change, but could not do anything about it, because the Fuehrer had stopped listening to me many years ago. ("Special Report on Walter Buch, *Reichsleiter* and *Oberster Parteirichter*", Special Detention Center "Ashcan"-Detailed Interrogation Report, in "Interrogation Records Prepared for War Crimes Proceedings at Nuernberg, 1945–1947/OCCPAC Interrogation Transcripts And Related Records: Buch, Walter Hans"; Publication Number M1270, Record Group RG238)

• After the war, Buch married a woman, the widow of a dentist, who had visited him during his postwar internment.

Sources:

Goebbels, Dr. phil. Paul Joseph: *The Goebbels Diaries 1942–1943* (edited by Louis P. Lochner). Doubleday, 1948.

Gordon, Harold J. Jr.: *Hitler and the Beer Hall Putsch*. Princeton University Press, 1972.

Hamilton, Charles: *Leaders and Personalities of the Third Reich, Volume I*. R. James Bender Publishing, 1984.

Höffkes, Karl: *Hitler's politische Generale: Die Gauleiter des Dritten Reiches*. Grabert-Verlag-Tübingen, 1986.

Krebs, Dr. phil. Albert: *The Infancy of Nazism: The Memoirs of Ex-Gauleiter Albert Krebs, 1923–1933*. New Viewpoints, 1976.

Lang, Jochen von: *The Secretary. Martin Bormann: The Man Who Manipulated Hitler*. Random House, 1979.

Lilla, Joachim; Döring, Martin; & Schulz Andreas: *Statisten in Uniform. Die Mitglieder des Reichstags 1933–1945*. Droste Verlag, 2004.

McKale, Donald M.: *The Nazi Party Courts: Hitler's Management of Conflict in His Movement, 1921–1945*. The University Press of Kansas, 1974.

Mendelsohn, S.: *The Polish Jews behind the Nazi Ghetto Walls*. Yiddish Scientiic Institute, 1942.

National Archives and Records Administration, College Park, Maryland: *SS-Personalakte* of Walter Buch. Microfilm document collection A3343SS

Office of United States Chief Counsel for Prosecution of Axis Criminality: *Nazi Conspiracy and Aggression (11 volumes). U.S. Government Printing Office, District of Columbia, 1946.*

Orlow, Dietrich: *History of the Nazi Party: 1933–1945*. University of Pittsburgh Press, 1973.

Reiche, Erich G.: *The Development of the SA in Nuremberg, 1922–1934*. Cambridge University Press, 2002.

Schirach, Baldur von: *Die Pioniere des Dritten Reiches*. Zentralstelle fur der deutschen Freiheitskampf, 1933.

Tell, Rolf: *Sound and Fuehrer*. Hurst & Blackett Ltd., 1939.

Weinreich, Max: *Hitler's Professors: The Part of Scholarship in Germany's Crimes against the Jewish People*. Yale University Press, 1999.

Franz Felix Freiherr von Pfeffer

[Known as Pfeffer von Salomon until 03.07.1941]

01.11.1926–29.08.1930	*Oberster SA-Führer (Reichs-SA-Führer)*

Born:	19.02.1888 in Düsseldorf.
Died:	12.04.1968 in München / Bayern. Gravesite: München, Waldfriedhof Solln (Plot 15 / Row 3 / Grave 71/72).
NSDAP-Nr.:	16 101 (Joined, 00.03.1925; Expelled from the Party, 14.11.1941)

Promotions

01.10.1910	*Einjährig-Freiwilliger*
16.02.1911	*Fahnenjunker*
00.00.1911	*Unteroffizier*
23.05.1911	*Fähnrich*
18.08.1911	*Leutnant* (mit Patent vom 20.08.1909)
18.08.1915	*Oberleutnant*
20.09.1918	*Hauptmann i. G.*
00.08.1926–01.11.1926	"mit der Führung der *SA* beauftragt"
01.11.1926–29.08.1930	*Oberster SA-Führer (Reichs-SA-Führer)*
01.04.1930–12.08.1930	*NSAK-Korpsführer*

Career

ca. 1894-ca. 1899	Attended *Volksschule*.
ca. 1899-ca. 1906	Attended *Gymnasium* (graduated and passed his *Abitur*).
ca. 1906-ca. 1907	Studied law at the University of Heidelberg.
ca. 1906–00.00.1907 (?)	Employed as a *Referendar* (junior barrister or law clerk) in Dühmen.
00.00.1907–00.00.1909	Attended a *Militärschule*.
01.10.1910–00.11.1918	Entered service as an *Einjährig-Freiwilliger*, assigned to *Infanterie-Regiment "Herwarth von Bittenfeld" (1.Westfälische) Nr. 13* (Münster). As of 1912, he was assigned to the Regiment's *6. Kompanie*.
00.11.1918–00.03.1920	*Führer* of the *Westfälische Freikorps von Pfeffer* (on the Baltic and in Lithuania, Poland, and Oberschlesien). He also participated in the *Kapp-Putsch* of March 1920.
00.03.1920–00.00.192_	Worked as a *Zivilangestellter* (civilian employee) with the *Wehrkreiskommando* in Münster.
00.00.1920–00.00.1921	Arrested and detained for his involvement in the *Kapp-Putsch*. Released on the grounds of an amnesty and the decision of a *Sondergericht* (special court) that he be acquitted.
00.00.1922–00.00.1924	Member of the *Rheinisch-Westfälische Treuebund*.
00.00.1923–00.00.1923	Member of an anti-French sabotage troop in the Ruhrgebiet.
00.00.1924–00.00.1925	Member of the *Völkisch-Sozialer Block* (VSB).
00.05.1924–27.03.1925	*Vorsitzender* of the *VSB Landesverband* in Westfalen.
00.03.1925	Joined the *NSDAP*.
27.03.1925–07.03.1926	*Gauleiter* of *Gau Westfalen der NSDAP*.

Franz Pfeffer von Salomon, ca. 1926.

Franz Pfeffer von Salomon
in 1924. (*Bundesarchiv Bild*)

10.09.1925–00.00.1926	Member of the *Arbeitsgemeinschaft der Nord- und Nordwestdeutschen Gaue der NSDAP* (Working Organization of the North and West German Districts of the Nazi Party).
Christmas 1925	Submitted an internal Party memorandum entitled "Zucht. Eine Forderung zum Programm" (Breeding: A demand in relation to the Party Program):

> First of all I share with Strasser and all revolutionaries the view that property in Germany is wrongly distributed: property, power, culture are in the wrong hands, misery and ruin are suffered by the wrong people, and this situation is growing and consolidating itself to such an extent that only the most drastic, ruthless intervention can force German life and Volkstum back into the right paths. Immediately after this common assumption, however, the thinking diverges.
>
> There are those who take as their starting-point the equality of human beings, or rather the equality of Germans. This premise leads naturally to the logical conclusion that there can be no reason why among equals a different distribution of property, state power, culture should prevail. If someone who is equal is forced to live in circumstances significantly lower than the mean, then this is naturally "unjust", or a "scandal" with respect to other

equals. If someone who is equal is treated markedly better than the average then this is likewise "unjust", for it can only happen "at the expense" of the share the other equals enjoy of property, power, and culture ...

In the last analysis I accuse Strasser's program of being rooted in this basic mentality (and I fear I must accuse him of coming out with far too many arguments which in our camp are called "socialist"). It is the Jewish-liberal-democratic-Marxist-humanitarian mentality. As long as there is even a single minute tendril which connects our program with this root then it is doomed to be poisoned and hence "to wither away to a miserable death ... All Germans are unequal. That is the starting-point ... The first logical conclusion to be drawn from inequality is the inequality of value. Some Germans are more valuable than others ("value" is a relative concept, which I use here naturally to refer to value measured in terms of the German, Nordic world-view, and within this measured in terms of what serves the interests of collective German well-being).

A logical consequence of this inequality must be the principle of unequal treatment, that is, unequal share of state power, property, culture. All these must be distributed to people on the basis of how valuable they are ...

A further logical consequence of inequality in what people are worth, and the continuous changes brought about by the development of the Volk, is the duty of the state to take charge of this development, which means influencing it in every way possible. Excellence must be increased and enhanced further. Inferiority must be reduced. In plain terms this is a question of breeding. Improving the stock of the race. Breeding human beings ... From the above considerations I can draw up the following programmatic demands of our future state, the "Third *Reich*":

To determine the degree of higher or lower value of all inhabitants of Germany. The value will be assessed as the function of four criteria:

(1)   actual performance in their professions
(2)   physical attributes, according to health and racial characteristics
(3)   spiritual, moral and cultural traits
(4)   hereditary traits evaluated by considering parents and grandparents ...

No pity is to be shown to those who occupy the lower categories of the inferior groups: cripples, epileptics, the blind, the insane, deaf and dumb, children born in sanatoria for alcoholics or in care, orphans (= children born out of wedlock), criminals, whores, the sexually disturbed, etc. [Krüppel, Epileptikern, Blinden, Irren, Taubstummen, Trinkerheilanstalten- (sic), Fürsorgezöglingen,

3.Reichsparteitag der NSDAP in Nürnberg, 1927. Pfeffer von Salomon
and Hitler salute the crowd, with Hess behind them.

Waisen, Verbrechern, Dirnen, Sexualgestörten u.s.w.] Everything
done for them not only means taking resources away from more
deserving causes, but counteracts the breeding selection process.
Nor should we mourn the dumb, the weak, the spineless, the
apathetic, those with hereditary diseases, the pathological, because
they go under "innocently" ...

This bottom category means destruction and death. Weighed
and found wanting. Trees which do not bear fruit should be cut
down and thrown into the fire. ("*NSDAP* Hauptarchiv", Hoover
Institution Microfilm Collection), reel 44, folder 896, i-ii.)

07.03.1926–20.06.1926	*Gauleiter* of *Großgau Ruhr der NSDAP*, together with Dr. Joseph Goebbels and Karl Kaufmann.
07.03.1926–20.06.1926	*Gau-SA-Führer Ruhr.*
00.08.1926–01.11.1926	*Mit der Führung der SA beauftragt* (Charged with leadership of the SA).
01.11.1926–29.08.1930	*Oberster SA-Führer (Reichs-SA-Führer).* Submitted his resignation, 12.08.1930 [see entry for that date below], and succeeded at the end

Pfeffer von Salomon and Walther Stennes overseeing an *SA* training exercise on 30.09.1928. (Roger Bender photo)

of the month by Hitler. Prof. Bruce Campbell describes Pfeffer von Salomon's leadership of the *SA* as follows:

> By all accounts he was a genuine leader, a strong-willed personality of great energy and ability, but of equally great arrogance and independence, who did not lack enemies within the *NSDAP* because of his ambition and lack of tact. Clearly he was a man who believed he was right and therefore did not like to compromise. Yet it was only through just such a combination of will, ability, and leadership that he was able to transform the *SA* into a national organization. And despite a certain inflexibility and brutal directness of character, von Pfeffer had learned to adapt to changing conditions and political contexts.... he could see beyond the world of guns and uniforms. It is this political side of von Pfeffer which enabled him to create a new form for the *SA,* different from that of the *Wehrverbände* and better adapted to current conditions. Critical of the "dilettantism" of the Beer Hall Putsch, he was able to draw the lesson from it that if the 'national revolution' were to succeed, what was needed was not a better organized putsch, which was the solution proposed by many who continued to think in exclusively military terms, but rather entirely new tactics. As *Gauleiter* of the Ruhr he gained the experience necessary to adapt the paramilitary form of the *Wehrverband* to the substance of a political movement. In this he

Pfeffer von Salomon with Hitler and Franz Ritter von Epp during
the *4.Reichsparteitag der NSDAP* in Nürnberg, August 1929.

was aided by Viktor Lutze. (Campbell, *The SA Generals and the Rise of Nazism*, p. 50)

03.11.1926   Issued an order on the "*SA* and the Public (Propaganda)", which included the following:

The only form in which the *SA* appears to the public is that of the closed formation. This is at the same time one of the most powerful forms of propaganda. The sight of a large number of inwardly and outwardly calm, disciplined men, whose total will to fight may be unequivocally seen or sensed, makes the most profound impression on every German and speaks to his heart a more convincing and inspiring language than writing and speech and logic can ever do. Calm composure and matter-of-factness underline the impression of strength- the strength of the marching columns and the strength of the cause for which they are marching. The inner strength of the cause leads the German emotionally to deduce its rightness: "For only the right, the honest, the good can release true strength." Where whole hosts purposefully (not in the welling up of sudden mass suggestion) stake life and limb and existence for a cause, the cause must be great and true!

... The *SA* man is the sacred freedom fighter. The Pg [Parteigenosse-Party Member] is the instructor and skilled agitator. Political propaganda seeks to enlighten the adversary, to dispute with him, to understand his viewpoint, to go into his ideas, up to a certain point to agree with him- but when the *SA* appear on the scene, this stops. They are out for all or nothing. They know only the motto (metaphorically): Strike dead! You or me. (Joachim C. Fest, "Ernst Röhm and the Lost Generation" in *The Face of the Third Reich*, pp. 142–143)

05.12.1926–29.08.1930   *Direkter Vorgesetzter der HJ.* His appointment to this post resulted from his successful demand that the *Hitler-Jugend* be directly subordinated to the *SA.*

27.10.1927–11.06.1930   *Vorsitzender des Jugendausschuß* (Chairman of the Youth Committee; redesignated *Jugendamt,* 00.00.1928) in the *Reichsleitung der NSDAP.*

19.08.1927–21.08.1927   *Organisationsleiter* (Director of Organization) for the *3. Reichsparteitag der NSDAP* in Nürnberg.

01.03.1928   Created seven new *SA-Oberführer* commands, placing them under the leadership of retired military officers, as follows:

SA-Oberführer Ost (Berlin)	*Hauptmann a. D.* Walter Stennes
SA-Oberführer Nord (Hannover)	*Major a. D.* Karl Dincklage
SA-Oberführer West (Kassel)	*Oberstleutnant a. D.* Curt von Ulrich

Oberster SA-Führer Pfeffer von Salomon with Hitler during a Party function, ca. 1926.

SA-Oberführer Mitte (Dresden)	*Kapitänleutnant a. D.* Manfred Freiherr von Killinger
SA-Oberführer Süd (München)	*Major a. D.* August Schneidhuber
SA-Oberführer Ruhr (Elberfeld)	*Oberleutnant a. D.* Viktor Lutze
SA-Oberführer Ostmark (Wien)	*Hauptmann a. D.* Hermann Reschny

01.08.1929–04.08.1929	Participated in the *4. Reichsparteitag der NSDAP* in Nürnberg.
02.09.1929	Attended the wedding of Martin Bormann and Gerda Buch.
Autumn 1928	In letters to an *SA* subordinate in Köln, wrote:

13.10.1928: We are of the opinion that the *SA,* as the nucleus of the future German army, must be so trained and organized that even today, slowly but surely, a State is formed within the non-State, so that when the National Socialist Greater Germany is created the internal structure will already exist, determined and strong, invincible in every respect.

28.11.1928: The *SA* is the militant force of the movement-it is the personification of the will to power of a political organization.... As our *SA* is the expression of a party machine's politically organized will to power which wishes-not with words, but with actions- to accomplish the national and social liberation of the German

Memorial service for Germany's 1914-18 war dead during the *4.Reichsparteitag der NSDAP*, Nürnberg, August 1929. From right to left: Rudolf Hess, Pfeffer von Salomon, Hitler, and Franz Ritter von Epp.

people, the concept of comradeship in its ranks takes on an entirely different meaning from that in the so called military and veteran organizations.... The comradeship in the *SA* formations must, therefore, become such a solid structure that all police bans or other underhand methods will recoil from its granite wall. (Robert M. W. Kempner, "Blueprint of the Nazi Underground-Past and Future Subversive Activities", in *Research Studies of the State College of Washington, Volume XIII, Number 2*, June 1945)

28.12.1929   In a letter to an *SA* leader in Köln, wrote:

Before you stands but one individual: der *Führer*. And you obey him. Faith in him is faith in the German people. He commands and you obey. You ask not why nor wherefore. You know the goal. (ibid)

01.04.1930–12.08.1930   *Korpsführer* of the *Nationalsozialistische Automobilkorps* (*NSAK*, the forerunner of the later *NSKK*). First holder of this post.

00.00.1930–00.10.1934   *Leiter* of the *Abteilung für den kulturellen Frieden* of the *NSDAP*.

12.08.1930   Submitted his resignation from the post of *Oberster SA-Führer*, and succeeded at the end of the month by Hitler himself (who retained Dr. Otto Wagener as Chief of Staff). According to Dietrich Orlow, his decision to step down was the culmination of a long history of

Franz Pfeffer von Salomon, ca. 1929.

dissatisfaction with the *Führer*'s leadership. It had become clear to him that Hitler's desire was to place further limitations on the autonomy of the *SA*. Whereas Pfeffer hoped to gain more *Reichstag* seats for his own subordinates in the *SA*, "Hitler", writes Dietrich Orlow, "either refused to allow *SA* leaders to become members of the *Reichstag*, or else turned down Pfeffer's demand that he, Pfeffer, should control the votes of the *SA* deputies [In a letter- marked in handwriting "not sent"- to Pfeffer dated 13.09.1930, Walter Buch stated the latter was true]. In either case, Pfeffer finally withdrew his request and even lamely attempted to explain why the failure to have *SA* members in the *Reichstag* was best for the organization." (Orlow, *History of the Nazi Party: 1919–1933*, p. 211) In a letter to *SA* leaders dated 29.08.1930, von Pfeffer wrote that the *SA* was justified in demanding "other visible and materially noticeable" confirmation that the *SA* was to benefit from the electoral successes of the *NSDAP*. Orlow continues:

> [Pfeffer] did not specify what proof he had in mind, but it appears that in addition to wanting more money, he demanded more freedom to disintegrate the *Reichswehr* and to train the *SA* as the core of the new German mass army. Since Hitler was particularly unwilling to embark on a campaign of subversion against the *Reichswehr*, Pfeffer had no choice but to resign his position. (ibid, pp. 211–212)

Franz Pfeffer von Salomon, Rudolf Hess, and Hermann Göring
during the *6.Reichsparteitag der NSDAP*, September 1934.

06.11.1932–01.02.1933	Member of the *Reichstag*.
06.12.1932–01.02.1933	*Stellvertretendes Mitglied* (alternate member) of the *2. Ausschuß (Auswärtige Angelegenheiten) der Reichstag* (2nd Committee [Foreign Affairs] of the *Reichstag*).
05.03.1933–14.10.1933	Member of the *Reichstag*.
12.11.1933–27.11.1941	Member of the *Reichstag (Wahlkreis 16, Südhannover-Braunschweig)*.
13.03.1934–00.00.193_	*Beauftragter und Verhandlungsführer in der Kirchenfrage der NSDAP* (Representative and Negotiation Leader for Church Questions of the *NSDAP*).
[00.00.1936]	Honorary member of the *Volksgerichtshof*.
00.09.1939–00.03.1940	Interned in a concentration camp for seven months (according to his own account given on 20.02.1953).
00.00.1941	Briefly arrested in connection with the unauthorized flight of Rudolf Hess to Scotland.
03.07.1941	In accordance with an order by the *Reichsminister* des Innern, Dr. Wilhelm Frick, removed "von Salomon" from his name and subsequently known only as "von Pfeffer". For many years previous, he had already gone by "von Pfeffer" due to the "Jewish" sound of "Salomon".
14.11.1941	Expelled from the *NSDAP*.
27.11.1941	Expelled from his seat in the *Reichstag*.
00.00.1942–21.07.1944	Lived on his estate in Pommern.

Pfeffer von Salomon with Rudolf Hess in Nürnberg's Hauptmarkt during
the *9.Reichsparteitag der NSDAP*, September 1937. (NARA photo)

21.07.1944	Arrested in connection with the 20. July 1944 assassination attempt on Hitler and, according to his own account, imprisoned for four months.
00.01.1945	Fled from Pommern to Bayern.
00.00.1945–00.00.1945	*Kommandeur* of a *Volkssturm* Division.

Postwar Confinement

Interned in Heilbronn, 00.00.1945–00.00.1946, then lived in Wiesbaden until 00.00.1960
when he moved to München.

Decorations and Awards

09.11.1918	*Ritterkreuz des Kgl. Hausordens von Hohenzollern mit Schwertern*
00.00.191_	*1914 Eisernes Kreuz I. Klasse*
00.00.191_	*1914 Eisernes Kreuz II. Klasse*
21.02.1916	*Ritterkreuz II. Klasse mit Schwertern des Kgl. Sächsischen Albrechts-Ordens*
ca. 1919	*Balten Kreuz I. Klasse*
ca. 1919	*Balten Kreuz II. Klasse*
ca. 1921	*Schlesischer Adler 1. Stufe*
ca. 1921	*Schlesischer Adler 2. Stufe*

ca. 1918	*Verwundetenabzeichen, 1918 in Schwarz*
ca. 1934	*Ehrenkreuz des Weltkrieges 1914–1918 mit Schwertern*
ca. 1933	*Goldenes Ehrenzeichen der NSDAP*
ca. 1929	*Nürnberger Parteitagsabzeichen 1929*
00.00.194_	*Dienstauszeichnung der NSDAP in Silber*
00.00.194_	*Dienstauszeichnung der NSDAP in Bronze*
00.00.1934	*Ehrendolch der SA* (mit Wirkung vom 03.02.1934)
00.02.1934	*Ehrenwinkel für alte Kämpfer*

Notes

- Parents:

   - Father: *Kgl. Preußischen Geheimen Regierungsrat Max* Karl Friedrich Ferdinand Pfeffer von Salomon (born 25.04.1854 in Friedeberg / Neumarkt, died 11.08.1918 in Münster as a *Hauptmann* der Landwehr. He was the son of the Prussian *Oberst* of cavalry Ferdinand Pfeffer von Salomon [1822–1901; ennobled in 1862] and his wife *Ida* Henriette, née Hoffmann-Scholtz [1834–1899]).

   - Mother: Anna, née von Clavé-Bouhaben (born 26.07.1862 in Simmern, died 28.01.1919 in Münster), daughter of the Prussian Appelationsgerichtsrat [appeals court counselor] Franz von Clavé-Bouhaben, an estate owner in Königswinter, and his wife Marie, née Coninx. She married Franz's father on 23.04.1887.

   - Younger brother: *SA-Obergruppenführer* and *Regierungspräsident* of Wiesbaden, Friedrich ("Fritz") Ludwig Ferdinand Felix Pfeffer von Salomon (born 19.05.1892 in Berlin-Charlottenburg, died 29.10.1961 in Straelen / Niederrhein), *NSDAP*-Nr. 77 889 (Joined 01.03.1929). Among his cousins were the famous *Freikorps* leader and writer (and co-assassin, on 24.06.1922, of German foreign minister Walther Rathenau) Ernst von Salomon (born 25.09.1902 in Kiel, died 09.08.1972 in Stoekte bei Winsen) and his brother Bruno von Salomon, a leading figure in the *KPD* (*Kommunistische Partei Deutschlands*).

- Married in Bakenhof bei Krefeld on 14.01.1922 to *Marie* Sophie Clara Freiin Raitz von Frentz (born 01.04.1895 in Altena / Westfalen, died 18.05.1984 in Altena / Westfalen). She was the daughter of Adolf Freiherr Raitz von Frentz (1843–1907], a Prussian *Premier-Lieutenant*, and his wife Sophie Freiin von Lüninck (1850–1934]. Former Hamburg *Gauleiter* Dr. Albert Krebs describes Frau Pfeffer as follows:

   Insofar as Pfeffer criticized the party, he was undoubtedly supported or even encouraged to do so by his wife. Her repugnance for many leading Nazis, especially Hitler himself, derived from her unhindered instincts and her still solid appreciation of real values and rules … (Dr. Krebs, *The Infancy of Nazism*, p. 282)

- The following children were born to Franz Pfeffer von Salomon and his wife:

   Ferdinand (born 1929), Dr. jur. and attorney in Lünen.

   Max (born 1932), state attorney in Cape Town, South Africa.

   Irmgard (born 1923), Dr. rer. nat., a biologist, married to Friedemann Freiherr von Wintzingerode [1913–1964], a farmer in South Africa.

   Kunigunde (born 1927), Dr. rer.nat., also a biologist, married to the geologist Dr. phil. Jobst Hülsemann.

Sources

Campbell, Bruce: *The SA Generals and the Rise of Nazism*. The University Press of Kentucky, 1998.

Fest, Joachim C.: *The Face of the Third Reich: Portraits of the Nazi Leadership*. Weidenfeld and Nicolson, Ltd., 1970.

Höffkes, Karl: *Hitlers politische Generale: Die Gauleiter des Dritten Reiches*. Grabert-Verlag-Tübingen, 1986.

Höhne, Heinz: *The Order of the Death's Head*. Martin Secker and Warburg, 1969.

Krebs, Dr. phil. Albert: *The Infancy of Nazism: The Memoirs of Ex-Gauleiter Albert Krebs, 1923–1933*. New Viewpoints, 1976.

Lilla, Joachim; Döring, Martin; & Schulz, Andreas: *Statisten in Uniform. Die Mitglieder des Reichstags 1933–1945*. Droste Verlag, 2004.

Stockhorst, Erich: *5000 Köpfe: Wer war was im 3. Reich* (3rd Edition). Arndt-Verlag, 1998.

Wagener, Dr. Otto: *Hitler: Memoirs of a Confidant* (edited by Henry Ashby Turner, Jr.; translated by Ruth Hein). Yale University Press, 1985.

Weiss, Hermann: *Neue Deutsche Biographie*. Duncker & Humblot, 2001.

Otto William (Wilhelm) Heinrich Wagener

Dr. phil. h.c.

01.10.1929–31.12.1930	*Stabschef der SA*
29.08.1930–31.12.1930	*Obersten SA-Führer*

Born:	29.04.1888 in Durlach / Baden.
Died:	08.08.1971 in Chieming-Stöttham.

NSDAP-Nr.:	159 203 (Joined 01.10.1929)

Promotions

09.07.1906	*Fahnenjunker*
22.03.1907	*Fähnrich*
18.11.1907	*Leutnant* (mit Patent vom 22.05.1906)
08.11.1914	*Oberleutnant*
18.12.1915	*Hauptmann*
01.10.1929–31.12.1930	*Stabschef der SA*
29.08.1930–31.12.1930	*Obersten SA-Führer* (m.d.W.d.G.b.)
18.12.1931	*SA-Gruppenführer*
01.04.1940	*Hauptmann z. V.*
15.03.1941	*Major z. V.*
01.06.1942	*Oberstleutnant z. V.*
01.08.1943	*Oberst z. V.*
01.12.1944	*Generalmajor z. V.*

Career

00.00.189_-00.00.1906	Attended *Gymnasium* in Karlsruhe (passed his *Abitur* at the *Kadettenanstalt Karlsruhe*, 09.07.1906).
09.07.1906	Entered service as a *Fahnenjunker*, assigned to *Infanterie-Regiment "Markgraf Ludwig Wilhelm" (3. Badische) Nr. 111* (Rastatt).
00.00.19_-30.09.1913	Adjutant of *I. Bataillon / Infanterie-Regiment "Markgraf Ludwig Wilhelm" (3. Badische) Nr. 111.*
01.10.1913–00.00.1914	Attended the *Kriegsakademie* in Berlin.
01.07.1914–31.07.1914	Attended an *Ausbildungskurs* (training course) as a *Flugzeugbeobachter* (aerial observer) with *Fliegerkommando Döberitz.*
01.08.1914–00.11.1914(?)	Deployed to the front with *Reserve-Infanterie-Regiment 55.*
00.11.1914–00.12.1915	Assigned as a *Brigade-Adjutant*, then as a *Kompanieführer.*
01.02.1916–11.07.1916(?)	Bataillonsführer in *3. Reserve-Infanterie-Regiment Nr. 110.*
11.07.1916–07.12.1916(?)	Attached to the *Generalstab* of *Armeegruppe von Stein.*
07.12.1916–00.00.1918	Assigned to the *Generalstab* of *5. Armee.*
06.05.1918	Following a hearing before a court of honor, dismissed from service.
00.00.1919–00.00.1919	Member of a *Freikorps* on the German-Polish border.
Summer 1919–00.00.1920	*Chef des Generalstabes* of the *Deutsche Legion* (*Freikorps*), also known as the *Baltische Legion*, in the Kurland region of Latvia and in Lithuania,

Hauptmann Wagener during World War I.

where it served as part of Colonel (later *Major* General) Prince Pavel Mikhailovich Avalov-Bermondt's *Freiwillige Russische Westarmee* (also known as the *Freiwillige Deutsch-Russische Westarmee*). The Legion, which had been formed in Mitau, Latvia, on 25.08.1919 (from *2.Infanterie-Brigade*, elements of *1.Garde-Reserve-Division*, and the *Baltische Landwehr*) was disbanded on 18.12.1919, shortly after its return to Germany.

16.11.1919–18.12.1919	*Führer* of the *Deutsche Legion*. Succeeded *Kapitän zur See* Paul Siewert, who had been killed in action on 16.11.1919.
00.00.1919–00.00.1919	Studied economics.
00.03.1920–00.03.1920	Arrested and briefly detained in connection with the *Kapp-Putsch*.
31.03.1920	Released from *Freikorps* service.
00.00.1920–00.00.1920	Assistant director of a sewing machine factory.
00.00.1920–00.00.1921	Landesleiter of the *Organisation Escherich* in Baden.
End of 1920–00.00.1924	Direktor of his father's and Vorstand of his father's sewing machine factory *Haid und Neu* in Karlsruhe (named for Georg Haid and Carl Wilhelm Neu, who had founded it on 14.04.1860).
00.00.1922–00.00.1924	Lecturer on ecomic and social policy matters in business college courses at the *Technischen Hochschule* in Karlsruhe and the University of Würzburg.
00.00.1923	Joined the *SA*.
00.00.1924	Received his doctorate (Dr. phil. h. c.) from the *Philosophischen Fakultät* (philosophical faculty) of the University of Würzburg.
00.00.1924–00.00.1925	Traveled abroad and lectured on matters of economic policy.
00.00.1925–00.00.1929	Part owner of a plywood and veneer business in Villingen.
00.00.1925–00.08.1925	*Vorsitzender* (chairman) of the *Badische Wahlausschuß* (Baden electoral committee) for the *Rechtsblock* in the *Reich* presidential elections.
00.07.1929	Established contacts with the *NSDAP*.

Dr. Otto Wagener, ca. 1933.

01.08.1929–04.08.1929	Participated in the *4. Reichsparteitag der NSDAP* in Nürnberg.
01.10.1929	Joined the *NSDAP* and the *SA*.
01.10.1929–13.07.1933	Member of the *Reichsleitung der NSDAP*.
01.10.1929–31.12.1930	*Stabschef der SA*. Succeeded by Ernst Röhm.
29.08.1930–31.12.1930	*Obersten SA-Führer* (m.d.W.d.G.b.). Succeeded by Adolf Hitler.
01.01.1931–00.06.1932	*Leiter* of the *Wirtschaftspolitische Abteilung* (Economic Policy Department) in the *Reichsorganisationsabteilung II der Reichsleitung der NSDAP*. Manfred Berg and Martin H. Geyer write:

> Wagener was a zealous proponent of economic corporativism, a position that identified him with the self-consciously fascist wing among National Socialist intellectuals ... Probably in late 1931 or early 1932, at a breakfast at the party headquarters in Munich, Wagener advised the *Führer* that the Jewish problem could be solved quite simply by forbidding any further Jewish immigration into Germany. Given the low birth rate among Jews, Wagener calculated, and without replenishment from abroad, they would be condemned to gradual extinction. Until that happened, he advised Hitler, a Nazi regime would need to do little more than bar Jews from positions in government and watch while the Jewish problem solved itself. No Nazi ever proposed a milder solution to the Jewish problem. Hitler obviously did not consider it adequate. (Berg and Geyer, *Two Cultures of Rights: the Quest for Inclusion*

Dr. Otto Wagener, ca. 1933.

and Participation in Modern America and Germany)

00.00.1931	Founder of the *Wirtschaftspolitische Pressedienst* (WPD, Economic Policy Press Service).
00.00.1931–00.00.193_	Editor of the National-Socialist periodical *Wirtschaftspolitischen Briefe.*
18.12.1931	Granted the rank of *SA-Gruppenführer* with assignment as an *SA-Führer z. V.*
00.06.1932–04.09.1932	*Leiter* of *Hauptabteilung IV (Wirtschaftspolitik,* Economic Policy) in the *Reichsorganisationleitung der NSDAP.*
04.09.1932–01.04.1933	Designated as *"z.b.V. im Stabe des Führers"* (at disposal of the Staff of the *Führer*).
01.04.1933–13.07.1933	*Leiter* of the *Wirtschaftspolitisch Hauptamt der NSDAP* (Main Office for Economic Policy of the *NSDAP*) and *Sachbearbeiter für Wirtschaftspolitik im Verbindungsstab der NSDAP* (Specialist for Economic Policy in the Liaison Staff of the *NSDAP*).
15.04.1933–13.07.1933	*Regierungskommissar* in the *Geschäftsführung* (business management) of the *Reichsverband der Deutschen Industrie.*
24.04.1933–30.06.1933	*Reichskommissar für die Wirtschaft* (*Reich* Commissioner for the Economy).
03.05.1933–00.00.1933	*Kommissar des Reiches für den Reichsverband der deutschen Industrie und für die übrige Wirtschaft mit Ausnahme der Landwirtschaft* (*Reich* Commissioner for the *Reich* Association of German Industry and for the remaining Economy with the exception of Agriculture).

Dr. Otto Wagener (far left) posing with Hitler and members of his cabinet and others early in 1933. Also present, from right to left: Rudolf Hess (mostly obscured), Heinrich Himmler, R. Walther Darré, Hermann Göring, Ernst Röhm, Hitler, Dr. Joseph Goebbels, Dr. Wilhelm Frick, Hanns Kerrl, and Wilhelm Kube.

00.06.1933–13.07.1933	*Ehrenvorsitzender* (honorary chairman) of the *Reichsarbeitsgemeinschaft der technisch-wissenschaftlichen Arbeit* (RTA, *Reich* Working Group for Technical-Industrial Work)".
13.07.1933	Dismissed from all offices by Hitler, having fallen into disgrace with the *Führer* the previous month.
12.11.1933–28.03.1938	Member of the *Reichstag (Wahlkreis 21, Koblenz-Trier)*.
30.06.1934	Arrested and briefly detained during the so-called "Röhm-Putsch", narrowly avoiding execution. He fled to the Erzgebirge region, working as a farmer in Hohenwedel über Wiesenbad), thereafter avoiding any contact with the *NSDAP* leadership.
01.04.1937–00.05.1945	Reentered the *SA* with the rank of *SA-Gruppenführer* and assigned as an *SA-Führer z. V.* to the *Stabes der Obersten SA-Führung*.
01.04.1940–12.05.1940	Assigned as *an Ordonnanzoffizier* to the staff of *Armeeoberkommando 6* (*6.Armee* HQ).
12.05.1940–31.08.1940	Attached as *Gehilfe* (assistant) to the Ia in the *Generalstab des Heeres* (Army General Staff).
01.09.1940–00.01.1943	Assigned as *1.Generalstabsoffizier* (Ia) to the staff of *332.Infanterie-Division*.

Generalmajor Wagener inspecting his troops on Crete, 1944/45.

00.01.1943–11.02.1944	*Kommandeur* of *Sicherungs-Regiment 177.*
12.02.1944–06.05.1944	*Führer* (m.d.F.b.) of a *Sicherungs-Division.*
06.05.1944–20.07.1944(?)	In *Führerreserve OKH.*
20.07.1944–08.05.1945	*Kommandant Ost-Ägäis* and *Militärgouverneur der italienischen Inseln der Dodekanes* (Military Governor of the Italian Islands of the Dodecanese [Rhodes, Kos, and Leros]). Under his command were 6,000 troops, half of them from the punitive *Bewährungseinheiten 999.* On Rhodes, he had established an "Internierungslager" (internment camp) and the concentration camp "Calitea", where Italian POW's were held. According to the 14.02.1951 issue of the West German magazine *Der Spiegel*, 1,300 death warrants were executed under Wagener's authority in April/May 1945 alone.

Postwar Prosecution

Held as a prisoner of war at a British camp in Wales, 00.00.1945–00.01.1947, he was turned over to Italian authorities on 07.01.1947. Tried by an Italian war tribunal and sentenced on 16.10.1948 to 15 years' imprisonment for the shooting of Italian POW's on Rhodes. On 04.06.1951 he was released from prison thanks to the intercession of Bishop Alois Hudal and West German Chancellor Konrad Adenauer. During the 1950's, he was *Vorsitzender* (chairman) of the *Seeckt-Gesellschaft* and affiliated with nationalist political circles in Bayern.

Generalmajor Wagener in Allied captivity, 1945.

Published Works

Das Wirtschafts-Programm der NSDAP (1932)

Hitler aus nächster Nähe. Aufzeichnungen eines Vertrauten 1929–1932 (Written in 1946, but
   published posthumously; Edited by Henry Ashby Turner, 1978)

Decorations and Awards

* 05.05.1945	*Ritterkreuz des Eisernen Kreuzes* as *Generalmajor z. V.*, *Kommandant Ost-Ägäis*, and *Militärgouverneur der italienischen Inseln der Dodekanes* *Award technically invalid as it was made by the acting *Reich* Government of *Großadmiral* Karl Dönitz, who did not have authority to make such awards. On 20.05.1945 at 2035 hours, *Generalmajor* Benthack (Kdr. of *Festungs-Division Kreta* and Kdr. of *Festung Kreta* transmitted a proposal for award of the *Ritterkreuz* to five men, including Wagener, to *OKW/ WFSt (Oberkommando der Wehrmacht / Wehrmachtführungsstab)*. On 21.05.1945 at 23.33 hours, *Großadmiral* Dönitz sent a telegram authorizing award of the *Ritterkeuz* to Wagener and four other soldiers "in appreciation of the excellent conduct and spirit of the brave garrison of Crete."
00.00.19__	*1939 Spange zum 1914 Eisernes Kreuz I. Klasse*
00.00.19__	*Spange zum 1914 Eisernes Kreuz II. Klasse*
00.00.191_	*1914 Eisernes Kreuz I. Klasse*
00.00.191_	*1914 Eisernes Kreuz II. Klasse*
25.05.1916	*Militärischer Karl-Friedrich-Verdienstorden (Baden)*

ca. 1934	*Ehrenkreuz des Weltkrieges 1914–1918 mit Schwertern*
00.00.194_	*Dienstauszeichnung der NSDAP in Silber*
00.00.194_	*Dienstauszeichnung der NSDAP in Bronze*
ca. 1929	*Nürnberger Parteitagsabzeichen 1929*
00.02.1934 (?)	*Ehrendolch der SA*
00.00.1934	*Ehrenwinkel für alte Kämpfer*

Notes

• Son of a factory director.

Sources

Berg, Manfred & Geyer, Martin H.: *Two Cultures of Rights: The Quest for Inclusion and Participation in Modern America and Germany*. Cambridge University Press, 2002.

Lilla, Joachim; Döring, Martin; & Schulz Andreas: *Statisten in Uniform. Die Mitglieder des Reichstags 1933–1945*. Droste Verlag, 2004.

Orlow, Dietrich: *History of the Nazi Party: 1933–1945*. University of Pittsburgh Press, 1973.

Wagener, Dr. Otto: *Hitler: Memoirs of a Confidant* (Henry Ashby Turner, Jr., ed). Yale University Press, 1985 (Translation of *Hitler aus nächster Nähe. Aufzeichnungen eines Vertrauten 1929–1932*. Ullstein, 1978).

Ernst Julius Günther Röhm

05.01.1931–01.07.1934	*Chef des Stabes der SA*
00.03.1925–01.05.1925	*Führer der SA*

Born:	28.11.1887 in München / Regierungsbezirk Oberbayern / Bayern.
Executed:	01.07.1934 in Cell 474 at Strafgefängnis München-Stadelheim (see the "Career" section, entry of 01.07.1934, for further details).

NSDAP-Nr.:	41 (First joined the DAP with Nr. 623, 00.01.1920, and remained in the Party until it was banned in the wake of the *München-Putsch* of 09.11.1923; Reenrolled in the *NSDAP* with Nr. 41 on 01.11.1930)

Promotions

15.07.1906	*Zweijährig-Freiwilliger und Fahnenjunker*
15.10.1906	*Unteroffizier*
16.02.1907	*Fähnrich*
09.03.1908	*Leutnant*
30.11.1914	*Oberleutnant*
17.04.1917	*Hauptmann* (mit Patent vom 17.01.1917)
00.00.1928	*Teniente Coronel* (lieutenant-colonel, Bolivian Army)
00.03.1925–01.05.1925	*Führer* der *SA*.
05.01.1931–01.07.1934	*Chef des Stabes der SA*
01.05.1931–30.04.1933	*Korpsführer des NSKK*
13.04.1933–01.12.1933	*Staatssekretär*
00.06.1933–01.07.1934	*Reichsleiter* der *NSDAP*
02.12.1933–01.07.1934	*Reichsminister ohne Geschäftsbereich*

Career

Autumn 1897-Spring 1906	Attended the *Königliches humanistisches Maximilians-Gymnasium* in München through graduation. Of his education, Eleanor Hancock writes:

> He studied German and classical literature, Latin, Greek, French, history, geography, nature studies, mathematics and physics, drawing and calligraphy, and gymnastics. Whether at school or privately, he also learned to play the piano very well.
>
> He himself described [in his memoirs] his achievements at school as "rather varied" and admitted that he was never a particularly good student. (Hancock, *Ernst Röhm: Hitler's SA Chief of Staff*, p. 8)

15.07.1906–00.00.1919	Entered service as a *Zweijährig-Freiwilliger* and *Fahnenjunker*, assigned to *Kgl. Bayerisches Feldartillerie-Regiment "Prinz Ludwig"* (later redesignated *"König Ludwig"*) *Nr. 10* (Ingolstadt).

Fahnenjunker Röhm in 1906.
(Röhm, *Die Geschichte eines
Hochverräters*, 1928)

Leutnant Röhm in 1908. (Röhm, *Die
Geschichte eines Hochverräters*, 1928)

25.01.1907   Passed his examinations for promotion to *Fähnrich*.

01.03.1907–07.02.1908   Attended the *Lehrkurs* (instructional course) at the *Kriegsschule* in München. He passed his examinations toward commissioning as an officer on 07.02.1908, and graduated 98th in a class of 124. A performance evaluation from this period describes *Fähnrich* Röhm as follows:

> A not yet completely firm character, who however works on himself with good will. His conception of the profession has become more serious. Intellectually of normal capacity, even if partially distracted. Zeal adequate. Physically strong and sufficiently dexterous. Attitude good: he still lacks assurance and certainty of bearing when facing formations. Off duty conduct and social manners are good. Needs supervision. Financial conditions are settled. (ibid, p. 10)

00.00.1908–01.08.1914   Assigned as a *Rekrutenausbilder* (recruit instructor) to *Kgl. Bayerisches 10. Infanterie-Regiment "König Ludwig"*. In an evaluation report written in 1912, his company commander, *Hauptmann* Schleicher, wrote:

Kompanieführer Röhm (third from left) at the front, ca. 1915.
(Röhm, *Die Geschichte eines Hochverräters*, 1928)

> [Röhm is] a vigorous, cheerful, capable officer with a well developed
> military spirit. Mentally of normal gifts and physically sufficiently
> dexterous that he is easily deployable in all aspects of service
> and forms a reliable support of his company chief. He fervently
> endeavors to fulfill his duties with diligence and conscientiousness.
> His military knowledge is appropriate for his time of service: his
> oral and written orders are good. Possesses in addition a good
> knowledge of French. He is very employable as a platoon leader,
> understands quickly and circumspectly, sensible and energetic. He
> treats his subordinates calmly and benevolently. ("Qualifikations-
> Bericht zum 1. Januar 1913 über den *Leutnant* Ernst Röhm im K.
> 10.*Infanterie-Regiment*", in ibid, p. 13)

Röhm experienced some trouble during this assignment, however, when
on 09.03.1911, he was punished with 3 days' confinement to quarters
(punishment later suspended) by a *6.Division* court for mistreating a
trainee. Eleanor Hancock writes:

> ... [Röhm] repeatedly push[ed the recruit] lightly in the side during
> musketry training and [gave] him a smack on the back of the head.
> (ibid, p. 10)

In another incident, on 22.04.1912, Röhm demanded that three
civilians in Ingolstadt's Hotel Adler relinquish their table to him and
his fellow officers. Because of their insulting response, Röhm brought
charges against the civilians for "defamation of officers". The two men
and one woman accused by Röhm were tried and found guilty of the
offense on 30.08.1912. It was later determined by his superiors, however,

that Röhm had provoked the civilians, and he was ordered to serve a confinement to quarters for two days.

01.06.1912–28.06.1912	Attended the *Pionierkurs* (combat engineer instructional course) of *1. Bayerische Pionier-Bataillon*.
25.09.1912–15.10.1912	Attached to *1. Kgl. Bayerisches Reserve-Infanterie-Regiment* under *III. Kgl. Bayerisches Armee-Korps*.
27.04.1914–16.05.1914	Attached for training in *Waffeninstandsetzungsgeschäft* (weapons repair) to the *Gewehrfabrik* (artillery factory) in Amberg.
02.08.1914–24.09.1914	Adjutant of *I. Bataillon / 10. Kgl. Bayerisches Feldartillerie-Regiment "König"* on the Western Front.
24.09.1914	Severely wounded in the face near Spada in the Chanot Wood, Lorraine, losing half of his nose. He was subsequently hospitalized in the following medical facilities: 26.09.1914–15.10.1914 *Festungslazarett* in Metz 15.10.1914–10.01.1915 (?) *Distriktkrankenhaus* in Kaiserslautern 10.01.1915–00.04.1915 *Sanatorium* in Reichenhall
00.04.1915–00.04.1915	Assigned to *Ersatz-Bataillon / 10. Kgl. Bayerisches Feldartillerie-Regiment*.
17.04.1915–02.06.1915(?)	Returned to active service with *10. Kgl. Bayerisches Feldartillerie-Regiment*, and again assigned as *Adjutant* of its *I. Bataillon*.
02.06.1915–23.06.1916	*Führer* of *10. Kompanie / 10. Kgl. Bayerisches Feldartillerie-Regiment*. On 20.06.1916, the commanding general of *III. Bayerische Armee-Korps*, *General der Kavallerie* Ludwig Freiherr von Gebsattel, issued the following *Korps-Tagesbefehl* (corps order of the day):

> In the name of his majesty the German Kaiser, I award the Iron Cross First Class to *Oberleutnant* Röhm, 10.*Infanterie-Regiment* König, in recognition of the extraordinary drive and courage he has exhibited on all occasions since the beginning of the war.
> [signed] Freiherr von Gebsattel. (ibid)

He received the award shortly before the regiment's capture of the French fortification of Thiaumont near Verdun and the occupation of most of the village of Fleury. The regiment's dead in this battle amounted to 150 men (10 officers and 140 NCO's and enlisted men), with 17 officers and nearly 1,000 NCO's and enlisted men wounded). Röhm later wrote:

> I bled from many wounds and was greatly weakened; but I felt full of pride, that I had been there. The 23rd of June, the regiment's greatest day of victory, is also the proudest day of my life. (ibid, p. 22)

On the basis of his involvement in this action, Röhm applied for award of the highest military distinction of the Bavarian Army, the Militär-Max-Joseph-Orden (MMJO), for actions above and beyond the call of duty. According to his own account, his former battalion commander suggested he submit his name. His application was strongly endorsed by his superiors, including *Generalmajor* Ludwig Freiherr von Tautphoeus (commander of *11.Königlich Bayerische Infanterie-Brigade*) who wrote that Röhm was "one of the 10th Infantry Regiment's best officers, whose sang-froid, bravery and constant dash had an exemplary effect on his subordinates." (ibid, p. 23) However, as Eleanor Hancock continues:

> Röhm was not awarded the MMJO. Although the order's chapter voted four to three in favor of the award, War Minister von Hellingrath followed the advice of the Chancellery Office of the order and recommended the king reject the award on the grounds that actions like Röhm's occurred frequently, and were therefore not over and above the dictates of duty. The king accepted this advice. (ibid)

Röhm's claim that reasons of finance influenced Bavarian War Ministry's rejection of the award may have been a cover story to assuage his wounded pride and anger.

23.06.1916–16.08.1916	Severely wounded during an attack on the French fortification at Thiaumont, and suffered a serious wound (caused by 14 pieces of shrapnel) to the chest. He was then hospitalized at the *Reservelazarett Kriegsschule* München.
16.08.1916–00.00.1916	Hospitalized at the *Reservelazarett Hohen-Aschau*, and classified as unfit for further frontline duty.
30.10.1916–27.05.1917	*Adjutant* of the *Armeeabteilung* in the *Bayerische Kriegsministerium* (Bavarian War Ministry).
29.05.1917–08.05.1918	*Ordonnanzoffizier* and *Nachschuboffizier* with the staff of *12. Kgl. Bayerisches Infanterie-Division* in Romania and France.
09.05.1918–05.11.1918	*2.Generalstabsoffizier (Ib)* of *12. Bayerische Infanterie-Division*.
00.10.1918	Contracted Influenza, from which he became very ill. After a lengthy convalescence, he returned to duty.
02.12.1918	Demobilized from *12. Bayerische Infanterie-Division*.
01.01.1919–31.03.1919	Adjutant of *11. Bayerische Infanterie-Brigade* (Ingolstadt, then München).
00.04.1919–00.05.1919	Assigned as *Stabsoffizier* (for *Verpflegung und Ausrüstung* [Provisions and Equipment]) to *Freikorps Epp*.
03.05.1919–00.07.1919	*Chef des Stabes* to the *Stadtkommandant* of München and *Offizier für die Abwehr und für politische Angelegenheiten* (Officer for Defense and Political Affairs) with the *Reichswehr* staff in München.
00.07.1919–31.12.1920	*Ib* (*Bewaffnung und Ausrüstung*, [staff officer] for armaments and equipment) with the staff of *Schützenbrigade 21 / Bayerisches Schützenkorps* (commanded by Franz Xaver Ritter von Epp).
00.00.1919	Joined the *Deutschnationale Volkspartei* (*DNVP*, German Nationalist Peoples Party).
00.07.1919	Cofounder of the *Eisernen Faust* ("Iron Fist") club, an organization of radical nationalist officers.
16.10.1919	Attended the first meeting of the *Deutsche Arbeiterpartei* addressed by Adolf Hitler. He was probably introduced to the eventual *Führer* by *Hauptmann* Karl Mayr, and the two soon became close friends; Röhm

was one of Hitler's few "Duz-Freunden" (a person he permitted to use the informal pronoun "Du").

00.01.1920	Joined the *DAP*.
01.10.1920–05.12.1922	*Adjutant* and *2.Stabsoffizier* to the *Infanterie-Führer VII* (Ritter von Epp); also *Leiter* of the *Feldzeugmeisterei* (Quartermaster's Office) in *Gruppenkommando 4* of the *Reichswehr*.
Late 1921–00.00.1923	*Leiter* of the *München Ortsgruppe* of the *Bund Reichsflagge*, a *Wehrverband* (paramilitary organization) established in the summer of 1919 by his friend and fellow *Reichswehr Hauptmann*, Adolf Heiß.
05.12.1922–03.05.1923	*Stabsoffizier* with *7.Reichswehr-Division* (München).
31.01.1923–00.04.1923	Member of the *Arbeitsausschuß der Landesleitung* (Working Staff of the State Leadership) of *Reichsflagge* and *bevollmächtigter Vertreter* (deputized representative) of *Reichsflagge's Führer, Hauptmann* Heiß, in München.
04.02.1923–00.04.1923	*Führer* (*spiritus rector*) of the *Arbeitsgemeinschaft der Vaterländische Kampfverbände* (Working Community of Patriotic Fighting Organizations); simultaneously assigned as a representative of the *Organisation Niederbayern* in München. Within the *Arbeitsgemeinschaft* were the following: *Reichsflagge, NSDAP, Vaterländische Bezirksvereinen Münchens* (*VVM*), *Zeitfreiwilligenkorps München, Bund Oberland*, and *Bund Unterland*. Röhm led the *Arbeitsgemeinschaft* together with Dr. Christian Roth (former Bavarian Justice Minister), who handled political affairs, and *Oberstleutnant* Hermann Kriebel, responsible for military matters.
03.05.1923–00.09.1923	*Kompaniechef* in Bayreuth.
00.09.1923–09.11.1923	Member of the *Deutsche Kampfbund-Kampfgemeinschaft Bayern*.
26.09.1923	Submitted his request for discharge from the *Reichswehr*.
12.10.1923	Founded the *Reichskriegsflagge* in München.
08./09.11.1923	Participated in the *München-Putsch*, directing the occupation of the *Bayerische Kriegsministerium* building on Schönfeldstraße. Several of his men fired shots at *Reichswehr* troops who surrounded the building, resulting in the death of two *Reichskriegsflagge* men, Martin Faust and Theodor Casella. Soon afterward, he surrendered his forces and was arrested by the München police.
16.12.1923	Discharged from the *Reichswehr*.
01.04.1924	Sentenced in the so-called "Hitler-Prozeß" to 15 months' imprisonment (with 4 months' credit for time served) as an accessory to high treason. He was in fact released on probation the same day.
01.04.1924	Charged by Hitler with the reconstruction of the *SA* and the *Wehrverbände*. In the following note to Röhm, the convicted *Führer* wrote:

> *Hauptmann* Röhm is military leader of the Kampfbund. I command, therefore, that his ordinances be obeyed by members and particularly leaders of the *NSDAP*'s *SA*. Those who can't unconditionally follow the commands of *Hauptmann* Röhm

Hauptmann Röhm, ca. 1918.

are to be considered as no longer belonging to the *SA*. (David Jablonsky, *The Nazi Party in Dissolution: Hitler and the Verbotzeit, 1923–1925*, p. 80)

00.04.1924–01.05.1925   *Kommandeur* of *Frontbann*, a cover organization for the *SA* and other right-wing extremeist *Wehrverbände* banned in the aftermath of the *München-Putsch*. By September 1924, it numbered some 30,000 men throughout Germany. One of his colleagues at this time, Kurt G. W. Lüdecke, later wrote of Röhm as:

... a brilliant leader of men, an excellent officer, fearless and straightforward. His massive, round head, battle-scarred and patched, looked like something hammered from rock. He was the living image of war itself, in contrast to his polished manner and exceptional and instinctive courtesy. That, with his naturalness, diplomatic tact and savoir faire distinguished him from leading Nazis then and afterwards, who for the most part were boorish and arrogant, or were bullies. For all his one-sided military mind, he was a passionate politician, having, for a soldier, a rare intelligence and understanding for politics. I liked his keen, open gaze and his firm hand clasp. (Lüdecke, *I Knew Hitler: The Story of a Nazi Who Escaped the Blood Purge*, p. 245)

München, 01.04.1924: Group photo of defendants in the "Hitler-Prozeß", on trial for their part in the "München-Putsch" of 09.11.1923. From left to right: Heinz Pernet, Dr. Friedrich Weber, Dr. Wilhelm Frick, Hermann Kriebel, Erich Ludendorff, Hitler, Wilhelm Brückner, Ernst Röhm, and Robert Wagner. (Hermann-Historica, Auctioneers, München)

00.00.1924	Member of the *Deutschvölkische Freiheitspartei* (*DVFP*, German Nationalist Freedom Party).
04.05.1924–20.10.1924	Member of the *Reichstag* (representing the *NS-Freiheitspartei*).
17.05.1924–00.03.1925	*(geschäftsführender) Stellvertretender Kommandeur der SA* ([acting] Deputy Commander of the [illegal] *SA*).
00.03.1925–01.05.1925	*Führer der SA.* He met with Hitler in München, 16.-17.04.1925, and exhibited the rebellious spirit which would eventually lead to his death by presenting the *Führer* with a memorandum reading: "I categorically refuse to allow the *SA* to become involved in Party matters; equally, I categorically refuse to allow *SA* commanders to accept instructions from Party political leaders." (Röhm, *Die Geschichte eines Hochverräters*, pp. 313–314) Hitler would have none of it; the *SA,* as a subordinate part of the *NSDAP*, would have no independence from the Party. On 18.04.1925, Röhm submitted his letter of resignation to Hitler; the *Führer* disregarded it. On 30.04.1925, attempting to salvage his relationship with Hitler, Röhm wrote another letter which ended with: "I take this opportunity, in memory of the fine and difficult hours we have lived together, to thank you and to beg you not to exclude me from your personal friendship." The following day, he resigned from his post

as leader of the *SA* and the *Frontbann* due to his differences of opinion with Hitler. In a conversation with Kurt Lüdecke soon after, he stated:

> No need to tell you what Hitler is like. Believe me, I didn't mince words when I last talked with him. But it's useless; if you try to tell him anything, he knows everything already. Though he often does what we advise, he laughs in our faces at the time, and later does the very thing as if it were all his own idea and creation. He doesn't even seem to be aware how dishonest he is. I've never seen a man so magnificently unaware that he's adorning himself with borrowed plumage. Usually, he solves suddenly, at the very last minute, a situation that has become intolerable and dangerous only because he vacillates and procrastinates ... But nobody is perfect, and he has his great qualities. Apparently there's nobody else who would do better than he. (Lüdecke, *I Knew Hitler*, p. 287).

00.00.1925–00.00.1928	Employed with varying degrees of success in a number of civilian occupations.
11.10.1926	Sentenced to 10 days' jail as a result of his behavior during a conference in München of the *Reichstagsausschuß 'Feme-Organisationen und Fememorde* (*Reichstag* Committee "Feme Organizations and Feme Murders"). This committee, chaired by Pal Levi, convened in München 05.10.1926–14.10.1926, and investigated murders of "traitors" by right-wing extremist "*Feme*" detachments. Among these was the 06.10.1920 killing of 19-year-old Maria Sandmeyer, a waitress. Eleanor Hancock writes that during the committee's hearings:

> Röhm and Epp both admitted to helping von Schweikhart [the alleged murderer] ... In the course of Röhm's interrogation, he refused to answer Levi's further questions and was fined three hundred marks by the committee for contempt [he refused to pay the fine / MdM]. It is not clear what the questions were that Röhm refused to answer, and whether he refused because Levi's questioning was getting close to something legally incriminating. Röhm's own account in his autobiography was marked by anti-Semitic attacks on Levi, which suggests that his anti-Semitism and consequent resentment of Levi's authority led him to refuse to answer the questions. (Hancock, *Ernst Röhm: Hitler's SA Chief of Staff*, pp. 84–85)

09.02.1927–20.02.1927	Imprisoned at München-Stadelheim.
ca. 1928–00.06.1934	Member of the *Reichsverband Deutscher Offiziere* (*RDO*, Reich Association of German Officers).
00.08.1928–00.11.1928	*Vorsitzender* of the reestablished *Reichskriegsflagge*, under the new name *Flaggenklub*. His purpose in resurrecting the *RKF* was not to resume paramilitary activities, but to bring its members into the *NSDAP*.

Mid-1928	Received an offer from *Major* Wilhelm ("Guillermo") Kaiser, a former Imperial German Army officer serving as temporary Bolivian military attaché in the Netherlands, to work for the Bolivian Army.
25.07.1928–00.11.1928	*Vorsitzender* of the *Wehrpolitische Vereinigung* (*WPV*, League for Defense Policy).
00.01.1929–00.10.1930	Service as a *Teniente Coronel* (lieutenant-colonel) with the Bolivian Army as a military instructor and inspector of two infantry regiments. His contract was to run from 01.01.1929 to 31.12.1930. On 14.12.1928, he had embarked for South America aboard the Hamburg-South America liner *SS Cap Polonio*. With him was a 19-year-old München art student, Martin Schätzl and the chief of the Bolivian General Staff (since 09.02.1921), General Hans Kundt. Per Dr. Hancock, "their [Röhm's and Schätzl's] friendship may have arisen through shared political views," however Schätzl did not share Röhm's sexual orientation. The *Cap Polonio* weighed anchor in Buenos Aires on 31.01.1929, and the trio then went by train to La Paz, arriving there on 05.01.1929. The Bolivians backdated Röhm's service to 05.12.1928.
00.00.1930	Recalled from South America by Hitler, who requested that he assume the post of *Chef des Stabes der SA* (Chief of Staff of the *Sturmabteilung*).
01.11.1930	Reenrolled in the *NSDAP*.
05.01.1931–01.07.1934	*Chef des Stabes der SA* (Chief of Staff of the *Sturmabteilung*). Röhm built the *SA* into a massive fighting organization. At the time of his appointment as *Chef des Stabes*, its membership stood at approximately 77,000; within three months, it exceeded 100,000; by 15.12.1931, it had risen to 260,438, a figure that more than doubled in 1932. By the time of Hitler's appointment as *Reichskanzler* on 30.01.1933, Röhm led a paramilitary army of over 500,000 men. Following Röhm's murder by the SS, Viktor Lutze took over as *Stabschef*.
01.05.1931–30.04.1933	*Korpsführer des NSKK* (*NS-Kraftfahrkorps*). Succeeded by his deputy, Adolf Hühnlein.
00.00.1931	Initiation of proceedings against Röhm for "wiedernatürlicher Unzucht" (unnatural sex acts), but these were ultimately discontinued. Hitler tolerated and greatly valued Röhm despite the many rumors regarding his lifestyle. He often asserted that the tales of homosexuality on Röhm's part were unproven, and were, in any case, "an entirely private matter". However, Röhm did little to keep his behavior private, using his position as *SA* Chief of Staff to procure partners, with *SA* men specially allocated as the procurers. His most active pimp was Peter Granninger, installed by Röhm in the *SA-Nachrichtenabteilung* (intelligence department) and intimately linked to the *Stabschef* since 1928.

> The head pimp was a shop assistant named Peter Granninger, who had been one of Röhm's partners since 1928 and was now given cover in the *SA* Intelligence Section. For a monthly salary of 200 marks he kept Röhm supplied with new friends, his main hunting ground being Gisela High School [in] Munich; from

Ernst Röhm as a defendant in the "Hitler-Prozeß".
Below, from left to right: Heinz Pernet, Röhm, Wilhelm Brückner, and Robert Wagner.

"Deutscher Tag" in Weimar, 15–17.08.1924, a meeting of the *NS-Freiheitsbewegung Großdeutschland*, with 12,000 people in attendance. Ernst Röhm and Wilhelm Brückner follow the retired *General der Infanterie* Erich Ludendorff. (Roger Bender photo)

this school he recruited no fewer than eleven boys, whom he first tried out and then took to Röhm... The general meeting-point for the Granninger circle and the *SA* homosexuals was Röhm's reserved table in the Bratwurstglöckl, Munich. The proprietor, Karl Zehnter, was a homosexual himself ... Röhm and his highly organized squad of informers generally succeeded in covering their tracks. In March 1932, however, Röhm's letters began to leak, several being published in the Social-Democrat Münchner Post. (ibid, p. 82)

Hitler's foreign press representative, Ernst "Putzi" Hanfstaengl, writes of Röhm's "perversion":

Fellow-officers who had known Roehm during the war always maintained that he had been completely normal and even described orgies in which he had taken part in the Army brothels. He had certainly acquired a syphilitic infection during this period and this may have had some effect on his subsequent development. The scandal started soon after he had returned in October 1930. Letters from his male companions in Bolivia somehow came into the hands of third parties and the accusations started. General von Epp, who had held a high opinion of Roehm's organizational abilities before the Ludendorff Putsch, even taxed him with the

Heinrich Hoffmann's studio portrait of Ernst Röhm, ca. 1925.

rumours at quite an early stage and received Roehm's completely
false word of honour that they were not true. Later, about 1932,
the scandal became public, and although it was somehow glossed
over, Roehm quite openly admitted his aberration to Toni Drexler
[Anton Drexler, a cofounder of the *DAP*], because he passed it on
to me. Hitler can have had no illusions at any time and his mock
horror when he found it necessary to shoot Roehm in 1934 was,
of course, pure invention. (Hanfstaengl, *Hitler: The Missing Years*,
p. 155)

13.04.1932–14.06.1932	*SA* banned by the government of Chancellor Brüning.
10.03.1933–13.04.1933	*Staatskommissar zur besonderen Verwendung* (State Commissioner for Special Assignment) in Bayern.
00.04.1933–02.08.1933	*Ministerial-Kommissar für die Hilfspolizeibeamten der Sicherheitspolizei* (Ministerial Commissioner for the Auxiliary Police Officials of the Security Police) in the *Preußische Ministerium des Innern*.
13.04.1933–01.12.1933	*Staatssekretär* to the *Reichsstatthalter* in Bayern (Franz Xaver Ritter von Epp).
25.04.1933	Appointed as *Staatskommissar z.b.V. in Bayern* (State Commissioner for Special Assignment in Bavaria).
01.05.1933–31.12.1933	*Kommandeur der Sicherheitshilfspolizei* (Commander of the Auxiliary Security Police) in Bayern.

Spring 1933

First of two meetings with Hermann Rauschning, President of the Danzig Senate, who later wrote:

> ... [S]oon after the party's accession to power, [*Gauleiter* Albert] Forster had introduced us to each other ... Roehm was dissatisfied. He had not been made a minister. The entire meaning of the National Socialist revolution seemed lost to him.
>
> "We've just beaten up the game for the generals," he grumbled.
>
> Could not Forster use his influence with the *Führer* on Roehm's behalf? The entire National Socialist revolution would be bogged if the S.A. were not given a public, legal function, either as militia or as a special corps of the new army. He was not inclined to be made a fool of. At a later date, I had the opportunity of speaking to him in greater detail at Kempinski's well-known wine-restaurant in the Leipzigerstraße in Berlin, where he usually lunched. We discussed the new defensive power of the State, and who ought to command it, who, in fact, ought to create it, the *Reichswehr* generals or he – Roehm, who had made the party possible in the first place. Apart from his special weakness [his homosexuality / MdM], Roehm was unquestionably a pleasant companion, gifted, and a competent organizer, but, generally speaking, an adventurer whose right place was in the colonies, as far away from Europe as possible. In his reproaches against the *Reichswehr* he was unjust and embittered. He resented the arrogant reserve of the *Reichswehr* officers. Ardently desirous of action, in the consciousness of being able to accomplish something great, he confided to me in a few disconnected sentences his dream of the future.
>
> We were sitting in the great windowed dining-hall. His scars were scarlet with excitement. He had drunk a few glasses of wine in quick succession.
>
> "Adolf is a swine," he swore. "He will give us all away. He only associates with the reactionaries now. His old friends aren't good enough for him. Getting matey with the East Prussian generals. They're his cronies now."
>
> He was jealous and hurt.
>
> "Adolf is turning into a gentleman. He's got himself a tail-coat now!" he mocked.
>
> He drank a glass of water and grew calmer.
>
> "Adolf knows exactly what I want. I've told him often enough. Not a second edition of the old imperial army. Are we revolutionaries or aren't we? Allons, enfants de la patrie! If we are, then something new must arise out of our élan, like the mass armies of the French Revolution. If we're not, then we'll go to the dogs. We've got to produce something new, don't you see? A new

discipline. A new principle of organization. The generals are a lot of old fogeys. They never have a new idea.

"Adolf has learnt from me. Everything he knows about military matters, I've taught him. War is something more than armed clashes. You won't make a revolutionary army out of the old Prussian N.C.O.s. But Adolf is and remains a civilian, an "artist," an idler.

"Don't bother me," that's all he thinks. What he wants is to sit on the hilltop and pretend he's God. And the rest of us, who are itching to do something, have got to sit around doing nothing." ...

"They expect me to hang about with a lot of old pensioners, a herd of sheep. I'm the nucleus of the new army, don't you see that? Don't you understand that what's coming must be new, fresh and unused? The basis must be revolutionary. You can't inflate it afterwards. You only get the opportunity once to make something new and big that'll help us to lift the world off its hinges. But Hitler puts me off with fair words. He wants to let things run their course. He expects a miracle. Just like Adolf! He wants to inherit an army all ready and complete. He's going to let the "experts" file away at it. When I hear that word, I'm ready to explode. Afterwards he'll make National Socialists out of them, he says. But first he leaves them to the Prussian generals. I don't know where he's going to get his revolutionary spirit from. They're the same old clods, and they'll certainly lose the next war. Don't try to tell me! This is where you're letting the heart of our movement rot." ...

I had very little to do with Roehm after that. This outpouring seemed to me, albeit partly inspired by strong drink, to reveal the tragedy of a man who had creative talents of a sort, a man who, in spite of everything, was a rebel, as he himself said, and knew how to die ... (Rauschning, *Hitler Speaks: A Series of Political Conversations with Adolf Hitler on His Real Political Aims*, pp. 153–156)

29.05.1933	Appointed as *Präsident* of the *Deutschen Akademischen Austauschstelle* (German Academic Exchange Office).
00.06.1933–01.07.1934	*Reichsleiter der NSDAP.*
01.07.1933	Appointed as *Präsident* of the *Landesverband Bayern der Deutschen Bühne e. V.* (Bavarian State Association of the German Stage).
14.09.1933–01.07.1934	*Preußischer Staatsrat.*
03.10.1933–01.07.1934	Member of the *Akademie für Deutsches Recht*, München.
00.11.1933–01.07.1934	Member of the *Führerrat der Akademie für Deutsches Recht* (Leadership Council of the Academy for German Law).
12.11.1933–01.07.1934	Member of the *Reichstag.*
02.12.1933–01.07.1934	*Reichsminister ohne Geschäftsbereich* (*Reich* Minister without Portfolio, sworn in 04.12.1933) and member of the *Reichsverteidigungsrat* (*Reich* Defense Council).

Ernst Röhm in Bolivian Army uniform, ca. 1930.

07.12.1933   Delivered his speech "Warum SA?" (Why SA?) to foreign diplomats in Berlin, from which the following is excerpted:

> The Storm Troops cannot be compared with any army, any militia or any other military system in the world ... This is absolutely not the case with the Storm Troops according to Adolf Hitler's expressed will. On the contrary, in all the proclamations which deal with relations between the *Reichswehr* and the Storm Troops he has clearly and unmistakably indicated the dividing line: the *Reichswehr* is the sole armed force in the state and the Storm Troops are the representatives of the will and ideas of the National Socialist German Revolution. The *Reichswehr* is charged with the defense of the frontiers and the protection of the interests of the *Reich* as against foreign countries. The task set the Storm Troops is to form the new German state in mind and will on the basis of National-Socialist ideas and to educate the individual German as a living member of this National-Socialist state. There is no connection whatever between the *Reichswehr* and the Storm Troops. Thus the German Army took no part whatever in the National-Socialist Revolution, a fact which is probably unique in the history of revolutions.
>
> In spite of their numerical strength of about 2,500,000 men, the Storm Troops are not concentrated in barracks and rationed in common, as is the case with all formations belonging to any

A postcard photo of Ernst Röhm,
giving the German equivalent of his
Bolivian Army rank. (*Oberstleutnant*,
or Lieutenant Colonel)

Chef des Stabes Ernst Röhm in 1931.
(Baldur von Schirach, *Die Pioniere
des Dritten Reiches*, 1933)

military system in neighboring states. They are not paid and are not provided with service clothing. To-day, as in the past, service in the Storm Troops is based on the absolutely volunteer system. The Storm Trooper pursues his civil vocation, and merely devotes his leisure hours, in the evening and at night, to Storm Troop service.

The Storm Troops were created as a protective and fighting force for dealing with the internal political opponents of National Socialism, namely Communism and Marxism. For fourteen years the Storm Troops waged the moral fight to obtain power in the state. Their prime task is now to secure the victory of the National-Socialist Revolution.

The Storm Trooper is the exponent of the National-Socialist conception of human existence and its apostle who conveys the principles of National Socialism to the remotest cottage and to all his fellow-countrymen without exception.

In the course of the years of struggle the Storm Troops, with enormous sacrifices, have given convincing proof of their absolute loyalty to the *Führer* and to the movement, and thus showed themselves qualified for their task. It will always be a glorious page in the history of Germany that in times of the greatest need hundreds of thousands of men came forward who were ready, from pure idealism and absolutely voluntarily, to defend their principles to the last ...

So far it was simply and solely due to the fact that the Storm Troops in the heart of Europe stood for years with consciously anti-Bolshevist aims as a bulwark protecting peace and order in the world, that Bolshevism was unable to lay hands on the western European countries as well. Hence it is absolutely in the interest of foreign countries to see order and discipline firmly established in the German nation. The world ought to be thankful for that, instead of distorting the facts and representing the Storm Troops as a menace to peace ...

Nothing is ... more natural than that the National-Socialist State should make use of the old and well-tried champions of this new political faith to educate the entire population to accept in their hearts and really live in accordance with the principles of National Socialism. It would be a contradiction of the totalitarian claim of the National-Socialist State if the State were not to include the party as representing the idea upon which the State is based. This applies even more to the Storm Troops who, as regards their ideas, organization and fighting qualities, are the strongest expression of the power of National Socialism as such. The Storm Troops have got rid of a form of government in spite of the most embittered resistance on the part of its supporters. They have replaced the vanquished state of the November revolution and the Weimar National Assembly by their own State, the National-Socialist state.

Adolf Hitler has now incorporated the Storm Troops in the State. The Storm Troops have thus become not only the representatives of authority but also the bearers of responsibility in the National-Socialist state. (Röhm, *Why SA? German Minister Ernst Röhm, Chief of Staff, Addressing the Diplomatic Corps in Berlin on December 7, 1933*)

00.12.1933–01.07.1934	*Bayerischer Staatsminister.*
00.12.1933–01.07.1934	*Bayerischer Staatsrat.*
31.12.1933	One year into his chancellorship, and just months before ordering the massacre of Röhm and other *SA* leaders, Hitler wrote the following letter to Röhm:

> My dear Chief of Staff,
> The fight of the National Socialist Movement and the National Socialist Revolution was made possible only by the consistent suppression of the Marxist terror by the *SA*.

Stabschef Röhm in 1933. (Laurens Hessels photo)

Röhm with Himmler and Daluege preside over the mass meeting of *SS-Gruppe Ost* in Berlin-Döberitz, August 1933. (Bundesarchiv Bild 102-14886)

SA-Stabschef Röhm marches beside *Kronprinz* Wilhelm von Preußen, oldest son of the former Kaiser Wilhelm II, in 1933. (NARA photo from the Heinrich Hoffmann collection)

Röhm chats with *SA* officers in 1933. (NARA photo)

Ernst Röhm and Rudolf Hess, ca. December 1933. (Michal Sika photo)

Stabschef Röhm in his lavishly decorated office, ca. 1933. (Michal Sika photo)

If the Army is to guarantee the protection of the nation abroad, it is the task of the *SA* to secure the victory of the National Socialist Revolution, the continued existence of the National Socialist State and the community of our Volk at home.

When I summoned you, my dear Chief of Staff, to your present position, the *SA* was going through a serious crisis.

It is primarily thanks to you that, in the space of only a few years, this political instrument was able to develop the force which made it possible for me to finally win the struggle for power by overcoming the Marxist opponent.

At the close of the Year of the National Socialist Revolution, I feel compelled to thank you, my dear Ernst Röhm, for the immortal service which you have done to the National Socialist Movement and the German Volk and to assure you how grateful I am to Fate to be able to call such men as you fighting comrades [*Kampfgenossen*].

In true friendship and grateful regard,

Yours, Adolf Hitler (Dr. Max Domarus, ed, *Hitler. Speeches and Proclamations, 1932–1945: The Years 1932–1934*)

05.04.1934	Appointed as *Ehrenführer* of the *Deutsche Reichskriegerbund "Kyffhäuser"*.
28.02.1934	Luncheon held for *SA* and *Reichswehr* leaders at Röhm's headquarters on the Standartenstraße, Berlin. This followed a meeting at the *Reichswehr* Ministry, during which Hitler officiated over a reluctant agreement between *Reichswehrminister* General Werner von Blomberg and Röhm in which the *Reichswehr* was deemed "the sole bearer of arms in the

Third *Reich*", and pre- and post-military training would be the province of the *SA* alone. During the luncheon, Röhm and von Blomberg sealed the agreement with a handshake. Röhm's real goal was more ambitious: to absorb the entire *Reichswehr*, limited by the Treaty of Versailles to just 100,000 men, into the *SA*, with himself ruling as *Reichswehr*minister. Toward this end, he exploited the discontent within the ranks of the *SA*, now numbering some 2,500,000 men, which had not benefited in any appreciable way since the seizure of power. Because these plans were contrary to those of the *Führer*, Hitler – under pressure from Göring and Himmler – ultimately decided to eliminate his old comrade Röhm and a number of other *SA* leaders (see entry of 30.06.1934, below). Almost immediately following the departure of Hitler and the *Reichswehr* leaders from the luncheon, Röhm launched into a tirade before his *SA* lieutenants, shouting:

> What that ridiculous Corporal [Hitler] says means nothing to us ... I have not the slightest intention of keeping this agreement. Hitler is a traitor and at the very least must go on leave ... If we can't get there with him, we'll get there without him. (Heinz Höhne, *The Order of the Death's Head*, p. 110)

In response to these treasonous utterances, the stunned Lutze, whose primary loyalty was to the *Führer*, sought to expose the mutinous spirit displayed by Röhm. In March 1934, he shared his concerns with Rudolf Hess. Making no headway with the deputy *Führer*, he then presented his case directly to Hitler at Berchtesgaden. The *Führer*, hearing of the dangerous rumblings of discontent throughout the *SA* and rising as high as its top leadership, responded with, "We must let the matter develop." Knowing of Röhm's bitter rivalry with the *Reichswehr*, Lutze later spoke of the matter to *Generalmajor* Walter von Reichenau during military exercises in Braunfels. Von Reichenau was grateful for the information as, unbeknownst to Lutze, he was already working out the details, together with Reinhard Heydrich, for a joint *Reichswehr-SS* action to solve the problems presented by the troublesome Röhm and his clique. Heydrich eventually persuaded Himmler of the need to decapitate the *SA*, and throughout the Spring of 1934 the two plotted the eventual "Night of the Long Knives". On 22.06.1934, Hitler met with Lutze and informed him of the impending purge to be carried out on the *SA* leadership, including the removal of Röhm. He advised Lutze that he was to accept orders only from him, and not from München.

07.06.1934   Issued an announcement that on the advice of his physician, he was taking several weeks' leave due to a painful nervous disorder, and that the *SA* would be going on its customary summer leave (to return to duty on 01.08.1934). He added:

If the enemies of the S.A. hope that the S.A. will not be recalled or will be recalled only in part after its leave, we may permit them to enjoy this brief hope. They will receive their answer at such time and in such form as appears necessary. The S.A. is and remains the destiny of Germany. (Frederick L. Schuman, *Hitler and the Nazi Dictatorship. A Social Pathology and the Politics of Fascism*)

30.06.1934   Personally arrested by Adolf Hitler at the Pension Hanselbauer in Bad Wiessee am Tegernsee and transported to Stadelheim Prison near München, where he was held- as prisoner 4034- in cell 474. According to the recollections of Viktor Lutze, who arrived on the scene just as Hitler approached the door of Röhm's hotel room, Hitler, brandishing a revolver, shouted abuse at the *Stabschef*, and called him a traitor. He informed Röhm that he was under arrest and ordered him to get dressed.

30.06.1934   Expelled from the *Preußische Staatsrat* by order of Hermann Göring.

30.06.1934   At 1500 hours on that date, Hitler issued a *Führererlaß* (*Führer* Decree) to the German press removing Röhm from the *SA* and *NSDAP*. The next day, the following article, beginning with Hitler's decree, appeared in a special edition of the *Völkischer Beobachter*:

> Röhm excluded from the Party and the *SA*. The *Reich* press bureau of the *NSDAP* reports the following decree issued by the *Führer*:
>
> "As of today I have relieved the Chief of Staff Röhm of his position and have ejected him from the Party and the *SA* ["Ich habe mit dem heutigen Tage den *Stabschef* Röhm seiner Stellung enthoben und aus Partei und *SA* ausgestossen."]. I am appointing *Obergruppenführer* Lutze as chief of staff. *SA* Leaders and *SA* men who do not obey his orders or who oppose him will be ejected from the *SA* and the Party and consequently will be arrested and convicted.
>> Signed: Adolf Hitler
>> *Oberst*er *Führer* der Partei und der SA"
>
> -----
>
> The *Führer* to the new Chief of Staff: München, 30. June 1934
> The *Führer* has sent the following letter to *SA-Obergruppenführer* Lutze:
> "To *SA-Obergruppenführer* Lutze
> My Dear *SA-Führer* Lutze!
> The gravest failures on the part of my former chief of staff have compelled me to relieve him of his position. You, my dear *Obergruppenführer* Lutze, have for many years in good and bad days been an always reliable and ideal *SA-Führer*. If I am appointing you today as chief of staff it is done with the conviction that you will succeed through your reliable and obedient work in making of my *SA* the instrument which the nation needs

and which I imagine. It is my wish that the *SA* be built up as a reliable and strong part of the national socialist movement. Full of obedience and blind discipline, they must help to create and form the new German citizens.

Signed: Adolf Hitler" (Translation of Document 2407-PS, in *Nazi Conspiracy and Aggression, Volume VIII*)

In a statement of 30.06.1934, the *Reich* Press Office of the *NSDAP* declared:

For many months now, individual elements have attempted to drive wedges between the *SA* and the Party and between the *SA* and the State and to create conflicts. More and more evidence arose in support of the suspicion that these attempts were attributable to a limited clique with a definite purpose.

Chief of Staff Röhm, in whom the *Führer* had placed a rare trust, did not combat these manifestations but unquestionably promoted them. His known unfortunate predisposition gradually became such an insupportable burden that the *Führer* of the Movement and Supreme Commander of the *SA* was driven into an extremely difficult moral dilemma.

Chief of Staff Röhm made contact with General Schleicher without the knowledge of the *Führer*. In doing so, he made use of the services of an obscure character from Berlin of whom Adolf Hitler most strongly disapproves as wellas those of another *SA* leader. Due to the fact that these negotiations—likewise, of course, without the *Führer*'s knowledge—ultimately involved a foreign power or, respectively, its respresentatives, an intervention was no longer avoidable, both from the standpoint of the Party and from the standpoint of the State.

Strategically initiated incidents culminated in the fact that the *Führer* left Westfalen after he had toured labor camps there, flying from Bonn to München at 2:00 a.m. this morning to order that the most seriously incriminated leaders be removed from office and placed under arrest. The *Führer* proceeded to Wiessee in person with a small escort in order to nip any attempt at resistance in the bud. The act of arresting the men was accompanied by such morally pitiful scenes that every trace of sympathy was necessarily banned. A number of the *SA* leaders had taken Lustknaben [lust boys] with them. One of them [Edmund Heines] was surprised in a most revolting situation and arrested.

The *Führer* issued the order to ruthlessly eradicate this plague spot. In the future he is no longer willing to tolerate that millions of decent people are incriminated and compromised by isolated persons with pathological leanings. The *Führer* issued the order to the Prussian Minister-President Göring to carry out a similar action in Berlin and particularly crack down on the reactionary accomplices to this political conspiracy.

At 12:00 noon, the *Führer* made a speech to the higher-ranking *SA* leaders who had convened in Munich, in which he stressed his unshakable alliance with the *SA,* but at the same time announced his decision to show no mercy from now on in exterminating and destroying undisciplined and disobedient characters and asocial or diseased elements. He pointed out that service in the *SA* was an honorary service for which tens of thousands of upright *SA* men had made the most difficult sacrifices. He expected from the leader of each *SA* division that he prove himself worthy of these sacrifices and be a living example to his organization. He also pointed out that he had defended Chief of Staff Röhm for years against the heaviest attacks but that the most recent development had forced him to place all personal feeling second to the welfare of the Movement and to that of the State, and that above all he would eradicate and nip in the bud any attempt to propagate a new upheaval by ludicrous circles of pretentious characters. (Dr. Max Domarus, ed, *Hitler. Speeches and Proclamations, 1932–1945: The Years 1932–1934*, pp. 472–473)

01.07.1934

According to the account included by Heinz Höhne in *The Order of the Death's Head*, p. 121, three *SS* officers arrived at Stadelhim Prison at approximately 1700 hours on that date, and demanded to see Röhm in his cell. They were *SS-Brigadeführer* Theodor Eicke (*Kommandant* of Dachau), his adjutant *SS Sturmbannführer* Michael Lippert, and *SS-Gruppenführer* Heinrich Schmauser (*Führer* of *SS-Gruppe Süd).* They were escorted to Cell 474 where Eicke confronted Röhm with the words, "You have forfeited your life. The *Führer* gives you one more chance to draw the consequences." He set a loaded revolver on a table in the cell and advised the *SA* leader that he had ten minutes to use it. The *SS* officers left the room and waited for the shot. After 15 minutes of silence, Eicke and Lippert returned to the cell with pistols drawn. Eicke shouted, "*Stabschef,* get ready." The *SS* offices each fired a shot into Röhm, who collapsed to the floor, mortally wounded and groaning, "My *Führer,* my *Führer.*" The cynical Eicke replied, "You should have thought of that earlier; it's too late now." At 1800 hours, the coup de grace was then fired into Röhm's heart by either Eicke or Lippert. A different account of the events was was presented in 1949 by Dr. Robert Koch, the former governor of the prison:

> ... Nobody was allowed to leave the prison that night either. Next morning (Sunday 1 July 1934) two *SS* men asked at the reception desk to be taken to Röhm. Zink, who was at the entrance, in view of the strict instructions he had been given, refused.

It was about 9.30 a.m. When the two tried to force their way in, Zink alerted the prison governor and the green police, who at once occupied the corridors and prevented any intrusion. The governor ascertained that neither of the *SS* men had proper authorization. It therefore took hours of telephoning to check their papers; even the *Reich* Chancellery was rung up. When at least it became clear that they had an order from Hitler, the two murderers had to be taken to Röhm in the new building.

There they handed over a Browning to Röhm, who once again asked to speak to Hitler. They ordered him to shoot himself. If he did not comply, they would be back in ten munutes and kill him … . When the time was up, the two *SS* men re-entered the cell, and found Röhm standing with his chest bared. Immediately one of them from the door shot him in the throat, and Röhm collapsed on the floor.

Since he was still alive, he was killed with a shot point-blank through the temple. The bullet not only penetrated his skull, but also the ceiling of the cell below … . (Translation from Hans-Adolf Jacobsen and Werner Jochmann, eds, *Ausgewählte Dokumente zur Geschichte des Nationalsozialismus 1933–1945, Bd. I*, in Jeremy Noakes and Geoffrey Pridham, eds, *Nazism: A History in Documents and Eyewitness Accounts, 1919–1945*, document 124, p. 180)

Eleanor Hancock writes:

Eicke told the prison officials that medical help for Röhm was forbidden. Eicke and Lippert left, taking four *SA* prisoners with them to Dachau. Prison records showed 6:00 p.m. on July 1 as the time of Röhm's death. His body was put in a coffin after twilight and buried toward 10:00 p.m. in section 1, plot number 24 at the Perlacher Forst cemetery...

On July 10, [*Gauleiter* Adolf] Wagner instructed the Munich police to exhume and cremate the corpses of the *SA* leaders buried at Perlacher Forst. The ashes were then released to their families with strict instructions as to how and when they could be buried. Röhm's ashes were buried in the same grave as his father on July 21, 1934 … (Hancock, *Ernst Röhm: Hitler's SA Chief of Staff*, pp. 161–163)

In August 1956, Bavarian police arrested Lippert and the former leader of the *Leibstandarte Adolf Hitler*, "Sepp" Dietrich (who had overseen the execution of *SA* leaders at Berlin-Lichterfelde), charging them with manslaughter for the killings of June/July 1934 (Lippert being charged specifically for the murder of Röhm). They were soon released on bail, and their trial finally began on 06.04.1957. Lippert claimed

Berlin, 31.01.1934: Röhm delivers a memorial address to *SA-Sturmführer* Hans Maikowski on the anniversary of his death. On 30.01.1933, the day of the Nazi seizure of power, Maikowski (as *Führer* of *SA-Sturm 33*) was mortally wounded by gunfire in a largely leftist area of Berlin-Charlottenburg. He died the following day. (Heinrich Hoffmann collection at NARA)

SA-Stabschef Röhm attending Party gatherings in 1933.
(Heinrich Hoffmann collection at NARA)

An autographed studio portrait of *SA-Stabschef* Röhm.
(Hermann-Historica, Auctioneers, München)

The *Stabschef* at work in 1934.
(Todd Gylsen photo)

A Hoffmann postcard of *Stabschef*
Röhm at his desk. (NARA photo)

he had waited in the corridor while Eicke (killed on the Eastern Front, 26.02.1943) committed the murder on his own. On 14.05.1957, Lippert and Dietrich were sentenced to 18 months' imprisonment, the President of the court stating that even 12 years after the war, Lippert was "filled with a dangerous and unrepentant fanaticism". In addition to Röhm, the following senior *SA* leaders were murdered during the purge:

SA Leader	Final Assignment	Execution
SA-Obergruppenführer Edmund Heines	*Führer* of *SA-Obergruppe VIII*, Breslau	30.06.1934
SA-Obergruppenführer Fritz Ritter von Kraußer	*Leiter* of the *Führungsamt der OSAF*	01.07.1934
SA-Obergruppenführer August Schneidhuber	*Führer* of *SA-Obergruppe VII*, München	30.06.1934
SA-Gruppenführer Georg von Detten	*Leiter* of the *Politischen Amt der OSAF*	02.07.1934
SA-Gruppenführer Karl Ernst	*Führer* of *SA-Obergruppe III*, Berlin	30.06.1934
SA-Gruppenführer Hans Hayn	*Führer* of *SA-Gruppe Sachsen*	30.06.1934
SA-Gruppenführer Peter von Heydebreck	*Führer* of *SA-Gruppe Pommern*	30.06.1934
SA-Gruppenführer Wilhelm Schmid	*Führer* of *SA-Gruppe Hochland*	01.07.1934
SA-Gruppenführer Konrad Schragmüller	*Führer* of *SA-Gruppe Mitte*	01.07.1934
SA-Brigadeführer Hans Koch	*Führer* of *SA-Gruppe Westmark*	01.07.1934
SA-Brigadeführer Hans Ramshorn	*Führer* of *Gruppenbefehlstellen Oberschlesien*	30.06.1934
SA-Brigadeführer Karl Freiherr von Wechmar	*Führer* of *Gruppenbefehlstellen Niederschlesien*	30.06.1934
SA-Brigadeführer Willi Klemm	*Führer* of *Gruppenstaffel Schlesien*	01.07.1934

02.07.1934	Posthumously expelled from the *NSDAP*.
03.07.1934	Issuance of the "Law Relating to National Emergency Defense Measures", which read:

> The *Reich* Government has enacted the following law, which is hereby promulgated:
>
> The measures taken on 30 June and 1 and 2 July 1934 to suppress attempts at treason and high treason are legal emergency measures in defense of the state.
>
> Berlin, 3. July 1934
>
> The *Reich* Chancellor: Adolf Hitler
>
> The *Reich* Minister of the Interior: Frick
>
> The *Reich* Minister of Justice: Dr. Gürtner (*Reichsgesetzblatt, 1934, I*, p. 529)

Published Works

Die Geschichte eines Hochverräters (memoirs, 1928)
Die nationalsozialistische Revolution und die SA (1934)

A formal portrait of the *SA-Stabschef* in full regalia, including his unique
SA-Führer dagger bearing the symbols of the *SA* and *SS* denoting his
leadership of both organizations. (Roger Bender photo)

Decorations and Awards

20.06.1916	*1914 Eisernes Kreuz I. Klasse*
00.00.191_	*1914 Eisernes Kreuz II. Klasse*
08.02.1920	*Kgl. Bayerischer Militär-Verdienstorden IV. Klasse mit der Krone und mit Schwertern*
29.12.1914	*Kgl. Bayerischer Militär-Verdienstorden IV. Klasse mit Schwertern*
00.00.1921	*Schlesisches Bewährungsabzeichen (Schesischer Adlerorden) 1. (?) Stufe* (per his own account, in *Die Geschichte eines Hochverräter*s, awarded for his role in shipping weapons and other military materials to the men of *Freikorps Oberland* who were engaged in fighting Polish insurgents in Oberschlesien)
00.00.1918	*Verwundetenabzeichen, 1918 in Silber*
00.00.1905 (?)	*Prinzregent-Luitpold-Medaille in Silber*
00.00.1933	*Goldenes Ehrenzeichen der NSDAP*
00.00.1934	*Ehrenzeichen des 9. November 1923 (Blutorden)* (Nr. 2, with effect from 09.11.1933)
00.00.193_	*Ehrendolch der SA*
00.02.1934	*Ehrenwinkel für alte Kämpfer*
26.05.1934	*Ehrenbürgerrecht der Stadt Stettin*
24.08.1933	*Ehrenbürgerrecht der Stadt Bad Reichenhall* (revoked 10.07.1947)
20.04.1933	*Ehrenbürgerrecht des Freistaates Bayern.*

08.03.1934: The *Internationale Automobil- und Motorrad-Ausstellung* (International Automobile and Motorcycle Exposition. From right to left: Wilhelm Brückner, Hitler, Göring, Röhm, Dr. Joseph Goebbels, Kurt Schmitt, Dr. Hans Heinrich Lammers, Paul von Eltz-Rübenach, and Adolf Hühnlein. (Roger Bender photo)

00.00.1933	*Ehrenbürgerrecht der Stadt München* (revoked after 1946)
00.00.1933	*Ehrenbürgerrecht der Stadt Ingolstadt*
00.00.193_	*Ehrenbürgerrecht der Stadt Magdeburg* (revoked after 1946)

Notes

• Parents:

- Father: Guido *Julius* Josef Röhm (born 08.07.1847 in Langenzenn near Nürnberg, died 03.03.1926), a *Kgl. Bayerisches Eisenbahnoberinspektor* (Royal Bavarian Senior Railway Inspector) and son of the forester Friedrich Wilhelm Bernhard Röhm (1824–1897) and his wife Meta Johanna, née Weigel (1827–1900). Eleanor Hancock writes that Julius Röhm

was ... the second oldest of eight children of an estate forester. As a middle level railway official, he had status to uphold, but little money on which to do so. His family had been officials in Thuringia, and then Franconia, for centuries. He traveled widely as a young man and made his own way in life. He eventually rose into the senior ranks of the railway service, retiring as a railway chief inspector...

In later life, Röhm indicated that he never felt particularly close to his father or brother. He described his father as strict, but conceded that- once Julius Röhm

Ernst Röhm and his mother.

realized that he achieved better without exhortation- he gave him much freedom and allowed full scope to his interests. (Hancock, *Ernst Röhm: Hitler's SA Chief of Staff*, pp. 7–8)

- Mother: Sofia Emilie, née Baltheiser (born 15.12.1857 in Wunsiedel, died of a heart ailment, 06.01.1935). Eleanor Hancock continues:

... Röhm was very close to his mother and his sister. Both in public and in private, for Röhm, his mother was "the best wife and mother in the world. As her youngest, who loves her above all, I can say no more. (ibid, p. 8, citing excerpt from Röhm's memoirs).

• Siblings:
   - Brother: Robert Heinrich Bernhard Röhm (born 29.04.1879, died 31.05.1974). He followed his father in the career of a railway worker, and like Ernst, was an army officer in World War I.
   - Sister: Meta Eleonore (Lore) Sofie (born 14.05.1880 in Schweinfurt), married to the Hofrat and Forstrat Adolf Lippert (ironically the same last name as one of the men, Michael Lippert, who murdered her brother).
• Religion: Protestant (baptized by the town curate of the Evangelical Church, Röhmeder, on 11.12.1887). Julius Baltheiser, senior district judge in Herieden and a relative of his mother, was his godfather.

• Foreign language proficiency: French, Latin, and Spanish.

Sources

Bayerisches Hauptstaatsarchiv, München, Abteilung IV Kriegsarchiv: Excerpts from various *Kriegsranglisten* containing data on the Royal Bavarian Army service of Ernst Röhm.

Bundesarchiv, Berlin-Lichterfelde (former Berlin Document Center): *Personalunterlagen von SA-Angehörigen: SA-Personalakte* of Ernst Röhm.

Campbell, Bruce: *The SA Generals and the Rise of Nazism*. The University Press of Kentucky, 1998.

Hancock, Eleanor: *Ernst Röhm: Hitler's SA Chief of Staff*. Macmillan, 2008.

Hanfstaengl, Ernst Franz Sedgwick ("Putzi"): *Hitler: The Missing Years*. Arcade Publishing, 1957.

Höhne, Heinz: *The Order of the Death's Head*. Martin Secker and Warburg, 1969.

Jablonsky, David: The Nazi Party in Dissolution: Hitler and the Verbotzeit, 1923–1925. Psychology Press, 1989.

Lang, Jochen von: *The Secretary. Martin Bormann: The Man Who Manipulated Hitler*. Random House, 1979.

Lilla, Joachim; Döring, Martin; & Schulz Andreas: *Statisten in* Uniform. *Die Mitglieder des Reichstags 1933–1945*. Droste Verlag, 2004.

Lüdecke, Kurt G. W.: *I Knew Hitler: The Story of a Nazi Who Escaped the Blood Purge*. Charles Scribner's Sons, 1937.

Office of United States Chief Counsel for Prosecution of Axis Criminality: *Nazi Conspiracy and Aggression* (11 volumes). U.S. Government Printing Office, District of Columbia, 1946.

Orlow, Dietrich: *History of the Nazi Party: 1933–1945*. University of Pittsburgh Press, 1973.

Rauschning, Hermann: *Hitler Speaks: A Series of Political Conversations with Adolf Hitler on His Real Political Aims*. Thornton Butterworth, 1940.

Röhm, Ernst: - *Die Geschichte eines Hochverräters* (memoirs). Franz-Eher-Verlag, 1928.
  - *Why SA? German Minister Ernst Röhm, Chief of Staff, Addressing the Diplomatic Corps in Berlin on December 7, 1933*. Reprint by "Sons of Liberty", a Louisiana-based American anti-Semitic organization, 1975.

Schirach, Baldur von: *Die Pioniere des Dritten Reiches*. Zentralstelle fur der deutschen Freiheitskampf, 1933.

Schuman, Frederick L.: *Hitler and the Nazi Dictatorship. A Social Pathology and the Politics of Fascism*. Robert Hale & Company, 1936.

Toland, John: *Adolf Hitler*. Doubleday, 1976.

Viktor Heinrich Lutze

Born: 28.12.1890 in Bevergern / Bezirk Münster / Westfalen.

Died: 02.05.1943 (at 2230 hours) during an operation in a Potsdam hospital, as a result of injuries sustained in an auto accident near Michendorf. This occurred at 1430 hours on 01.05.1943 as Lutze, his wife, his 17-year-old daughter Inge, oldest son Viktor Jr., and chauffeur were returning on the *Reichsautobahn* from Bevergern to Berlin. Viktor Jr. was reportedly driving and fell asleep at the wheel, the car then careening over an embankment. Jill Halcomb, referencing a 03.05.1943 press report issued by Dr. Sommer, describes the accident as follows:

> Suddenly, from the grass median several pedestrians stepped into the path of Lutze's speeding car. The driver slammed on the brakes, and apparently hit some of the pedestrians. The sudden braking caused the car to change direction and roll with great velocity down the highway for more than fifty meters. Lutze was thrown against the windshield of the car. At first Lutze was not thought to be seriously injured, but late in the day he grew increasingly worse; a blood clot had formed in his lungs. He was rushed into surgery, but Lutze died soon after. His daughter, who had been riding in the rear seat of the car, died earlier, from her severe injuries. (Halcomb, *The SA: A Historical Perspective*)

Arved Fredborg writes:

> A scandal which lately did enormous harm to the Party concerned Viktor Lutze ... On their way back to Berlin, with Lutze's son at the wheel, a pedestrian crossed the road and the son twisted the wheel to avoid hitting him. The car skidded and turned over – at 70 m.p.h .... What annoyed people most was not that the pleasure trip was made in an official car, but that, after the accident, geese, hams, eggs and packages of butter were strewn along the highway. (Fredborg, *Behind the Steel Wall: A Swedish Journalist in Berlin, 1941–43*, p. 232)

Lutze was honored with a state funeral in the new *Reichskanzlei*. In his diary entry of 08.05.1943, *Reichsminister* Dr. Goebbels wrote of Lutze's funeral:

> At noon the state funeral for Viktor Lutze was held in the new *Reichskanzlei*. It developed into a huge demonstration of the solidarity of the Party and especially of the *SA*. Never since the

Einjährig-Freiwilliger Viktor Lutze during his training at Münsterlager in 1913.

Christmas 1914. Lutze (marked "X") with his comrades early in World War I.

Viktor Lutze (left) in the trenches on the Western Front, December 1916.

Viktor Lutze in civilian attire, ca. 1924.

founding of our *Reich* have such high honors been accorded to a man at his burial as were accorded Viktor Lutze. The ceremony itself was a very dignified one. My memorial address created a deep impression. The *Führer* himself then took the floor and in a brief tribute honored Lutze. He concluded by conferring upon him the highest category of the German Order. Thereby an honor was accorded Lutze and the *SA* that far exceeds the customary. But I am glad Viktor Lutze received it. He was such a decent and fine fellow that this decoration at least had to be awarded to him posthumously. I escorted Paula Lutze and the other members of the family out of the Mosaic Hall. Outside, in the Wilhelmstraße, a very impressive funeral parade was held. (Louis P. Lochner, ed, *The Goebbels Diaries 1942–1943*)

On 07./08.05.1943, the late *SA-Stabschef*'s remains were transferred from Berlin to his home town of Bevergern for interment.

NSDAP-Nr.: 84 (First joined 21.02.1922 with a different number; Party banned following the *München-Putsch* of 09.11.1923; Reenrolled, 22.03.1926, and granted the very low membership number of 84 "in Anerkennung seiner Verdienste" [in recognition of his meritorious service])

Promotions

[12.04.1916]	*Vizefeldwebel*
00.00.191_	*Leutnant d. R.*
00.00.191_	*Oberleutnant d. R.*
00.00.1923	*SA-Mann*
00.07.1925–26.09.1925	*Gauleiter* der *NSDAP*
00.06.1926–00.00.1927(?)	*Stellvertretender Gauleiter der NSDAP*
01.03.1928	*SA-Oberführer*
14.10.1931	*SA-Gruppenführer* (reinstated at that rank, 01.07.1932, when the government ban on the *SA* was lifted)
01.01.1933	*SA-Obergruppenführer* (formally granted this rank by Adolf Hitler, 04.01.1933)
25.03.1933–28.03.1941	*Oberpräsident*
30.06.1934–02.05.1943	*Stabschef der SA*
20.07.1934–02.05.1943	*Reichsleiter* der *NSDAP*
30.01.1938	*DRK-Generalhauptführer*

Career

ca. 1897-ca. 1901	Attended *Volksschule* in Bevergern.
ca. 1901–00.00.190_	Attended the *Rektoratsschule* in Ibbenbüren.
00.00.190_-00.00.1907	Attended *Gymnasium* in Rheine.
00.00.1907-ca. 1912	Served an internship with the *Reichspost* in Rheine, assigned as a *Wochenendfahrer* (weekend driver). He then underwent training with the *Oberpostdirektion* in Münster as an *Anwärter für die gehobene Beamtenlaufbahn* (civil service candidate).

Essen, 1925: Viktor Lutze (behind Goebbels) as *Gau SA-Führer* in the Ruhr.
(*SA-Obersturmführer* Karl H. W. Koch, *Das Ehrenbuch der SA*, 1934)

Viktor Lutze and *Gauleiter* Dr. Goebbels with *SA* men in the Ruhr, 1926.

Hitler consecrates new party colors by touching them to the *Blutfahne* (blood banner, spattered with the blood of Nazi "martyrs" during the *München-Putsch*). Viktor Lutze, third from right, next to the Austrian *SA-Oberführer* Hermann Reschny, looks on. (Darrel English photo)

01.10.1912–30.09.1913	Military service as an *Einjährig-Freiwilliger* with *Infanterie-Regiment "Graf Bülow von Dennewitz" (6.Westfälisches) Nr. 55* (Höxter).
00.08.1914–00.11.1918	War service, assigned as *Zugführer*, then *Kompanieführer* in *Infanterie-Regiment 369* and *Reserve-Infanterie-Regiment 15*; final assignment as *Adjutant* of *Reserve-Infanterie-Regiment 15* in Unna. He was lightly wounded on 12.04.1916 while serving with *6. Kompanie / Infanterie-Regiment 369*, and was wounded a further three times by war's end (including the loss of his left eye to shrapnel).
00.04.1919	Discharged from military service at his own request.
00.00.1919–00.00.1921	Moved from Unna to Elberfeld, resuming work as a postal employee for the local Hauptpostamt.
00.00.1921–00.00.192_	Geschäftsführer (business manager) of the firm *Schwelmer Messingwerk GmbH*, then worked as a salesman in Elberfeld and later Hannover.
00.00.192_-00.00.1922(?)	Member of the *Deutschvölkischen Schutz-und Trutzbund*.
00.00.192_-00.00.192_	*Führer* of the *Freischar Schill* (the Elberfeld *Stoßtrupp* [shock troop] of Heinz Hauenstein's *Organisation Heinz*, which resisted the French occupation of the Ruhr in 1923).
00.00.192_-00.00.192_	Founder and leader of *Kameradschaft Schill*.

The Gasthof Märker in Hattingen, 26.11.1926: Hitler, Goebbels, Karl Kaufmann
(3), and Lutze (1)as guests at a meeting of NSDAP *Ortsgruppenführers* in
the Ruhr. Also present is future *SA-Stabschef* Wilhelm Schepmann (2) (*SA-
Obersturmführer* Karl H. W. Koch, *Das Ehrenbuch der SA*, 1934).

21.02.1922	Joined the *NSDAP / Ortsgruppe Elberfeld*. He did so against the wishes of his father.
00.00.1923	Joined the *SA* as an *SA-Mann*.
00.00.1923–00.00.1923	Participated in underground resistance to the French occupation of the Ruhrgebiet.
22.03.1925	Reenrolled in the *NSDAP / Ortsgruppe Elberfeld*.
27.05.1923	Participated as a pallbearer in the funeral, at Elberfeld, of the Nazi saboteur Albert Leo Schlageter (executed by a French firing squad at Golzheimer Heide, 26.05.1923). The other pallbearers were Erich Koch, Karl Kaufmann, Hans Hayn, and two *SA* leaders Jürgens and Hugenell.
00.07.1925 -26.09.1925	*Gauleiter* of *Gau Rheinland-Nord der NSDAP* (jointly, as *SA-Führer*, with Dr. Joseph Goebbels and Paul Schmitz). In his diary entry of 31.08.1925, Dr. Joseph Goebbels wrote: "Lutze, our storm troop leader, has become my friend. A splendid fellow."
27.09.1925–01.03.1928	*Gau-SA-Führer Ruhr*.
00.06.1926–00.00.1927(?)	Stellvertretender *Gauleiter* of *Großgau Ruhr der NSDAP*. Succeeded by Erich Koch, 00.00.1927 (?).
03.07.1926–04.07.1926	Participated in the *3. Reichsparteitag der NSDAP* in Weimar.
00.00.1927–01.03.1928(?)	*Gausturmführer Ruhr*.
01.03.1928–00.10.1930	*Selbständiger* [independent] *SA-Oberführer Ruhr* (Wuppertal-Elberfeld).

Viktor Lutze in civilian attire, ca. 1929. (Igor Karpov photo)

Hitler and Lutze inspect the *SA* in Oldenburg. From left to right: Otto Herzog, Hitler, Gerret Korsemann, Lutze, and Rudolf Hess. (*SA-Obersturmführer* Karl H. W. Koch, *Das Ehrenbuch der SA*, 1934)

The Braunschweiger Hauptfriedhof (main cemetery), 18.10.1930: Lutze and Hitler attend the memorial service for Karl Dinklage (deputy to the *Obersten SA-Führer* and *Oberster SA-Führer Nord*). (*SA-Obersturmführer* Karl H. W. Koch, *Das Ehrenbuch der SA*, 1934)

SA-Obergruppenführer Lutze in 1933. (Left: Igor Karpov – Right: *SA-Obersturmführer* Karl H. W. Koch, *Das Ehrenbuch der SA*, 1934.)

SA-Oberführer Lutze in 1931. (Baldur von Schirach, *Die Pioniere des Dritten Reiches*, 1933)

SA-Obergruppenführer Lutze and *SA-Gruppenführer* Josef Seydel accompany
Hitler during a Party function in 1933. (Roger Bender photo)

08.02.1929	Appointed to the *Stab der Obersten SA-Führung* during a conference involving Hitler and senior *SA* leaders (all of whom were likewise appointed to the *Stab OSAF*).
14.09.1930–02.05.1943	Member of the *Reichstag (Wahlkreis 23, Düsseldorf-West*; after 31.07.1932, *Wahlkreis 16, Südhannover-Braunschweig*; after 29.03.1936, *Wahlkreis 7, Breslau*).
00.10.1930–01.04.1931	*Oberster SA-Führer Nord* (Hannover).
02.04.1931–13.04.1932	*Führer* (m.d.F.b. until 13.10.1931, then permanent from 14.10.1931) of *SA-Gruppe Nord* (Hannover).
01.07.1932–14.09.1932	*Inspekteur Nord der Obersten SA-Führung (SA-Gruppen Nordmark, Nordsee, Niedersachsen,* and *Mitte*) in Hannover.
15.09.1932–30.06.1933	*Führer* of *SA-Obergruppe II (SA-Gruppen Niedersachsen, Westfalen, Niederrhein,* and *Nordsee*) in Hannover.
08.10.1932–16.02.1933	*Führer* (m.d.F.b.) of *SA-Gruppe Nordsee* (Bremen).
15.02.1933–25.03.1933	*Polizeipräsident* of Hannover. Succeeded *Oberregierungsrat* Johannes Habben. Placed "in den Ruhestand" (in retirement), 25.03.1933, with Habben again assuming the post.
01.03.1933	Entered active service with the *Deutsches Rotes Kreuz* (DRK, German Red Cross).
25.03.1933–28.03.1941	*Oberpräsident* of *Provinz Hannover* ("vertretungsweise mit der Verwaltung beauftragt" [temporarily charged with administration until 29.05.1933, then officially assigned to the post). Succeeded Friedrich Carl Ludwig von Velsen. After formal assumption of the office, he was additionally assigned as *Vorsitzender* of the *Spruchkammer für Siedlung und Auseinandersetzung der Provinz Hannover* (Arbitration board for settlement and argument in the Province of Hannover). Resigned from this post at his own request and succeeded by Hartmann Lauterbacher.
00.00.1933	Appointed as *Leiter* of the *Hannoverschen landwirtschaftlichen Berufsgenossenschaft* (Agricultural Professional Association of Hannover).
00.00.1933	Appointed as *Vorsitzender* of the *Provinzial-Beratungsstelle für Kriegerehrungen* and *Chef* of the *Wasserstraßendirektion* (directorate for canals) in Hannover.
00.00.1933	Appointed as *Leiter* of the *Vermögensverwaltung des Herzogs von Cumberland* (Administration Office for the Property of the Duke of Cumberland).
00.00.1933	Appointed as *Vorsitzender des Aufsichtsrates* of *Hannoverschen Siedlungsgesellschaft mbH*, Hannover.
01.07.1933–30.06.1934	*Führer* of *SA-Obergruppe VI (SA-Gruppen Niedersachsen, Nordsee, Niederrhein,* and *Westfalen*) in Hannover. The *SA-Gruppen Niederrhein* and *Westfalen* were transferred to *SA-Obergruppe X* in May 1934.
00.00.1933	Appointed as *Ehrenvorsitzender* (honorary chairman) of the *Kunstverein Hannover* (artists' association of Hannover).
00.00.1933	Appointed as *Staatlicher Kommissar der Technische Hochschule und der Tierärztlichen Hochschule in Hannover* (Government Commissioner of the Technical University and Veterinary College in Hannover).
30.08.1933–03.09.1933	Participated in the *5. Reichsparteitag der NSDAP* in Nürnberg.

A Heinrich Hoffmann postcard of *SA-Stabschef* Lutze. The collar insignia has been modified in this photo; Lutze was an *SA-Obergruppenführer* when it was originally taken. (NARA)

An autographed portrait of *SA-Stabschef* Lutze, dated 11.08.1935. (Hermann-Historica, Auctioneers, München)

20.08.1934: Lutze and Hitler with the film director Leni Riefenstahl during preparations for the *6.Reichsparteitag der NSDAP*.

Hitler and Lutze inspect *SA* troops in
Berlin, 1935. (Michal Sika photo)

SA-Stabschef Lutze, ca. 1935.

München, 14. July 1938: The state visit of Lutze's Italian counterpart, *Luogotenente Generale*
Luigi Russo, Chief of Staff of the *Milizia Volontaria per la Sicurezza Nazionale* (MVSN,
Volunteer Militia for National Security, or "Black Shirts") from 04.10.1935 to 29.10.1939.

An exceptional pencil portrait
of Viktor Lutze, ca. 1938.

SA-Stabschef Lutze, ca. 1938.

Viktor Lutze, ca. 1938. (NARA photo)

Heinrich Hoffmann portrait of Viktor
Lutze, ca. 1939. (NARA photo)

14.09.1933–02.05.1943	*Preußischer Staatsrat.*
23.09.1933	Attended the *Stahlhelmtag* (in honor of the *Stahlhelm* veteran's organization) in Hannover.
00.00.1934–01.04.1941	*Präsident der Provinzialrat* (President of the Provincial Council) for *Provinz Hannover.*
30.06.1934	At 0630 hours, accompanied Hitler and his entourage during the arrest of Ernst Röhm, Edmund Heines, and other *SA* leaders at the Pension Hanselbauer at Bad Wiessee. According to Lutze's recollections, Hitler called Lutze a traitor and personally arrested the *Stabschef.* Following the discovery of *SA-Obergruppenführer* Heines sharing his bed with an 18-year-old *SA-Obertruppführer*, Heines pleaded "Lutze, I've done nothing. Can't you help me?" to which Lutze replied, "I can do nothing... I can do nothing." Later in the day, at the Braunes Haus in München, Lutze attended a meeting with Hitler and other leaders concerning the anticipated fate of the arrested *SA* leaders. Rudolf Hess and Max Amann were bickering over the "privilege" of murdering Röhm. Lutze, despite his important role in bringing about the *SA* purge, had not expected things to proceed in such a violent way. He expressed his ignorance when Hitler asked which *SA* leaders were most incriminated in Röhm's alleged conspiracy, and could give no answer to the *Führer*'s question of who should be put to death. In his *Reichstag* speech of 13.07.1934, in which he offered his justifications for the purge of the *SA*, Hitler gave special recognition to those leaders of the *SA* "who made no secret of their inner disgust and reprobation [at the activities of Ernst Röhm and his followers] and were in consequence in part removed from responsible posts, in part thrust aside, and in many respects left out of at. At the head of this group of *SA* leaders, who because of their fundamental decency had been hardly treated, stood the present Chief of Staff, Lutze, and the leader of the SS, Himmler." (for further information on Lutze's role in the *SA* purge, see biography of Röhm, above)
30.06.1934–02.05.1943	*Chef des Stabes der SA.* Succeeded Ernst Röhm. As far back as 05.10.1932, Martin Bormann – in a letter to Hitler's personal secretary Rudolf Hess – had proposed Lutze as a candidate to replace the embarrassing renegade Röhm; he wrote: " ... I am convinced that the post of [*SA*] chief of staff could be filled by any *SA* leader with an understanding of people and a talent for organization (Lutze) ... " (Jochen von Lang, *The Secretary*, p. 388) On 01.07.1934, the following article appeared in a special edition of the *Völkischer Beobachter*:

> Röhm excluded from the Party and the *SA*. The *Reich* press bureau of the *NSDAP* reports the following decree issued by the *Führer*:
>
> As of today I have relieved the Chief of Staff Röhm of his position and have ejected him from the Party and the *SA*. I am appointing *Obergruppenführer* Lutze as chief of staff. *SA* Leaders and *SA* men who don't obey his orders or who oppose him will

be ejected from the *SA* and the Party and consequently will be arrested and convicted.

Signed: Adolf Hitler
*Oberst*er *Führer* der Partei und der SA

The *Führer* to the new Chief of Staff: München, 30. June 1934

The *Führer* has sent the following letter to *SA-Obergruppenführer* Lutze:

To *SA-Obergruppenführer* Lutze

My Dear *SA-Führer* Lutze!

The gravest failures on the part of my former chief of staff have compelled me to relieve him of his position. You, my dear *Obergruppenführer* Lutze, have for many years in good and bad days been an always reliable and ideal *SA-Führer*. If I am appointing you today as chief of staff it is done with the conviction that you will succeed through your reliable and obedient work in making of my *SA* the instrument which the nation needs and which I imagine. It is my wish that the *SA* be built up as a reliable and strong part of the national socialist movement. Full of obedience and blind discipline, they must help to create and form the new German citizens.

Signed: Adolf Hitler (Translation of Document 2407-PS, in *Nazi Conspiracy and Aggression, Volume VIII*)

After Lutze's death, his deputy, *SA-Obergruppenführer* Max Jüttner was designated "mit der Wahrnehmung der Geschäfte des *Stabschef*s der *SA* beauftragt", holding this post from 03.05.1943 until 17.08.1943 when Wilhelm Schepmann was appointed as Lutze's permanent successor.

In his testimony before the tribunal at Nürnberg on 13.06.1946, Jüttner spoke of Lutze's minimal power in the hierarchy of the Third *Reich*:

Chief of Staff Lutze was a *Reichsleiter* in the Party. In spite of that fact he had no influence on the Party leadership. In the last few years, already before the war, he avoided *Gau* – and *Reichsleiter* meetings. Lutze did not become a *Reich* Minister, therefore he had no influence whatsoever on the conduct of Government affairs. (*Trial of the Major War Criminals Before the International Military Tribunal, Nuremberg, 14 November 1945 – 1 October 1946, Vol. XXI*)

20.07.1934–02.05.1943	*Reichsleiter der NSDAP.*
[00.10.1934]	Member of the *Präsidialrat der Deutscher Roten Kreuz.*
26.06.1935–02.05.1943	Member of the *Akademie für Deutsches Recht*, München.

Heinrich Hoffmann portraits of Viktor Lutze, ca. 1939. (NARA photo)

SA-Stabschef Lutze in Prague, 28.10.1939. (Roger Bender photo)

05.09.1934–16.09.1934	Participated in the *6. Reichsparteitag der NSDAP* in Nürnberg. On the second day of the celebrations, Lutze declared:

> Comrades! Many of you here this night know me from those early years of our movement when I marched rank and file with you as an *SA* man. I am still very much an *SA* man today just as I was then. We *SA* men have known only one thing: loyalty, and fighting for the *Führer*!

27.01.1935	Attended the *SA und SS-Skiwettkämpfe* (skiing competition) in Garmisch-Partenkirchen.
10.09.1935–16.09.1935	Participated in the *7. Reichsparteitag der NSDAP* in Nürnberg.
00.00.1935–02.05.1943	Member of the *Große Rat der Nordische Gesellschaft* (Supreme Council of the Nordic Society).
04.04.1936	Appointed as a *Reichskultursenator*.
15.04.1936–02.05.1943	Member of the *Reichsarbeitskammer* (with membership number 3).
25.07.1936	Involved in an auto accident near Genthin.
08.09.1936–14.09.1936	Participated in the *8. Reichsparteitag der NSDAP* in Nürnberg.
22.11.1936–29.11.1936	Participated in the *4. Reichsbauerntag* (4th *Reich* Farmers Rally) in Goslar.
00.00.193_–01.04.1941	Landesführer of *Landesstelle XI des DRK* in Hannover.
06.09.1937–13.09.1937	Participated in the *9. Reichsparteitag der NSDAP* in Nürnberg. Hamilton T. Burden writes:

> [Lutze's] speech contained the party's first public attack on the church and its followers in Germany. Although his threats were by no means veiled, they were extremely vague. Christians were accused of not serving their country as devotedly as they should. A good Christian was defined by Lutze as a man who served his country well. (Burden, *The Nuremberg Party Rallies*, p. 145; speech reported in the *Völkischer Beobachter* of 12.09.1937)

27.10.1937–29.10.1937(?)	In Rome as a member of the German delegation, led by Rudolf Hess, to ceremonies marking the 15th anniversary of the fascist "March on Rome".
24.02.1938	Attended the celebration marking the anniversary of the establishment of the *NSDAP* at the Hofbräuhaus, München.
11.03.1938	Together with other senior *NSDAP* leaders, including *Reichsführer-SS* Himmler, visited *KL-Dachau*.
26.05.1938	Attended the foundation-stone laying ceremony for the *Volkswagenwerke* in Fallersleben (Wolfsburg). There he approached *General der Artillerie* Wilhelm Ulex, the commanding general of *XI.Armee-Korps*; during their private meeting, Lutze stated that the recent *SS*-engineered purging of Army chief of staff Werner Freiherr von Fritsch, on false charges of homosexuality, should be used as the basis for a *Wehrmacht* revolt against Himmler. According to Ulex's postwar recollections, Lutze

proclaimed himself and the *SA* firmly on the side of the *Wehrmacht*. Ulex advised Lutze that, provided the *SA* leader obtain proof that the charges against von Fritsch had been manufactured by the *SS*, he would convey Lutze's plan for a revolt to von Fritsch and *Generaloberst* Walther von Brauchtisch (von Fritsch's successor). One or two weeks later, *SA-Gruppenführer* Erich Reimann reported to Ulex with news that Lutze could now provide supporting documentation that the statements of von Fritsch's accusor, Otto Schmidt, had been false and the result of coercion by the *Reichsführer-SS*. When Ulex shared this information with von Brauchitsch, however, the senior officer refused to play any part, stating: "If these gentlemen want to do it, they must do it alone.'" (Höhne, *The Order of the Death's Head*, pp. 470) Interestingly in light of Lutze's attempts to forge a conspiracy with von Brauchitsch, the 03.02.1939 issue of *Der SA Mann* contained a photograph of Lutze and von Brauchitsch, reviewing a unit of the SA; it was captioned "We [the *SA*] will be the bridge between the Party and the *Wehrmacht*." Despite von Brauchitsch's failure to take action against the *SS*, Lutze continued to impress upon the Army C-in-C the dangers posed by Himmler's powerful and sinister order, making good on his vow in Stettin of August 1935. On 12.03.1940, Gottlob Berger, a staunch Himmler supporter, wrote to Himmler's chief of staff, Rudolf Brandt:

> The attitude of Lutze... is gradually becoming a danger to the SS, if not to the Party. His "communal evenings", particularly those attended by the *Wehrmacht*, are invariably used for anti-SS propaganda and in a form unworthy of any decent person.... [In the company of *Wehrmacht* officers, Lutze] adopted an intolerable attitude toward the Reichsführer.... In my view it is necessary to place Lutze under observation. (ibid, p. 471)

Lutze, who remained under constant *SS* surveillance due to his outspoken criticism of Himmler's organization, eventually turned to *Generalgouverneur* Dr. Hans Frank – incidentally the most senior *SA* man in occupied Poland – for help. Dr. Frank was having his own difficulties with the SS, and particularly with his independent-minded and brutal *HSSPF*, Friedrich-Wilhelm Krüger. Late in 1942, the *Generalgouverneur* had attempted to establish an *SA* auxiliary police force, "clearly in the hopes that the *SA* might counter-balance the *SS* and police formations" (ibid, citing 10.10.1942 letter from Krüger to Himmler). Per Himmler's instructions, Krüger refused to permit this. Other attempts by the *SA* to gain a stronger role in police and security duties in the *Generalgouvernement* and the occupied eastern territories were also halted by the *SS*. Himmler still perceived Lutze as a threat, however, particularly after learning that the *Stabschef* had gone, in February 1943, for a cure to Frank's villa and spa "Bad Kryniza"; in a letter to Martin Bormann dated 26.02.1943, Himmler wrote: "I regard

SA-Stabschef Lutze in Pommern, 06.08.1942. (Roger Bender photo)

10.09.1942: Viktor Lutze and General Enzo Galbiati (head of the MVSN) visit Führer HQ "Werwolf" at Vinnitsa in the Ukraine, escorted by *SS-Hauptsturmführer* Richard Schulze.

Lutze's visit to a spa in the *Generalgouvernement* as unfortunate; I consider it would be better if he took his cure in some German spa." In his prison memoirs, Alfred Rosenberg wrote of Lutze's conflicts with Himmler as follows:

> Lutze, staff chief of the *Sturmabteilung*, was also watching Heinrich Himmler expand his influence, not so much by means of positive accomplishments as by sly pressure made possible by executive power entrusted to him by the state. It may be that Lutze lacked the gift of finding new tasks for the slighted *Sturmabteilung* that really would have given it a new lease on life; but, historically speaking, it is also possible that the *Sturmabteilung* was so exclusively identified with our battle for power that it would have taken a long time before such new tasks could have been found for it. Many *Obergruppenführer* complained to Lutze. The sports insignia of the *Sturmabteilung* seemed inadequate, nor did the plans for a large German sports competition to be held during the Party Day at Nürnberg satisfy them. Thus the *Sturmabteilung* lost a number of its old leaders to the foreign diplomatic service, and others to the *SS*. Himmler countered Lutze's aversion, which he considered a purely personal prejudice, by stating that it was he who had suggested Lutze as chief of staff to succeed Röhm. The *Führer*, he claimed, had also offered him the leadership of the *Sturmabteilung*, but he hadn't wanted to fall heir to the leadership of an organisation that others had led during battle. Now Lutze showed his gratitude by making these unjust accusations. Later on, secure in his personal victory, Himmler declined even to see Lutze. However, this antagonism was by no means personal but, rather, a matter of principle. How justified it was, now becomes apparent at Nürnberg. When we discussed this matter, Lutze told me that every young state needed a harsh police executive which occasionally had to use methods running counter to accepted morality, quite without regard to party affiliations. Thus, it was bad for the National Socialist movement that a Party organisation should be so intimately tied through a personal union with the *Sicherheitsdienst*, the *Staatspolizei* [*Gestapo*], and the rest, and that a man like Himmler should draw these ties even tighter. This was necessarily harmful to the Party's reputation among the people, who began to identify certain necessary steps taken by the state with the party and its political identity. In addition, Lutze said, instances of vile corruption in *SS* leadership as well as indefensible chicanery on the part of the police had occurred. Just then -- and this was already during the war -- he couldn't bother the *Führer* with such things. But later on he would submit all his material to the *Führer* together with an either-or. But even noble Viktor Lutze couldn't possibly guess at that time what Himmler and Heydrich

11.09.1942: Lutze hosts General Galbiati during his visit to Berlin.

Lutze and *Reichsleiter* Dr. Ley at the *Arbeitstagung der DAF* at
Berlin's *Haus der Flieger*, 00.11.1942. (NARA photo)

Viktor and Frau Lutze attend the Summer Olympic Games in Berlin, 1936. Dr. Hans Frank and his wife are seated behind them, and Julius Schaub sits in the foreground.

were really up to. (Rosenberg, *Memoirs of Alfred Rosenberg*)

17.07.1938	With Lt. Gen. Luigi Russo, Chief of Staff of the fascist *Milizia Volontaria per la Sicurezza Nazionale* (*M.V.S.N.*, Volunteer Militia for National Security) as his guest of honor, attended the *Reichswettkämpfen der SA* on the *Reichssportfeld* in Berlin.
05.09.1938–12.09.1938	Participated in the *10. Reichsparteitag der NSDAP* in Nürnberg.
09./10.11.1938	Attended the annual commemoration of the *München-Putsch* at the Bürgerbräukeller. Dr. Goebbels announced during this meeting of "Old Fighters" that anti-Semitic outbreaks were occurring in various German cities as a "spontaneous" protest to the murder of German diplomat Ernst vom Rath in Paris. He further stated that the *Führer* had decided such demonstrations should not be discouraged by German authorities, but permitted to occur without the appearance of government or Party involvement. While many of those present raced to the telephone to relay orders for the inconspicuous organization and carrying-out of anti-Jewish agitation, Lutze took aside his *SA* leaders and "ordered them not to participate in any actions against the Jews." In many cases, Lutze's orders were ignored, with *SA* units playing a leading role in what became known as "Reichskristallnacht". It is likely that the manner in which the

Hitler gives a final salute to his *SA* Chief of Staff, 07.05.1943.

pogroms were carried out offended the disciplined officer in Lutze; he was certainly no friend to the Jews, writing (ca. 1937) that *Der Giftpilz* ("The Poison Mushroom", a virulently anti-semitic "children's book" by *Der Stürmer* editor Ernst Hiemer) was "a unique book, well calculated to clarify the Jewish question". (ibid, p. 372) In 1937, he declared:

> Der Stürmer has an essential role in seeing that each German today views the Jewish question as the crucial question of the nation, and the honor of having put racial thought in popular language. *(Der Stürmer, 1937/Nr. 11)*

01.02.1939	Attended, as Benito Mussolini's guest of honor, the ceremony marking the 16th anniversary of the establishment of the fascist MVSN militia on the Piazza Venezia in Rome.
00.02.1939	Visited the Italian colony of Libya.
17.02.1939	Accompanied Hitler during his visit to the *Internationale Automobil-Ausstellung* in Berlin.
09..06.1939–10.06.1939	Two-day visit to Danzig, during which he inspected the *SA* in that city together with *Gauleiter* Albert Forster, *Senatspräsident* Arthur Greiser, and *SA-Obergruppenführer* Heinrich Schoene.
22.09.1939–01.04.1941	Member of the *Verteidigungsausschuß* for *Wehrkreis XI*.

19.07.1940	In his *Reichstag* speech of that date, Hitler gave special recognition to various high-ranking Party leaders (Rudolf Hess, Heinrich Himmler, Konstantin Hierl, Dr.-Ing. Fritz Todt, Dr. Joseph Goebbels, and Lutze) "… who, among innumerable others, have won the greatest merit in the struggle to make it possible for the new Germany to celebrate victory." In summarizing Lutze's contributions, Hitler declared:

> Lutze, Chief of Staff of the *Sturmabteilung*, who has organized the millions of Storm Troopers in the spirit of the greatest sacrifice to the State and assured their preliminary training and their subsequent careers …

00.10.1940–00.10.1940	Visited Marienburg / Westpreußen, where he laid a wreath at the gravesite of *SA-Obergruppenführer* Joachim Meyer-Quade (killed in action during the Polish Campaign on 12.09.1939), then proceeded to *Posen* where he inspected the newly-formed *SA-Standarte "Hans von Manteuffel"* (also known as *SA-Standarte* "Posen"). He then participated in celebrations of the "Tag der Freiheit" in Litzmannstadt before proceeding to a ceremony at *Generalgouverneur* Dr. Frank's headquarters at Wawel Castle in Kraków.
15.05.1941–24.05.1941	Visit to France together with other high-ranking *SA* leaders. During this tour, he and Admiral Karl Dönitz, inspected the crew of *U-94* (commanded by *Kapitänleutnant* Herbert Otto Kuppisch, decorated with the Knight's Cross the following month) at Lorient on the French Atlantic coast.
06.06.1941	Delivered a speech during the swearing in ceremony for newly established *elsässischer SA-Einheiten* (Alsatian *SA* units) in Strassburg.
09.06.1942	Attended the funeral of Reinhard Heydrich in the *Mosaiksaal* of the *Reichskanzlei*, Berlin.
23.07.1942	Accompanied by *SA-Obergruppenführer* Max Jüttner, met with Hitler at the *Führerhauptquartier "Werwolf"* near Vinnitsa in the Ukraine.
21.11.1942	Visit to Norway as guest of the *Reichskommissar, Gauleiter* Josef Terboven, in Oslo.

Decorations and Awards

08.05.1943	*Goldenes Kreuz des Deutschen Ordens* (posthumous award)
00.00.191_	*1914 Eisernes Kreuz I. Klasse*
00.00.191_	*1914 Eisernes Kreuz II. Klasse*
31.03.1941	*Kriegsverdienstkreuz I. Klasse mit Schwertern*
00.00.1940 (?)	*Kriegsverdienstkreuz II. Klasse mit Schwertern*
19.01.1919	*Fürstlich Lippisches Kriegsehrenkreuz*
30.05.1919	*Fürstlich Lippisches Offiziersehrenkreuz mit Schwertern*
00.00.1918	*Verwundetenabzeichen, 1918 in Silber*
ca. 1939	*Medaille zur Erinnerung an den 1. Oktober 1938*
ca. 1938	*Medaille zur Erinnerung an den 13. März 1938*
ca. 1934	*Ehrenkreuz des Weltkrieges 1914–1918 mit Schwertern*

00.00.1933	*Goldenes Ehrenzeichen der NSDAP*
17.05.1939	*Goldenes Hitler-Jugend Ehrenzeichen mit Eichenlaub* (presented by *Reichsjugendführer* Baldur von Schirach)
30.01.1942	*Dienstauszeichnung der NSDAP in Gold*
30.01.1940	*Dienstauszeichnung der NSDAP in Silber*
30.01.1940 (?)	*Dienstauszeichnung der NSDAP in Bronze*
00.00.1929	*Nurnberger Parteitagsabzeichen 1929*
00.00.1931	*Abzeichen des SA-Treffens Braunschweig 1931*
00.00.1933	*Schlageter Gedächtnisabzeichen* (Nr. 7)
00.00.193_	*Ehrenzeichen des Deutschen Roten Kreuzes I. Klasse*
20.10.1940	*Kreuz von Danzig I. Klasse*
24.08.1936	*Deutsches Olympia-Ehrenzeichen I. Klasse*
13.12.1939	*Deutsche Schutzwall-Ehrenzeichen*
ca. 1936	*Ehrenplakette für die Mitglieder des Reichskultursenats*
00.00.193_	*Deutsche Reiterabzeichen in Bronze*
31.03.1941	*Dankschreiben des "Führers"* (Letter of Appreciation from the *Führer,* Adolf Hitler)
28.12.1940	*Ehrendolch des Deutschen Heeres* (Honor Dagger of the German Army; presented on the occasion of his 50th birthday by *Generalfeldmarschall* Walther von Brauchitsch)
00.00.1934	*Ehrendolch der SA* (mit Wirkung vom 03.02.1934)
28.12.1938	*Luftwaffen-Zivilabzeichen*
ca. 1932	*Tyr-Rune*
00.02.1934	*Ehrenwinkel für alte Kämpfer*
26.06.1937	*Ehrenbürgerrecht der Stadt Göttingen* (title revoked by decision of the "Hauptausschuß der Stadt Göttingen", 25.08.1952)
08.10.1934	*Ehrenbürgerbrief der Stadt Bevergern*
00.00.1937	Order of Saints Maurice and Lazarus, Grand Cross (*Gran Croce dell'Ordine dei Santi Maurizio e Lazzaro*)
01.11.1937	Honor Dagger of the *Milizia Volontaria per la Sicurezza Nazionale* (M.V.S.N., Volunteer Militia for National Security)
00.00.193_	*Österreichische Kriegserinnerungsmedaille*
16.07.1918	Commander's Cross of the Royal Bulgarian Military Service Order with War Decoration
30.09.1940	Grand Cross of the Royal Bulgarian Order of Civil Merit

Notes

- Son of the master shoemaker and smallholder August Lutze (born 00.00.1859, died 00.00.1947).
- Married on 09.11.1918 to Paula Lehmann (born 07.09.1900, died 23.06.1953). From the marriage of Viktor and Paula Lutze were born two sons (Viktor, Jr., born 29.02.1924 in Wuppertal-Elberfeld, killed in action as a *Leutnant* in Normandy, 20.06.1944, and Addi, born 23.07.1936, killed in a car crash on 17.08.1957) and one daughter (Inge, born 23.11.1925, killed in the accident that mortally injured her father on 01.05.1943). During a private conversation on the night of 03./04.01.1942, Hitler made the following remarks concerning Viktor Lutze, Jr.:

Young Lutze has gone off to the front as a volunteer. Let's hope nothing happens to

Viktor Lutze and his family, ca. 1935.

him. He's truly a pattern of what a young man should be—perfect in every way. When he has had a long enough period of training at the front, I'll take him onto my staff. He has plenty of breeding. On one occasion, Inge and he had come to Obersalzberg. They must have been thirteen and fourteen years old. Inge had done something that was not too well-behaved, no doubt. He turned to us and made the observation: "What young people are coming to, nowadays!" (Dr. Henry Picker, ed, *Hitler's Table Talk 1941–1944*)

Sources

Burden, Hamilton T.: *The Nuremberg Party Rallies*. Frederick A. Praeger, Publishers, 1967.

Campbell, Bruce: *The SA Generals and the Rise of Nazism*. The University Press of Kentucky, 1998.

Fredborg, Arvid: *Behind the Steel Wall: A Swedish Journalist in Berlin 1941–43*. Viking Press, 1944.

Gisevius, Hans Bernd: *To the Bitter End*. Houghton Mifflin Company, 1947.

Halcomb, Jill: *The SA: A Historical Perspective*. Crown/Agincourt Publishers, 1985.

Höffkes, Karl: *Hitlers politische Generale: Die Gauleiter des Dritten Reiches*. Grabert-Verlag-Tübingen, 1986.

Höhne, Heinz: *The Order of the Death's Head*. Martin Secker and Warburg, 1969.

Lang, Jochen von: *The Secretary. Martin Bormann: The Man Who Manipulated Hitler*. Random House, 1979.

Lilla, Joachim; Döring, Martin; & Schulz, Andreas: *Statisten in Uniform. Die Mitglieder des Reichstags 1933–1945*. Droste Verlag, 2004.

Orlow, Dietrich: *History of the Nazi Party: 1933–1945*. University of Pittsburgh Press, 1973.

Rosenberg, Alfred: *Memoirs of Alfred Rosenberg*. Ziff-Davis, 1949.

Schirach, Baldur von: *Die Pioniere des Dritten Reiches*. Zentralstelle fur der deutschen Freiheitskampf, 1933.

Toland, John: *Adolf Hitler*. Doubleday, 1976.

Max Paul Wilhelm Werner Jüttner

SA-Obergruppenführer

02.05.1943–08.08.1943	mit der *Wahrnehmung der Geschäfte des Stabschefs der SA beauftragt*

Born:	11.01.1888 in Saalfeld/Saale / Thüringen.
Died:	14.08.1963 in München / Bayern.

NSDAP-Nr.:	2 039 331 (First joined 28.01.1923 with one of the following Party Nr's: 19 487, 19 499, or 19 500; Party banned following the *München-Putsch* of 09.11.1923; Reenrolled, 01.07.1933, with effect from 01.05.1933)

Promotions

22.03.1906	*Fahnenjunker*
18.08.1907	*Leutnant* (mit Patent vom 14.02.1906 A6a)
00.10.1914	*Oberleutnant*
18.12.1915	*Hauptmann*
01.11.1933	*SA-Brigadeführer*
20.04.1935	*SA-Gruppenführer*
09.11.1937	*SA-Obergruppenführer*
00.00.19__	*Major z. V.*
02.05.1943–18.08.1943	*mit der Wahrnehmung der Geschäfte des Stabschefs der SA beauftragt*

Career

00.00.1894–00.00.1897	Attended *Vorschule* in Saalfeld,
00.00.1897–*00.03.1906*	Attended a *Reformgymnasium* in Saalfeld (passed his *Abitur*, 00.03.1906).
22.03.1906–*00.00.191_*	Entered service as a *Fahnenjunker*, assigned to *2.Thüringische Feldartillerie-Regiment Nr. 55* (Naumburg/Saale).
00.10.1906–00.06.1907	Attended the *Kriegsschule* in Glogau.
[00.00.1914]-10.12.1914(?)	Adjutant of *II.Abteilung / Feldartillerie-Regiment 55*. Deployed to the front from 01.08.1914.
10.12.1914–25.08.1916	Assigned to the staff of *4.Landwehr-Division*.
26.08.1916–06.12.1916	Assigned to the *Generalkommando* (Corps HQ) of the *Landwehr-Korps*.
07.12.1916–06.03.1917	Assigned to the staff of *225. Infanterie-Division*.
13.07.1917–00.00.1917	Assigned to the *Generalstab der Armee*.
00.00.1917–03.08.1917	Assigned to the staff of *Armee Woyrsch*.
04.08.1917–20.11.1917	Attached to the staff of the German *Gouverneur* in Metz.
21.11.1917–08.01.1918	Attached to the *Generalkommando of XIII.* and *XIV.Armee-Korps*.
09.01.1918–20.09.1918	Attached to *Armeeoberkommando 6 (AOK 6)*.
21.09.1918–09.12.1918	*1.Generalstabsoffizier (Ia)* of *2.Infanterie-Brigade*.
09.12.1918–00.01.1919(?)	*1.Generalstabsoffizier (Ia)* of *119.Infanterie-Division*.
00.01.1919–01.08.1919	*Führer* of the *Zeitfreiwilligen-Verband Mitteldeutschland* of the *Landesjägerkorps* in *Regierungsbezirk Halle-Merseburg*.
01.08.1919	Discharged from army service.

SA-Obergruppenführer Max Jüttner, ca. 1938. (Igor Karpov photo)

00.00.1919–00.00.1921	Studied law and economics (5 semesters) at the Universities of Halle and Jena.
01.09.1920–00.00.1921	Coalminer in the lignite mines of Halle.
15.09.1919–31.10.1933	Member of the *Stahlhelm. Bund der Frontsoldaten / Ortsgruppe Halle/ Saale.*
00.00.1920–00.01.1923	Member of the *Deutschnationale Volkspartei* (*DNVP*).
00.00.1921–30.11.1933	Employed- first as a clerk, later as *Abteilungsleiter (Prokurist) für Sozial- und Wirtschaftspolitik-* with the *Deutschen Braunkohlen-Industrie-Verein* in Halle, and ultimately as *Prokurist* of *Anhaltische Kohlenwerke.*
30.01.1923–12.04.1933(?)	*Führer* of the *Ortsgruppe Halle/Saale* of the "Stahlhelm"-Bund.
12.04.1933–31.10.1933	Landesführer of the *Landesverband Mitteldeutschland der "Stahlhelm".*
01.11.1933	Joined the *SA* at the request of Ernst Röhm.
00.05.1921–00.06.1921	*Führer* of the *Studentenbataillon Jena* in Oberschlesien, with which he participated in fighting against Polish insurgents.
28.01.1923	Joined the *NSDAP / Ortsgruppe München.*
00.01.1923	Participated in the *1. Standartenweihe der NSDAP* (first ceremonial consecration of Nazi Party flags).
30.01.1923–31.10.1933	*Gauführer* of the "Stahlhelm"-Bund in *Gau* Halle.
00.00.1926–00.00.1929	Member of the *Provinziallandtag* for *Provinz Sachsen.*
00.00.19__-31.10.1933	*Stellvertreter des 2. Bundesführers des "Stahlhelm"* (deputy to the 2nd *Bundesführer* of the *Stahlhelm-Bund*, Theodor Duesterberg).
12.04.1933–31.10.1933	*Landesführer des "Stahlhelm"-Bund in Mitteldeutschland.*
01.07.1933	Reentered the *NSDAP / Ortsgruppe Halle an der Saale* (transferred to the *Ortsgruppe Braunes Haus der NSDAP* on 01.12.1935).
01.11.1933	Joined the *SA.*
01.11.1933–30.11.1933	*Wehrstahlhelmführer der SA* (m.d.W.d.G.b.).
01.11.1933–30.11.1933	*Beauftragter der Wehrstahlhelm* to *SA-Obergruppe IV* (comprising *SA-Gruppen Sachsen* and *Thüringen*) in Halle an der Saale.

12.11.1933–08.05.1945	Member of the *Reichstag (Wahlkreis 11, Merseburg)*.
01.12.1933–20.09.1934	Appointed as a *hauptamtlicher SA-Führer* and assigned as *Leiter* of the *Abteilung Ausbildung und Organisation* (Department for Training and Organization) in the *Führungsamt im Stab der Obersten SA-Führung* (München). In testimony before the International Military Tribunal at Nürnberg on 13.08.1946, Jüttner stated:

> My appointment into the *SA* leadership was connected with the incorporation of the *Stahlhelm* into the *SA*. The Central German *Stahlhelm* enjoyed a good reputation even among its political opponents. My especially good relations with the miners and also with the trade unions were well known to Röhm. The Central German *Stahlhelm* was especially successful in the social field. All this might have contributed to my appointment. I left the mining industry voluntarily and became a professional *SA Führer* ... (*Trial of the Major War Criminals Before the International Military Tribunal, Nuremberg, 14 November 1945 – 1 October 1946, Vol. XXI*)

01.12.1933–20.09.1934	*Verbindungsführer der SA zur Reichswehr* (Head of Liaison between the *SA* and the *Reich* Armed Forces).
27.07.1934–01.11.1937	*Chef der Führungsamt* (Chief of the Leadership Office) in the *Obersten SA-Führung* (m.d.W.d.G.b. until 19.09.1934, then permanent from 20.09.1934). Upgraded to *Hauptamt* status, 01.11.1937.
30.04.1936–00.09.1936	Charged with the *Gesamtleitung des Aufmarsches der SA, SS, NSKK und Flieger beim Reichsparteitag der Ehre 1936* in Nürnberg.
00.03.1937–00.09.1937(?)	Charged with *Leitung einschließlich der Festsetzung der Stärken der Marschsäulen sowie der Kommandoführung beim Appell der SA, SS, des NSKK und des DLV* (Management, including the fixing of the strengths of the marching columns and the command leadership, at the roll call of the *SA, SS, NSKK,* and *DLV*) for the *9. Reichsparteitag der NSDAP* in Nürnberg.
01.11.1937–08.05.1945	*Chef* of the *Führungshauptamt* in the *Obersten SA-Führung*.
[00.00.1938]	*Abwehrbeauftragter der Obersten SA-Führung*.
05.03.1938–08.05.1945	*Führer* of the *Deutsche Schützenverband* in the *Nationalsozialistischer Reichsbund für Leibesübungen* (*NSRL*, National Socialist *Reich* League for Physical Exercise under *SA-Obergruppenführer* Hans von Tschammer und Osten).
01.10.1938–30.10.1938	*Verbindungsführer der SA* to the *Sudetendeutsche Freikorps*.
07.02.1939–00.00.1939	Member of the *Deutsche Olympische Ausschuß* (German Olympic Committee) for Germany's participation in the planned 1940 Olympic Games in St. Moritz and Helsinki.
15.06.1939–08.05.1945	*Stabsführer* of the *Obersten SA-Führung* and *ständiger Stellvertreter des Stabschefs der SA* (Permanent Deputy of the Chief of Staff of the SA). Succeeded Otto Herzog.

November 1933: Newsreel stills of Max Jüttner during the ceremonial transfer of his men from the *Stahlhelm-Bund* to the *SA*.

02.05.1943–08.08.1943	*Mit der Wahrnehmung der Geschäfte des Stabschefs der SA beauftragt.* Assumed the duties of *SA-Stabschef* following the death of Viktor Lutze, and until the appointment of a permanent *Stabschef* (Wilhelm Schepmann).
00.11.1944–00.05.1945	*Mit Führungsaufgaben beim Aufbau des Volkssturms beauftragt* (Charged with managerial responsibilities for the creation of the German People's Army).
00.04.1945–00.04.1945	*Führer* of the *Volkssturm-Kampfgruppe "Jüttner"*, numbering approximately 160 men, in München.

Postwar Confinement and Activities

Arrested by U.S. Army troops while commanding his *Volkssturm* unit in the Oberhaushammer Hütte south of Schliersee, 11.05.1945. He was then held successively in the camps of Bad Aibling, Neu-Ulm, Heilbronn, Ludwigsburg, Camp 74, Seckenheim and Kornwestheim (Camp 75). He lost 65 lbs while in U.S. custody. From 13.08.1946 to 16.08.1946, he appeared before the International Military Tribunal at Nürnberg as a defense witness on the subject of the *SA*. The following are excerpts from his testimony:

13.08.1946:

JÜTTNER: I know the Leadership of the *SA,* its aims, and the *SA* leaders, especially the higher *SA* leaders, very well. I do not propose to gloss over anything. A small fraction of *SA* leaders who had turned out to be mere troopers was eliminated. Even those *SA* leaders had in the past, during the first World War as brave soldiers, and later as members of the Free Corps under the governmeht of Ebert and Nosske, deserved well of their country. Their attitude and their way of life, however, were opposed to the principles of the *SA,* therefore they had to leave. But the rest, that is the bulk of the *SA* Leadership Corps, were decent and clean, and irreproachable in their sense of justice and duty.

HERR BÖHM: Tell us about the professional Leadership Corps.

JÜTTNER: As to the active leaders, the *Obergruppenführer* and the *Gruppenführer,* I know their history, their way of life and their political and ethical attitude. Apart from the insignificant number who had to leave, these *SA* leaders were irreproachable. Not one of them had a police record, not one of them was what one might call a failure, all of them had a civil profession before they were taken into the Leadership Corps of the *SA.* Their way of life was simple and modest. They received, however, in relation to comparable positions of civil servants or business men, extremely low salaries. All incomes from other sources were charged against them; there was no one in the *SA* who was allowed more than one source of income; no one could enrich himself personally owing to his position, and only he could spend money on social activities who had means of his own. Of the *Gruppenführer* and *Obergruppenführer* who in 1939 were active in the *SA* Leadership Corps or with the *SA* Gruppen, half the number lost their lives in the war. They gave their lives in the belief that they had fought for a just cause. They were patriots, and they committed no wrong or ungodly acts. And even today, I pride myself on having belonged to such an upright leadership corps.

HERR BÖHM: Were the *SA* leaders paid?

JÜTTNER: Up to 1933 there were no paid *SA* leaders. Only the leaders of the so-called *Untergruppe*n, of which there was one in each *Gau*, received a remuneration of about 300 marks a month. After 1933 a wage scale was established. In 1940 there was a small increase in pay. The maximum basic salary for an *Obergruppenführer* was 1,200 marks a month. From Scharführer up to *Obersturmbannführer* inclusive, all *SA* leaders, with the exception 'of the auxiliary personnel, were honorary workers. Of the entire Leadership Corps, including the nominal leaders, roughly two percent were paid.

HERR BÖHM: How was the *SA* Leadership Corps organized?

JÜTTNER: In the *SA* we differentiated between: *SA* leaders, *SA* administrative leaders, *SA* medical leaders. The *SA* leaders formed the leadership staffs and led the units. The *SA* administrative leaders handled the budget, financial matters, and the audit. Together with the administrative leaders of the other branches and of the Party they formed a special leadership body and had to follow the directives of the *Reich* Treasurer. The medical leaders were physicians and pharmacists; they were charged with the medical care of the *SA.* The administrative and medical leaders had no influence whatsoever on the running of the *SA,* and they had no right to that. Besides, the *SA* had leaders for special purposes, the so-called "z. V." leaders and honorary leaders, some of whom are among the main defendants here.

HERR BÖHM: Was not one of the main defendants an honorary leader?

JÜTTNER: Yes, I believe several of them were honorary leaders, such as Göring, Frank, Sauckel, Von Schirach, Streicher, and, to my knowledge, perhaps Hess and Bormann. I might add in this connection that the honorary leaders were never informed about the business

SA-Brigadeführer Jüttner, 1934. (Igor Karpov photo)

affairs of the *SA*. They had neither the opportunity nor the authority to exert any influence on training, leadership, or use of the *SA*. They had merely the right to wear the *SA* uniform and, at meetings and festivities, to take their positions in the ranks of the *SA* leadership. Even Hermann Göring-who in 1923 headed the *SA* temporarily when it numbered but a few thousand men-no longer exerted any influence on the *SA* after that time, nor did he have any time to do so. His nomination as chief of the "Standarte Feldherrnhalle" was only a formal honor, similar to the honors that were extended in the days of the Kaiser to military leaders of merit, or members-even feminine members- of royal families. Herr Frank was appointed leader of the *SA* for the former Government General by Chief of Staff Lutze. That too was and remained only a formal honor, because the administration itself was carried out by a special administrative staff under *Brigadeführer* Peltz, and later Kühnmunde. He did not receive any orders concerning the administration of the *SA* in that region from the Chief of Staff. Such orders went to the administrative staff who, in turn, were responsible to the Supreme *SA* Leadership. The 'ZV' (*z. V.*, leaders for special purposes whom I have mentioned could temporarily be called in for duty if they were willing. They were advisory duties, for example on legal and social questions.

HERR BÖHM: Of what types of people did the *SA* in general consist?

JÜTTNER: From the beginning, the *SA* was made up of former soldiers of the First World War, that is, soldiers and young idealists who loved their country above all. The *SA* was not, as the witness Gisevius asserted, a mob of criminals or gangsters, but rather, as Sinclair Lewis is said to have written, pure idealists. Many clergymen, many students of theology, belonged to the *SA* as active members, some until the very end. Each and every *SA* man will be able to confirm that never at any time were criminal actions demanded of him, and that the *SA* leadership never pursued criminal aims"

[…]

HERR BÖHM: What was the opinion and the attitude of the *SA* on the Jewish question?

JÜTTNER: The *SA* demanded that the influence of the Jews in national affairs, in the economy, and in cultural life, be reduced in accordance with their position as a minority in Germany. It advocated a numerus clausus.

HERR BÖHM: And what was the reason for this demand or this attitude?

JÜTTNER: This demand, which was not only that of the *SA,* became general in Germany when after the first World War, in 1918 and 1919, great numbers of Jewish people emigrated from Poland to Germany and entered into the economic and othy spheres of life, where they gained considerable influence in an undesirable manner. Through certain large judicial proceedings all this profiteering and this disintegrating influence had become known, and it caused much ill-will and resulted in a movement of opposition. Even Jews who had lived in Germany for a long time, and societies of German citizens of the Jewish faith, took position against these influences in a decided manner. So one can readily see that the demand of the *SA* was well-grounded.

HERR BÖHM: Did the *SA* incite, others to active violence against the Jews?

JÜTTNER: No, in no way. Never did, the Chiefs of Staff, Röhm, Lutze, or Schepmann treat the Jewish question in their speeches, or issue any directives in that respect, much less incite others to violence. The concept of a so-called "master race" was never fostered in, the SA; that would have been quite contrary to reason, for the *SA* received its replacements from all strata. The extermination of a people because of its type was never given any support by the *SA,* and actions of violence against Jews were not favored by the *SA.* Quite the contrary, the leadership always objected most strongly to actions of that kind.

14.08.1946:

HERR BÖHM: Witness, yesterday we left off in your examination with the manner in which the Jewish question was handled by the *SA.* Now I should like to ask you how the participation of members of the *SA* in actions lagainst the Jews in November 1938 can be explained?

JÜTTNER: The participation of *SA* members in this action consisted of irresponsible actions by individuals which were in gross contradiction to the directive of Staff Chief Lutze's executives. Staff Chief Lutze was in Munich in the old city hall. There, in connection with the speech made by Dr. Goebbels, he immediately assigned the chief of the administrative office, *Obergruppenführer* Matthes [sic, Mappes], to go to the Hotel Rheinhof, where a part of the *SA* leaders present had already retired, in order to give these *SA* leaders strict orders not to participate in any action against the Jews. About an hour later, when he received the news that the synagogue in Munich had been set on fire, Lutze, in my presence, repeated this order to the *SA* leaders who were still present in the Munich city hall and said that it was to be passed on to all units immediately. This was actually done, which is confirmed by the fact that in many places no actions were carried out at all, and numerous *SA* men state under oath that they received this order.

HERR BÖHM: Then how did it come about that, in spite of that, members of the *SA* participated in the destruction of Jewish establishments?

Hans Jüttner in 1938 as *Chef* of the *SA Führungshauptamt* with two of his principal assistants, *SA-Brigadeführers* Rolf Michaelis and Walter Nibbe (heads of the *Amt Körperliche Ertüchtigung* [Office of Physical Training] and *Amt Organisation und Einsatz* [Office of Organization and Operations], respectively).

JÜTTNER: As was ascertained afterwards, certain individuals let themselves be misled by agencies which were undoubtedly under the influence of Dr. Goebbels. As an actual fact, compared with the *SA,* relatively few real members of the *SA* participated in this action, although public opinion later blamed the *SA* for this entire action. And here again it so happened that everyone in a brown shirt was considered an *SA* man. That the *SA* was in no way the sponsor of this action may also be seen from the fact that, as I have read in the press in the last few months, in certain trials, for example in Bamberg, Stuttgart, and, I believe, in Hof, people were convicted who had destroyed synagogues and yet did not belong to the *SA*. The fact also that in many places *SA* men upon instructions from the leadership offered to afford protection to Jewish installations against plundering by shady elements, et cetera, created a popular impression that the *SA* had committed these misdeeds. In any event, Staff Chief Lutze one or two days later gave voice to his indignation to Dr. Goebbels about the action itself and the unjustified accusation against the *SA,* and strongly condemned the irresponsible way in which the *SA* 'men had been incited to1 commit these misdeeds. Soon after he issued an order that in the future *SA* men were not to place themselves at the disposal of other agencies for any tasks or actions unless he himself had given express approval. Staff

SA-Obergruppenführer Jüttner, ca. 1939.

Chief Lutze punished the guilty ones whom he discovered, and if the case warranted it, they were turned over to the regular courts for judgment.

HERR BÖHM: Had things been different up to that time when Lutze took this particular line? Was the Political Leadership in a position to use *SA* members for its own purposes?

JÜTTNER: The Political Leadership only had authority to use the *SA* for certain tasks, which included the following: participation in *Gau* and *Kreis* rallies; demands for the use of the *SA* in cases of disaster, and also for propaganda purposes; for collection drives for the Winter Relief, for collecting clothing and the Like. These were the usual demands which the Political Leadership made on the *SA* in the course of the year. So far as I know, at no time did the Political Leadership make any other demands of an illegal nature of the *SA*. But Lutze issued this order to prevent those offices which were under Dr. Goebbels' influence from leading *SA* men astray in the future.

[...]

HERR BÖHM: Did the *SA,* leadership have any influence on politics?

JÜTTNER: After the death of Röhm none at all. The *SA* was completely unsuited for exerting any influence on politics, both by

its organization and its leadership. Even the misuse of the *SA* for war-mongering purposes was quite out of the question. Militarism such as the glorification of military activities, uniforms, drilling, jingoism, or the creation of a warlike spirit, was never approved by the SA; Röhm's attitude toward neighboring countries and Lutze's attitude toward war in general, in themselves speak for that.

[...]

HERR BÖHM: ... Herr Jüttner, now I should like to put: my final question to you. Did the political aims of the *SA* have a criminal character?

JÜTTNER: The things which the *SA* did and the aims which its leaders pursued need never fear the light of day. The *SA* leadership did not pursue any criminal aims and did not even know of any criminal aims of any other agencies. The *SA,* as an organization, never carried out any actions which could justify its defamation as a criminal organization. The *SA,* Mr. President, had many followers in the *Reich*, that is, in the former *Reich*, and even beyond its boundaries. The *SA* had opponents as well. Many of these opponents raised their voices, and out of hate or envy created prejudices against the *SA*. Not the truth-only prejudices of the kind which, as is well known in history, have caused the downfall of many a brave man, could lead to a situation where five or six million men who belonged to the *SA* in the last two and a half decades would be stamped as criminals ... . For these men, for these five or six million men and for the many millions in their families, I can solemnly declare under oath that the *SA* never had a criminal character. Mr. President, my entire life has been guided by the rule that one should stand by whatever one has done, whatever the risk may be, and fear nothing, not even death itself, save only dishonor.

I consider it to be dishonorable to evade responsibility by putting an end to one's Life, or to become untruthful. In this respect, Mr. President, my consciencel is clear. Therefore, with my declaration on the blamelessness of the *SA* I can stand in front of the Highest Judge."
15.08.1946:
JÜTTNER: I shall be very brief, Your Lordship. To conclude the questions put to me I should like to assure you upon my oath that we of the *SA* did not do anything bad. We did not want a war and we did not prepare for a war. We of the *SA,* the leadership and the

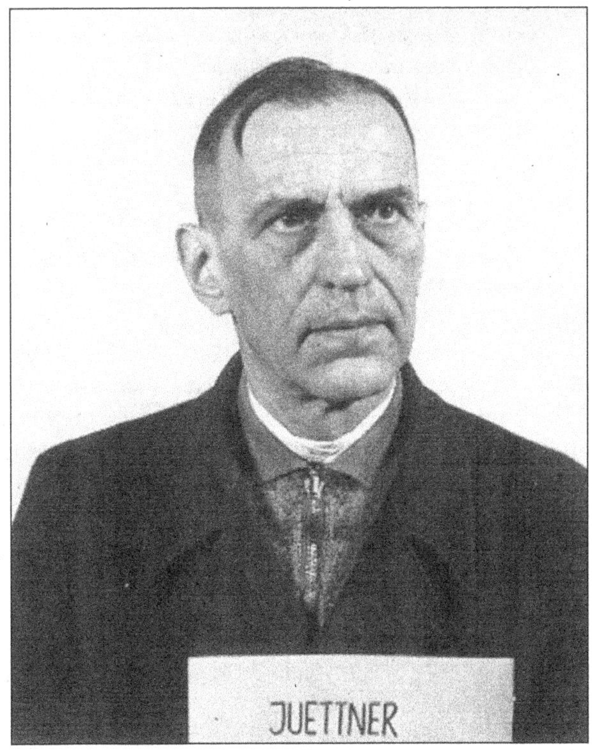

Witness identification photo of Max Jüttner, taken in Nürnberg in 1946. (Ian Sayer photo)

organization itself, did only those things which in other countries are expected of the men of the nation as their moral duty, which Mr. Truman or Marshal Stalin or the statesmen of England and France expect of their men, namely, to do everything to protect the home country and to maintain peace. We of the *SA* did not commit any crimes against humanity, either. The leadership did not decree them, nor did they tolerate them, nor allow the organization to become guilty of any of them. When individuals committed misdeeds they should be punished and it is our will, too, that they should be brought to just punishment. We therefore do not ask for mercy or sympathy by portraying our domestic distress. We ask only for justice; for nothing else, for our conscience is clear. We acted as patriots. If patriots are to be labelled as criminals, then we were criminals.

Jüttner later returned to work in the mining industry, and by 1957 was working as a trade representative in München-Solln. In May of that year he appeared as a witness in the trial of "Sepp" Dietrich for overseeing the killing of *SA* leaders during the "Röhm-Putsch".

Decorations and Awards

18.06.1915	*1914 Eisernes Kreuz I. Klasse*
01.10.1914	*1914 Eisernes Kreuz II. Klasse*
12.07.1918	*Bayerisches Militar-Verdienstkreuz IV. Klasse mit Schwertern*
15.12.1916	*Hamburgisches Hanseatenkreuz*
20.03.1915	*Sachsen-Meiningensches Ehrenkreuz fur Verdienste im Kriege*

09.03.1915	*Osterreichisches Militar-Verdienstkreuz III. Klasse mit Kriegsdekoration*
13.06.1918	*Verwundetenabzeichen, 1918 in Schwarz*
22.03.1935	*Ehrenkreuz des Weltkrieges 1914–1918 mit Schwertern*
11.01.1938	*Goldenes Ehrenzeichen der NSDAP*

Notes

- Son of the printing company owner August Jüttner (born 02.01.1838 in Stendal / Altmark, died 26.10.1903) and his wife Anna, née Franke (born 16.06.1856 in Saalfeld, died 04.01.1931, daughter of the tannery owner Hermann Franke).
- Religion: Lutheran.
- Married on 10.02.1913 to Erna Nies (born 23.11.1899 in Zobtau). One son (Klaus, born 29.10.1915, killed in Stalingrad, 21.01.1943) two daughters (born 24.02.1914 and 01.09.1920) were born to this marriage. Both of his sons-in-law lost their lives in World War II (Walther Rohde, born 18.10.1906 in Loitz, killed in action as a *Leutnant* 2.5 km. west of Ludwigshöhe near Salla / Karelia, and ______ Wiedemann, killed in action in Budapest, 09.11.1944).

Sources

Bundesarchiv, Berlin-Lichterfelde (former Berlin Document Center): *Personalunterlagen von SA-Angehörigen: SA-Personalakte* of Max Jüttner.

Campbell, Bruce: *The SA Generals and the Rise of Nazism.* The University Press of Kentucky, 1998.

Lilla, Joachim; Döring, Martin; & Schulz Andreas: *Statisten in Uniform. Die Mitglieder des Reichstags 1933–1945.* Droste Verlag, 2004.

Wilhelm Hans Schepmann

18.08.1943–08.05.1945	*Stabschef der SA* [m.d.F.d.G.b. until 09.11.1943]

Born:	17.06.1894 in Baak bei Hattingen / Ruhr.
Died:	26.07.1970 in Gifhorn / Landkreis Gifhorn / Niedersachsen.

NSDAP-Nr.:	26 762 (First joined 00.00.1922; Party banned following the *München-Putsch* of 09.11.1923; Reenrolled, 28.12.1925)

Promotions

00.01.1915	*Leutnant d. R.*
00.00.1931	*SA-Oberführer*
01.11.1932	*SA-Gruppenführer*
09.11.1936	*SA-Obergruppenführer*
ca. 1940	*Hauptmann d. R.*
18.08.1943–09.11.1943	*mit der Führung der Geschäfte des Stabschefs der SA beauftragt*
09.11.1943–08.05.1945	*Stabschef* der SA

Career

ca. 1900-ca. 1904	Attended *Volksschule*.
ca. 1904-ca. 1912	Attended *Gymnasium*.
ca. 1912–00.00.1914 (?)	Attended a *Lehrerseminar* (teachers' college).
00.00.1914–00.11.1918	War service with *7. Westfälische Jäger-Bataillon* and *Infanterie-Regiment 56*. After commissioning in January 1915, assigned successively as *Bataillonsadjutant* of *7.Jäger-Bataillon* (in which assignment he was severely wounded on two occasions), then as *Gerichts- und Unterichtsoffizier*, *Zugführer*, and *Führer* of *4. Kompanie*.
00.00.1920–00.00.1929	Employed as a *Volksschule* teacher in Hattingen.
ca. 1920–00.00.1922	Member of the *Deutschvölkische Schutz- und Trutzbund*.
00.00.1922	Joined the *NSDAP*.
ca. 1922–00.00.1923 (?)	*Propagandaleiter* of the *Ortsgruppe Hattingen der NSDAP*.
00.00.1923–00.00.1927	*Führer* of the *SA-Sturm Hattingen*.
09.05.1924	Arrested and allegedly mistreated by French occupation troops.
28.12.1925	Reenrolled in the *NSDAP*.
00.00.192_-00.00.19__	*Vorsitzender* of the *Gau-Uschla des Gaues Ruhr der NSDAP*.
00.00.1927–00.00.1930 (?)	*Führer* of *SA-Standarte "Ruhr"*.
00.00.1928	Appointed as a *Parteiredner der NSDAP*.
20.05.1928	Unsuccessful candidate of the *NSDAP* for election to the *Preußische Landtag* and *Reichstag*.
01.-04.08.1929	Participated in the *4. Reichsparteitag der NSDAP* in Nürnberg.
17.11.1929–00.00.1933	*Stadtverordneter* in Hattingen.
0.01.1930–00.00.1933	*Fraktionsführer* (parliamentary group leader) of the *NSDAP* in Hattingen.
00.00.1930–00.00.193_	*Gausturmführer* of *Gausturm Essen* (Bochum).

00.00.1930–00.00.1931	*Gauorganisationsleiter* of *Organisationen I und II* in the *Gauleitung Westfalen-Süd der NSDAP.*
00.00.1930–00.00.1931	Gaupropagandaleiter and *Gaufachberater für Kommunalpolitik"* of the *Gauleitung Westfalen-Süd der NSDAP".*
00.00.1931	Dismissed from the public school service with loss of pension rights due to his *NSDAP* activities.
00.00.1931	Appointed as a *hauptamtlicher SA-Führer.*
00.00.1931–31.10.1932	*Führer* of *SA-Untergruppe Westfalen-Süd.*
24.04.1932–14.10.1933	Member of the *Preußische Landtag (Wahlkreis 18, Westfalen-Süd).*
01.11.1932–14.03.1934	*Führer* of *SA-Gruppe Westfalen* (Westfalen-Nord und -Süd, Lippe, Schaumburg-Lippe; HQ: Dortmund; initially m.d.F.b., then permanent). Succeeded Werner von Fichte.
15.02.1933–10.11.1934	*Polizeipräsident* in Dortmund (assumed the post on 17.02.1933). Resigned from this post at his own request and succeeded by Otto Schramme.
12.11.1933–08.05.1945	Member of the *Reichstag (Wahlkreis 18, Westfalen-Süd;* after 29.03.1936, *Wahlkreis 28, Dresden-Bautzen).*
00.10.1933–00.07.1934	*Sonderbevollmächtigter des Obersten SA-Führers für die Provinz Westfalen* (Special Plenipotentiary of the Supreme *SA* Leader for the Province of Westfalen).
05.12.1933–00.07.1934	*Sonderbevollmächtigter des Obersten SA-Führers für den Freistaat Lippe-Detmold und Schaumburg.*
15.03.1934–30.06.1934	*Führer* of *SA-Obergruppe X* (comprising *SA-Gruppen Westfalen* and *Niederrhein;* HQ: Dortmund; m.d.F.b. until 04.04.1934, then permanent).
10.07.1934–00.08.1943	*Führer* of *SA-Gruppe Sachsen* (Dresden)(m.d.F.b. until 15.09.1935, then permanent). He actually assumed the post on 17.07.1934, succeeding Hans Hayn (who had been murdered in the purge of the *SA* leadership).
30.07.1934	Indicted by the *Obersten Parteigericht der NSDAP* in connection with the so-called "Röhm-Affäre", charged *with* violation of § 4 Abs. 2 b of the *Satzung* [Statute] *der NSDAP*, at the initiative of *Gauleiter* Josef Wagner. Also charged was his subordinate, *SA-Brigadeführer* Paul Giesler (*Führer* of *SA-Gruppe Westfalen*). They were specifically accused of planning an armed *SA* rebellion on 30.06.1934 and of acting against the authority of *Gauleiter* Wagner and the *SS.* Proceedings against him were opened before *OPG* on 08.08.1934, but vacated on 24.10.1934. He was fully exonerated and acquitted by the *II. Kammer* (2nd Chamber) of the *OPG* on 10.04.1935.
30.03.1936–01.01.1939	*Kreishauptmann* of the *Kreishauptmannschaft Dresden-Bautzen* ("kommissarisch mit der Verwaltung beauftagt" until 01.06.1936, then permanent).
01.01.1939–00.08.1943	*Regierungspräsident* in *Dresden-Bautzen* (with effect from 22.03.1936).
00.00.1939–00.00.1940	Army service with successive assignment as *Kompanieführer, Ordonnanzoffizier,* and *Regimentsadjutant* in an *Infanterie-Regiment.*

Wilhelm Schepmann in civilian attire,
ca. 1932 (Baldur von Schirach, *Die
Pioniere des Dritten Reiches*, 1933)

An official portrait of *SA-Gruppenführer*
Schepmann, April 1934.

Wilhelm Schepmann and Viktor Lutze during the 1928 *Kundgebung der SA* in
Düsseldorf (*SA-Obersturmführer* Karl H. W. Koch, *Das Ehrenbuch der SA*, 1934)

SA-Oberführer Schepmann and an unidentified senior *SS* officer inspect an *SA* unit in 1932. *(SA-Obersturmführer* Karl H. W. Koch, *Das Ehrenbuch der SA, 1934)*

Hitler visits Dortmund in 1933, escorted by Schepmann as Führer of *SA-Gruppe* Westfalen. From left to right: "Sepp" Dietrich (with Wilhelm Brückner behind), Julius Schaub, Hans-Joachim Riecke, Hitler, and Schepmann (NARA, Heinrich Hoffmann collection)

SA-Obergruppenführer Schepmann, ca. 1937 (Igor Karpov photo)

SA-Obergruppenführer Schepmann just before his appointment as *SA-Stabschef*, August 1943.

Formal portraits of *SA-Stabschef* Schepmann, 1943. (NARA, Heinrich Hoffmann collection)

16.08.1943–08.05.1945	*Stabschef der SA* (m.d.F.d.G.b. until 09.11.1943, then permanent). Succeeded Max Jüttner, who attended to the duties of *Stabschef* in the months following the death of Viktor Lutze.
00.09.1944–00.05.1945	*Inspekteur für die Schießausbildung* (Inspector for Marksmanship Training) of the *Deutschen Volkssturm.*

Postwar Prosecution and Activities

After the war, Schepmann lived in Gifhorn under the alias "Schuhmacher" and worked as a materials administrator in the local district hospital. Arrested by British security forces, 00.04.1949, then handed over to German authorities. Joined the *Bund der Heimatvertriebenen und Entrechteten* (BHE), 00.00.1949. Sentenced by the Dortmund *Schwurgericht* to 9 months' imprisonment, 00.06.1950, due to coercion in office (in the case of the liquidation of the newspaper *Dortmunder Generalanzeiger,* when his *SA* men occupied its offices on 20.04.1933 in retaliation for their publishing a caricature of Hitler), but this sentence was later annulled. In April 1952, a further de-Nazification proceeding placed him in Gruppe V, but without a specific indication of grounds. The same year, he was elected as a member (representing the BHE) of the *Stadtrat* (city council) in Gifhorn / Niedersachsen, a seat he was to hold from 28.11.1952 to 09.06.1961. The following article appeared in the 17.11.1952 issue of *Time* magazine:

> In Lower Saxony, an old friend of Hitler's emerged from a wooden hut where he is living, unemployed, on a dole of $6.90 a week, to win a seat on both the town and county councils of Gifhorn. He was Brownshirt Wilhelm Schepmann, 58, last chief of staff of Hitler's Storm Troopers. Schepmann won easily, without even bothering to campaign. In other local elections in Lower Saxony the neo-Nazis campaigned on the slogan: "Stand fast. Remain German ... We shall return.'" The Refugee Party, which had the Nazis' support, won 17% of the total vote.

Also on 28.11.1952, he became a Beigeordneter (delegate) to the *Verwaltungsausschuß* (administrative council) of Gifhorn. In 1954, he was tried but acquitted of charges of coercion in office. The same year, he was elected to serve as a teacher at a Volsschule in Gifhorn, however the *Kultusministerium* of Niedersachsen prohibited him from teaching. From 09.11.1956 to 17.05.1961, he served as *1. Beigeordneter* (i.e., *stellvertretender Bürgermeister* [deputy mayor]) of Gifhorn, and was reelected on 18.04.1961. However, amid public protests, he was removed from office the next month, on 17.05.1961. On 27.08.1961, the *Amtsgericht* of Niedersachsen, in Hannover, removed Schepmann from the pension rolls, ruling that despite the fact that he had committed no crimes against humanity, he had nonetheless "violat[ed] the principles of the legal state".

Decorations and Awards

00.00.1940	*1939 Eisernes Kreuz I. Klasse* (awarded for his part in the breakthrough of the Maginot Line)
00.00.1940	*1939 Spange zum* Eisernen Kreuz II. Klasse
00.00.191_	*1914 Eisernes Kreuz II. Klasse*
00.00.194_	*Kriegsverdienstkreuz I. Klasse ohne Schwerter*
00.00.194_	*Kriegsverdienstkreuz II. Klasse ohne Schwerter*
31.05.1917	*Schaumburg-Lippisches Kreuz für treue Dienste 1914*

00.00.1918	*Verwundetenabzeichen, 1918 in Schwarz*
ca. 1939	*Medaille zur Erinnerung an den 1. Oktober 1938*
ca. 1938	*Medaille zur Erinnerung an den 13. März 1938*
ca. 1934	*Ehrenkreuz des Weltkrieges 1914–1918 mit Schwertern*
00.00.193_	*Goldenes Ehrenzeichen der NSDAP*
00.00.194_	*Dienstauszeichnung der NSDAP in Gold*
00.00.194_	*Dienstauszeichnung der NSDAP in Silber*
00.00.194_	*Dienstauszeichnung der NSDAP in Bronze*
00.12.1943 (?)	*Gau-Ehrenzeichen des Gaues Sudetenland der NSDAP*
ca. 1929	*Nürnberger Parteitagsabzeichen 1929*
00.00.19__	*Silbernes Treudienst-Ehrenzeichen*
00.00.193_	*SA-Sportabzeichen in Bronze*
00.00.1934	*Ehrendolch der SA*
00.02.1934	*Ehrenwinkel für alte Kämpfer*
22.06.1944	RM100,000 Dotation from the *Führer* (for his 50th birthday)
00.00.193_	Commemorative Medal of the War of 1914–1918 with Swords (Hungary)
00.00.193_	Medal for Participation in the European War 1915–1918 (Bulgaria)

Notes
- Married to Gertrude Schumann (1898–1959), 00.00.19__.
- Schepmann's essay "Task and Role of the SA", published in *Der Politische Soldat*, Nr. 12, October 1944)

As *Stabschef* responsible to the *Führer* for the *SA*, I consider it right that I should once more define explicitly the task and role of the *SA* within the Party and the *Reich*, in order to clear up once and for all any obscurity concerning the task of the *SA*.

During the early years of the movement it was the *SA* which, in accordance with the will of the *Führer*, as part of the Party brought together the activist forces as the sword-arm of the National-Socialist movement. For the National-Socialist who was ready for action, membership in the *SA* was a matter of course. It was just as much a matter of course that the *SA* man became a party member and considered himself first and always a National-Socialist. As an *SA* man, it was his pride and his honor always to be in the forefront of the fight. Where propagandists, collaborators within the inner circle of the Party, dare-devils, helpers for Party members in trouble, everywhere where men ready for action were needed, the *SA* was there and naturally the *SA* was called upon. In his calling the *SA* man was at all times a fanatical representative of the National-Socialist Weltanschauung. The Party was unimaginable without the *SA*, just as the *SA* was never an independent power outside the Party, but exclusively a part and an organ of the movement. From this role of the *SA* as the organization of the activist men of the Party, from its indispensability for the carrying out of National-Socialist activities, and not from the training of the *SA* or any other branch of their self-education, is derived the justified self-confidence of the *SA* man. From this role resulted also the esteem of the *SA* within the whole movement. Tested *SA* men, capable as political leaders, were continually transferred to the political directorate of the Party, to advance there to leading positions. Almost every one of the old men in authority [Hoheitsträger] in the Party was also an *SA* man in the years of struggle.

Formal portraits of *SA-Stabschef* Schepmann, 1943. (Top: NARA, Heinrich Hoffmann collection. Bottom: Jerry Luisi)

SA-Stabschef Schepmann smiles for
the camera (Roger Bender photo)

Rastenburg / Ostpreußen, 07.10.1943:
SA-Stabschef Schepmann and *Reichsminister*
Alfred Rosenberg during the conference
of Reichs- and *Gauleiter* at Führer
HQ *Wolfsschanze* (NARA photo)

Prerequisite to this success of the *SA* in the party struggle was its construction according to military principles, was the *SA* man's unconditional duty to obey, his education in discipline, was above all the constant schooling of the *SA* man to become a conscious and fanatical political soldier of the *Führer*. In the narrow sense this is the goal which the work of the *SA* served. Just as, however, the peace-time training of a soldier does not have significance except as future or at least possible action is in view for the actual defense of the life of the nation, so work within the *SA* was only a means to an end. The end was the completion of the manifold tasks of the *SA* in support of the Party's struggle in the field of internal politics, was the victory of the National-Socialist movement.

After Hitler came to power the *SA* was at first in danger of developing into a defense organization with a military foundation, as a result of a misconception of the order given by the *Führer* from the beginning that it should educate all German men to have a military, National-Socialist attitude. This led to a loosening of its spiritual ties to the Party and thus necessarily to serious setbacks. Open and veiled alienation between political leaders and *SA* leaders was the inevitable consequence. The result was unsatisfactory for both sides and disadvantageous to the Party. Many a Party endeavor would have attained easier success if the right use had been made of the largest and strongest formation of the Party, the SA— tightly organized and disciplined down to the last detail, and comprising millions of zealous men from all classes of the people. By and large, however, the *SA* got stuck in its own field

Führerappell der SA-Gruppe Pommern, 22.11.1943 (NARA photos)

of service. *SA* work in the narrow sense, the training of units in the evening and on Sundays, became too exclusively the sum and substance of *SA* activities in general. Above all, this work determined, to too great an extent, the conception of the present task of the *SA* which was held by the German people, by the Party, and finally even by many *SA* men.

To be sure, the *Führer* himself again and again provided fresh impetus, for example, by his order for the military education of German manhood by means of *SA* training for the Wehrabzeichen and by other assignments. These assignments, however, because they were often not approached in the right way, did not come to fruition. Partly in the Party, partly in the *SA* Itself, they were regarded too much as independent individual tasks of secondary importance and top little as important constituents of the entire Party's work of educating and leading the whole German people. Thus for a long time the position of the *SA* and the self-confidence of the *SA* men remained shaken, they remained so, even though the *SA,* in this time of outward stagnation and self-imposed limitation achieved results of which they are fully entitled to be proud.

One could not judge as well in peace-time, and even the *SA* man himself was often never aware, what a great work of education had been done. Especially in the psychologically difficult conditions in which he had to lead his men, the *SA* leader gained almost unequalled experience in the field of human leadership. In a wide measure, as well; he gained organizational knowledge in the course of the numerous *SA* internal undertakings. The entire *SA,* however, became a body of men who had learned how to be strict with themselves and to stand by the *Führer* and Party in the most difficult circumstances. Now, during the war, it has been shown how valuable this work of the *SA* was in itself. The *SA* men in the units on the fighting front have stood the test, like all real National-Socialists, and have proved to be the backbone of the forces. The *SA* leaders who remained at home, however, put their knowledge and experience at the disposal of the Party and justified, by their achievements, the *Führer*'s order to instill the spirit of National-Socialism in the German men eligible for military service. Thus we have every reason to do away with the old inhibitions. It is the will of the *Führer* and my task as *Stabschef* of the *SA* to clear up these questions conclusively. Every *SA* man and every Party member, no matter what his position, must come to realize:

As in the period of struggle, it remains the task of the *SA* to stand to the fore in all matters in which the Party needs an active representation among the people of its will and its measures. The *SA* man must be the most active propagandist of the Party, the most courageous warrior of the Party in the air war, the most active collaborator with the political leaders in war-time welfare work, the most active National-Socialist in the *Wehrmacht*, and also, if must be, again the most active fighter against grumblers and defeatists.

The *SA* is nothing outside the Party's field of duty, and the Party cannot deny itself the use of an army, millions strong, of staunch National-Socialists, comprising the best men of all classes of our people. The National-Socialist leadership of the *Wehrmacht* knows that the *SA* men in its units with the other men of the National-Socialist Party represent a nucleus of determined warriors. It cannot refuse to call upon these men— these political soldiers, in the best sense, educated to display unswerving zeal even in difficult situations it cannot fail to call upon them for leadership, education, and spiritual strengthening of the troops.

The *SA* man himself must know, however, that he is always committed, wherever he may be, as an *SA* man, as the most active champion of the will of the Party. It makes no difference whether he does *SA* service in his unit in continuation of his peace-time work, whether he helps with the pre-military training of untrained citizens, whether he conducts

SA-Stabschef Schepmann with, at right, *SA-Hauptsturmführer* Wittfoth (Roger Bender photo)

19.-22.01.1945: Schepmann speaks with highly decorated officers of Panzer-Division *"Feldherrnhalle"*, including, from left to right, *Oberst d. R.* and *SA-Sanitäts-Standartenführer* Dr. Franz Bäke, *Hauptmann d. R.* and *SA-Oberführer* Ewald Bartel, and *Oberleutnant* Otto Erdmann (NARA photo)

military sports contests or military shooting contests, whether in groups or alone, in uniform or mufti, he works in the air war, in the auxiliary police, in the NSV relief work, or in one of the thousand other fields of war service: he is always committed, first of all, as an *SA* man, as a warrior of the Party, and performs definite tasks in this capacity.

To a special degree all those *SA* men who have been called up as soldiers today must consider themselves as *SA* men at work—they more than the rest of the *SA*. It does not matter whether they are used as generals or staff officers, as leaders of military units or National-Socialist guidance officers, as non-commissioned officers or common soldiers. They all have, along with their military tasks, the political assignment of being examples of the National-Socialist will-to-fight, of sternness and comradeship, the embodiment of the will of the movement. They all have to carry out this assignment as part of their *SA* service.

Thus every *SA* man, down to the last one, gains the self-confidence which he absolutely must have to perform his great and wonderful, timeless task. The *SA* man will, then, always be judged by all other members of the Party and by the whole German nation in the way which the *SA* and he himself deserve and must claim. The heart of the matter is end remains: The work of the *SA* can only be understood correctly as an integral part of the Party's task of leading the German people and infusing into them the National-Socialist ideology. In this

19.04.1944: *SA-Stabschef* Schepmann on a firing range in Reichsgau Tirol-Vorarlberg, with *Gauleiter* Franz Hofer observing (NARA photo)

it is not if a number of subordinate auxiliary tasks which is involved, but the I utilization of the units of the Party in all fields of the Party's work. The *SA* man is everywhere and always on duty as an *SA* man.

I demand this attitude of every *SA* man, no matter whether he be a general (SA) or a private, no matter whether active or not, no matter whether honorary or paid. But I also ask all other National-Socialists, from *Reichsleiter* and *Gauleiter* down to Blockleiter and even to inactive Party members, from National-Socialist guidance officer to unit leader, to make this attitude the basis of their co-operation with the *SA*.

This will only be of benefit to the National-Socialist movement.

Nothing of this will change after the war, for what is involved is the lasting task of the *SA* within the framework of the Party. In our work one task will come to the fore: the *SA* will be charged with the extra-military National-Socialist education for defense. It will have to bring together the mass of men capable of bearing arms after their discharge from the *Wehrmacht*, for the preservation and education of their National-Socialist readiness for defense and for the preservation and training of their valiant capacity for action. Naturally special consideration will be given psychologically to the soldiers returning from the front. The service in the reserve corps of ex-service men will be so arranged that it can be performed by everybody. This assignment, too, has to be regarded in the first instance as an essential part of the National-Socialist educational duty of the Party to the German nation. The aim of making every German man into a resolute warrior for the National-Socialist *Reich*, into a thoroughly convinced and loyal follower of the *Führer*, is our first and most important aim. The means to this end is training in defense. Detached from the political goal, however, this training would be only a job half-done.

19.-22.01.1945: Schepmann speaks with Knight's Cross holder *Unteroffizier* Josef Fink during a reception for men of *Division "Feldherrnhalle"* (NARA photo)

That is how we *SA* men regard our task, that is how we value it, and that is how we wish, thereby, to serve the Party, our people, and our beloved *Führer*. In order that cowardice and treachery may never again be able to creep into the German nation and that our children and grandchildren may be able in the future to lay the final stone of the mighty, storm-defying edifice of the National-Socialist Greater German *Reich* of Adolf Hitler!

Sources

Bundesarchiv, Berlin-Lichterfelde (former Berlin Document Center): *Personalunterlagen von SA-Angehörigen: SA-Personalakte* of Wilhelm Schepmann.

Campbell, Bruce: *The SA Generals and the Rise of Nazism*. The University Press of Kentucky, 1998.

Lilla, Joachim; Döring, Martin; & Schulz Andreas: *Statisten in Uniform. Die Mitglieder des Reichstags 1933–1945*. Droste Verlag, 2004.

Orlow, Dietrich: *History of the Nazi Party: 1933–1945*. University of Pittsburgh Press, 1973.

Schirach, Baldur von: *Die Pioniere des Dritten Reiches*. Zentralstelle fur der deutschen Freiheitskampf, 1933.

Part 2

SA-Obergruppenführer (B – J)

Adolf Heinz Beckerle
SA-Obergruppenführer

Born:	04.02.1902 in Frankfurt-am-Main / Hessen.
Died:	03.04.1976 in Frankfurt-am-Main.

NSDAP-Nr.:	80 983 (First joined 29.08.1922 with Nr. 7 197; Party banned following the *München-Putsch* of 09.11.1923; Reenrolled with Nr. 80 983, 01.09.1928)

Promotions

01.04.1925	*Offiziersanwärter der preußischen Schutzpolizei*
01.09.1928	*SA-Mann*
01.04.1929	*SA-Sturmführer*
01.01.1930	*SA-Standartenführer*
10.12.1930	*SA-Brigadeführer* (the old form of this rank)
01.03.1931	*SA-Oberführer*
01.03.1933	*SA-Gruppenführer*
14.09.1933-00.05.1941	*Polizeipräsident* ("vertretungsweise mit der Verwaltung der Stelle beauftragt" [charged with acting administration of the position] until 31.12.1933, then permanent)
09.11.1937	*SA-Obergruppenführer*
01.04.1940	*Leutnant d. R.*

Career

ca. 1908-ca. 1912	Attended *Volksschule* in Frankfurt-am-Main (presumably).
ca. 1912 -00.03.1921	Attended the *Wühlergymnasium* in Frankfurt-am-Main (passed his *Abitur*).
00.00.1919	Joined the *Jungdeutschen Orden*.
00.04.1921-00.00.1927	Studied political economy, philology, and medicine (with breaks in studies) at *Johann Wolfgang Goethe-Universität*, Frankfurt-am-Main. He was also a member of the *Landsmannschaft "Nibelungia" D.L.*, and employed as a nurse and commercial trainee during this period.
00.00.1922	Joined the *Technische Nothilfe* (*TN*).
00.00.1922	Joined the *Verband nationalgesinnter Soldaten* (Association of Nationalist-Minded Soldiers).
00.00.1922	Joined the *Wikingbund*.
29.08.1922	Joined the *NSDAP / Ortsgruppe München*.
00.00.1924-00.00.1925	Worked in commerce, banking, and industry, visiting North and South America (spending one year in Argentina).
01.04.1925-00.06.1926	Service as an *Offiziersanwärter* in the *Preußischen Schutzpolizei*, assigned to the *Polizeischule* in Hannoversch Münden.
00.00.1925	Joined the *Eiserne Schar. Bund deutscher Männer und Frontkämpfer*.

SA-Oberführer Beckerle, ca. 1932. (Igor Karpov photo)

00.00.1927	Obtained his *volkswirtschaftlicher Diplom* (*Diplom-Volkswirt*, Diploma in political economy) from the Johann Wolfgang Goethe-Universität, Frankfurt-am-Main.
01.09.1928	Reenrolled in the *NSDAP*.
01.09.1928-01.08.1929(?)	Joined the *SA*, assigned to *Gausturm Frankfurt-am-Main*.
01.08.1929-01.01.1930	Assigned to the *SA-Reserve Frankfurt-am-Main*, then to *SA-Sturm 68* (Frankfurt-am-Main).
01.01.1930-05.02.1931	*Führer* of *SA-Standarte II* (Frankfurt-am-Main).
00.00.19__-00.00.1932	Studied law at *Johann Wolfgang Goethe-Universität*, Frankfurt-am-Main (did not obtain a degree).
05.02.1931-13.04.1932	*Gausturmführer* of the *SA* in Hessen-Nassau-Süd.
01.07.1931-13.04.1932	*Führer* (m.d.F.b.) of *SA-Untergruppe Hessen-Nassau-Süd* (Frankfurt-am-Main) (*SA* banned by the government of *Reichskanzler* Heinrich Brüning, 13.04.1932-01.07.1932). Succeeded by Emil Steinhoff.
18.10.1931	Participated in the *SA-Aufmarsch in Braunschweig*.
06.12.1931-19.12.1931	Participated in the *7. Lehrgang* (7th instructional course) of the *Reichsführerschule der SA* in München.
24.04.1932-00.09.1932	Member of the *Preußische Landtag* (*Wahlkreis 19*, Hessen-Nassau).
01.07.1932-01.07.1933	*Führer* of *SA-Untergruppe Hessen-Nassau-Süd*.
31.07.1932-08.05.1945	Member of the *Reichstag (Wahlkreis 19, Hessen-Nassau)*.
00.04.1933	Appointed as *Beauftragter der Reichssportkommissar* (Representative of the *Reich* Sport Commissioner, Hans von Tschammer und Osten) for *Gau XIII* of the *Reichsbund für Leibesübungen* (*Reich* Physical Training League).

01.07.1933-31.01.1942	*Führer* of *SA-Gruppe Hessen* (Frankfurt-am-Main; m.d.F.b. until 30.09.1933, then permanent from 01.10.1933).
14.09.1933-00.05.1941	*Polizeipräsident* in Frankfurt-am-Main ("vertretungsweise mit der Verwaltung der Stelle beauftragt" [charged with acting administration of the position] until 31.12.1933, then permanent). Briefly suspended from office in 1934 due to the so-called "Röhm-Putsch".
30.10.1933-03.07.1934	*Sonderbevollmächtigter der OSAF* (Special Plenipotentiary of the *Obersten SA-Führer*) to the *Oberpräsident* of *Provinz Hessen-Nassau*.
00.00.193_-00.00.193_	Member of the *Deutsche Olympische Komitee*.
00.00.1934	Appointed as a *Preußischer Provinzialrat* for *Provinz Hessen-Nassau*.
00.00.193_-00.00.19__	*Vorsitzender des Aufsichtsrates* of *Treuhand-Vereinigung AG*, Frankfurt-am-Main.
00.00.193_-00.00.19__	Member of the *Vorstand, Hippodrom AG*, Frankfurt-am-Main.
0.12.1935	Appointed as *Gauführer* of *Gau XIII (Südwest)* in the *Reichsbund für Leibesübungen*.
20.04.1936	Appointed as *ehrenamtlicher Beauftragter des Reichssportführers* (honorary representative of the *Reich* Sport Leader [Hans von Tschammer und Osten]) for the *Provinzen Starkenburg, Bayerische Pfalz,* and *Saargebiet*, and the Prussian *Kreise* of *Frankfurt-am-Main, Wiesbaden, Rheingau,* and *Main-Taunuskreis*.
07.02.1939-00.00.1939	Member of the *Deutsche Olympische Ausschuß* (German Olympic Committee tasked with preparations for Germany's participation in the 1940 Olympic Games in St. Moritz, Switzerland and Helsinki, Finland).
26.09.1939-00.11.1939	*(kommissarischer) Polizeipräsident in Lodsch* (Łódź). On 25.10.1939, he met with the *Generalgouverneur* of occupied Poland, Dr. Hans Frank, as indicated by the following entry in Frank's official diary:

> The *Generalgouverneur* received the *Polizeipräsident* of Lodsch, *SA-Obergruppenführer* Beckerle. Questions pertaining to the committment of Police and civil service employees were discussed. *Obergruppenführer* Beckerle pointed out that the former fear psychology present among the people had passed and that there were more Jews in Lodsch than before. This is one of the reasons for the continued deterioration of the food situation. (Partial Translation of Document 2233-G-PS, in *Nazi Conspiracy and Aggression, Volume IV*)

He was informed of his appointment by Kurt Daluege, who had arranged it through an agreement with *SA-Stabschef* Lutze. Beckerle arrived in the city within two days of the end of hostilities, accompanied by *Oberstleutnant der Schutzpolizei* Willi Hartmann and several other police officers from Frankfurt-am-Main. To assist him in organizing a police force, the military administration allocated to him one army *Reserve-Bataillon* and one *Sturm* of the *NSKK*. He also requested several members of the *Schutzpolizei* from Berlin, receiving 30 men. Soon after his arrival, he came into conflict with *Kreisleiter* and *SS-Standartenführer*

An illustration of *SA-Gruppenführer* Beckerle from the cigarette card album *Männer des Dritten Reich*. (Orientalischen Cigaretten-Compagnie "Rosma", 1934)

SA-Gruppenführer Beckerle in 1933. (Laurens Hessels)

Ludwig Wolff over the Polish "Freedom Monument" in the city; Wolff offered to blow it up, in order to demonstrate to Poles the power and superiority of the German race. Beckerle, however, wanted to dismantle it and use the scrap metal for the German war industry. Nevertheless, the monument was blown, which resulted in severe damage to surrounding houses and casualties among the local population. Beckerle summoned the representative of the Łódź bishop, and advised that the slightest attempt to cause unrest among the faithful would result in the arrest of the bishop and all the higher clergy of Łódź. Likewise the Jewish elders were threatened and instructed to organize work parties to clean the streets of debris. *Reichsführer-SS* Himmler inspected the city early in October 1939, and sharply criticized *Polizeipräsident* Beckerle because of the lack of discipline among the Polish police, his failure to organize demonstrative executions of the Polish intelligentsia and Jews, the use of sidewalks (rather than roadways) of Polish pedestrians, and the fact that Jews had not been provided with yellow identification badges. In mid-November 1939, Beckerle filed a request to be relieved of his office as *Polizeipräsident*, because he was offended by Himmler's attacks on him, and because of his subordination to an *SS- und Polizeiführer*, which he considered to be beneath his dignity and contrary to the *Führerprinzip*.

00.00.1939-00.00.1940   War service on the Western Front.

SA-Gruppenführer Beckerle at a sporting event in 1934.

17.06.1941-18.09.1944   Employed in the *Auswärtige Amt* (German Foreign Office), based on a recommendation by *Unterstaatssekretär* and *SA-Brigadeführer* Martin Luther.

28.06.1941-18.09.1944   *Deutscher Gesandter* (German Ambassador) in Sofia. Succeeded Herbert von Richthofen. While holding this post, he played a role in the deportation of 12,000 Jews from the Bulgarian-occupied regions of Macedonia and Thrace to Auschwitz and Treblinka. The following is a letter sent by Beckerle to the *Auswärtige Amt*:

Deutsche Gesandtschaft [German Legation]
A4318/42
On edict Nr. 1769 of 16. October 1942 and following edict Nr. 1839 of 2. November 1942
Subject: Umsiedlung der Juden [Relocation of the Jews]
3 double
1 attachment (triplicate)
Sofia, 16. November 1942

[Document Stamp on the letter, confirming receipt:
Auswärtige Amt
D III 1039. g
ring. 21st November 1942]

A charcoal portrait of *SA-Gruppenführer* Beckerle, ca. 1936.

Official portrait of *SA-Obergruppenführer* Beckerle, ca. 1937.

The Prime Minister [Bogdan Filov] spoke to me today regarding the Judenfrage [Jewish Question]. He principally welcomed the possibility to transport the Jews towards the east but he also pointed out that some of the male Jews are needed as workforce for roadworks and are currently indispensible. He would be thankful if a German advisor could be sent here to help carrying out the corresponding measures prior to the deportation.

The verbal note regarding this question of the Bulgarian government arrived today, too. As I already pointed out in the afore mentioned edict, I only raised this topic with the Prime Minister verbally. At the request of the secretary-general of the foreign ministry, only informal notes were provided as reminders for individual points during the consultations in the council of ministers. Romania is mentioned here, because I hinted that the deportation could be brought into connection with the Romanian Jews.

In summary, it is possible to register and deport most of the Bulgarian Jews, with the exception of a number of male Jews which are needed as workforce.

[Signed] Beckerle
To the
Auswärtige Amt
Berlin

On 07.06.1943, he wrote to the *Reichskanzlei*:

> I beg you to believe me that my service is doing everything possible
> to reach a final solution to the Jewish question in conformity with
> decisions taken. I am deeply convinced that the Prime Minister
> [Bogdan Filov] and the Bulgarian government also desire a final
> solution of the Jewish problem and are trying to reach it. But they
> are forced to take into account the mentality of the Bulgarian
> people, who lack the ideological conceptions that we have. Having
> grown up among Turks, Jews, Armenians, the Bulgarian people
> have not observed in the Jews faults which would warrant these
> special measures against them. (Guy H. Haskell, *From Sofia to
> Jaffa: The Jews of Bulgaria and Israel*, p. 109)

Marshall Lee Miller writes:

> The attitude of ... Beckerle, toward the Jewish question is not
> completely clear. A dedicated Nazi ... it ... seems inconceivable
> that he would have had any but the most orthodox Nazi views
> concerning the Jews, although not surprisingly he has written
> to the author [in 1966] that he was responsible for saving the
> Bulgarian Jews from deportation. Normally this claim would not
> be taken seriously. However, both Hannah Arendt and Gerald
> Reitlinger have stated that Beckerle's reports to Berlin discouraged
> Germany from putting more pressure on Bulgaria to persecute
> the Jews. It is true that many of the cables he sent during this
> period were the barest minimum that he could do and still keep
> the confidence of his superiors ... (Miller, *Bulgaria in the Second
> World War*, p. 105)

00.00.1941	Suceeded in mobilizing 40,000 Bulgarian workers to perform work in German enterprises.
01.02.1942-00.05.1945	Assigned as an *SA-Führer z. V.* to the *Stab der Obersten SA-Führung*.

Postwar Prosecution and Activities

Arrested by Soviet troops – together with 36 other German diplomats- at Svilengrad, Bulgaria, on the border with Turkey, 18.09.1944. On 17.10.1951, he was sentenced by the Soviet MGB-Special Authority *Osoboe Soveščanie* (OSO) to 25 years' imprisonment (in accordance with Control Council Order #10 of 20.12.1945) and held in Soviet captivity until 13.10.1955 when he was repatriated to West Germany. Classified as a "Hauptschuldiger" (Major Offender) by the *Zentralkammer* (Central de-Nazification Court) Hessen-Süd, 22.03.1950. On 26.07.1956, Beckerle was granted a remuneration of 5,600 DM by a Frankfurt court. The *London Times* reported on 27.07.1956: "The court had to overlook his Nazi career in order to approve payment, and managed this by arguing that he had been arrested and sentenced by the Russians, not on political grounds, but because he was a member of the diplomatic service ... When he first applied for compensation his request was rejected on the grounds that it was against the spirit of

A formal portrait of Beckerle, taken after his 01.02.1933 promotion
to *SA-Gruppenführer*. (Laurens Hessels photo)

the compensation laws to pay money to leading ex-Nazis." Employed as a *Prokurist* (authorized representative) of a firm in Neu-Isenburg, 00.00.1956-00.00.1966, he was terminated due to his indictment on war crimes charges. On 25.09.1959, Beckerle was arrested at the direction of the Generalstaatsanwalt in Frankfurt-am-Main, Fritz Bauer, following issuance of an arrest warrant by the *Amtsgericht* in Frankfurt. Released in October 1959, but following a complaint by Bauer, he was arrested again on 27.05.1960. Charged again, 17.03.1966, together with former Foreign Office *Legationsrat*, Fritz von Hahn, in connection with the deporation of 11,343 Jews from Macedonia and Thrace to the Treblinka extermination camp (von Hahn was additionally accused of involvement in the deportation of 20,000 Greek Jews to Auschwitz). Beckerle and von Hahn were, from November 1967 through 21.06.1968, tried in the so-called "Judenmord-Prozess" before the *Schwurgericht* of the *Oberlandesgericht* in Frankfurt-am-Main. Hospitalized, 21.06.1968, due to sciatica (prosecutor Bauer died of a heart attack two days later). The proceedings against Beckerle were terminated due to his poor health on 30.06.1968, and he was not charged again.

Published Work

Wir wollen arbeiten: Erlebnisse deutscher Auswanderer in Südamerika (1942, written under the
pseudonym "Heinz Edelmann" [Edelmann being his wife's maiden name])

Decorations and Awards

00.00.1940	*1939 Eisernes Kreuz II. Klasse*
00.00.193_	*Goldenes Ehrenzeichen der NSDAP*
ca. 1929	*Nürnberger Parteitagsabzeichen 1929*

ca. 1931	*Abzeichen des SA-Treffens Braunschweig 1931*
00.00.194_	*Dienstauszeichnung der NSDAP in Silber*
00.00.194_	*Dienstauszeichnung der NSDAP in Bronze*
01.10.1935	*Ehrenzeichen der TN (Nr. 1 376)*
30.03.1935	*SA-Sportabzeichen in Gold*
17.05.1936	*Deutsches Reichssportabzeichen in Silber*
02.04.1936	*Deutsches Reiterabzeichen in Silber*
30.06.1934	*Abzeichen der Deutschen Lebensrettungsgesellschaft- DLRG- in Bronze*
00.00.1934	*Ehrendolch der SA* (mit Wirkung vom 03.02.1934)
19.12.1931	*Tyr-Rune*
00.02.1934	*Ehrenwinkel für alte Kämpfer*

Notes

- Son of the postal official (*Postoberamtsmann*) Hans Beckerle (born 03.01.1875 in Eltville) and his wife Margarethe, née Kammerer (born 03.10.1875 in Frankfurt-am-Main).
- Religion: Roman Catholic (his father's faith; mother was Lutheran).
- Married on 27.02.1935 to Silke Edelmann (born 13.10.1906 in München, Suicide 00.00.1951).
- Foreign language proficiency: English, French, and Spanish.

Sources

Bundesarchiv, Berlin-Lichterfelde (former Berlin Document Center): *Personalunterlagen von SA-Angehörigen: SA-Personalakte* of Adolf Heinz Beckerle.

Haskell, Guy H.: *From Sofia to Jaffa: The Jews of Bulgaria and Israel.* Wayne State University Press, 1994.

Lilla, Joachim; Döring, Martin; & Schulz Andreas: *Statisten in Uniform. Die Mitglieder des Reichstags 1933-1945.* Droste Verlag, 2004.

Miller, Marshall Lee: *Bulgaria in the Second World War.* Stanford University Press, 1975.

Office of United States Chief Counsel for Prosecution of Axis Criminality: *Nazi Conspiracy and Aggression* (11 volumes). U.S. Government Printing Office, District of Columbia, 1946.

Dr. phil.
Hans *Heinrich* Bennecke
SA-Obergruppenführer

Born:	08.02.1902 in Dresden / Sachsen.
Died:	29.01.1972 in Stuttgart-Vaihingen.

NSDAP-Nr.:	4 840 (First joined, 00.05.1922; Party banned following the *München-Putsch* of 09.11.1923; Reenrolled, 15.05.1925)

Promotions

10.07.1927	*SA-Standartenführer*
01.07.1932	*SA-Oberführer*
01.07.1933	*SA-Brigadeführer*
01.03.1934	*SA-Gruppenführer*
09.11.1937	*SA-Obergruppenführer*
01.03.1939	*Leutnant d. R.*
01.07.1941	*Oberleutnant d. R.*
01.10.1943	*Hauptmann d. R.*

Career

00.00.1908-00.00.1912	Attended a *Bürgerschule.*
00.00.1912-00.00.1921	Attended a *Real-Reformgymnasium* (passed his *Abitur*).
Summer 1917 & 1918	Worked as an *Erntehelfer* (harvest assistant).
00.00.1919-00.00.1919	Service with *Zeitfreiwilligen-Regiment Dresden.*
15.03.1920-15.05.1920	*Reichswehr* service as a *Zeitfreiwilliger* in 2. *Kompanie / Regiment 23* (in Dresden, Leipzig, and the Vogtland).
00.05.1921-00.06.1921	Service with *Freikorps Hassfurther* in Oberschlesien.
00.00.1921	Passed his *Abitur.*
00.00.1921-00.00.1922	Studied economics and political science at the *Technische Hochschule* in Dresden.
00.00.1922-00.00.1923	Attended the University of München.
01.05.1922-09.11.1923	Joined the *SA*, assigned to *SA-Regiment "München".*
00.05.1922	Joined the *NSDAP / Ortsgruppe München.* He was fined, in 1922, 1923, and 1924, for his *NSDAP* activities.
00.08.1922-00.00.1923	Part time assistant in the *Geschäftsstelle der SA* in München.
15.10.1923-08.11.1923	Service as *Zeitfreiwilliger* in *Pionier-Bataillon 7* (München).
00.00.1923-09.11.1923	Adjutant of *SA-Regiment "München".*
08./09.11.1923	Participated in the *München-Putsch*, for which he was subsequently expelled from Bayern.
01.02.1924-31.12.1925	Employed as a shop assistant with a paper business in Dresden.
00.00.1924-00.12.1925	Assigned as a *Zugführer* with the *Frontbann* in Dresden.
15.05.1925	Reenrolled in the *NSDAP / Ortsgruppe Dresden.*

SA-Gruppenführer Dr. Bennecke, ca. 1936, wearing three of the highest decorations of the *NSDAP* (the *Goldenes Ehrenzeichen*, *Blutorden*, and the *Coburger Abzeichen*).

Autumn 1925-Autumn 1926	Member of the *Großdeutsche Jugend* (Greater German Youth, laterabsorbed into the *Hitler-Jugend*).
00.04.1926-00.00.1929	Studied journalism, history, and philosophy at the University of Leipzig.
00.00.1926-00.00.1928	Member of the *NS-Deutschen Studentenbund* (*NSDStB*, National Socialist German Students' League).
00.07.1926-00.12.1926	*Führer der HJ Ostsachsen* (Leader of the Hitler-Jugend in Eastern Saxony).
00.07.1927	Reentered the *SA* in Leipzig.
10.07.1927-31.08.1929	*Führer* of *SA-Standarte IV* (Leipzig).
01.09.1929-30.05.1930	Adjutant to the *OSAF-Stellvertreter Mitte* (Dresden), Manfred Freiherr von Killinger.
01.09.1929-30.05.1930	Editor of the *Sächsischer Beobachter*, the official newspaper of the *Gauleitung Sachsen der NSDAP*.
00.00.1930-00.12.1930	Editor of the National-Socialist daily *Freiheitskampf.*
00.05.1930-31.01.1934	Member of the *Sächsische Landtag* (1. Wahlkreis).
01.06.1930-30.06.1931	*Führer* of *SA-Brigade V* in *Gausturm Sachsen* (Dresden).
01.07.1930	Entered *hauptamtliche SA-Dienst* (full-time *SA* service).
00.00.193_-00.00.193_	Member of the *Haushaltsausschuß A* (budget committee A) and the *Büchereiausschuß* (library committee) of the *Sächsische Landtag*.
10.06.1930	Received his doctorate (Dr. phil.) from the University of Leipzig.
00.00.1930-00.00.1930	*Gaupressewart* of the *Gauleitung Sachsen der NSDAP*.
01.07.1931-31.12.1931	*Führer* of *SA-Standarte 5* (Dresden).

SA-Gruppenführer **Dr. Bennecke at his desk, ca. 1934. (NARA photo)**

01.01.1932-25.02.1932	*Führer* of *SA-Standarte 100* (Dresden).
26.02.1932-28.02.1933	*Führer* of *SA-Untergruppe Dresden*.
29.01.1933-04.02.1933	Participated in the *15. Lehrgang* (15th instructional course) at the *Reichsführerschule der SA*, München.
01.03.1933-15.09.1933	*Führer* of *SA-Brigade 33* (Dresden).
15.09.1933-00.00.193_	Assigned to the *Stab der Obersten SA-Führer*.
00.09.1933-00.09.1933	Participated in the *5. Reichsparteitag der NSDAP* in Nürnberg.
16.09.1933-01.03.1934	*Führer* (m.d.F.b.) of the *Reichs-SA-Hochschulamt* in Berlin, subordinated to the *Chefs des Ausbildungswesens der SA (Chef AW)*, Friedrich Wilhelm Krüger, and responsible for *vormilitärische Ausbildung* (pre-military training) of students who had joined the *SA*.
31.07.1934-31.12.1936	*Führer* of the *Reichsführerschule der SA* in München (m.d.F.b. until 14.09.1935, then permanent from 15.09.1935).
29.03.1936-08.05.1945	Member of the *Reichstag (Wahlkreis 18, Westfalen-Süd*; after 10.04.1938, he represented *Wahlkreis 6, Pommern*).
15.12.1936	Appointed as a member of the *Verfassungsausschuß der Reichsstudentenführung* (Constitutional Committee of the *Reich* Student Leadership).
01.01.1937-31.12.1944	*Führer* of *SA-Gruppe Pommern* (m.d.F.b. until 31.03.1937, then permanent from 01.04.1937) in Stettin. Succeeded by Walter Nibbe.
00.00.1937-31.12.1944	*Preußischer Provinzialrat* for *Provinz Pommern*.
15.11.1938-23.01.1939	Participated in military training exercises with the *Wehrmacht*.

The 48th birthday of Viktor Lutze, 28.12.1938: *SA-Obergruppenführer* Dr. Bennecke presents the *SA-Stabschef* and Frau Lutze with a handcrafted wooden trunk. (Scherl Bilderdienst photo)

& 31.07.1939-14.08.1939	
00.09.1939-00.00.1940	War service as a reserve officer in Poland and France.
00.01.1940-00.06.1941	Continued *Wehrmacht* service.
00.06.1941	Left the *Wehrmacht* and was briefly considered for assignment as a *Generalkommissar* to *Reichskommissariat Ukraine*.
00.00.1943-00.00.1943	Continued *Wehrmacht* service.
01.12.1944-00.05.1945	*Führer* of *SA-Gruppe Südmark* (kommissarisch until 31.12.1944, then permanent from 01.01.1945). in Graz. Succeeded Walter Nibbe.

Published Works

Hitler und die SA (1962)

Die Reichswehr und der 'Röhm-Putsch' (1964)

Wirtschaftliche Depression und politischer Radikalismus. 1918-1938 (1970)

Decorations and Awards

30.05.1940	*1939 Eisernes Kreuz II. Klasse*
30.01.1943	*Kriegsverdienstkreuz I. Klasse ohne Schwerter*
30.01.1943	*Kriegsverdienstkreuz II. Klasse ohne Schwerter*
ca. 1939	*Medaille zur Erinnerung an den 1. Oktober 1938*

SA-Gruppenführer Dr. Bennecke inspects SA troops, ca. 1936.

ca. 1938	*Medaille zur Erinnerung an den 13. März 1938*
ca. 1933	*Goldenes Ehrenzeichen der NSDAP*
00.00.1934	*Ehrenzeichen des 9. November 1923 (Blutorden)* (Nr. 28; with effect from 09.11.1933)
ca. 1932	*Coburger Abzeichen 1922*
00.00.194_	*Dienstauszeichnung der NSDAP in Gold*
00.00.194_	*Dienstauszeichnung der NSDAP in Silber*
00.00.194_	*Dienstauszeichnung der NSDAP in Bronze*
00.00.193_	*SA-Sportabzeichen in Gold*
00.00.1932 (?)	*Frontbann-Abzeichen (presumably)*
00.00.1934	*Ehrendolch der SA* (mit Wirkung vom 03.02.1934)
04.02.1933	*Tyr-Rune*
00.02.1934	*Ehrenwinkel für alte Kämpfer*

Notes

- Son of the neurologist *Generaloberarzt a. D.* Dr. med. Kurt Bennecke (Born 19.08.1869 in Eisenach / Thüringen) and his wife Margarethe, née Sputh (Born 24.08.1879 in Sebnitz / Sachsen).

- Married on 06.02.1930 to Ruth Heyde (Born 24.02.1905 in Dresden / Sachsen). No children.
- Religion: Lutheran until 00.11.1936, then left the church and declared himself "gottgläubig".

Sources

Bundesarchiv, Berlin-Lichterfelde (former Berlin Document Center): *Personalunterlagen von SA-Angehörigen: SA-Personalakte* of Heinrich Bennecke.

Lilla, Joachim; Döring, Martin; & Schulz Andreas: *Statisten in Uniform. Die Mitglieder des Reichstags 1933-1945*. Droste Verlag, 2004.

Joseph Berchtold
SA-Obergruppenführer

Born: 06.03.1897 in Ingolstadt / Regierungsbezirk Niederbayern / Bayern.
Died: 23.08.1962 in Herrsching am Ammersee / Landkreis Starnberg / Regierungsbezirk Oberbayern / Bayern.

NSDAP-Nr.   964 (First joined 00.02.1920 with *NSDAP*-Nr. 750; Left, 29.07.1921; Reenrolled, 07.03.1922; Party banned following the *München-Putsch* of 09.11.1923; Reenrolled with Nr. 36 003, 07.04.1926; Later granted Nr. 964)

Promotions

01.01.1916	*überzähliger Gefreiter* (i.e., granted the rank but not the pay thereof)
11.06.1916	*überzähliger Unteroffizier*
21.06.1916	*etatmässiger Unteroffizier* (i.e., receiving all the pay and benefits of the rank)
15.10.1916	*Offizier-Aspirant* (until 22.07.1917, when he was stripped of his officer candidacy)
12.03.1917	*Vizewachtmeister d. R.*
10.03.1918	Reinstated as an *Offizier-Aspirant*
22.06.1918	*Leutnant d. R.* (ohne Patent; later granted Patent vom 13.06.1918)
15.04.1926-31.10.1926	*Oberleiter [Führer] der Schutzstaffel* (redesignated *Reichsführer-SS*, 01.11.1926)
18.12.1931	*SA-Standartenführer*
01.01.1933	*SA-Oberführer*
01.01.1936	*SA-Brigadeführer*
01.05.1937	*SA-Gruppenführer*
29.04.1940	*Hauptmann d. R.*
30.01.1942	*SA-Obergruppenführer*

Career

00.00.1903-00.00.1915	Attended *Volksschule* and *Realgymnasium* in München.
01.02.1915	Entered military service as a *Kriegsfreiwilliger*.
03.02.1915-25.07.1915	Assigned as a Kanonier to *4. Ersatz-Batterie / 1. Kgl. Bayerisches Feldartillerie-Regiment*. He was sworn into service on 24.02.1915.
25.07.1915-17.11.1915	Assigned to frontline duty with *5.Batterie / 1. Kgl. Bayerisches Feldartillerie-Regiment*.
11.11.1915-16.11.1915	Hospitalized in a *Krankenrevier* due to enteritis.
17.11.1916-29.11.1915	Hospitalized in *Feldlazarett 6* of the *I. Kgl. Bayerisches Armee-Korps*.
29.11.1915	Returned to service with *5.Batterie / 1. Kgl. Bayerisches Feldartillerie-Regiment*.
16.03.1916-22.03.1916	Hospitalized in a *Krankenrevier* due to a contusion on the left foot.
19.06.1916-25.06.1916	Again in *Krankenrevier* due to a nervous disorder and exhaustion.

Josef Berchtold in civilian attire, ca. 1933.

11.02.1917-08.03.1917	Attended the *Artillerie-Schießschule* of *Armee-Abteilung "C"*.
19.03.1917-19.11.1917(?)	Assigned to *8.Batterie / 1. Kgl. Bayerisches Feldartillerie-Regiment*.
19.11.1917	Designated as under medical care for gas poisoning, but "dienstfähig" (fit for duty).
22.11.1917-01.01.1918(?)	Assigned to *4. Ersatz-Batterie / II. Ersatz-Abteilung / 1. Kgl. Bayerisches Feldartillerie-Regiment*.
01.01.1918-09.07.1918	Assigned to *8.Batterie / 8. Kgl. Bayerisches Reserve-Feldartillerie-Regiment* (redeployed to the Western Front, 31.01.1918).
09.07.1918-15.12.1918	Assigned to *9.Batterie / 8. Bayerische Reserve-Feldartillerie-Regiment*.
15.12.1918	Discharged from military service at the *Bezirkskommando München II*.
00.00.19__-00.00.19__	Studied economics at the University of München, then employed as a journalist.
00.02.1920	Joined the *NSDAP*. In 1933, Berchtold was interviewed by the journalist Heinz A. Heinz, who wrote of the encounter as follows:

> Herr Berchtold was bending over a paper as I entered, doing busy things with a big blue pencil, but he looked up at once, extended his hand, and indicated a seat.
>
> I got this impression instantly- no sedentary individual, Herr Berchtold, and no one bound down by red tape. A practical man; agile and decisive both in mind and body. Full of temperament,

Berchtold (far left) as *Führer* of *III.Bataillon / SA-Regiment "München"*, 1923. (Max Williams photo)

but no neuropath. "Lots of people," he began, "look upon that so-called Putsch of November 9th, 1923, as a bolt from the blue which split up the Movement. But it wasn't really anything of the sort. It was a perfectly logical and necessary outcome of all that had gone before. We National Socialists knew well enough that some such gesture was bound to come; what we didn't know, and what we weren't prepared for in the least, was the betrayal of the whole thing by sheer reactionary cowardice.' I will mostly confine myself to telling you what I know about it at first hand and content myself with a mere sketch of what happened when I wasn't, myself, actually present. I had been attracted to the 'Deutsche Arbeiter-Partei' so far back as 1920, not so much for itself as on account of this Adolf Hitler. To hear him was to be convinced that he had got hold of the right end of the stick. But for all that, there weren't many who believed then any use could be made of the fact. What exactly was Hitler's message? He appealed to reason when nobody else had hope of things being reasonable any more. He took his stand on re-establishing order in Germany, since without it the whole country trembled on the brink of final collapse and ruin.

I felt that no Party could achieve such an end. It was not the job so much for a political Party as for a sort of superman. Only he could set himself to such a task who possessed a strong hand, great decisive power and an incorruptible purpose. The last quality seemed to be remarkable for its absence in all parties at that time.

As for me, I no sooner heard Hitler in the Sterneckerbrau than I became a convinced adherent. I resolved to follow this man no matter whither he led. I joined up with the Party in January, 1920. As an active sort of man I didn't want to be merely a passive member, but an energising one. From this it follows I was pretty well mixed up with all the frays and conflicts in the beer halls and the streets which marked the rise of National Socialism- particularly did I lend a hand on the famous occasion of the Battle of the Hofbrauhaus. I was one of the Guards that night. We laid about us as we did, simply because our opponents left us no choice. It may seem pretty brutal to have to defend ideas with beer mugs and broken chair legs, but necessity knows no law. We weren't pummelling people out of an irresponsible love of fighting, but to gain a hearing for our Leader ..." (Heinz, *Germany's Hitler*, pp. 148-149)

00.06.1920-29.07.1921	Member of the *Arbeitsausschuß der NSDAP* (Working Committee of the Nazi Party).
00.06.1920-21.01.1921	*2. Kassierer der NSDAP* (Second Treasurer, after Franz Xaver Schwarz) of the Nazi Party.
21.01.1921-29.07.1921	*1. Kassierer der NSDAP.*
29.07.1921	Resigned from the *NSDAP.*
00.00.1921	Co-founder of the *Freien Nationalsozialistische Vereinigung München.*
07.03.1922	Reentered the *NSDAP* and *SA*, reappointed as *2. Kassierer der NSDAP.*
00.00.1922-00.01.1923	*Führer* of *III. Bataillon / SA-Regiment "München".*
00.00.1923-00.00.192_	Owner of a tobacco and stationery business at Im Tal Straße 54 München.
00.03.1923	Cofounder, with Julius Schreck, of the *Stabswache Hitler*, an *SA* company tasked with maintaining Hitler's personal security. The forerunner of the later *SS*, it was redesignated as the *Stoßtrupp Adolf Hitler* in May 1923. At that time it numbered just 20 men.
00.08.1923-00.11.1923	*Führer* of the elite *Stoßtrupp Adolf Hitler*. During his interview with

Heinz A. Heinz in 1933, Berchtold described the origins of the *Stoßtrupp* and Hitler's close involvement with these elite troops:

By the spring of 1923 [the SA] already amounted to a respectable number of men. Among these, however, were many who, despite all the goodwill in the world, were not fitted for the rougher work for which the Abteilung was constantly called upon. So it gradually became necessary to form a small body of specially-picked men on whom the most absolute dependence could be placed . They had to be ready to deal with the worst elements of the underworld under all and instant circumstances . Such a detachment was organised in May, 1923 .

It went by the name of "Strosstrupp Hitler" and came under the *Führer*'s immediate orders. Later these men were commanded by Captain Goering. One of the units was placed under me . In a few weeks' time my detail "Sturm Abteilung-Storm Troops." reached full company strength of 120 men. For the most part they were all old front-line soldiers, thoroughly dependable men.

The Storm Troops at that time wore grey water-proof; and grey cloth caps, like the former Austrian Infantry. The uniform of the Shock Troops was the old field grey uniform of the War ; with the addition of high black boots or gaiters, and a black cloth cap with a silver death's head on a red ground. Our shoulder straps exhibited the initials "St. Tr. H." We had arm-bands with the hooked cross in red. On special occasions of danger we exchanged our cap for a helmet with death's head and swastika upon it.

The Shock Troops were designed for specially dangerous work at specially dangerous places. Whenever the S.A. went on the march we led the van, or followed up the rear. We were put through continuous courses of training especially with a view to street fighting and fighting in assembly rooms. When, on our National Socialist expeditions, the spitting or stone-throwing of the onlookers (the Reds) waxed beyond a joke, I blew my whistle, and in no time at all the Shock Troops had cleared the streets . Their efficiency and discipline soon won the admiration of the populace. After a while the enemy learnt to let us alone : only those who had not yet had a taste of our quality still ventured to dispute our passage .

Hitler had his own method of attaching each and every man to himself. He would appear unannounced in our quarters, here in these offices in the Schellingstraße, on a drill night, and after a word or two with me he would address the men in the most comradely way possible . Then he'd inspect the Company, but not so much like an officer as like a friend . He would shake each man's hand, and look him squarely in the eyes. It was this glance, more than anything, which made every trooper Hitler's man to the death! (ibid, pp. 149-150)

08./09.11.1923   Participated in the *München-Putsch* at the head of his *Stoßtrupp*, leading its 125 men in a number of brutal and criminal actions. Shortly before the "operation" to overthrow and supplant the Bavarian government of Gustav von Kahr, Berchtold addressed his men in the bowling alley of the café-restaurant Torbräu:

Comrades, the hour we have all yearned for is at hand. Tonight a

Josef Berchtold (leaning on truck cab at left) with men of the *Stoßtrupp Adolf Hitler* in München, 1923. Standing near him, with goggles, is Julius Schreck.

new *Reich* government will be formed by our *Führer* Adolf Hitler and Herr von Kahr. Before I tell you more, is there anyone who for any reason wants to withdraw from the ranks? Now is your opportunity- no questions asked [no one accepted this offer].

All right. I will now swear you to absolute loyalty and unquestioning obedience to me and to the *Führer*, no matter what our orders. (John Dornberg, *Munich 1923: The Story of Hitler's First Grab for Power*, p. 34)

Following this ceremony, Berchtold led his men to the corner of St. Martin and Balin streets. Trucks full of weapons and ammunition awaited them there. During his 1933 interview with Heinz A. Heinz,, Berchtold spoke of his and the *Stoßtrupp*'s role in the *Putsch*:

... About eleven o'clock on the morning of the 8th of November I got the order to stand ready for the National Revolution. My men took a last oath, to serve true to death if needs be, and we got our final instructions from Captain Goering.

I busied myself all day with preparations, and then at six in the evening assembled the troops, in instant readiness for action in the Torbrau, opposite the Sterneckerbrau. I harangued my fellows, "Any one of you," I said, "who isn't going into this thing heart and soul had better get out right now." As no one budged by so much as an inch, I pursued, "It's our job, as Shock Troops, to bear the brunt

14./15.08.1926: *Deutscher Tag* in Starnberg. Hitler and Berchtold at the
Gasthaus *Zur Eisenbahn*. Hitler's photographer, Heinrich Hoffmann sits at
far left, with Jakob Grimminger holding up a flag in the background.

of what's coming. We're going to run the Government out. Hitler
and Kahr are united over this, they are going to set up another one."
Everyone of us gripped hands, and we were ready.

We marched off and took up a position in the neighbourhood
of the place where von Kahr's great gathering was to be held. It was
already crammed to bursting; a group of police near us could find
no standing room within, so hung about outside.

I glanced at my watch. Now for it!

Shouting orders to the Troop I sprang forward brandishing
my revolver.

"Out of the way-you there!" I yelled to the police, who
incontinently fell back amazed and unprepared, and burst our way
into the entrance. I ordered my men to cover every window and
exit, but pressed on myself with the rest into the hall. Hitler was
already there. Catching sight of us, he at once placed himself at our
head, and led us quickly and quietly right down the main gangway
until we reached the platform. The audience numbered several
thousand. It was listening to von Kahr who was addressing them
on the subject of "The People and the Nation." ... " Our sudden
appearance in their midst caused consternation.

Von Kahr was struck dumb.

People began shouting to know what was the matter. Women
fainted; here and there panic was imminent.

Meantime we had mounted the platform, and Hitler made an

SA-Oberführer Berchtold with the lawyer *SA-Gruppenführer* Prof. Dr. Walter Luetgebrune, senior legal counsel for the SA and SS, in 1933.

> attempt to speak. But the excitement was so great he could neither make himself heard nor understood. So he drew his revolver and a loud report rang out. He had pointed it upwards, to the ceiling.
>
> An instant silence fell in which his voice could be heard proclaiming the end of the Red regime. (Heinz A. Heinz, *Germany's Hitler*, pp. 154-155)

The *SA-Führer*, Hermann Göring, then took the stage and made a short speech, which ended with:

> We've taken this step because we're convinced that the men who stand at the head of us here in Bavaria will help us wrench free from Berlin and the Jews. The new Government will form itself round Hitler, Ludendorff, Pohner and von Kahr! (ibid)

Berchtold then received new orders, to:

> ... clear out the offices of the Munich Post, a much-detested Social Democratic paper, often styled the "Munich Pest." We forced the doors of this place, ransacked the building, and flung all the

printed stuff we could lay hands upon out into the street where it was promptly burnt. This was about eleven o'clock at night.

... I reassembled my men before the main doors of the premises and ordered them to wait there until I should return, with a convoy of perhaps twenty men, from a visit I intended to pay to the house of the notorious Social-democrat Auer.

We drove off thither in an armoured car, but Auer was not at home, and there was nothing for it but to return to the Munich Post, fall in the rest of the men, and return to the Burgerbraukeller ... On the morning of November 9th I received orders to occupy Police Headquarters ... When my men and I arrived we found the police fully armed and prepared to resist us, so avoiding immediate conflict I went on to the Rathaus.

I was aware that a Session was in progress. I flung open the doors and, cocking my revolver, informed the assembled Councillors, Social Democrats and Communists to a man, that they must consider themselves under arrest. Alarmed and startled, they sprang to their feet. We shepherded them from the chamber, and the building, and down the wide flight of steps without.

Here the rest of my men took charge. Each member, accompanied by two troopers, was assisted into a truck.

So we went on to the Burgerbrau and shut the whole lot up in the room to the rear from which Ludendorff had released von Kahr and his companions overnight ... (ibid)

The following day, Berchtold and his men joined in the march- led by Hitler, Ludendorff, and Göring- to the *Feldherrnhalle*:

Twelve ranks of Shock Troops followed with myself in command; then came the S.A, the United Troops and hundreds of civilians, workmen, students, all with brassards showing the hooked cross, by way of a great mass demonstration. The appeal was to be to the streets, to the people, to Munich. No arms to be used. Everything was to be put to the test of popular feeling...

A police cordon had been drawn across the bridge head. The police themselves were armed, and helmeted. We were within a stone's throw of them when they raised their rifles. Ulrich Graf, Hitler's bodyguard, shouted, "Don't fire; Ludendorff is with us," whereupon they lowered them again, and I sprang forward at the head of some ten of my fellows and promptly disarmed them. They were despatched forthwith to the Burgerbrau. I myself convoyed them. Having safely bestowed them in custody I made the utmost speed to rejoin the column, which forged ahead meanwhile through the Marienplatz to the Feldherrnhalle end of the Odeonsplatz. Taking the shortest cut I sprinted down the Maximilianstraße to the Max Joseph Platz, and arrived on the spot just as the head of the column debouched between the Feldherrnhalle and the Residenz. (ibid, pp. 154-160)

The police had fired on the marchers, killing 16 Nazis for the loss of three of their own. Berchtold continued:

I was in the Odeonsplatz and saw it all. When the crowd broke I made off as hard as I could go, back to the Burgerbrau where lots of the fellows soon came trooping in. When most of them seemed to have turned up we released the Councillors from the room, where they had been shut up all this time, and, loading them again into trucks, drove them right out of the city and away into the woods towards Rosenheim. When we'd gone some considerable distance I called a halt.

We all got out, and leaving half my fellows behind I led the other half and our prisoners deep into the wood.

The Councillors thought for a certainty that their last hour had struck. But I had a far less ruthless purpose.

I merely wanted to change an entire suit of clothes with a couple of them. Each one contributed this, that or the other garment, and got the corresponding one of mine in return. In a few moments I had transformed myself from a Shock trooper into a very passable likeness of the ordinary man in the street. Thereupon we returned to the cars, and began our journey back to Munich.

I dumped the City Councillors at the first convenient village and left them to shift for themselves.

But, dressed as I now was, I could make my way about without attracting attention. I ventured first to seek out the hospital to which Goering had been taken. I even saw him. He was in such pain it was all he could do to keep his wits about him, but he managed to give me some details about the end of the affair. Warrants were out for the lot of us, Hitler, Ludendorff, Goering and all of us, especially we of the Shock Troops, as the prime delinquents.

As a matter of fact they caught me that same night. But I managed to slip through their hands, and made off again in the direction of Rosenheim. In Traunstein they all but had me a second time, but I doubled back and returned to Munich where I lay low for a time. There I heard the end of the story.

As for me, ... I remained on, more or less in hiding, in Munich until the following February (1924), when 1 managed to get away into Austria. I kept in touch with Hitler as best I could, and when I heard his sentence would gladly have given myself up for a chance of sharing it with him. But by his own wish I stuck where I was, in Austria, until such time as I might return without danger. I stayed until 1926 when Hindenburg's general amnesty for political

Hitler poses with Berchtold and other former members of the *Stoßtrupp Adolf Hitler*, 1933. (NARA, Heinrich Hoffmann collection)

offenders allowed so many of us, Goering included, to return to Germany.

The outlook, however, was pretty miserable. Our great Movement, everything, was suppressed, 'verboten.' The only thing that remained was fidelity in many a breast to Hitler and his Idea. When the hour struck, months and months afterwards, for the resurrection of this idea, we were all ready. (ibid, pp. 162-164)

23.04.1924	While still in the Tirol region of Austria, tried *in absentia* by the *Volksgericht für den Landgerichtsbezirk München I* and sentenced to a term of imprisonment for his role in the *München-Putsch*.
00.00.1924-00.04.1926	*Gaugeschäftsführer* (regional business leader) of the *NSDAP* and *SA-Führer* in Kärnten (Carinthia).
00.04.1926	Amnestied and returned to Germany.
07.04.1926	Reenrolled in the *NSDAP*.
15.04.1926-31.10.1926	*Oberleiter [Führer] der Schutzstaffel* (50 Schellingstraße, München).
01.11.1926-00.03.1927	*Reichsführer-SS* (confirmed in this post by *Oberster SA-Führer* Pfeffer von Salomon, who had issued an order on 04.11.1926 declaring the *SS* as an independent organization under the authority of the *Obersten SA-Führung*; Berchtold resigned due to the humble position to which the *SS* was relegated). Succeeded by his deputy, Erhard Heiden.

A jovial *SA-Gruppenführer* Berchtold collecting donations, ca. 1937.

01.01.1927-00.00.19__	Editor of the official Party newspaper, the *Völkischer Beobachter,* München.
00.00.1928-27.01.1934 (?)	Assigned to *Stab der Obersten SA-Führung* in München.
00.00.1928-00.01.1938	Founder and *Hauptschriftleiter* (Editor in Chief) of the *Obersten SA-Führung* newspaper *SA-Mann.*
01.-04.08.1929	Participated in the *4. Reichsparteitag der NSDAP* in Nürnberg.
00.00.1934-00.00.1943	*Chef vom Dienst* (permanent deputy to the Hauptschriftleiter, Wilhelm Weiss) of the *Völkischer Beobachter,* München.
27.01.1934-00.05.1945	Assigned as an *SA-Führer z. V.* to the *Obersten SA-Führung.*
30.06.1934	Member of Hitler's entourage in carrying out the arrests of Ernst Röhm, Edmund Heines, and other *SA* leaders at Bad Wiessee.
00.00.193_-00.00.19__	*Ratsherr der Hauptstadt der Bewegung München* (Councilman of the Capital City of the [National-Socialist] Movement, München).
ca. 1936-00.05.1945(?)	Member of the *Reichskultursenat* (*Reich* Senate of Culture).
06.03.1936	Appointed as a member of the *Kulturkreis der SA* (Cultural Circle of the SA).
00.00.1936-08.05.1945	Member of the *Reichstag (Wahlkreis 32, Baden).*
00.01.1938-00.02.1943	*Stellvertretender Hauptschriftleiter* (deputy editor in chief) of the *Völkischer Beobachter,* under Wilhelm Weiss.
29.04.1940-00.00.194_	Army service as a reserve officer.

Published Works
Adolf Hitler über Deutschland (1932).

Postwar Internment
Arrested and interned in Oberpfaffenhofen, 1945.

Decorations and Awards
12.11.1916	*1914 Eisernes Kreuz II. Klasse*
21.10.1917	*Bayerisches Militär-Verdienstkreuz II. Klasse mit Schwertern.*
06.12.1918	*Bayerischer Militär-Verdienstorden IV. Klasse mit Schwertern*
ca. 1934	*Ehrenkreuz des Weltkrieges 1914-1918 mit Schwertern*
ca. 1933	*Goldenes Ehrenzeichen der NSDAP*
00.00.1934	*Ehrenzeichen des 9. November 1923 (Blutorden)* (Nr. 9; with effect from 09.11.1933)
00.00.1932	*Coburger Abzeichen 1922*
00.00.1929	*Nürnberger Parteitagsabzeichen 1929*
00.00.194_	*Dienstauszeichnung der NSDAP in Gold*
00.00.194_	*Dienstauszeichnung der NSDAP in Silber*
00.00.194_	*Dienstauszeichnung der NSDAP in Bronze*
ca. 1936	*Ehrenplakette für die Mitglieder des Reichs-Kultur-Senats*
00.00.1934	*Ehrendolch der SA* (mit Wirkung vom 03.02.1934)
00.00.193_	*Ärmelband der Stoßtrupp Hitler*
00.02.1934	*Ehrenwinkel für alte Kämpfer*

Notes
- Son of the conservator Joseph Berchtold (born 14.02.1863 in Dorffen, died 29.04.1935 in München; joined the *NSDAP*, 00.00.1920, and participated in the *München-Putsch* on 09.11.1923) and his wife Maria, née Schmidt.
- Religion: Roman Catholic (?). Declared himself "gottgläubig", 19__.

Sources
Bayerisches Hauptstaatsarchiv, München, Abteilung IV Kriegsarchiv: Excerpts from various Kriegsranglisten containing data on the Royal Bavarian Army service of Joseph Berchtold.

Bundesarchiv, Berlin-Lichterfelde (former Berlin Document Center): *Personalunterlagen von SA-Angehörigen: SA-Personalakte* of Joseph Berchtold.

Dornberg, John: *Munich 1923. The Story of Hitler's First Grab for Power.* Harper & Row, 1982.

Flood, Charles Bracelen: *Hitler: The Path to Power.* Houghton Mifflin Company, 1990.

Heinz, Heinz A.: *Germany's Hitler.* Hurst and Blacket, 1934.

Lilla, Joachim; Döring, Martin; & Schulz Andreas: *Statisten in Uniform. Die Mitglieder des Reichstags 1933-1945.* Droste Verlag, 2004.

Miller, Michael D.: *Leaders of the SS & German Police, Volume I (Reichsführer-SS-SS-Gruppenführer, Georg Ahrens to Karl Gutenberger).* R. James Bender Publishing, 2006.

Franz Bock
SA-Obergruppenführer

Born:	28.06.1905 in Kaltenbrunn / Bezirksamt Neustadt an der Weinstraße / Oberpfalz.
Died:	10.05.1974 in Köln.

NSDAP-Nr.:	33 014 (first joined, 03.01.1923 with Nr. 15 381; Party banned following the *München-Putsch* of 09.11.1923; Reenrolled 06.03.1926 with Nr. 33 014)

Promotions

24.11.1922	*SA-Mann*
00.00.1925	*SA-Mann*
01.06.1929	*SA-Truppführer*
26.01.1930	*SA-Sturmführer*
01.05.1931	*SA-Sturmbannführer* (01.07.1932: Reconfirmed in that rank with the lifting of the ban on the *SA*)
17.12.1932	*SA-Standartenführer*
01.07.1933	*SA-Oberführer*
20.04.1935	*SA-Brigadeführer*
09.11.1938	*SA-Gruppenführer*
01.09.1940	*Unteroffizier d. R.*
01.02.1943	*Oberleutnant d. R.*
09.11.1944	*SA-Obergruppenführer*

Career

00.00.1911-00.00.1914	Attended *Volksschule*.
00.00.1914-00.00.1920	Attended a humanistic *Gymnasium* (graduated *Obersekunda*).
00.00.1920-00.00.1922	Served an apprenticeship in banking.
00.00.1922-00.00.1927	Worked as a commercial employee and authorized representative in the tobacco industry, München.
24.11.1922	Joined the *SA* (with *SA*-Nr. 3502).
24.11.1922-09.11.1923	Assigned to the *Technischen Kompanie* of *SA-Regiment München*.
03.01.1923	Joined the *NSDAP / Ortsgruppe München*.
09.11.1923	Participated in the *München-Putsch*.
00.00.1924	Joined the *Großdeutsche Volksgemeinschaft/Sektion* in München-Schwabing.
00.00.1925	Reentered the *SA*.
00.00.1925-20.08.1927	Assigned to *SA-Sturm Schwabing*.
06.03.1926	Reenrolled in the *NSDAP / Ortsgruppe München*.
00.00.1927-00.00.1932	Commercial employee (balance account and examiner) in the ceramics industry, Worms.
20.08.1927-01.06.1929	Assigned to *SA-Sturm 58 Worms*.
01.06.1929-26.01.1930	Assigned as a *Truppführer* to *SA-Sturm 58 Worms*.

SA-Brigadeführer Bock, ca. 1936.

26.01.1930-01.05.1931	*Führer* of *SA-Sturm 58 Worms*.
01.05.1931-09.09.1932	*Führer* of *SA-Sturmbann I/117*.
18.10.1931	Participated in the *SA-Aufmarsch in Braunschweig*.
01.01.1932	Appointed as a *hauptamtlicher SA-Führer*.
00.03.1932-00.03.1932	Participated in the *10. Lehrgang* at the *Reichsführerschule der SA, München*.
03.07.1932-16.07.1932	Participated in *Lehrgang 10a* at the *Reichsführerschule der SA*, München.
09.09.1932-20.02.1933	Adjutant of *SA-Gruppe West* (Koblenz) (with effect from 01.07.1932).
20.02.1933-14.04.1933	Adjutant of *SA-Obergruppe III (SA-Gruppen West, Thüringen*, and *Südwest)* in Koblenz.
15.04.1933-28.04.1933	Adjutant of *SA-Obergruppe IV (SA-Gruppen Sachsen* and *Mitte)*.
28.04.1933-01.07.1933	Adjutant of *SA-Obergruppe VII (SA-Gruppen Bayerische Ostmark, Franken*, and *Hochland)* in Ingolstadt.
01.07.1933-04.10.1934	*Stabsführer* of *SA-Gruppe Bayerische Ostmark*.
13.09.1933-01.10.1934	*Stellvertreter* (deputy) to the *Sonderbevollmächtigten des Obersten SA-Führers bei der Kreisregierung Oberpfalz-Niederbayern* (Special Plenipotentiary of the Supreme *SA* Leader to the District Government of Oberpfalz-Niederbayern) in Regensburg.
05.10.1934-28.02.1935	*Führer* of *SA-Jägerstandarte J 3* (Traunstein).
01.03.1935-31.12.1936	*Führer* of *SA-Brigade 75* (Düsseldorf) in *SA-Gruppe Niederrhein*.
01.10.1935-01.04.1937	*Ratsherr der Stadt Düsseldorf* (Councilman of the City of Düsseldorf).
01.11.1935-09.11.1935	Attended a *Lehrgang* at the *Reichsführerschule der SA*, München.
29.03.1936-08.05.1945	Member of the *Reichstag (Wahlkreis 22, Düsseldorf-Ost)*.

SA-Brigadeführer Bock in a posed image with *SA-Obergruppenführer* Dr. Horst Raeke. Future *Obergruppenführer* Leopold Damian stands at center.

01.03.1937-31.10.1937	Abteilungschef in the *Personalamt der Obersten SA-Führung* (Personnel Office of the Supreme *SA* Leadership) (m.d.W.d.G.b. until 31.03.1937, then permanent from 01.04.1937).
01.11.1937-30.11.1938	*Amtschef* and Chef of the *Amt Soziale Fürsorge* (Social Welfare Office) in the *Personalhauptamt der Obersten SA-Führung*.
01.12.1938-31.01.1942	Chef of the *Amt Gruppenschulen* in the *Erziehungshauptamt der Obersten SA-Führung* (Educational Main Office of the Supreme *SA* Leadership).
10.05.1940-05.05.1944	*Wehrmacht* service, assigned to *Infanterie-Lehrregiment Döberitz-Elsgrund* and later to *Infanterie-Regiment 409* with which he participated in the French Campaign as a *Bataillons-Adjutant*. Discharged from active military service, 05.05.1944.
01.02.1942-00.05.1945	*Führer* of *SA-Gruppe Niederrhein* (Düsseldorf). Succeeded Heinrich Knickmann, who had been killed in action in Russia on 05.08.1941.
00.00.1943-00.00.194_	*Preußischer Provinzialrat* in the *Rheinprovinz*.
17.09.1944-00.00.194_	Unknown assignment with the *Wehrersatzinspektion Köln*.
30.09.1944	Appointed as *Gehilfe für die Organisation des Volkssturmes* (Assistant for the Organization of the German Home Defense Army) attached to the *Gauleiter* and *Reichsverteidigungskommissar* in Düsseldorf (Friedrich Karl Florian).

00.10.1944-00.00.1945   *Gaustabsführer des Volkssturms im Gau Düsseldorf* (Chief of Staff of the *Volkssturm* in *Gau* Düsseldorf).

Postwar Activities

Appeared as a witness before the International Military Tribunal, Nürnberg, 12.08.1946, speaking on the development and organization of the *Sturmabteilung*.

Decorations and Awards

00.04.1942	*1939 Eisernes Kreuz I. Klasse*
00.00.194_	*1939 Eisernes Kreuz II. Klasse*
31.07.1943	*Kriegsverdienstkreuz I. Klasse mit Schwertern*
26.05.1943	*Kriegsverdienstkreuz II. Klasse mit Schwertern*
00.00.194_	*Verwundetenabzeichen, 1939 in Schwarz*
ca. 1942	*Medaille "Winterschlacht im Osten 1941/42"*
05.10.1940	*Medaille zur Erinnerung an den 1. Oktober 1938*
ca. 1938	*Medaille zur Erinnerung an den 13. März 1938*
00.00.193_	*Goldenes Ehrenzeichen der NSDAP*
ca. 1934	*Ehrenzeichen des 9. November 1923 (Blutorden)*
30.01.1942	*Dienstauszeichnung der NSDAP in Gold*
30.01.1940	*Dienstauszeichnung der NSDAP in Silber*
30.01.1940	*Dienstauszeichnung der NSDAP in Bronze*
00.00.193_	*SA-Sportabzeichen in Bronze*
16.07.1932	*Tyr-Rune*
00.00.1934	*Ehrendolch der SA* (mit Wirkung vom 03.02.1934)
00.02.1934	*Ehrenwinkel für alte Kämpfer*

Notes

- Son of the orchestra director Karl Bock (born 17.04.1880 in Kaltenbrunn) and his wife Elisabeth, née Hösl (born 24.02.1875 in Kaltenbrunn).
- Religion: Roman Catholic until 18.11.1937, then left the church and declared himself "gottgläubig".
- Married on 05.08.1933 to Anneliese Finninger (born 09.12.1905 in Worms).

Sources

Bundesarchiv, Berlin-Lichterfelde (former Berlin Document Center): *Personalunterlagen von SA-Angehörigen: SA-Personalakte* of Franz Bock.

Lilla, Joachim; Döring, Martin; & Schulz Andreas: *Statisten in Uniform. Die Mitglieder des Reichstags 1933-1945*. Droste Verlag, 2004.

Arthur Böckenhauer
SA-Obergruppenführer

Born:	13.09.1899 in Hamburg.
Died:	18.04.1953 in Hamburg.

NSDAP-Nr.:	12 815 (First joined 01.10.1922; Party banned following the *München-Putsch* of 09.11.1923; Reenrolled 01.07.1925; Expelled, 08.05.1928; Reenrolled 01.06..1930)

Promotions

00.00.1918	*Unteroffizier*
01.11.1930	*SA-Standartenführer*
15.04.1931	*SA-Oberführer*
01.03.1933	*SA-Gruppenführer*
09.11.1936	*SA-Obergruppenführer*
27.08.1939	*Unteroffizier d. R.*
01.12.1939	*Feldwebel d. R. und Reserve-Offizier-Anwärter*
01.05.1940	*Leutnant d. R.*
02.02.1941	*Oberleutnant d. R.*
01.02.1943	*Hauptmann d. R.*
01.05.1944	*Major d. R.*

Career

00.00.1906-00.00.1913	Attended *Volksschule* in Hamburg.
00.00.1913-00.00.1915	Attended a *kaufmänische Fortbildungsschule* in Hamburg.
00.00.1913-00.00.1916	Served a business apprenticeship in the paper export business.
00.00.1914-00.00.1916	Underwent pre-military training with *Kompanie 28* of *IX. Armee-Korps*.
00.00.1917-00.00.1917	Briefly worked as a laborer in a gunpowder factory, Hamburg.
23.05.1917-01.01.1918	Entered service as a *Kriegsfreiwilliger*, assigned to *3. Kompanie / Ersatz-Bataillon / Reserve-Infanterie-Regiment 96*.
00.12.1917	Tried by a *Militärgericht* (military court), charged with "privater Sachbeschädigung" (damage to private property) and sentenced to 3 days' confinement (amnestied on 27.01.1918).
02.01.1918-30.03.1918	Assigned to *2. Kompanie / Ersatz-Bataillon / Reserve-Infanterie-Regiment 86*.
30.03.1918-02.08.1918	Assigned to with *11. Kompanie / Infanterie-Regiment 31* on the Western Front.
03.08.1918-30.11.1919	Assigned to *5.(Genesenden-)Kompanie / Ersatz-Bataillon / Infanterie-Regiment 31* and *Freikorps Bade*.
01.12.1919-07.09.1920	Assigned to *9. Kompanie / Reichswehr-Infanterie-Regiment 17*.
08.09.1920-01.07.1923	Assigned to *3. Kompanie / Infanterie-Regiment 6*.
00.00.1920-00.00.1922	Member of the *Deutschvölkischische Schutz- und Trutzbun*d.
00.00.1921-00.00.1922	Attended a *Heeresfachschule* (arrmy specialists' school).

Arthur Böckenhauer, ca. 1930.

01.10.1922	Joined the *NSDAP / Ortsgruppe Hamburg*.
01.10.1922-12.02.1923	Joined the *SA,* assigned as *Führer der SA* in Hamburg.
12.02.1923-01.03.1925	*Vorsitzender* of the *Blücher-Turn-, Sport- und Wandervereinigung von 1923* (Blücher Gymnastic, Sport, and Hiking Club). At the time of its establishment, it had only five members (Böckenhauer, Walter Ballhorn, Gustav Dahlhaus, Paul Schlange, and Erich Seiler); by 1924, it numbered 50. The early Hamburg *Gauleiter* Dr. Albert Krebs writes:

> This was originally a cover organization of the Stormtroopers [*SA*], made necessary by the repeated prohibitions of the party and its affiliates in Hamburg since 1922, but over the years it had become so accustomed to its organizational independence that one could hardly speak of it as party of the party any more. Very few members of the Blücher League were also members of the party. Böckenhauer himself belonged but took no orders for the party for the Blücher League. It would have been completely unthinkable to dismiss him from his position of leadership: the Blücher people followed him, not the party. (Krebs, *The Infancy of Nazism*, p. 41)

00.00.1923-00.00.1933	Served as a *Gauredner der NSDAP* in Hamburg.
00.04.1923-01.07.1923	On leave.
01.07.1923	Discharged from the *Reichswehr*.
00.00.1923-00.00.1924	Employed as a bank clerk in Hamburg.

29.06.1924-12.09.1924	Trained former *SA* members as *Zeitfreiwilligen* (temporary volunteers) for service in *Küstenabwehr-Abteilung VI* (Coastal Defense Detachment 6) of the *Reichsmarine*.
01.03.1925	Reenrolled in the *NSDAP*.
01.03.1925-00.06.1927	Reentered the *SA*, assigned as *Führer der SA* in Hamburg. Dr. Krebs, his nominal superior as *Ortsgruppenleiter* of the independent *Ortsgruppe Hamburg*, provides the following character study of Böckenhauer:

> A representative of [the] more intellectual sort [of Nazi], noted for their strongly individualistic traits, was the first leader of the Hamburg Stormtroopers, Arthur Böckenhauer. The tension between emotion, will, and reason led him, after reason had triumphed, to [the] worship of force and power... But it was not this alone that indicated this spiritual relationship; it could be seen also in the way in which Böckenhauer substantiated his opinions, in his method of subtly and pitilessly dispatching his opponents, in the way in which he was driven by a constant restlessness, the real motive for which was often unclear, and which he tried to justify or to hid by calling it other names. At such times he spoke of his "duties," which left him no time for a private life. In fact his duties were his private life. This life denied him entry into certain middle-class circles, just as he prevented himself from giving in to certain middle-class emotions.
>
> It was remarkable how much Böckenhauer, who certainly knew nothing of the theories of nihilism or those of the fate conscious supermen, incorporated those theories in his attitude. Within the bounds of his effectiveness he was the complete representative of the breed of man who finds only in politics the opportunity to acquire and exercise power. In this connection he was never troubled for an instant by the question of the moral justification of power, as far as I could see.
>
> Full of primal vitality, this product of a small artisan family and the noncommissioned officers' school grasped for power in order to satisfy his burning but wholly egotistical ambition. To him Nazism meant little more than a chance to develop and use his own powers and abilities. Once, when Hitler refused to accede to his wishes, Böckenhauer made the following significant statement in a public meeting: 'What right has this man from München to give us orders in Hamburg?' But at the same time he resembled Hitler in his ability to inspire a fanatical following and in his daredevil courage, which was partly a matter of temperament, partly the result of the sober reflection that nothing ventured means nothing gained.
>
> ... It was his ambition, rather than any ideological doubts or reservations, that led to his repeated quarrels with officials of the civilian sector of the party and to consequent disciplinary

SA-Obergruppenführer Böckenhauer shakes hands with Kurt Schmalz, deputy *Gauleiter* of Südhannover-Braunschweig, ca. 1938, as his fellow *SA* leaders Waldemar Geyer and Georg Oberdieck look on.

measures against him. I had to expel him from the party once myself. Nevertheless, he always found his way back into the party again and in increasingly higher positions, even though he was not looked on with any particular favor by the national leadership. Rudolf Hess, for example, believed Böckenhauer to have Jewish blood in his veins, because of his physiognomy. (ibid, pp. 48-49)

03.03.1925	Cofounder of the *Landesverband Hamburg* (Hamburg State Association) of the *NSDAP*.
18.03.1925-03.11.1926	Police official (assigned as a training instructor and staff official) with the *kasernierten Ordnungspolizei* (barracks-housed Order Police) in Hamburg; ultimately assigned to *4.(Radfahr-) Hundertschaft*. He was immediately discharged from police service due to his political activities on 03.11.1926.
12.06.1925	Organizer of the *Sportfest* (sports festival) in Groß-Borstel (first march of the Hamburg *SA* since the lifting of the government ban on the organization).
01.07.1925	Officially reenrolled in the *NSDAP* and the *SA*.
01.04.1926-14.09.1927	*Führer der SS* in Hamburg. Establishing the first *SS* unit in the city was an act of open defiance against the elderly *Gauleiter*, Josef Klant.

Official portrait of Arthur Böckenhauer, ca. 1936. (Laurens Hessels photo)

Böckenhauer's first 50 *SS* men had earlier been expelled from the *NSDAP* by Klant, who experienced much difficulty in dealing with the *SA* during his tenure as *Gauleiter*. Hamburg police reports from 1926 state:

> Klant [had] fallen into disfavor among many members of his party for not being radical enough, and especially for resisting every expansion of the *Sturmabteilungen*. He has recently expelled a great number of members for the party.

and

> In National Socialist circles the opinion reigns that Klant has harmed the further development of the National Socialist movement, especially through his handling of the *SA,* whose expansion he opposes. The *SA,* furthermore, seeks to protect its independence from the regional and local leadership [*Gau- und Ortsgruppenleitung*]. (StAH [*Staatsanwaltschaft* Hamburg] 331-3 / 1097, reports of 17.12.1926 and 18.11.1926, respectively, in Andrew Wackerfuss, M.A., *The Stormtrooper Family: How Sexuality, Spirituality, and Community Shaped the Hamburg SA* [dissertation], p. 80)

04.11.1926	Declined to accept leadership of the *selbständigen Ortsgruppe Hamburg der NSDAP* (independent local group of the *NSDAP* in Hamburg), which instead went to Dr. Albert Krebs.
04.01.1927-03.05.1928	*Gauverwalter* (regional administrator) of the *Gausturm Nordmark*.
00.00.1927-00.00.1930	Employed in various professions, including work as a warehouse laborer.
01.03.1927-03.05.1928	*Gau-SS-Führer Nordmark*.
00.03.1927-21.08.1927	*Führer* of *SA-Standarte II* (Hamburg). Succeeded by Paul Ellerhusen.
19.08.1927-21.08.1927	*Führer* of the *SA-Transportzug* sent from northern Germany to the *3. Reichsparteitag der NSDAP* in Nürnberg.
21.08.1927-20.04.1928	*Gau-SA-Führer Nordmark* (Schleswig-Holstein, Hamburg, Lüneburg-Stade).
16.02.1928-20.04.1928	*Gau-SA-Führer* for Hamburg.
19.02.1928	Unsuccessful candidate for election to the *Hamburger Bürgerschaft*.
00.00.1928-00.00.1928	*Wahlleiter der NSDAP* for the *Wahlkreis Hamburg-Land* in the elections for the *Hamburgische Bürgerschaft* (the state parliament for Hamburg).
08.05.1928	Expelled from the *NSDAP* and the *SA*. As with Klant, Böckenhauer had difficulty working with Dr. Krebs, whom he saw as a rival, and actively conspired against the new *Ortsgruppenleiter* (soon to be appointed as *Gauleiter*). Andrew Wackerfuss writes:

> [Alfred] Conn [a Hamburg *SA* leader] reported that Böckenhauer had spied on Krebs and sought to review, "in secret", his every decision as *Ortsgruppenleiter*. Böckenhauer spread rumors that the poor financial condition of the party was caused by the corruption and incompetence of Krebs and his business manager, Edgar Brinkmann. Furthermore, Böckenhauer charged that the two had failed to turn over the proceeds from the Hitler rallies. He also declared that the party leadership had turned into a "boss system" ("Bonzenwirtschaft"), an attack aimed at subverting the very same younger and working-class youths whom Krebs sought to attract. When Krebs declined to remove Brinkmann from his post, Böckenhauer, as he had against Klant, ordered the *SA* to stop protecting meeting halls. (ibid, p. 89)

As a result of this infighting, *Reichsorganisationsleiter* Gregor Straßer convened a meeting with Böckenhauer, Dr. Krebs, and *SA-Standartenführer* Paul Ellerhusen (who had succeeded Böckenhauer as Hamburg *SA-Führer*).

00.00.1928	Sentenced by a *Standgericht* (summary court) to 3 months' imprisonment for assault. He was released early due to an amnesty.
01.06.1930	Reenrolled in the *NSDAP*.
00.10.1930-15.04.1931	Reentered the *SA*, assigned as *Sachbearbeiter Hamburg* and *Stabszahlmeister* (staff paymaster) to the Stab Gausturm Nordmark.
01.11.1930	Entered full-time service as an *SA-Führer* in Hamburg.
01.11.1930-15.04.1931	*2. Adjutant of Gausturm Nordmark*, simultanously "beauftragt mit der Wahrnehmung der Stabszahlmeister-Geschäfte" (charged with the

Arthur Böckenhauer as a reserve officer of the German Army, ca. 1941.

duties of staff treasuer) and assigned as *Sachbearbeiter für die Bri IV, V, XV* on the staff of *Gausturm Nordmark*.

00.00.1930-30.04.1931	*Führer* of *SA-Sturmbann I/31* (Altona).
15.04.1931-30.06.1933	*Führer* of *SA-Untergruppe Hamburg*.
27.09.1931-00.03.1933	Member of the *Hamburger Bürgerschaft*.
17.10.1931-18.10.1931	Participated in the *SA-Aufmarsch in Braunschweig*.
06.12.1931-19.12.1931	Participated in the *7.Lehrgang* (7th Instructional Course) at the *Reichsführerschule der SA* in München.
00.00.1932-00.00.1932	*Führer* of the *Freiwilligen-Partei-Arbeitsdienst* in Hamburg.
31.07.1932-14.10.1933	Member of the *Reichstag (Wahlkreis 34, Hamburg)*.
ca. 1933-28.02.1934	*z. V. [zur Verfügung] des Preußischen Ministeriums des Innern* (at disposal of the Prussian Ministry of Interior).
14.03.1933-01.03.1934	Sonderkommissar of the *Obersten SA-Führung* in Hamburg.
01.07.1933-14.09.1933	*Führer* (m.d.F.b.) of *SA-Gruppe Hansa* (Hamburg), formed 01.07.1933 from his *SA-Untergruppe Hamburg* as well as *SA-Untergruppe Mecklenburg*. Succeeded by Herbert Fust.
15.09.1933-28.02.1934	Assigned as an *SA-Führer z. V.* to *SA-Obergruppe II* (Stettin).
25.09.1933-22.12.1933	Participated in an *Offiziers-Einweisungslehrgang* (introductory instructional course for officers) at the *Höheren Polizeischule* in Eiche bei Potsdam, and passed his examinations toward becoming a police officer.
01.03.1934-31.03.1935	Transferred to the active *SA-Führerkorps* and assigned as *Abteilungschef (Sachbearbeiter Feldjägerkorps)* to the *Politische Amt* of the *Stab der Obersten SA-Führung*.

01.08.1934-31.03.1935	*Vorsitzender* of an *SA-Sondergericht* (special court) tasked with investigating the so-called "Röhm-Affäre". In this assignment, Böckenhauer was tasked with conducting an investigation- conducted by commissions assigned to each *SA-Gruppe-* and purge of the *SA*. Hitler's order for the formation of the court, dated 09.08.1934, stated that it was "to investigate 'all circumstances by which *SA* leaders have rendered themselves unworthy of membership of the *SA* corps of leaders, such as mode of life, immorality, place-seeking materialism, embezzlement, drunkenness, snobbery and debauchery.'" (Höhne, *The Order of the Death's Head*, p. 147)
01.04.1935-31.10.1937	Chef of the *Gerichts- und Rechtsamt der Obersten SA-Führung* (Courts and Law Office of the Supreme *SA* Leadership).
01.04.1935-00.04.1936	*Inhaber der gesamten Dienstaufsicht des Stab der Obersten SA-Führer* (Holder of the Entire Administrative Supervision of the Staff, Supreme *SA* Leader).
00.03.1936	Appointed as *Vorsitzender* of the *I. Disziplinarkammer* (1st Disciplinary Chamber).
29.03.1936-18.04.1945	Member of the *Reichstag (Wahlkreis 18, Westfalen-Süd*; after 10.04.1938 *Wahlkreis 16, Südhannover-Braunschweig).*
01.05.1936-30.11.1936	Chef (m.d.W.d.G.b.) of the *Personalamt* (personnel office) in the *Obersten SA-Führung.* Succeeded Dr. Otto Marxer.
05.01.1937-00.00.194_	Honorary member of the *Volksgerichtshof* (5-year appointment, possibly extended on 07.11.1941).
01.11.1937-31.05.1938	*Führer* of *SA-Gruppe Niedersachsen* (Hannover). Succeeded Siegfried Kasche. Resigned at his own request due to illness, and succeeded by Max Linsmayer.
01.06.1938-01.09.1939(?)	Assigned as an *SA-Führer z. V.* and *Hauptamtschef* to the *Stab der Obersten SA-Führung.*
01.09.1939-18.04.1945(?)	Assigned as an *SA-Führer z. V.* to the *Adjutantur* of the *SA-Stabschef* in the *Stab der Obersten SA-Führung.*
27.08.1939-00.00.194_	Entered Army service as an *Unteroffizier d. R.*, assigned to an as yet unknown unit with *Feldpost-Nr.* 39 253, Hamburg (later in France).
[01.05.1944]-00.00.194_	Assigned as *Major d. R.* to an as yet unknown unit (with *Feldpost-Nr.* 24210; possibly one of the following: *Kommando* (headquarters) *550.Grenadier-Division*; *Divisions-Nachrichten-Führer 1550*; or *Feldgendarmerie-Trupp 1550*).

Published Work

10 Jahre SA Hamburg in Bildern (1932)

Decorations and Awards

00.00.1940	*1939 Eisernes Kreuz I. Klasse*
00.00.1940	*1939 Spange zum* Eisernen Kreuz II. Klasse
18.10.1919	*1914 Eisernes Kreuz II. Klasse*
23.05.1935	*Ehrenkreuz des Weltkrieges 1914-1918 mit Schwertern*
30.01.1937	*Goldenes Ehrenzeichen der NSDAP*

20.04.1940	*Dienstauszeichnung der NSDAP in Silber*
20.04.1940	*Dienstauszeichnung der NSDAP in Bronze*
ca. 1931	*Abzeichen des SA-Treffens Braunschweig 1931*
08.10.1932	*Deutsches Turn- und Sportabzeichen in Silber* (Nr. 9 579)
20.11.1925	*Deutsches Turn- und Sportabzeichen in Bronze* (Nr. 34 582)
00.00.1934	*Ehrendolch der SA*
19.12.1931	*Tyr-Rune*
00.02.1934	*Ehrenwinkel für alte Kämpfer*
24.04.1931	Commemorative Medal of the War of 1914-1918 with Swords (Hungary)

Notes

- Son of the tailor Jochen Christian *Friedrich* Theodor Böckenhauer (born 22.03.1870 in Sterley / Kreis Lauenburg) and his wife Theresia Franziska Henriette, née Sontag (born 10.11.1870 in Hamburg).
- Married on 04.10.1934 to *Nina* Gerda Christina Egbera (born 31.08.1911 in Wilhelmsburg/ Elbe). Divorced, 01.02.1939. One daughter (born 02.09.1936).
- Religion: Lutheran until 23.12.1936, then left the church.

Sources

Bundesarchiv, Berlin-Lichterfelde (former Berlin Document Center): *Personalunterlagen von SA-Angehörigen: SA-Personalakte* of Arthur Böckenhauer.

Krebs, Dr. phil. Albert: *The Infancy of Nazism: The Memoirs of Ex-Gauleiter Albert Krebs, 1923-1933*. New Viewpoints, 1976.

Lilla, Joachim; Döring, Martin; & Schulz Andreas: *Statisten in Uniform. Die Mitglieder des Reichstags 1933-1945*. Droste Verlag, 2004.

Wackerfuss, Andrew, M.A.: "The Stormtrooper Family: How Sexuality, Spirituality, and Community Shaped the Hamburg SA. A Dissertation submitted to the Faculty of the Graduate School of Arts and Sciences of Georgetown University" Washington, DC, 15.12.2008.

Johann *Heinrich* Adolf Böhmcker
SA-Obergruppenführer

Born:	25.07.1896 in Bosau-Braak bei Eutin.
Died:	16.06.1944 near Hannover (suffered a heart attack during a railway trip from Berlin).

NSDAP-Nr.:	27 601 (Joined 11.01.1926)

Promotions

ca. 1918	*Unteroffizier und Offizieraspirant*
26.12.1925	*SA-Mann*
01.01.1931	*SA-Standartenführer*
09.09.1932	*SA-Oberführer* (mit Wirkung vom 01.07.1932)
01.07.1933	*SA-Brigadeführer*
09.11.1934	*SA-Gruppenführer*
06.10.1940	*SA-Obergruppenführer*

Career

00.00.1903-00.00.1906	Attended *Dorfschule* in Braak.
00.00.1906-00.08.1914	Attended the *Voß-Gymnasium* in Eutin (Passed his Notabitur, 06.08.1914 [or 13.09.1915 according to another source]).
00.08.1914-29.10.1918	Entered service as *Kriegsfreiwilliger*, initially assigned to a Lüneburg-based *Dragonen* unit, then to *Reserve-Kavallerie-Abteilung 78* and *1.Garde-Fußartillerie-Regiment*. Deployed to the Eastern Front, 00.12.1914, and later to the Western Front.
29.10.1918	Placed on leave, however he did not return to active service.
25.03.1919	Discharged from military service.
00.03.1919-00.07.1921	Studied law at the Universities of Kiel, Göttingen, Münster, and again in Kiel. He was a member of the student corps *Brunsviga Göttingen* and *Suevia Strassburg* during this period.
00.07.1921-00.01.1927	*Referendar* (junior lawyer) on various courts.
00.00.192_-00.00.192_	Member of *Organisation Consul*.
00.05.1923-00.05.1923	Arrested and briefly detained, presumably due to his activities with *Organisation Consul*.
26.12.1925-31.10.1928	Joined the *SA*, serving as an *SA-Mann* in Eutin.
11.01.1926	Joined the *NSDAP / Ortsgruppe Eutin*.
00.00.1927	Passed his *1. juristische Staatsprüfung (Referendar-Examen)* (1st state legal examinations [junior lawyer exams]) in Kiel.
00.01.1927	Passed his *2. juristische Staatsprüfung (Assessor-Examen)* on his third attempt.
00.00.1927-00.00.1931	Employed as an attorney and partner in the legal firm of Ernst Evers in Eutin.

Johann Heinrich Böhmcker as an *SA-Brigadeführer*, 1934. (Laurens Hessels photo)

00.00.1928	Appointed as a *Gauredner der NSDAP* (regional speaker of the Nazi Party).
01.11.1928-30.11.1928	*Führer der SA* in Eutin.
01.12.1928-01.06.1929	*Führer* of *SA-Sturm 30* (Eutin).
02.06.1929-31.10.1930	*Führer* of *SA-Standarte XI* (in *Landesteil Lübeck* and *Kreis Plön*).
02.06.1929-31.10.1930	*Führer* of *SA-Standarte III* (Kiel).
00.00.1930-00.00.1932	Bezirksleiter of the *NSDAP* for the *Kreise* of *Oldenburg, Plön*, and *Fürstentum Lübeck*.
00.00.1930-00.00.1933	*Stadtrat* and member of the *Kreistag* (district legislature) in Eutin.
00.00.1930-00.00.1933	Member of the *Landesausschuß für der oldenburgischer Landesteil Lübeck* in Eutin.
01.11.1930-30.06.1931	*Führer* of *SA-Brigade XV*.
01.01.1931-01.07.1932(?)	*Führer* of *SA-Standarte 163*.
17.05.1931-00.10.1933	Member of the *Oldenburgischer Landtag*.
01.07.1931-30.06.1932	*Führer* of *SA-Standarte 163* (with effect from 01.01.1931).
05.11.1931	Unsuccessful candidate of the *NSDAP* for the office of *Ministerpräsident in Oldenburg* (as part of a failed attempt at forming a coalition *NSDAP / DNVP* government).
16.06.1932-00.00.1933	Stellvertretender *Vorsitzender* (deputy chairman) of the *Ausschuß II (Verwaltung) der Oldenburgischer Landtag* (Committee II [Administration] of the Oldenburg State Parliament).
01.07.1932-30.06.1933	*Führer* of *SA-Untergruppe Ostholstein*.

SA-Gruppenführer Böhmcker dines with *SA-Stabschef Lutze* and
German military officers, ca. 1934. (Roger Bender photo)

15.07.1932-01.04.1937	*Regierungspräsident* of the *oldenburgischer Landesteil* (region) Lübeck (Seat: Eutin). The appointment of Böhmcker to this post elevated a hard-drinking street brawler (nicknamed "Latten-Böhmcker" for his choice of weapon – a wooden board) to a significant government post. He proved to be quite capable in office, however, and worked diligently to reform community organization and improve the economic situation in his area through the creation of new jobs. His abuse of power and heavy drinking led to several regular and Party court proceedings, none of which were successful.
00.01.1933	Appointed as a member of the *Oberversicherungsamt* (higher insurance office) in Oldenburg and *Vorsitzender* of the *Spruchkammer der Oberversicherungsamt* (arbitration board of the higher insurance office) established in Eutin (appointment for the duration of his service as *Regierungspräsident*).
Summer 1933	Ordered the establishment of *KL-Eutin*, one of the so-called "wild" concentration camps, which was in operation from July 1933 to May 1934. The majority of its inmates were Social Democrats, Communists, and labor union officials.
01.07.1933-09.07.1934	*Führer* of *SA-Brigade Ostholstein* (redesignated *SA-Brigade 14 "Ostholstein"*, 15.09.1933).
12.11.1933	Unsuccessful candidate for election to the *Reichstag (Wahlkreis 13, Schleswig-Holstein)*.

SA-Gruppenführer Böhmcker, ca. 1935.

10.07.1934-16.06.1944	*Führer* (m.d.F.b. until 14.09.1935, then permanent from 15.09.1935) of *SA-Gruppe Nordsee* (Bremen). Succeeded Wilhelm Freiherr von Schorlemer. Succeeded by Dr. Hans-Joachim Fischer.
30.01.1935-16.06.1944	*Bremischer Staatsrat.*
29.03.1936	Unsuccessful candidate for election to the *Reichstag.*
01.04.1937-16.04.1937	*(kommissarischer) Landrat* of *Landkreis Eutin.*
16.04.1937-16.06.1944	*Regierender Bürgermeister* of the Freien Hansestadt Bremen, appointed by *Reichsstathalter* Carl Roever on 25.06.1937 ("mit der kommissarischen Wahrnehmung der Geschäfte" [charged with acting supervision of business] until 22.06.1937, then permanent). Succeeded Otto Heider. Succeeded by Richard Duckwitz.
22.06.1937-16.06.1944	*Vorsitzender* of the *Bremische Landesregierung* (State Government of Bremen).
22.06.1937-06.07.1938	*Senator der inneren Verwaltung* (Senator for Internal Administration) in the *Bremische Landesregierung.*
00.00.193_-16.06.1944(?)	*Vorsitzender der Verwaltungsrat* (Chairman of the Board of Directors) of the *Bremer Landesbank*, Bremen.
00.00.193_-16.06.1944(?)	Member of the *Aufsichtsrat* of *Bremer Baumwoll AG*, Bremen and the *Eutin-Lübecker Eisenbahngesellschaft*, Lübeck.
01.12.1937-16.06.1944	Member of the *Kuratorium* (board of trustees), *Hamburgische Welt-Wirtschafts-Institutes e. V.* (*HWWI*, Foundation Committee for the Institute for World Economy).

Böhmcker with *Gauleiters* Otto Telschow and Carl Röver, after conferring
honorary citizenship (*Ehrenbürgerrechte*) of Bremerhaven upon them,
June 1937. (*Bremer Nachrichten mit Weser Zeitung*, 19.06.1937)

00.02.1938	Appointed as a member of the *Kolonialrat der Reichskolonialbund* (Colonial Council of the *Reich* Colonial League).
10.04.1938	Unsuccessfully proposed for the *Liste des Führers zur Wahl des Großdeutschen Reichstages am 10. April 1938* (List of the *Führer* for election to the Greater German *Reichstag* on 10. April 1938).
09./10.11.1938	While in München for the memorial celebration of the 09.11.1923 *Putsch*, issued the following explicit orders, via telephone, to *SA-Oberführer* Werner Römpagel, his *Stabsführer* in Bremen, concerning the German "retaliation" for the murder of the diplomat Ernst vom Rath, later known as *Reichskristallnacht*:

> All Jewish shops are immediately to be destroyed by *SA* men in
> uniform ... Jewish synagogues are to be immediately set on fire
> ... The fire brigade is not allowed to interfere. Only residential
> buildings of Aryan Germans are to be protected by the fire
> brigade. Adjacent Jewish residential buildings are also to be
> protected by the fire brigade; however the Jews in them must leave,
> since Aryans will move in there in the next few days ... The police
> are not permitted to interfere. The *Führer* wants the police not
> to interfere ... All Jews are to be disarmed. In case of resistance,
> immediately shoot them down. (Richard J. Evans, "David Irving,
> Hitler and Holocaust Denial: Electronic Edition", at http://www.
> hdot.org/en/trial/defense/evans/430ciiG.html; translated from

SA-Gruppenführer Böhmcker and Viktor Lutze attend a sporting
event in Bremen, 06.06.1936. (Roger Bender photo)

K. Pätzold and I. Runge, *Kristallnacht. Zum Pogrom 1938* [1988])

00.00.1941-16.06.1944   *Vorsitzender des Aufsichtsrates* (Chairman of the Supervisory Board) of
Stadtwerke Bremen AG, Bremen.

Decorations and Awards

21.06.1944   *Ritterkreuz des Kriegsverdienstkreuzes ohne Schwerter* (posthumously)
00.00.191_   *1914 Eisernes Kreuz I. Klasse*
00.00.191_   *1914 Eisernes Kreuz II. Klasse*
24.02.1943   *Kriegsverdienstkreuz I. Klasse mit Schwertern*
01.09.1942   *Kriegsverdienstkreuz I. Klasse ohne Schwerter*
00.00.194_   *Kriegsverdienstkreuz II. Klasse [ohne Schwerter?]*
ca. 1934   *Ehrenkreuz des Weltkrieges 1914-1918 mit Schwertern*
00.00.193_   *Goldenes Ehrenzeichen der NSDAP*
00.00.1929   *Nürnberger Parteitagsabzeichen 1929*
00.00.194_   *Dienstauszeichnung der NSDAP in Gold*
00.00.194_   *Dienstauszeichnung der NSDAP in Silber*
00.00.194_   *Dienstauszeichnung der NSDAP in Bronze*
00.00.193_   *Deutsches Reichssportabzeichen in Bronze*
28.09.1936   *SA-Sportabzeichen in Bronze* (Nr. 955 583)
00.00.1934   *Ehrendolch der SA*
00.02.1934   *Ehrenwinkel für alte Kämpfer*
00.10.1942   Grand Officer's Cross of the Order of the Crown of Italy (*Grande
Ufficiale dell'Ordine della Corona d'Italia*)

SA-Gruppenführer Böhmcker poses with SA-*Standartenführer* Werner Schäfer, *Führer* of *229. SA-Standarte,* ca. 1935. (Laurens Hessels photo)

Notes

- Only child of the farmer Adolf *Hermann Friedrich* Böhmcker (born 29.08.1851 in Bosau, died 15.04.1927 in Eutin) and his wife Elise Christine, née Sachs (born 20.05.1854 in Braak, died 09.05.1926 in Eutin).
- His cousin was Dr. jur. Hans Böhmcker (born 06.11.1899, suicide 19.10.1942), *Senator für Justiz* of the Hansestadt Lübeck (appointed 08.06.1933) and, from 1940 to 1942, *Beauftragter des Deutschen Reiches für die Stadt Amsterdam* (Representative of the German *Reich* for the City of Amsterdam, on the staff of the *Reichskommissar* for the occupied Netherlands, Dr. Artur Seyss-Inquart). In the latter capacity, he played a key role in the persecution and deportation of Dutch Jews.
- Married in Bremen on 29.08.1941 to Frieda Marie Kreide (born 12.08.1912 in Newcastle upon Tyne, England, a secretary to the attorney Dr. Evers and later to Johann Heinrich Böhmcker; widow of *SA-Oberführer* Johannes Valsechi [born 1904, killed in action, 1940], with whom she apparently had four children). Two children were born to the marriage of Heinrich and Frieda Böhmcker. She later married a third time, to Wilhelm Estorff, the founder, in May 1925, of the automotive dealership *Wilhelm H. P. Estorff* in Eutin.

Source

Bundesarchiv, Berlin-Lichterfelde (former Berlin Document Center): *Personalunterlagen von SA-Angehörigen: SA-Personalakte* of Heinrich Böhmcker.

Lilla, Joachim; Döring, Martin; & Schulz Andreas: *Statisten in Uniform. Die Mitglieder des Reichstags 1933-1945.* Droste Verlag, 2004.

Fritz Bracht

SA-Obergruppenführer

Born:	18.01.1899 in Heiden / Lippe.
Suicide:	09.05.1945 (?) in Bad Kudowa bei Glatz / Niederschlesien, by *Blausäure-Kapsel-* hydrogen cyanide capsule, together with his wife, to avoid capture by Soviet troops (date of death uncertain; he was later declared dead with effect from 31.12.1945).
NSDAP-Nr.:	77 890 (Joined 01.04.1927)

Promotions

ca. 1918	*Gefreiter*
01.04.1927-30.09.1928	*Ortsgruppenleiter* der *NSDAP*
01.04.1927	*SA-Sturmführer*
01.10.1928-28.02.1931	*Bezirksleiter der NSDAP*
01.03.1931-30.04.1935	*Kreisleiter* der *NSDAP*
15.10.1933	*SA-Sturmbannführer*
09.04.1935-27.01.1941	*Stellvertretender Gauleiter der NSDAP*
30.01.1938	*SA-Brigadeführer*
29.01.1941-08.05.1945	*Gauleiter* der *NSDAP*
30.01.1941	*SA-Gruppenführer*
20.04.1944	*SA-Obergruppenführer*

Career

ca. 1905-ca. 1909	Attended *Volksschule*.
ca. 1909-00.00.1913	Attended a *Fortbildungsschule*, then underwent training as a gardener.
00.02.1917-00.08.1918	Entered service as a *Kriegsfreiwilliger*, and after 10 weeks' training, assigned to *Infanterie-Regiment* 446. He saw combat on the Western Front, and was promoted to the rank of *Gefreiter* (lance corporal) for "Tapferkeit vor dem Feind" (bravery before the enemy).
00.08.1918-00.12.1919	In British captivity.
ca. 12.1919-00.00.1927	Upon his return from captivity, worked as a gardener and as a fitter for various firms.
01.04.1927	Joined the *NSDAP / Ortsgruppe Plettenberg*.
01.04.1927-30.09.1928	*Ortsgruppenleiter* of the *Ortsgruppe Plettenberg der NSDAP*.
00.00.192_-00.00.193_	*Bezirksleiter* of the *Bezirk Olpe der NSDAP*.
01.04.1927-00.07.1929	Joined the *SA*, assigned as *Führer* of an *SA-Sturm*.
01.10.1928-28.02.1931	*Bezirksleiter* of *Bezirk Sauerland der NSDAP*.
00.00.1929-00.00.1933	*Stadtverordneter* in Plettenberg.
14.09.1930	Unsuccessful candidate for election to the *Reichstag*.
01.03.1931-30.04.1935	*Kreisleiter* of *Kreis Altena-Lüdenscheid der NSDAP* (redesignated *Kreis Altena der NSDAP*, 16.08.1933). Ending date per Andreas Schulz;

Fritz Bracht at his desk in Kattowitz, ca. 1941. (Michal Sika photo)

	Michael Rademacher, in *Handbuch der NSDAP-Gaue 1928-45*, gives 30.03.1935).
24.04.1932-14.10.1933	Member of the *Preußischer Landtag (Wahlkreis 18, Westfalen-Süd)*.
00.00.1933	Appointed as a member of the *Kreistag* (District Legislature) of *Kreis Altena*.
12.11.1933-08.05.1945	Member of the *Reichstag (Wahlkreis 18, Westfalen-Süd*; after 29.03.1936, represented *Wahlkreis 7, Breslau)*.
09.04.1935-27.01.1941	*Stellvertretender Gauleiter* of *Gau Schlesien der NSDAP*, under *Gauleiter* Josef Wagner (assumed this post on 01.05.1935). Succeeded Walter Gottschalk. Ending date per Andreas Schulz; Joachim Lilla gives 09.02.1941 in *Die Stellvertretenden Gauleiter und die Vertretung der Gauleiter der NSDAP im Dritten Reich*. He was the final holder of this post, which was dissolved when *Gau Schlesien* was divided into two separate *Gaue- Oberschlesien* and *Niederschlesien*. In a 1939 fitness report, *Gauleiter* Wagner wrote:

> Bracht has great organizational ability, which I already know from his activity in *Gau* Westfalen-Süd. He proved that he is able to successfully coordinate the Party in such a large *Gau* as Schlesien. Has the ability to speak; in terms of political philosophy he is a National Socialist in the best sense of the word. He stands two feet firmly on the ground with political action in practical sense.

Deputy *Gauleiter* Bracht, ca. 1938. Note the cross on his pocket- the *Lippisches Kriegsehrenkreuz für heldenmütige* Tat (War Honor Cross for Heroic Conduct). (NARA photo)

In terms of character traits, trustworthy ... .; in certain situations, is a little soft in the right sense of the word, which by itself cannot speak of weakness. Diligent and thorough in work; has a great capacity for journalism. In his relations with colleagues and political leaders, tries to establish loyalty and trust, which basically succeeds. His basic modesty is impeccable and must be treated as a positive character trait. It has led here and there to disappointment, which, however, was a good [lesson] for him, and as a result he has become harder. As my closest collaborator, for me he is the best in the hard work of reorganization which was entrusted to me in Schlesien. (Translated and contributed by Igor Karpov)

In a letter of 10.11.1939 to Heinz Lindhorst, Martin Bormann pointed to the difficulties in cooperation between Wagner and Bracht, suggesting an urgent need for the replacement of Bracht. Proposed in Bracht's place as Deputy *Gauleiter* was Georg Tesche, Deputy *Gauleiter* of *Halle-Merseburg*, who wanted to move to Silesia and was considered as a person who could be able to find a common language with Wagner.

ca. 1935   Appointed as a *Preußischer Provinzialrat* for *Provinz Niederschlesien*.

Fritz Bracht in civilian attire, ca. 1939.

01.08.1936-15.10.1936	*(kommissarischer) Stellvertretender Gauleiter* of *Gau Westfalen-Süd der NSDAP*, under Josef Wagner. Succeeded Emil Stürtz. Succeeded by Heinrich Vetter.
30.01.1938-08.05.1945	Assigned as an *SA-Führer z. V.* to *SA-Gruppe Schlesien*.
27.04.1940-28.01.1941	*Mit der verantwortlichen Führung des Gaues Schlesien der NSDAP beauftragt* (charged with the responsible leadership of *Gau Schlesien der NSDAP*). Attended to the duties of this post while *Gauleiter* Josef Wagner was away on other duties, including the running of *Gau Westfalen-Süd*.
28.01.1941-08.05.1945	*Gauleiter* of *Gau Oberschlesien der NSDAP*, formed when *Gau Schlesien* was divided into the separate *Gaue* of *Ober-* and *Niederschlesien* (Karl Hanke was appointed to lead the latter region). Officially appointed to this post, 10.02.1941. From May 1944 to Autumn 1944 and 23.01.1945 (?) to 00.05.1945, the duties of the post were held by deputy *Gauleiter* Rudolf Metzner, due to Bracht's illness.
28.01.1941-08.05.1945	*Oberpräsident der Provinz Oberschlesien.* He was officially appointed to this post by Hermann Göring on 28.01.1941.
28.01.1941-08.05.1945	*Beauftragter des RKF im Gau Oberschlesien.*
01.02.1941	Appointed to the *Sektion Reichsleitung der NSDAP*.
01.03.1941	Greeted Heinrich Himmler and his entourage in Gleiwitz / Oberschlesien, proceeding to the town of Auschwitz and the nearby concentration camp. He then participated in an inspection of the camp,

together with Himmler, the local *HSSPF* Erich von dem Bach, the *Inspekteur der Konzentrationslager* Richard Glücks, Lagerkommandant Rudolf Höss, *Regierungspräsident* Walter Springorum, and leading representatives of *IG Farben*.

31.01.1942 — Meeting in Berlin with Heinrich Himmler, Ernst Heinrich Schmauser (von dem Bach's successor as *HSSPF Südost*), *SS-Oberführer* Rudolf Creutz (*Stabshauptamt RKFDV*), and *SS-Sturmbannführer* Dr. Fritz Arlt.

06.04.1942-00.05.1945 — *Beauftragter des GBA für den Gau Oberschlesien* (Representative of the *Generalbevollmächtigter für den Arbeitseinsatz* [Plenipotentiary General for Labor Allocation, *Gauleiter* Fritz Sauckel] for *Gau Oberschlesien*).

17.07.1942-18.07.1942 — Met with Himmler and his entourage on their arrival at the Kattowitz airport at 1515 hours, 17.07.1942, then drove to *KL-Auschwitz*. After tea in the quarters of *Kommandant* Höss, the group proceeded to inspect agricultural operations within the camp, the *Auschwitz Stammlager* (main camp [*Auschwitz I*]), the *Frauenkonzentrationslager* (*FKL*, concentration camp for women), and dined again in Höss's quarters (itinerary based on *Der Dienstkalender Heinrich Himmlers 1941/42*, pp. 491-493, and the agenda of Himmler's adjutant, Werner Grothmann). Himmler dined that evening in Bracht's residence. The following day, the *Reichsführer-SS* had breakfast with *Gauleiter* and Frau Bracht at 0900 hours before returning with Bracht to Auschwitz and meeting, at 1000 hours, with Schmauser, Dr. Joachim Caesar, and the *Leiterin* of the *FKL*, Johanna Langenfeld. Himmler was subsequently shown the factory grounds of *Auschwitz III* (Monowitz / *Bunawerk*) by representatives of *IG-Farben*. According to the unconfirmed postwar account of Rudolf Höss, during this inspection of *Auschwitz* facilities, Himmler, along with *Gauleiter* Bracht and the latter's deputy, Albert Hoffmann, observed the entire extermination process at Birkenau, beginning with the "selection" of Jews (from a recently arrived transport originating at Westerbork in the Netherlands) to their murder with *Zyklon B* gas in "Bunker II".

16.11.1942-00.05.1945 — *Reichsverteidigungskommissar für den Gau Oberschlesien.*

07.02.1943 — At *Führer* HQ "Wolffschanze" in Rastenburg / Ostpreußen, during which Hitler addressed most of the Reichs- and *Gauleiter* on the subject of the recent German defeat at Stalingrad.

26.07.1944 — In a speech to local leaders of the peasantry, delivered in Kattowitz, declared:

> There should be no anxiety in our *Gau* about events at the front. Oberschlesien remains firm and ready for any sacrifice, standing behind Hitler in the knowledge that the fate of the German people is in good and strong hands. (*Oberschlesische Zeitung*, Kattowitz, 26.07.1944; translated and contributed by Richard Hargreaves)

25.09.1944-00.05.1945 — *Führer des Deutschen Volkssturms im Gau Oberschlesien.* The following article appeared in the *Oberschlesische Zeitung* on Sunday, 22.10.1944:

Official portraits of *Gauleiter* Bracht.

The birth of the *Volkssturm* in the *Gau* of Upper Silesia
Kattowitz, 22. October
Around 3pm yesterday in front of the *Gau* headquarters in the presence of tens of thousands of people, the entire Party leadership and its formations, Generals, a delegation of the *Wehrmacht*, the police, every state and city authority, in front of our *Gauleiter* the historic roll-call of the first Upper Silesian *Volkssturm* battalions took place. The armed battalions of mature *Volkssturm* men radiated strength and determination. Like the huge crowd, they experienced this impressive, historic hour under the battle-hardened banners of the National Socialist *Reich* full of inner tension.

The speech by our *Gauleiter*, evocative and rich in content, every word stressing the importance of the hour, was often interrupted by enthusiastic cries of approval. At the same time that the *Volkssturm* men were proclaiming their loyalty to our homeland before history in the capital of the *Gau*, a delegation from the *Gauleiter* laid a wreath at the *Freikorps* memorial on the Annaberg in memory of the courageous deeds of those men and an expression of gratitude for their loyalty.

The speech by our *Gauleiter*
My *Volkssturm* soldiers and German men and women. At the same time as we are gathered here and have proclaimed our determination to fight to the last, to fight fanatically with all means at our disposal, persevering in the face of doubt and temptation, steadfast and on guard until our flags our victorious, I have ordered a *Volkssturm* unit led by my deputy [Rudolf Metzner] to go to the memorial for fallen *Freikorps* warriors on the Annaberg.

There this unit, as a delegation of the German *Volkssturm* in Upper Silesia, is bound in reverent gratitude with those who remained true to the German Volk and *Reich* in Germany's most shameful hour, who remained undaunted and full of belief, who obeyed only their inner conscience in the struggle for native German soil, for German honor and freedom, sacrificing their lives for German women and children. Whether they were members of the *Freikorps* in Upper Silesia, in the Baltic, in the Ruhr or somewhere else, whether they fought against Polish insurgents or Bolshevik hordes, these men-soldiers from within-were what we now wish to be as *Volkssturm* soldiers: German men and youths in action to the very end, prepared, if necessary, to make the ultimate sacrifice as long as the Fatherland, as long as the homeland is in danger...

Numerous battalions, hundreds of companies will form in the coming days and weeks in the *Gau* of Upper Silesia as part

Vienna, 22.09.1941: *Gauleiter* Bracht receives the *Goldenes Hitler-Jugend Ehrenzeichen mit Eichenlaub* from Baldur von Schirach. (NARA photo)

of the German *Volkssturm*. First of all they will receive infantry training and, especially, anti-tank warfare using close-combat weapons.

At the beginning, however, we men of the *Upper Silesian Volkssturm* affirm and make the holy pledge that in our ranks and in each one of us only the spirit which inspired our *Freikorps* warriors on the Annaberg, the spirit which made them and all the others capable of rising above themselves to make the manliest sacrifice for Germany shall live. This spirit of 21. May 1921 and similarly the spirit of the spring of 1813 will arise again even more fanatically, even more complete in the German *Volkssturm* in 1944. So it should be. We promise that, we vow it...

We know that now we have truly begun to wage total war and that in the shortest possible time we will be a nation in arms-in the truest sense of the word-the laurels of victory can no longer be snatched from us...

Now is it too late for our enemies. Now it has truly become a total, a holy war, one we have been forced to wage. It will end one day, suddenly perhaps. And the bells which will then announce peace to the world, they will be the bells of German victory. (Translation courtesy of Richard Hargreaves)

In a telegram to Hitler of 22.10.1944, Bracht wrote:

In keeping with the immortal tradition of soldiers and *Freikorps* men from my Gau, the soldiers of the Upper Silesian *Volkssturm* will fulfill their duty as soon as the future demands they do. My *Führer*! Many centuries of fate in this borderland rest upon the people of this Gau. It has made them strong and heroic. If Fate should bring the enemy to the borders of the Fatherland, the

Rastenburg / Ostpreußen, 07.10.1943: *Gauleiter* Bracht chats with *Reichsleiter* Walter Buch during the conference of *Reichs-* and *Gauleiter* at *Führer* HQ "Wolfsschanze". Also present, from right to left: Gauleiter Karl Hanke (Niederschlesien), Karl Holz (Franken), and Karl Weinrich (Kurhessen). (NARA photo)

> relentless struggle for every village and for every house will be a sign of our fanatical belief in Germany and you, my *Führer*. Heil, my *Führer*!" (Translation courtesy of Richard Hargreaves)

21.12.1944	In his capacity as *Reichsverteidigungskommissar* in Oberschlesien, issued guidelines for the evacuation of concentration camp inmates and prisoners of war from his Gau, on the basis of which "Todesmärsche" (death marches) to the west were organized by the *SS*.
14.01.1945	Articles appearing in the *Oberschlesischer Beobachter* of that date:

> • Upper Silesia – the *Reich*'s shield in the East
> Powerful declaration of will by all armaments workers to raise their output
> The *Gauleiter*'s speech in front of 12,000 workers
> In a mass rally in front of 12,000 Upper Silesian workers in Königshütte, which we report on this same page under the

headline "Our slogans: Heroic, faithful and tenacious", the *Gauleiter* stressed from the outset that through this rally we wanted to demonstrate-in unison with all the other German workers in the *Gau* of Upper Silesia, nearly one and a half million men in all-that we here in Upper Silesia know what is at stake and what our duties are...

It was extremely admirable, the *Gauleiter* continued, that after air raids, which he himself had experienced, in affected areas which had been attacked a dozen times or more, just ten minutes after the attack work had resumed and production continued under cardboard and corrugated iron roofs, something which was unimaginable.

"I want to take this opportunity to express gratitude from the bottom of my heart and wholeheartedly appreciate all the work of Upper Silesians. It is not only the miner, not only the worker in the mill, it is all of Upper Silesia, it is the farmer as well as the baker, and all the craftsmen and all the workers involved in this enormous front of workers in the *Gau* of Upper Silesia."

• Our slogans: Heroic, faithful and tenacious
A torrent of emotion, one single thought, like a mighty heartbeat, went through the many thousands of Upper Silesia's workers at a rally of the *Deutsche Arbeitsfront* in an armaments factory in Königshütte as *Gauleiter* Fritz Bracht graphically described the difficulties of the past year. The attitude of a Volk, who courageously accept the heaviest burdens to serve their nation was demonstrated not only by the huge procession of flags, which surrounded the platform, but also by the sound of the songs which were sung with gusto by the workers at this solemn hour.

At the beginning of this rally, *Kreisobmann* Poganatz told the *Gauleiter* that the 12,000 comrades had come to once again hear confirmation of their faith in victory in his words and that in the procession of eighty-six flags there were some from model National Socialist plants in Upper Silesia and model Upper Silesian war industries. (Translation courtesy of Richard Hargreaves)

00.01.1945-00.05.1945   With the start of the Soviet offensive, moved his headquarters to Neisse on the Oder and forbade the evacuation of the German civilian population. According to the 20.01.1945 diary entry of *Reichsminister* Dr. Goebbels, *Gauleiters* Bracht and Greiser, together with *Generalgouverneur* Dr. Hans Frank, were transported to Oppeln by *Generaloberst* Ferdinand Schörner on 17.01.1945. Bracht then suffered a severe heart attack (per Goebbels' entry of 24.01.1945) and was hospitalized in Neisse, with his deputy (Rudolf Metzner) taking the reins in *Gau Oberschlesien*. The tubercular Bracht later proceeded to Bad Kudowa bei Glatz / Niederschlesien to convalesce.

Decorations and Awards

00.09.1918	*1914 Eisernes Kreuz II. Klasse*
20.04.1941	*Kriegsverdienstkreuz I. Klasse ohne Schwerter*
00.00.194_	*Kriegsverdienstkreuz II. Klasse ohne Schwerter*
ca. 1918	*Lippisches Kriegsehrenkreuz für heldenmütige Tat*
ca. 1939	*Medaille zur Erinnerung an den 1. Oktober 1938*
ca. 1934	*Ehrenkreuz des Weltkrieges 1914-1918 mit Schwertern*
00.00.193_	*Goldenes Ehrenzeichen der NSDAP*
22.09.1941	*Goldenes Hitler-Jugend Ehrenzeichen mit Eichenlaub* (Presented to him in Wien by Baldur von Schirach)
00.00.194_	*Dienstauszeichnung der NSDAP in Silber*
00.00.194_	*Dienstauszeichnung der NSDAP in Bronze*
00.00.1934	*Ehrendolch der SA*
00.00.1934	*Ehrenwinkel für alte Kämpfer*

Notes

* Son of the laborer Ferdinand Friedrich Christoph Bracht (born 16.08.1871 in Distelberg bei Detmold) and his wife Johanna Wilhelmine Charlotte, née Bücker (born 05.10.1876 in Heiden / Lippe).
* Religion: Protestant until 00.10.1937, then left the church and declared himself "gottgläubig".
* Married to Paula Heseler (born 07.05.1907 in Plettenberg, committed suicide with her husband, 09.05.1945 [?]). One daughter (E., born 13.09.1922).

Sources

Aly, Götz & Herbert, Ulrich (Editors): *National Socialist Extermination Policies: Contemporary German Perspectives and Controversies*. Berghahn Books, 2000.

Höffkes, Karl: *Hitlers politische Generale: Die Gauleiter des Dritten Reiches*. Grabert-Verlag-Tübingen, 1986.

Hüttenberger, Peter: *Die Gauleiter: Studie zum Wandel des Machtgefüges in der NSDAP*. Deutsche Verlags-Anstalt GmbH, 1969.

Lilla, Joachim: *Die Stellvertretenden Gauleiter und die Vertretung der Gauleiter der NSDAP im "Dritten Reich". Materialien aus dem Bundesarchiv, Heft 13*. Koblenz, 2003.

Lilla, Joachim; Döring, Martin; & Schulz, Andreas: *Statisten in Uniform. Die Mitglieder des Reichstags 1933-1945*. Droste Verlag, 2004.

Miller, Michael D. & Schulz, Andreas: *Gauleiter: The Regional Leaders of the Nazi Party and Their Deputies, Volume I*. R. James Bender Publishing, 2012.

Seidler, Franz: *Deutscher Volkssturm: Das letzte Aufgebot 1944-1945*. Bechtermunz Verlag, 1999.

Hermann Max-Gustav **Brauneck**
Dr. med.
SA-Sanitäts-Obergruppenführer

Born:	19.12.1894 in Sulzbach / Kreis Saarbrücken / Rheinprovinz.
Killed:	27.07.1942 during a Soviet air raid on the Strait of Kertsch, eastern Crimea.
NSDAP-Nr.:	496 265 (Joined 01.04.1931)

Promotions

01.04.1913	*Seekadett*
03.04.1914	*Fähnrich zur See*
18.09.1915	*Leutnant zur See*
21.02.1920	*Oberleutnant zur See a. D.*
01.01.1932	*SA-Standartenarzt*
15.12.1932	*SA-Gruppenarzt* (mit Wirkung vom 01.01.1933) (26.05.1933: Redesignated *SA-Sanitäts-Gruppenführer*)
00.00.1935	*Marinestabsarzt d. R.* (mit RDA vom 19.12.1934)
00.06.1936	*Reichshauptstellenleiter der NSDAP*
26.06.1936	*Ministerialrat*
09.11.1937	*SA-Sanitäts-Obergruppenführer*
00.08.1939	*Marinestabsarzt d. R.*
01.02.1941	*Marineoberstabsarzt d. R.*
00.00.1942	*Geschwaderarzt d. R.* (posthumous promotion)

Career

00.00.1901-00.00.1903	Attended *Volksschule*.
00.00.1903-00.00.1904	Attended a *Rektoratsschule*.
00.00.1904-00.00.1913	Attended *Realgymnasium* (graduated and passed his *Abitur*, 00.00.1913).
01.04.1913	Entered the *Kaiserliche Marine* as a *Seekadett*.
01.04.1913-14.08.1913	Underwent naval training at the *Marineschule Mürwick*.
15.08.1913-13.03.1914	After completion of training, assigned to the crew of the heavy cruiser (and *Seekadettenschulschiff* [naval cadets' school ship]) *S.M.S. Hertha*, aboard which he participated in a cruise with the following itinerary: Wilhelmshaven → England → Spain → Halifax, Canada → Vera Cruz, Mexico → the Antilles (West Indies) → Kiel.
00.08.1914-00.07.1917	*Wachoffizier* (watch officer) aboard *S.M.S. Deutschland* (participated in the "Skagerrakschlacht" (Battle of Skagerrak).
00.08.1917-00.04.1918	*Adjutant* and *Funken-Telegraphie Offizier* (*FT-Offizier*, wireless telegraphy officer) *S.M.S. Friedrich der Große* (00.09.1917-00.04.1918), and with the staff of *IV.Linienschiffsgeschwader* (battleship squadron) (00.04.1918-00.11.1918).
00.04.1918 – 00.11.1918	*Flaggleutnant im Stabe* of *IV.Linienschiffsgeschwader*.

SA-Sanitäts-Obergruppenführer **Dr. Hermann Brauneck, ca. 1941. (NARA photo)**

21.02.1920	Discharged from the *Reichsmarine* at the *Marinestation der Ostsee* (Baltic Sea naval station).
00.00.1919-00.00.1922	Studied medicine (passed state medical examinations). His doctoral thesis was entitled *Beiträge zur Aetiologie der Rachiti* (Contributions to the Etiology of Rickets).
00.00.1922	Received his doctorate (Dr. med.).
00.00.1922-00.00.1930	Underwent training as an *Assistenzarzt*.
00.00.1930-00.00.1933	Worked as a surgeon in Bremen, also obtaining his *Sportarztdiploma* (sports physician certificate) during this period.
01.04.1931	Joined the *NSDAP / Ortsgruppe Bremen*.
01.08.1931-31.12.1931	Joined the *SA,* assigned as *Sturmbannarzt* of *SA-Sturmbann II/75* (Bremen).
17.10.1931-18.10.1931	Participated in the *SA-Aufmarsch in Braunschweig*.
01.01.1932-31.12.1932	Standartenarzt of *SA-Standarte* 75 (Bremen).
15.12.1932-14.02.1936	Gruppenarzt of *SA-Gruppe Nordsee* (Bremen), with effect from 01.01.1933.
00.00.1933	Appointed as *Leiter* of the *Gesundheitsamt* (health office) in Bremen.
00.00.1933-24.02.1936	*Präsident* of the *Gesundheitsbehörde* (health authority) in Bremen.
00.00.1934-24.02.1936	*Gauobmann* of the *NSDÄB* in *Gau Weser-Ems*.
00.00.1934-24.02.1936	*Gauamtsleiter* of the *Amt für Volksgesundheit* (public health office) for *Gau Weser-Ems der NSDAP*.
00.00.1934-24.02.1936	*Gauamtsleiter* of the *Rassenpolitisch Amt* for *Gau Weser-Ems der NSDAP*.

SA-Sanitäts-Obergruppenführer Dr. Brauneck, ca. 1937.

ca. 1934	Appointed as a *Gauredner der NSDAP* in *Gau Weser-Ems*.
01.11.1935-28.11.1935	Participated in a military training exercise at the *Marinelazarett Wilhelmshaven*.
00.00.1936-00.00.1937	*Referent Sanitätswesen* in the *Organisationsleitung der Reichsparteitag* (Organizational leadership of the *Reich* Party Congress).
08.02.1936	Appointed as a member of the *Reichsausschußes zum Schutze des deutschen Blutes* (*Reich* Committee for the Protection of German Blood).
15.02.1936-31.01.1937	Assigned as an SA-*Sanitäts-Gruppenführer z. V.* to *SA-Gruppe Berlin-Brandenburg*.
24.02.1936	Appointed as *stellvertretender Vorsitzender* of the *Reichsausschuß zum Schutze der deutschen Blut*, Berlin.
03.05.1936-30.05.1936	Instructor for the *Zweiter Jungärztelehregang* at the *Führerschule der deutschen Ärzteschaft* in Alt-Rehse. The topic of his lecture was "Die Nürnberger Gesetze" (The Nürnberg [Racial] Laws).
00.06.1936	Appointed as a *Reichshauptstellenleiter* in the *Hauptamt für Volksgesundheit der NSDAP*.
26.06.1936	Appointed as a *Ministerialrat im Reichsdienst* with the *Reichsministerium des Innern*.
00.00.1936	Appointed as a *Mitarbeiter* (work colleague) to the *Rassenpolitische Amt der Reichsleitung der NSDAP* (Racial Policy Office of the *Reich* Leadership of the Nazi Party).

16.09.1936-24.09.1936	Instructor for the *Erster Ärztinnenkurs* (first female physicians' course) at the *Führerschule der deutschen Ärzteschaft* in Alt-Rehse. The topic of his lecture was "Praktische Anwendung der Nürnberger Rassengesetze" (Practical Application of the Nürnberg Racial Laws).
01.02.1937	Appointed as a *hauptamtlicher SA-Führer*.
01.02.1937-31.10.1937	*Chef* (m.d.W.d.G.b.) of the *Sanitätsamt* (Medical Office) in the *Obersten SA-Führung* (m.d.W.d.G.b. until 30.04.1937, then permanent from 01.05.1937).
01.11.1937-27.07.1942	*Chef des Sanitätswesens der SA* (Chief of *SA* Health Services).
01.11.1937-31.01.1942	Chef of the *Gesundheitshauptamt* (Main Office for Health) in the *Obersten SA-Führung*.
10.04.1938	Unsuccessful candidate for election to the *Reichstag* (on the *Liste des Führers zur Wahl des Großdeutschen Reichstages am 10. April 1938*).
00.08.1939	Reentered military service as a *Marinestabsarzt d. R.*
01.09.1939-15.02.1940	War service as a surgeon at the *Marinelazarett Stralsund* / Pommern.
15.05.1940-04.07.1940	Service aboard the battleship *Gneisenau*.
Autumn 1940	Assigned as a surgeon to *Marinelazarett Rotterdam*.
00.00.1941-00.00.1941	Assigned as a surgeon to *Marinelazarett Bordeaux*.
01.02.1942-27.07.1942	Inspekteur of the *Inspektion Sanitäts-SA und Sanitätswesen der Reichsparteitage* (Medical Inspectorate of the Medical *SA* and Medical Services for the *Reich* Party Congresses).
Spring 1942-27.07.1942	Chefarzt of *Marine-Zeltlazarett* (tent hospital) *Nr. I*.

Published Articles

"Konservative Magenoperationsmethoden", in *Deutsche Zeitschrift für Chirurgie*, July 1929, Volume 217, Issue 3-4, pp. 228-252

"IV. Mesenterialfibrom", in *Deutsche Zeitschrift für Chirurgie*, April 19266, Volume 195, Issue 4-5, pp. 345-350.

Decorations and Awards

27.09.1919	*1914 Eisernes Kreuz I. Klasse*
01.07.1916	*1914 Eisernes Kreuz II. Klasse*
00.00.194_	*Kriegsverdienstkreuz II. Klasse ohne Schwerter*
15.08.1935	*Ehrenkreuz des Weltkrieges 1914-1918 mit Schwertern*
30.01.1938	*Ehrenzeichens des Deutschen Roten Kreuzes I. Klasse*
30.01.1935	*Verdienstkreuz des Deutschen Roten Kreuzes I. Klasse*
20.04.1940	*Dienstauszeichnung der NSDAP in Bronze*
ca. 1931	*Abzeichen des SA-Treffens Braunschweig 1931*
14.10.1937	*SA-Sportabzeichen in Bronze* (Nr. 1 223 162)
00.00.1934	*Ehrendolch der SA*
00.02.1934	*Ehrenwinkel für alte Kämpfer*

Notes

- Son of the *Sanitätsrat Dr. med.* Hermann Brauneck (born 04.02.1861 in St. Wendel, died 04.05.1918) and his wife Agnes, née Bergmann (born 27.06.1875 in Koblenz). His mother remarried, to Albert Kühme, in 1922.

- Married on 16.12.1922 to *Dr. med.* Elisabeth Lose (born 18.02.1892 in Bremen). Two sons (He. O., born 03.02.1924 and Ha., born 01.04.1935) and three daughters (Gi., born 16.05.1926, Gu., born 02.07.1933, and I., born 01.05.1936).
- Religion: Protestant until 11.06.1936, then left the church and declared himself "gottgläubig".

Sources

Bundesarchiv, Berlin-Lichterfelde (former Berlin Document Center): *Personalunterlagen von SA-Angehörigen: SA-Personalakte* of Hermann Brauneck.

Wilhelm Karl Friedrich ("Owambo") Brückner
SA-Obergruppenführer

Born:	11.12.1884 in Baden-Baden.
Died:	20.08.1954 in Herbstdorf (today known as Nußdorf) / Kreis Traunstein Chiemgau, of a heart attack.
NSDAP-Nr.:	298 623 (first joined 00.00.1922; Party banned following the *München-Putsch* of 09.11.1923; Reenrolled, 01.09.1930)

Promotions

01.10.1904	*Einjährig-Freiwilliger*
20.05.1905	*überzähliger Gefreiter*
01.09.1905	*überzähliger Unteroffizier und Offizier-Aspirant*
06.08.1914	*Unteroffizier*
15.11.1914	*Vizefeldwebel der Landwehr I* (mit Wirkung vom 01.11.1914)
15.11.1914	*Offizierstellvertreter* (mit Wirkung vom 01.11.1914)
13.07.1915	*Feldwebelleutnant*
02.06.1916	*Leutnant der Landwehr I* (mit Patent vom vom 31.10.1912)
15.03.1918	*Oberleutnant der Landwehr I* (mit Patent vom 15.03.1918)
18.12.1931	*SA-Oberführer* (01.07.1932: Reconfirmed in that rank when the government ban on the *SA* was lifted)
01.03.1933	*SA-Gruppenführer*
09.11.1934	*SA-Obergruppenführer*
00.00.193_	*Oberdienstleiter der NSDAP*
01.06.1941	*Major d. R.*
00.00.1943	*Oberstleutnant d. R.*
01.12.1944	*Oberst*

Career

ca. 1891-ca. 1895	Attended *Volksschule* in Baden-Baden (?).
ca. 1895-00.00.1904.	Attended *Realgymnasium* in Baden-Baden (passed his *Abitur*).
01.10.1904-30.09.1905	Entered service as an *Einjährig-Freiwilliger*, assigned to 1. *Kompanie / Kgl. Sächsische 6. Infanterie-Regiment Nr. 105* (Strassburg).
30.09.1905	Granted a *Befähigungszeugnis zum Reserveoffizier* (certificate of fitness as a reserve officer).
30.09.1905	Transferred to the reserves.
00.00.1906-00.00.1907(?)	Volontär (business trainee) in a plumbing supply factory in Baden-Baden, preliminary to studies in engineering.
14.03.1906-08.05.1906	Participated in *Übung A* (training exercise "A") with 6. *Kompanie / Kgl. Sächsische 6. Infanterie-Regiment Nr. 105.*
00.00.1907-00.00.1908	Studied engineering at the *Technische Hochschule* in München.
29.11.1907	At his own request, stricken from the *Liste der Offizier-Aspiranten* (officer candidates' list).

Wilhelm Brückner as *Führer* of *SA-Regiment München*, 1923. He inscribed this photo to a fellow Party member in May 1924. (Hermann-Historica, Auctioneers, München)

00.00.1908-00.00.1909	Underwent training in a furniture manufacturing business in Baden-Baden.
00.00.1909-00.00.1910	Studied economics and law at the University of Munchen.
00.00.1911-00.00.1912	Worked as an administrative clerk for an insurance firm, and later for *Dun Co.* in Frankfurt-am-Main, to earn money to continue his university studies.
31.07.1911-23.09.1911	Participated in *Übung B* with *1. Kompanie / 2. Kgl. Bayerisches Feldartillerie-Regiment* (München).
00.00.1912-00.00.1914	Continued his studies at the University of München, while simultaneously working for *Dun Co.* of Frankfurt-am-Main.
01.04.1912	Transferred to the *Landwehr I. Aufgebot* (1st Levy of the Bavarian State Militia).
07.07.1914-17.07.1914	Participated in a *Landwehr-Übung* with the *Landwehr-Kompanie / 15. Kgl. Bayerisches Feldartillerie-Regiment*.
06.08.1914-15.11.1914	Mobilized as an *Unteroffizier* and assigned to *16. Kompanie / IV. Bataillon / Bayerische Landwehr-Infanterie-Regiment Nr. 1* (the later *I. Bataillon / Bayerische Landwehr-Infanterie-Regiment 15*).
15.11.1914-00.00.1915	Assigned to *7. Kompanie / Bayerische Landwehr-Infanterie-Regiment Nr. 15* (Augsburg).
09.05.1915-18.05.1915	Admitted to *Krankenrevier* (sick bay) with hemorrhoids.
15.08.1915	Ordered confined to quarters for three days due to improper conduct during a patrol.

Three views of Wilhelm Brückner wearing the insignia of an
SA adjutant, ca. 1931. (Roger Bender photos).

Bad Harzburg, 11.10.1931: Wilhelm Brückner follows his Führer during
the "Harzburger Front" meeting. From left to right: Wilhelm Friedrich
Loeper, Brückner, Julius Schaub, Hitler, and Rudolf Hess.

20.10.1915-22.11.1915	Hospitalized due to hemorrhoids in the *Reservelazarett* in Duss.
22.11.1915-25.06.1916(?)	Following discharge from hospital, assigned to *Ersatz-Bataillon / Bayerische Landwehr-Infanterie-Regiment Nr. 1*.
25.06.1916-25.02.1917(?)	Assigned to the *Gefangenenlager* (prisoner of war camp) in Lechfeld.
25.02.1917-00.03.1917	Assigned to 2. *Kompanie / Ersatz-Bataillon / Bayerische Landwehr-Infanterie-Regiment 1*.
04.03.1917-11.03.1917	Participated in a *Granatwerferkurs* at Lager Lechfeld.
11.03.1917-25.03.1917	Again assigned to the *Gefangenenlager Lechfeld*.
25.03.1917-19.05.1917	Attached to 2. *Ersatz-MG-Kompanie*.
19.05.1917-26.09.1917	Attached to 1. *Ersatz-MG-Kompanie / I. Bayerische Armee-Korps*.
26.09.1917-19.10.1917	Deployed to the front with 3. *Kgl. Bayerisches Feldartillerie-Regiment*.
19.10.1917-02.01.1918	Acting *Kompanieführer* of 3. *Kompanie / I. Bataillon / 3. Kgl. Bayerisches Feldartillerie-Regiment*.
02.01.1918-28.01.1918	Acting *Kompanieführer* of 2.*MG-Kompanie / II. Bataillon / 3. Kgl. Bayerisches Feldartillerie-Regiment*.
17.01.1918	Participated in the *14.MG-Kurs für Fliegerbekämpfung* (an anti-aircraft machine gun training course) in Razoy.
28.01.1918-31.01.1918	Briefly held the post of acting *Kompanieführer* of 3. *Kompanie / 3. Kgl. Bayerisches Feldartillerie-Regiment*, then appointed to command a *Zug* (platoon) of the same unit.

Wilhelm Brückner in civilian attire. (Left: Roger Bender photo; Right:
Baldur von Schirach, *Die Pioniere des Dritten Reiches*, 1933)

Wilhelm Brückner (far right) poses with his close colleague Julius Schaub
(also a senior adjutant to Hitler) and *Reich* Minister of Agriculture
Walther Darré. (NARA, Heinrich Hoffmann collection)

A relaxed and refined-looking Wilhelm Brückner in 1933.

31.01.1918-15.04.1918	*Zugführer* in *3. Kompanie / 3. Kgl. Bayerisches Feldartillerie-Regiment.*
15.04.1918	Wounded in action near Zwartemolenhoek (bullet wound to the buttocks and right wrist).
15.04.1918-21.04.1918	Hospitalized in *Bayerisches Feldlazarett 21.*
21.04.1918-25.04.1918	In the *Reservelazarett Essen.*
26.04.1918-24.06.1918	In the *Reservelazarett München Station A.*
27.06.1918-31.10.1918(?)	Assigned to *2. Ersatz-MG-Kompanie / I. Bayerische Armee-Korps.*
31.08.1918-01.09.1918	Treated in a *Krankenrevier* for diarrhea.
21.09.1918-04.10.1918	Participated in a *Kurs für indirektes Schießen* (indirect fire course) at *Lager Hamelburg.*
31.10.1918-16.11.1918	*Kompanieführer* of *MG-Kompanie Brückner* (a *Grenzschutz* [border defense] unit). On 16.11.1918, it was subordinated to *Ersatz-Bataillon / 3. Kgl. Bayerisches Feldartillerie-Regiment* as *3. Ersatz-Kompanie,* Brückner retaining command of the unit.
16.11.1918-13.01.1919	*Kompanieführer* of *3. Ersatz-Kompanie / Ersatz-Bataillon / 3. Kgl. Bayerisches Feldartillerie-Regiment* (redesignated as *3. Demobilisierungs-Kompanie,* with Brückner retaining command, on 13.01.1919).
11.01.1919-11.05.1919	On paid leave in Geisenhofen.
13.01.1919-01.03.1919	*Kompanieführer* of *3. Demobilisierungs-Kompanie / Ersatz-Bataillon / 3. Kgl. Bayerisches Feldartillerie-Regiment.*
25.03.1919	Discharged from military service with effect from 31.03.1919.

SA-Oberführer Brückner marches past Hitler and other senior
NSDAP, *SA*, and *SS* figures, ca. 1931. (Roger Bender photo)

SA-Gruppenführer Brückner inspects a contingent of *NSDAP* political
leaders wearing steel helmets in 1933. Erich Koch, *Gauleiter* of
Ostpreußen, walks beside him. (Roger Bender photo)

00.00.1919-00.10.1919	Service with *Freikorps Epp* and *Schützen-Regiment 42* of the *Reichswehr*, participating in the suppression of the *Münchner Räterepublik* (the short-lived communist government of Bayern).
00.10.1919	Discharged from military service.
00.10.1919-00.00.1921	Resumed his studies at the University of München.
00.00.1919	Joined the *Münchner Einwohnerwehr* (München Home Guard).
00.00.1920-00.00.1923	Employed as a film cameraman and producer with the firm of *Arnold & Richter* in München.
00.00.1922	Joined the *NSDAP* and the *SA*.
00.00.1922-01.03.1923	*Kompanieführer*, then *Bataillonsführer*, in the München *SA*.
01.01.1923-31.12.1924	Full-time clerical employee of the *NSDAP*.
01.03.1923-09.11.1923	*Führer* of *SA-Regiment München*, comprising three battalions (*I.*, *II.*, and *III.*, commanded by Karl Beggel, Edmund Heines, and Hans Knauth, respectively). Brückner led this regiment of some 1,500 men in the *München-Putsch* of 08./09.11.1923, and marched in the first row (with Hitler, Ludendorf, Göring, and other Putsch leaders) toward the Feldherrnhalle.
01.04.1924	Sentenced by the *Landgericht* München to 1 year, 3 months' imprisonment for "aiding and abetting high treason" for his role in the *München-Putsch*. The sentence was reduced to 4 years' probation and he was immediately released, together with Dr. Wilhelm Frick and Ernst Röhm. He was additionally fined 100 RM and court costs totaling 769,10 RM.
00.00.1924-00.00.1925	Studied economics and history at the University of München (4 semesters).
00.00.1924-00.00.192_	Again assigned as *Führer* of *Regiment München* (now a part of the *Frontbann*).
00.08.1924-01.05.1925	*z.b.V.* (at disposal) of *Stab / Oberkommandos des Frontbann*.
End of 1924	Charged with "Geheimbündelei" (conspiracy) and sentenced to two months' imprisonment.
01.05.1925	Left the *NSDAP*.
00.00.1924-00.00.1928	*3. Generalsekretär* of the *Verein für das Deutschtum im Ausland* (VDA, Association for Germans Abroad), leading the München branch of that organization. He was primarily concerned with maintaining the "cultural consciousness" of Germans living in border areas ceded to France by the Treaty of Versailles.
00.00.1927-00.00.1929	Employed as a tennis and skiing instructor and commercial representative for sporting goods in München.
00.00.1929-00.00.1930	Clerical employee of the *Deutsche Auslandsinstitut* (*DAI*, German Foreign Institute) in Stuttgart. In this capacity, he organized a "Danzig Exposition" in 1929.
00.00.1930-00.00.1930	Worked as a tennis instructor in München.
01.09.1930	Reenrolled in the *NSDAP*.
01.09.1930	Joined the *SA*.
01.09.1930-18.10.1940	Member of the *persönlicher Stab* (personal staff) of Adolf Hitler.
01.09.1930-00.04.1933	*2. Persönlicher Adjutant* (2nd Personal Adjutant) to Hitler.

Wilhelm Brückner and the *Führer* during the *6.Reichsparteitag der NSDAP* in Nürnberg, September 1934. (Roger Bender photo)

Heinrich Hoffmann's postcard of Wilhelm Brückner, 1933. (Roger Bender photo)

Brückner and Julius Schaub follow the *Führer* during the *Heldengedächtnisfeier* (Heroes' Remembrance Festival) in München, 17.03.1935. (Roger Bender photo).

01.09.1930	Appointed as *SA Adjutant* to the *Führer*.
00.00.1931-00.00.1933	*Führer* of the *SA-Reichsleitung-Sturm* and *Leiter* of *Unterabteilung F* in *Amt III* of the *Quartiermeisterstab beim Stab der OSAF*.
15.08.1933	While driving with his fiance, Sophie Stork, behind Hitler's car, lost control of his vehicle. Brückner's car came to rest in a ditch, and he suffered a broken leg, fractured skull, and injury to his eye. Luckily for Brückner, the driver of the vehicle behind him was Dr. Karl Brandt, a talented young surgeon on holiday with his own fiance in the area of Berchtesgaden. Ulf Schmidt writes of this incident, which led to Brandt's appointment as Hitler's escorting physician, as follows: "Brandt immediately provided first aid and drove Brückner ... to the nearest hospital in Traunstein. Brandt ... operated on Brückner's fractured skull and removed one of his badly injured eyes ... [He] spent the next six weeks at Brückner's bedside until his condition had clearly improved ... " (Schmidt, *Karl Brandt: The Nazi Doctor. Medicine and Power in the Third Reich*, pp. 53-55)
20.02.1934-18.10.1940	*Chefadjutant* to the *Führer und Reichskanzler* (Hitler). Former Hamburg *Gauleiter* Dr. Albert Krebs provides the following character study of Hitler's chief adjutant:

> Brückner always came across to me as the harmlessly snappy lieutenant from a Balkan operetta, a man who combined a healthy natural boyishness with born cordial courtesy and well-trained good manners ... He was totally unproblematical and completely without personal plans or intentions, excepting a naturally patriotic viewpoint. But precisely for these reasons he was superbly suited for his office. Hitler needed a few hands to work for him and a few mouths to speak for him in his immediate staff, but not heads that might think. Beyond that, Brückner, with his sparkling external appearance, made a respectable receptionist as well as a messenger for the orders of his lord and master ... (Dr. Krebs, *The Infancy of Nazism*, p. 253)

00.00.193_-08.05.1945(?)	Assigned as an *SA-Führer z. V.* to the *Stab der Obersten SA-Führung*.
30.06.1934	Accompanied Hitler on his trip from München to Bad Wiessee, during which *SA-Stabschef* Röhm and other *SA* leaders were arrested.
29.03.1936-08.05.1945	Member of the *Reichstag (Wahlkreis 3, Berlin Ost)*.
00.00.1940	Falling out with Hitler, resulting in his dismissal. The incident that precipitated this was Brückner's speaking out on behalf of Hitler's ordnance officer, Max Wünsche, after complaints against him by the *Führer*'s butler, Arthur Kannenberg. In the course of a visit to Obersalzberg by Crown Prince Umberto of Italy, Kannenberg complained, several of Wünsche's subordinate ordnance officers had behaved improperly. Hitler then ordered Wünsche's immediate removal from his staff and return to duty with the *Leibstandarte-SS "Adolf Hitler"*. Brückner officially departed from Hitler's personal staff on 01.04.1941.

Wilhelm Brückner wearing native American headdress, ca. 1936. (Roger Bender photos)

Brückner speaks with the locals at the Berghof, ca. 1936. (Roger
Bender photo, from Brückner's personal collection)

A group of Hitler's most valued subordinates, photographed by Heinrich
Hoffmann, 1936. From left to right: Brückner, Julius Schaub, Dr. Otto
Dietrich, Dr. Joseph Goebbels, "Sepp" Dietrich, Hans Baur, and Dr. Karl
Brandt. (Roger Bender photo, from Brückner's personal collection)

Left to right: *Reichspräsidialrat* Dr. Werner Kiewitz, Gerhard Engel,
Brückner, Otto Meissner, Rudolf Schmundt, and Ludwig Bahls.
(Roger Bender photo, from Brückner's personal collection)

"To *Pg.* [*Parteigenoß*] [*Gauleiter* August] Eigruber with heartfelt comradeship. Heil Hitler! Wilhelm Brückner, Dec. 40." (Snyder's Treasures)

An autographed photo of *SA-Obergruppenführer* Brückner, ca. 1940. (Roger Bender)

01.06.1941-00.00.1942	Called up for *Wehrmacht* service as a *Major d. R.*, assigned to the staff of the *Feldkommandantur* in Nevers / Loire, France.
00.00.1942-00.00.1942	*Kreiskommandant* of Dijon, France. Retreated into the Vosges, 00.09.1944.
00.00.1942-00.10.1944	*Kreiskommmdant* of Macon, France.
00.10.1944-24.11.1944	*Kampfkommandant, Regimentskommandeur,* and *Kommandeur* of *Kampfgruppe* "Brückner" in Oberschulheim and Storzheim.
24.11.1944-00.00.1945	Severely injured (double fracture to the right ankle), then hospitalized in the *Lazaretten Holzkirchen* and *Traunstein.*

Postwar Prosecution

Arrested and interned by troops of the U.S. 7th Army at the *Lazarett Traunstein*, 04.05.1945, then transferred to the prisoner of war camp at Mundenheim. On 10.06.1945, he was delivered to the U.S. 7th Army Interrogation Center (SAIC) in Augsburg. From 01.08.1945 to 02.09.1945, he was held at Seckenheim near Mannheim; 02.09.1945 to 10.09.1945 in Ulm; 10.09.1945-18.09.1945 in Altenstadt; 18.09.1945 to 13.10.1945 at 3rd U.S. Army Interrogation Center in Freising, and on 13.10.1945 he was sent for further interrogation to U.S. Forces European Theater-Military Intelligence Service Center (USFET-MISC). A "Preliminary Interrogation Report" by Major Paul Kubala (commander of SAIC), dated 22.06.1945, stated:

12.01.1940: Hitler congratulates Hermann Göring on his 47th
birthday, as *SA-Obergruppenführer* Brückner looks on.

PW [Prisoner of War] is not very talkative and gives the impression that he either has no special knowledge about HITLER, or that he does not care to divulge any special knowledge he may have. (*Interrogation Records Prepared for War Crimes Proceedings at Nuernberg, 1945-1947/OCCPAC Interrogation Transcripts And Related Records: Brueckner, Wilhelm Friedrich*; Publication Number M1270, Record Group RG238)

A later report by 1st Lt. Arthur D. McKibbin, dated 31.10.1945, states:

BRUECKNER appears to be honest and entirely cooperative, anxious to impress the interrogator with the idealistic nature of his connection with National Socialism. (ibid)

On 14.09.1948, the *Lagerspruchkammer* in Garmisch classified him as a "Hauptschuldiger" (major offender) and sentenced him to 3½ years' labor camp and ordered him to forfeit 5000 DM in property. Following an appeal of this sentence, he was released from custody on 22.09.1948.

Decorations and Awards

16.11.1917	*1914 Eisernes Kreuz II. Klasse*
· 14.03.1915	*Bayerisches Militär-Verdienstkreuz II. Klasse mit der Krone und mit Schwertern*
ca. 1918	*Verwundetenabzeichen, 1918 in [Schwarz?]*
ca. 1939	*Medaille zur Erinnerung an die Heimkehr des Memellandes*

The Berghof, New Years' Day, 1940. Hitler poses with members of his inner circle and personal staff (NARA photo) Front row, left to right: Wilhelm Brückner, Christa Schröder, Eva Braun, Hitler, Gretl Braun; *Gauleiter* Adolf Wagner; and Dr. Otto Dietrich; Second Row, from left: Gerda Daranowski, Margarete Speer, Martin Bormann, Dr. Karl Brandt, and Heinrich Hoffmann; Last row, left to right: Dr. Theo Morrell, Hannelore Morell, Karl-Jesko von Puttkammer, Gerda Bormann, Max Wünsche; and Heinrich Heim. (NARA photo)

ca. 1939	*Spange "Prager Burg" zur Medaille zur Erinnerung an den 1. Oktober 1938*
ca. 1939	*Medaille zur Erinnerung an den 1. Oktober 1938*
ca. 1938	*Medaille zur Erinnerung an den 13. März 1938*
ca. 1934	*Ehrenkreuz des Weltkrieges 1914-1918 mit Schwertern*
00.00.192_	*Steckkreuz Deutscher Frontkriegerbund*
30.01.1938	*Goldenes Ehrenzeichen der NSDAP*
00.00.1934	*Ehrenzeichen des 9. November 1923 (Blutorden)* (Nr. 7, with effect from 09.11.1933)
00.00.1936	*Coburger Abzeichen 1922*
00.00.194_	*Dienstauszeichnung der NSDAP in Silber*
00.00.194_	*Dienstauszeichnung der NSDAP in Bronze*
00.00.193_	*Ehrenzeichen des Deutschen Roten Kreuzes I. Klasse*
00.00.192_	Badge of the *Ring der nationalen Kraftfahrt- und Luftfahrtbewegung* (Association of the National Motoring and Aviation Movement" [grade unknown])
00.00.1934	*Ehrendolch der SA* (mit Wirkung vom 03.02.1934)
00.02.1934	*Ehrenwinkel für alte Kämpfer*

An autographed photo of *Major*
Brückner, inscribed and dated
02.03.1942. (Snyder's Treasures)

14.09.1948: Wilhelm Brückner
during his de-Nazification
proceedings before the
Lagerspruchkammer in
Garmisch-Partenkirchen.

15.01.1936   *Ehrenbürgerbrief der Stadt Detmold* (status revoked by the *Stadtrat* of
Detmold, 09.11.1945)

Notes:

- Son of the musician (member of the *Städtischen Kurorchester*) Wilhelm Brückner and his wife Friederike Henriette, née Niebecker.
- Religion: Protestant.
- From 1929 to 1936, Brückner was engaged to the painter and ceramicist Sophie (aka Sofie) Stork (born 05.05.1903 in München, died 21.10.1981 in Seeshaupt), a member of the *NSDAP* from 1931 and a close friend of Eva Braun. They may have married in 1937, and had two children, although these details have not been confirmed.

Sources

Bayerisches Hauptstaatsarchiv, München, Abteilung IV Kriegsarchiv: Excerpts from various Kriegsranglisten containing data on the Royal Bavarian Army service of Wilhelm Brückner.

Bundesarchiv, Berlin-Lichterfelde (former Berlin Document Center): *Personalunterlagen von SA-Angehörigen: SA-Personalakte* of Wilhelm Brückner.

Hamilton, Charles: *Leaders and Personalities of the Third Reich, Volume I*. R. James Bender Publishing, 1984.

Lilla, Joachim; Döring, Martin; & Schulz Andreas: *Statisten in Uniform. Die Mitglieder des Reichstags 1933-1945*. Droste Verlag, 2004.

National Archives and Records Administration, College Park, Maryland: "Interrogation Records Prepared for War Crimes Proceedings at Nuernberg, 1945-1947/OCCPAC Interrogation Transcripts And Related Records: Brueckner, Wilhelm Friedrich"; Publication Number M1270, Record Group RG238.

Schirach, Baldur von: *Die Pioniere des Dritten Reiches*. Zentralstelle fur der deutschen Freiheitskampf, 1933.

Schmidt, Ulf: *Karl Brandt: The Nazi Doctor. Medicine and Power in the Third Reich*. Hambledon Continuum, 2007.

Josef Bürckel
SA-Obergruppenführer

Born: 30.03.1895 in Lingenfeld / Pfalz.

Died: 28.09.1944 (at approximately 1104 hours) in Neustadt-an-der-Weinstraße. A report from Bürckel's personal physician (since 1936), *Gaugesundheitsführer* Prof. Dr. med. Ewig, dated 28.09.1944, stated that Bürckel was physically and mentally worn out, spending all of his time at work because of the deteriorating situation in his Gau. He suffered an inflammation of the intestine with diarrhea, eventually becoming too ill to continue. Ewig was called in on 26.09.1944. Bürckel soon contracted pneumonia which hastened his death two days later. The death certificate gave the cause of death as pneumonia and blood failure. Dr. Josef Rowies stated on 23.10.1944 that the report of his death sent to the *SS-Personalhauptamt* by Himmler's personal staff office on 09.10.1944 had been "doctored" to conceal his mental breakdown. The consensus is that he gave up the will to live, dying at 1100 hours on 28.09.1944 basically of the symptoms shown in the death certificate plus exhaustion. A memorial ceremony was held on 03.10.1944 at the *Gauhaus* in Neustadt–an-der-Weinstraße on 03.10.1944, with *Reichsleiter* Alfred Rosenberg officiating as representative of the *Führer*. Also in attendance were *Reichsleiter*s Dr. Robert Ley, Karl Fiehler, and Baldur von Schirach; *Gauleiters* Wilhelm Murr, Karl Wahl, Jakob Sprenger, Fritz Sauckel, Dr. Gustav Adolf Scheel, Gustav Simon, and Friedrich Karl Florian; and *SA-Stabschef* Wilhelm Schepmann. In his eulogy for Bürckel, Rosenberg declared:

> Josef Bürckel was a devoted nationalist and a passionate socialist. In all of his life, he symbolized the unity that, from the standpoint of our worldview, reflects outwardly that great inner experience, that great idea, for which we fought ...
>
> The *Führer* has authorized me, party comrade Bürckel, to express to you his thanks for your complete loyalty to him and to the movement. More than ever before, the *Führer* remembers the loyal support of one of his oldest fellow fighters, who never grew weary during the years of struggle, who always followed the *Führer* and his banner. In particular recognition of this exemplary National Socialist life, and as a continuing reminder for coming generations, the *Führer* awards you, Joseph Bürckel, the highest level of the German Order with Swords. This will be a symbol of your beloved *Gau,* and of your loyalty to the whole National Socialist movement.
>
> We now take leave of you. the flags of the Greater German *Reich* will flutter over your grave, and the soldiers of the German

Josef Bürckel as a young teacher, ca. 1926.

people will pass by you as they march west, where they will protect German territory and realize what you gave your whole life for, along with the great loyalty of your heart. (*Berliner Morgenpost*, 05.10.1944; Translation at Prof. Randall Bytwerk's German Propaganda Archive- http://www.calvin.edu/academic/cas/gpa/mp01.htm)

NSDAP-Nr.:	33 979 (First joined, 00.00.1921; Party banned following the *München-Putsch* of 09.11.1923; Reenrolled, 09.04.1925)
SS-Nr.:	289 230 (Joined 09.11.1937)

Promotions

15.08.1915	*überzähliger Gefreiter*
04.12.1915	*etatmässiger Unteroffizier*
13.03.1926-28.09.1944	*Gauleiter* der *NSDAP*
25.09.1933	*SA-Gruppenführer*
11.02.1935-17.06.1936	*Reichskommissar*
23.02.1935	*NSKK-Gruppenführer*
09.11.1936	*SA-Obergruppenführer*
09.11.1937	*SS-Gruppenführer*
20.04.1938	*NSKK-Obergruppenführer*
23.04.1938-31.03.1940	*Reichskommissar*

30.01.1942	*SS-Obergruppenführer*

Career

00.00.1900-00.00.1909	Attended the Catholic *Volksschule* in Lingenfeld.
00.00.1909-00.00.1914	Attended the *Lehrerseminar* (teacher training institute) in Speyer.
00.08.1914	Volunteered for service with *17. Kgl. Bayerisches Feldartillerie-Regiment*. During the follow-up examination, he was rejected due to a recent operation. On the same day, he was re-examined and posted to the *12. Kgl. Bayerisches Feldartillerie-Regiment* in Landau, and soon after, transferred to *6.Batterie / II.Abteilung / 20. Kgl. Bayerisches Feldartillerie-Regiment*.
03.11.1914-01.01.1915	Reported as a *Freiwilliger mit Einjährigs Berechtigung*, assigned to *II.Rekruten-Depot / II. Ersatz-Abteilung / 12. Kgl. Bayerisches Feldartillerie-Regiment* (Landau / Pfalz).
01.01.1915-03.04.1915	Assigned to *4.Batterie / II. Ersatz-Abteilung / 12. Kgl. Bayerisches Feldartillerie-Regiment*.
03.04.1915-30.09.1915	Assigned to *6.Batterie / 20. Kgl. Bayerisches Feldartillerie-Regiment*. On 03.05.1915, he passed his final teaching examinations at the *Landesbildungsanstalt* in Speyer (while on leave), then returned to the field with his unit, in the area of Belgian-French Flanders southwest of Péronne, on 06.04.1915. He then took part in heavy defensive fighting near La-Bassée-Arras and Fronelles. From 12.05.1915 to 01.06.1915, he was deployed south of Oosttaverne and after 01.06.1915, near Deniercourt and on the Somme. From 12.08.1915 to 15.09.1915, he was hospitalized due to a heart ailment at the *Kriegslazarett* in Péronne / Departement Somme, then returned to duty with 6.Batterie on 16.09.1915.
00.00.1915	Passed his *Seminarabschlussprüfung* (teaching seminary graduation examination) in Speyer.
01.10.1915-05.11.1915	Assigned to *4. Kompanie / I. Ersatz-Bataillon / 17. Kgl. Bayerisches Feldartillerie-Regiment*.
05.11.1915 -06.12.1915	Assigned to *4.Batterie / II. Ersatz-Abteilung / 12. Kgl. Bayerisches Feldartillerie-Regiment* (Landau / Pfalz).
06.12.1915-07.02.1916	Assigned to *Kgl. Bayerisches Feldartillerie-Regiment Nr. 896*.
07.02.1916-17.05.1916	Assigned *to 4. Ersatz-Batterie / II. Ersatz-Abteilung / 12. Kgl. Bayerisches Feldartillerie-Regiment*. Honorably discharged from service, 17.05.1916.
16.05.1916-28.07.1919	Employed as a *Schuldienstanwärter* (candidate for the school service, with an annual salary of 849 *Reichsmarks*) to Catholic *Volksschule*n in the following towns: 16.05.1916-17.10.1917 Ramberg 17.10.1917-01.08.1918 Lingenfeld 01.08.1918-01.10.1918 Bellheim 01.10.1918-28.07.1919 Minfeld
18.05.1916	Determined unfit for further military service and discharged from the army.
00.01.1919	Passed a teacher employment examination in Speyer.

Führertagung der NSDAP in Weimar, part of a larger group photo. Back row, left to right: Gregor Strasser, Emil Holtz, Franz Ritter von Epp, Franz Xaver Schwarz. Middle row, left to right: Otto Erbersdobler, Heinz Haake, Josef Bürckel, and Dr. Joseph Goebbels. In the next row are, second from left, Helmuth Brückner and, at right, Adolf Hitler.

01.03.1919-01.08.1927	Employed as a *Hilfslehrer* (assistant teacher) to *Volksschulen* in the following towns:
01.03.1919-31.01.1920	Bobenheim-Roxheim
01.02.1920-07.01.1924	Rodalben bei Pirmasens / Pfalz (appointed as a "Lehrer aus Lebenszeit" [teacher for life] with an annual salary of 6200 RM, 01.04.1921)
10.05.1924 (?)-01.08.1927(?)	Resumed his teaching position in Rodalben
00.00.1920	Passed his state teaching examinations.
00.01.1919	Passed a teacher employment examination in Speyer.
01.03.1919-01.02.1920	Assistant teacher in Roxheim.
00.00.1921	Joined the *NSDAP*.
	Participated in actions against the French-supported separatist movement in the Pfalz, which had as its goal the establishment of a "rheinischen Republik" [Rhenish Republic]. On 07.01.1924, he fled to Heidelberg, then to Niederbayern. On 12.02.1924, he took part- pistol in hand- in an assault on the separatist-occupied *Bezirksamt* (district office) in Pirmasens.
18.03.1924-10.05.1924(?)	Appointed by the government in Niederbayern, worked as an *Aushilfslehrer* in Höhenstadt/Passau.
09.04.1925	Reenrolled in *NSDAP*.

Josef Bürckel, ca. 1933.

09.04.1925-13.03.1926	*Leiter* of the *Pfälzisch-Saarländische Nationalsozialisten* (National-Socialists in the Palatinate and Saar regions).
13.03.1926-28.02.1935	*Gauleiter* of *Gau Rheinpfalz der NSDAP* (Seat: Rodalben, then Neustadt an der Weinstraße). He succeeded Fritz Wambsganß. According to Franz Maier, Bürckel's deputy, Fritz Hess, served as acting *Gauleiter* from 30.03.1927 to 19.11.1927 (Maier, *Biographisches Organisationshandbuch der NSDAP und ihrer Gliederungen im Gebiet des heutigen Landes Rheinland-Pfalz*). When the Saarland (administered as a *Gau* by Bürckel-see entry of 31.01.1933-28.02.1935, below) was returned to Germany by the Plebiscite of 01.03.1935, it was merged with *Gau Rheinpfalz* to form *Gau Pfalz-Saar*, also under Bürckel.
01.08.1927-00.09.1930	*Schulleiter* (headmaster) of the Catholic *Volksschule* in Nußbach bei Neustadt. Went on leave due to his *Reichstag* candidacy, 22.08.1930, and advised the *Bezirksschulrat* on 15.10.1930 of his appointment as a member of the *Reichstag*.
00.03.1926	Together with his Party comrade Fritz Hess, founded the weekly newspaper *Der Eisenhammer* (with the subtitle *Pfälzische Wochenschrift zum Kampf um die Wahrheit und das Recht auf Arbeit und Brot* [later *Kampfblatt der NSDAP Gau Pfalz*]). Hess was editor of the newspaper until 00.08.1927, when Bürckel assumed the post. The paper was a scandal sheet, filled with anti-democratic, primitive racist, and Jew-baiting rhetoric.
00.08.1927-00.00.1930	Editor of *Der Eisenhammer*. Succeeded Fritz Hess.

20.01.1929	Participated in the *Führertagung der NSDAP* in Weimar.
00.00.1930-28.09.1944	Editor of the *Gau Rheinpfalz* newspaper *NSZ-Rheinfront* (renamed *NSZ-Westmark* on 01.12.1940).
14.09.1930-28.09.1944	Member of the *Reichstag* (*Wahlkreis 27, Pfalz, later Rheinpfalz-Saar*).
00.00.1931-00.00.1933	Embroiled in a violent feud with Theodor Eicke, who was then serving as *Führer* of *10.SS-Standarte "Pfalz"* in Neustadt an der Weinstraße. According to Eicke, the quarrel began when Bürckel supported the local police in prohibiting an *SS* parade in Ludwigshafen. In fact, it began because Bürckel, in his capacity as *Gauleiter*, saw himself as *de facto* commander of all *SA* and *SS* units in his *Gau*. Eicke refused to obey orders issued by Bürckel but was too proud and arrogant to complain to Himmler. During Eicke's absence on other duties in Italy, Bürckel spread the rumor that he had been dismissed from the *NSDAP*. In letters to his *SS* comrades in Ludwigshafen (written during 1932/1933), Eicke asked that they inform Bürckel of his vow that "I shall protect my honor with every means should he [Bürckel] continue his Jesuit tactics against me in our movement." He stated that, if necessary, he would use "the old methods" [i.e. assassination] against Bürckel. Upon his return from Italy to Ludwigshafen in March 1933, and despite his promises to Himmler that he would have no further contact with Bürckel, Eicke and his men stormed the *Gau* headquarters in Neustadt, locking Bürckel in a janitor's closet and arresting the local *Kreisleiter*. Bürckel's loyalists called the local *Schutzpolizei* and Eicke was arrested and imprisoned in Ludwigshafen with his men on 21.03.1933. The angry *Gauleiter* had Eicke arrested and found "mentally ill and a danger to the community." Eicke was then sent for an indefinite period of psychiatric observation to the *Nervenklinik* in Würzburg on 23.03.1933. The furious Himmler dismissed him from the *SS* because he had "not kept the word of honor he had given the *Reichsführer-SS*". While confined, Eicke befriended his psychiatrist, Prof. Dr. Werner Heyde (a leading figure in the later T-4 ["euthanasia"] program and future *SS-Standartenführer*); Dr. Heyde wrote to Himmler on 22.04.1933, stating that "[Eicke] has conducted himself in an exemplary manner and by his quiet, self-controlled character has left a very pleasant impression and by no means appears to be an intriguing [i.e. troublemaking] personality." Eicke was released from the clinic in Würzburg on 26.06.1933, and reinstated in the *SS* with no loss of seniority at his original rank of *Oberführer*. A Party court raised no objections to his release, provided he not return to *Gau Saarpfalz*.
00.08.1932	Appointed as *Sachberater "Besetzte Gebiete" der NSDAP-Reichstagsfraktion* (Specialist for Occupied Regions [i.e., areas of Germany under French control, such as the Saarland] of the *NSDAP* Faction in the *Reichstag*).
31.01.1933-28.02.1935	*(kommissarischer) Gauleiter* of *Gau Saarland der NSDAP* and *Beauftragter der NSDAP im Saarland* (Representative of the Nazi Party

Josef Bürckel, ca. 1934. (*Wiener Bild*, 05.02.1939)

Josef Bürckel, ca. 1935.

in the Saar region). Assumed *de facto* control of the post on 31.01.1933, with Karl Brück retaining it on paper until 06.05.1933.
Berater beim Beauftragten der SA und SS bei der Regierung der Pfalz (Advisor to the Representative of the *SA* and *SS* to the Government in Pfalz, *SA-Oberführer* Fritz Schwitzgebel).

25.09.1933-26.01.1934	Granted the rank of *SA-Gruppenführer* and attached to the staff of *SA-Gruppe Westmark*.
15.01.1934	Founder of the charities *Josef-Bürckel-Stiftung* (Josef Bürckel Endowment) and *Volkssozialistische Selbsthilfe* (VS, People's Socialist Self Help; redesignated *Volkssozialistische Gemeinschaft e.V.* in April 1934).
27.01.1934-00.00.193_	Assigned as an *Ehrenführer der SA* to *SA-Standarte* 18 (Landau) in *SA-Gruppe Westmark*.
11.04.1934	Appointed as *Politischer Beauftragter der Bayerische Staatsregierung bei der Regierung der Rheinpfalz* (Political Representative of the Bavarian State Government attached to the Government of Rheinpfalz).
10.08.1934-28.02.1935	*Reichsbevollmächtigter (m.d.W.d.G.b.) für die Saar-Angelegenheiten* (*Reich* Plenipotentiary for Saarland Matters). Succeeded Franz von Papen.
27.11.1934	Officially discharged from the public school service, having been appointed – with effect from 01.11.1934 – as a *Hauptlehrer ohne höhere Dienstbezüge*.

Bürckel greets Hitler in Saarbrücken, late March 1936. At right is then-*SA-Brigadeführer* Fritz Schwitzgebel. (*Hamburger Fremdenblatt*, 27.03.1936)

11.01.1935	Attended a conference regarding the upcoming *Saarabstimmung* (Saar Plebiscite) at the *Reichskanzlei* in Berlin, meeting with Hitler, Dr. Wilhelm Frick, and Rudolf Hess. During the plebiscite, conducted by Bürckel under the auspices of the League of Nations from 13.01.1935-15.01.1935, 90.67% of voters (477,119 of 228,005 votes cast) asserted their demand for a reunion with Germany, thus rejecting union with France or a continuation of the League of Nations' administration of the Saarland. The vote became effective on 01.03.1935, and the Saarland was formally transferred to the *Reich* in a ceremony (presided over by Bürckel and attended by *Reichsministers* Dr. Frick and Dr. Goebbels) at the *Kreisständehaus* in Saarbrücken on that date.
30.01.1935-01.03.1935	*Oberster Repräsentant der Reichsaufsicht* (Senior representative for *Reich* Supervision) and *Regierungschef* (Government Chief) of the Saarland.
11.02.1935-17.06.1936	*Reichskommissar für die Rückgliederung des Saarlandes* (*Reich* Commissioner for the Return of the Saarland). He established his office in Saarbrücken on 01.03.1935.
00.00.193_-00.00.19__	Chef of the *Landwirtschaftsamt* (agricultural office) in Saarbrücken.
01.03.1935-28.09.1944	*Gauleiter* of *Gau Pfalz-Saar der NSDAP* (redesignated *Gau Saarpfalz*, 00.00.1937, and *Gau Westmark* on 07.12.1940). This *Gau* was formed

Josef Bürckel delivering a speech, ca. 1935.

by combining the *Gaue* of *Rheinpfalz* and *Saarland* and including the recently incorporated [by the Plebiscite of 01.03.1935] Saarland region. The duties of the post were handled by his deputy, Ernst Ludwig Leyser, when Bürckel was away in Austria (00.03.1938-00.04.1940). After his death, he was succeeded by Willi Stöhr.

20.01.1936	In a decree of that date, ordered the application in the Saarland, with effect from 01.03.1936, of the Nürnberg Laws and all other orders already in force relating to the Jews in German territory.
17.06.1936-08.04.1940	*Reichskommissar für die Saarland* in Saarbrücken.
00.00.193_-28.09.1944	Assigned as an *SA-Führer z. V.* to the staff of *SA-Gruppe Kurpfalz*.
15.03.1937	Appointed as a member of the *Landesjagdrat für Bayern einschließlich Saarland* (State Hunting Council for Bavaria, including the Saarland).
06.10.1937	Cofounder of the *Saarpfälzische Erzbergbaugesellschaft AG*, Neustadt/Weinstraße.
09.11.1937-28.09.1944	Joined the *Allgemeine-SS*, assigned as an *ehrenamtlicher SS-Führer* to the staff of the *Reichsführer-SS*.
09.11.1937	Sworn in by the *Reichsführer-SS* as a *Hüter des Bluts- und Lebensgesetzes der Schutzstaffel* (Guardian of the Blood and Life Statute of the SS).
13.03.1938-00.05.1938	*(kommissarischer) Leiter der NSDAP von Österreich* (acting Head of the NSDAP for Austria), appointed via the following order by Hitler:

Josef Bürckel delivering a speech, ca. 1937.

1. I assign *Gauleiter* Bürckel, Saarpfalz, to reorganize the *NSDAP* in Austria.

2. *Gauleiter* Bürckel is entrusted with the preparations for the plebiscite in his capacity as acting leader of the *NSDAP* in Austria.

I have granted *Gauleiter* Bürckel full power of attorney to report and order all measures necessary to allow for a responsible completion of his assignment.

Linz, 13. March 1938

Adolf Hitler. (Dr. Max Domarus, ed, *Hitler. Speeches and Proclamations, 1932-1945: The Years 1935-1938*, p. 630)

13.03.1938-00.04.1938	*Beauftragter des Führers für die Volksabstimmung in Österreich* (Representative of the *Führer* for the Plebiscite in Austria) and *Leiter* of *Abteilung VII (Abstimmungs-Propaganda*, Plebiscite Propaganda) in that office. In this capacity he supervised the 10.04.1938 plebiscite held in Austria which resulted in a 99.67% vote favoring union with Germany.
24.03.1938	Delivered a speech in the *Wiener Konzerthaus*, from which the following is an excerpt:

March 1938: *Reich* Interior Minister Dr. Frick and Dr. Hans Heinrich Lammers visit Austria shortly after the *Anschluß*. *SS-Gruppenführer* Bürckel stands at far left.

Our people, our *Reich*, our honor require for their protection solely German freedom and this will also be guaranteed here in Austria, even to the extent that the Jews will be forced of their own democratic freedom to get out of Wien. [Applause] I allow myself this remark, because we are all of the opinion that the Jewish workers in Wien represent far too strong a factor with their characteristic freedom, than the Viennese population really deserves .... .

Each system has the Jews it deserves!

I'd like to expand on that and say:

Each nation deserves as much Jewish freedom, as it will put up with! [Thunderous applause]

I promise that the relation between Jewish freedom and arrogance on the one hand and German right to assert themselves on the other will be settled-settled by all means -, in order to balance things out.

23.04.1938-31.03.1940   *Reichskommissar für die Wiedervereinigung Österreichs mit dem Deutschen Reich* (*Reich* Commissioner for the Reincorporation of Austria into the German *Reich*). In this post, Bürckel- with the support of his superiors in the *Reich*- sought to achieve "the thoroughgoing partification and Nazification of the society. As a result, Bürckel soon pushed aside the more conservative 'native' Austrian Nazis." (Dietrich Orlow, *History of the Nazi Party, 1933-1945*, p. 237)

Regarding Bürckel's church policy in the *Ostmark*, Karl R. Stadler writes:

> ... Abolition of all church schools, the practical ban on church youth guilds, the dismissal of all ordained teachers, the closure of three out of the four theological faculties, and much more. The driving force behind this policy was *Gauleiter* Bürckel, of whom Schirach was to say at the Nuremberg Trial that his policy vis-à-vis the Church amounted to persecution where his Schirach's policy, had been to stop all anti-Church propaganda and demonstrations. (Stadler, *Austria*, p. 196)

In a report to the staff of the Deputy *Führer* (Rudolf Hess), submitted in June 1939, an Austrian Party official revealed the average Austrian appraisal of Bürckel and his staff/clique:

> ... Bürckel and his Pfälzer don't understand us and won't meet with us. Bürckel breakfasts with his Pfälzer in the morning, he sits with them at lunch, in the evening he goes drinking with them, and at night he is again in the same house with them. I don't need to say anything about the inadequacies of most Pfälzer ... . (ibid, p. 187)

In his table talk of 20.05.1942, Hitler justified his appointment of Bürckel as follows: " ... a man of unbending hardness, not a Viennese muddler." (ibid, pp. 246-247)

The German annexation of Austria brought some 190,000 Jews under Nazi control. Persecution of this large Jewish community began immediately, with countless instances of physical assault, public humiliation, and theft. Already on 17.03.1938, Reinhard Heydrich wrote to Bürckel that "those National Socialists who in the last few days allowed themselves to launch large-scale assaults [against Jews] in a totally undisciplined way" would be arrested by the *Gestapo*. "[However], it was only on April 29," writes Saul Friedländer, "when Bürckel announced that the leaders of *SA* units whose men took part in the excesses would lose their rank and could be dismissed from the *SA* and the party, that the violence started to ebb." (Friedländer, *Nazi Germany and the Jews, Volume I*, p. 242)

12.06.1938-19.06.1938	Participated in *Reichstheaterfestwoche 1938* (*Reich* Theater Festival Week) in Wien.
00.07.1938-00.12.1939	Editor of the monthly newsletter *Der Ostmark Brief, Mitteilungsblatt und Schulungsbrief.*
26.08.1938-04.09.1938	Attended the *6.Reichstagung des Auslandsdeutschen* at the Stadthalle in Stuttgart.

Formal portrait of Josef Bürckel by Heinrich Hoffmann, ca. 1939. (NARA photo)

20.08.1938   Ordered the establishment of the *Zentralstelle für jüdische Auswanderung* (Central Office for Jewish Emigration) in Wien, under the leadership of Dr. Franz Walther Stahlecker (who delegated the actual duties of the post to Adolf Eichmann). In an autumn 1938 discussion with subordinates concerning the "Jew Problem", Bürckel declared:

> Let's not forget that if you want to Aryanize and to deprive the Jews of their means to live, then you've got to solve the Jewish problem in an absolute fashion. (Hans Safrian, *Die Eichmann Männer*, p. 36)

09./10.11.1938   "Reichskristallnacht" in Wien, carried out primarily by the *SA*. According to the *Völkischer Beobachter*, 19 synagogues were burned. Following this mass arson, Jewish-owned businesses were destroyed, and 5,000 Jewish shops damaged and closed. 8 Jews were murdered, at least 21 committed suicide, and many more were physically assaulted. Bürckel then ordered Jewish residences to be raided and searched (for weapons and proscribed literature); this resulted in many arrests.

30.01.1939-02.08.1940   *Gauleiter* of *Gau Wien der NSDAP*. Succeeded Odilo Globocnik. Globocnik had "resigned", but was in fact directed to step down due to various financial irregularities and his "work for something akin to Austrian autonomy." While in Wien, with whose population he proved extremely unpopular (being viewed as a foreign interloper with a handpicked staff of like-minded outsiders), he was scornfully nicknamed

"Bierleiter Gaukel" and "Bauleiter Gürckel". He was succeeded by Baldur von Schirach.

06.03.1939	Attended a meeting of the *Arbeitsgemeinschaft für die deutsch-italienischen Rechtsbeziehungen* (Working Organization for German-Italian Legal Relations) at the Belvedere Palace, Wien.
00.03.1939-00.04.1939	*Chef des Zivilverwaltungs in Mähren* (Moravia), attached to *Heeresgruppe 5*.
01.05.1939-31.03.1940	*Beauftragter der NSDAP für die Stadt Wien* and *Führer des Gaues Wien* (*de facto Oberbürgermeister*). Succeeded Hermann Neubacher. Succeeded by Prof. Dr.-Ing. Johannes "Hanns" Blaschke.
04.05.1939-02.08.1940	*Reichsstatthalter für die Ostmark*, Wien.
00.05.1939	Accompanied Hitler during his inspection tour of the *Westwall* defenses (14.-19.05.1939).
01.09.1939-02.08.1940	*Reichsverteidigungskommissar für den Wehrkreis XVII*.
22.09.1939-02.08.1940	*Vorsitzender* of the *Verteidigungsausschuß* (defense committee) for *Wehrkreis XVII*.
22.09.1939-16.11.1942	Member (and *Vorsitzender* until 02.08.1940) of the *Verteidigungsausschuß* of *Wehrkreis XII* (Wiesbaden).
05.01.1940	In a press notice appearing in Viennese newspapers, the Jews of Wien were informed of the following ordinance:

> *Gauleiter* Bürckel has directed that in all retail establishments of Wien produce may be sold to Jews only during fixed hours. Hours during which Jews are permitted to shop are set as follows until further notice:
>
> A. In food store... between 11:00 A.M. and 1:00 P.M., in the *Kreis II* (Leopoldstadt) and in *Kreis XX* (Brigittenau) also between 4:00 and 5:00 P.M. It is forbidden to deliver foodstuffs to Jews in their homes.

08.02.1940-02.08.1940	*Präsident* of the *Südosteuropa-Gesellschaft (SOEG*, Southeastern Europe Society). Succeeded by Baldur von Schirach. The Society had been established under the patronage of *Reich* Economics Minister Walther Funk, "to coordinate Eastern studies and research with the long-term objective of taking over the Balkans and simultaneously exploiting their economy." (Ingo Haar and Michael Fahlbusch, *German Scholars and Ethnic Cleansing, 1920-1945*, p. 220)
01.04.1940-02.08.1940	*Reichsstatthalter des Reichsgaues Wien*. On 12.04.1940, he and the other newly appointed *Reichsstatthalter* of the *Ostmark* (former Austria) were officially sworn in by Hitler at the *Reichskanzlei* in Berlin. He was succeeded by Baldur von Schirach.
08.04.1940-11.03.1941	*Reichskommissar für die Saarpfalz*.
00.05.1940-07.08.1940	*Chef der Zivilverwaltung* (*CdZ*, Chief of the Civil Administration) attached to the staff of *Armeeoberkommando 1* (*AOK 1* [*1.Armee* HQ, under *Generaloberst* Erwin von Witzleben]).

A very rare photo of Josef Bürckel (left, with Dr. Artur Seyss-Inquart and
Konrad Henlein) wearing the collar tabs of a Gauleiter. The occasion is the
11.Reichsparteitag der NSDAP in Nürnberg, September 1938. (Mark Costa photo)

Führer HQ "Tannenberg" in Freudenstadt, 01.07.1940: Left to right:
Bürckel, Dr. Hans Heinrich Lammers, Martin Bormann, Robert Wagner,
Hitler, and Baldur von Schirach (in Army uniform). (NARA photo)

Metz, 1940: *Gauleiter* Bürckel and his deputy, Ernst August Leyser celebrate the return of Lorraine to the *Reich*. (NARA photos)

07.08.1940-28.09.1944   *Chef der Zivilverwaltung in Lothringen* (Chief of the Civil Administration in Lorraine). He officially undertook the duties of this post on 21.09.1940. Succeeded by Willi Stöhr. Bürckel promptly and zealously pursued a policy of "Germanization" aimed at expelling all the French-speaking people from the territory in order to make way for Volksdeutsche settlers. In late November or early December 1940, Bürckel declared: "The situation on Germany's western borders has been defined for all time … . The centuries-long battle over the Rhein is ended." At a conference attended by Hitler, Martin Bormann, Bürckel, and *Gauleiter* Robert Wagner of Baden (the *CdZ* in neighboring Elsaß [Alsace]) on 25.09.1940, the *Führer* ordered the two *Gauleiter* to completely Germanize their territories, by whatever means necessary, within the next decade. At a meeting with his assigned *HSSPF*, Theodor Berkelmann, in Metz on 08.10.1940, he admitted that the policy of forced expulsion he was implementing in Lorraine had not been explicitly ordered by Hitler, but that it had been his own innovation, submitted to and approved by the *Führer*. On 22./23.10.1940, *Gauleiter* Bürckel and Wagner ordered the sudden and extremely brutal deportation (code-named "Wagner-Bürckel-Aktion") of approximately 6,504 Jews from their respective Gaue to Lyon in unoccupied France (from there they were sent by French authorities to French concentration camps at Rivesaltes, Gurs, and Les Milles); 7 trains left Baden on 22. and 23.10.1940 and two trains departed Saarpfalz on 22.10.1940. This operation, organized by Adolf Eichmann of the *Gestapo*, was an extension to another that had occurred earlier the same month when, in accordance with an agreement between *General der Infanterie* Carl Heinrich von Stülpnagel (head of the German Armistice Commission) and French Minister of War / Chief of the French Armistice Commission General Charles Huntziger in which it was decided to deport all Jews from Alsace and Lorraine to unoccupied France. As a result, 22,000 Jews were removed from Elsass alone. Bürckel and Wagner took this opportunity to clear out the Jews from their own *Gaue*. During his tenure as *CdZ,* Bürckel was to expel some 100,000 "undesirable" people from Lothringen. The expulsion actions of Bürckel and Wagner outraged *Reichsführer-SS* Himmler, who believed the *Gauleiter* were recklessly throwing away the "racially valuable" populations of Alsace and Lorraine. Himmler ordered them to limit the expulsions to Jews, Gypsies, Africans, the mentally ill, criminals, and "all other trash that does not belong to us on the basis of blood."

15.11.1940 -28.09.1944   *Gauwohnungskommissar* for *Gau Westmark*. Succeeded by Willi Stöhr.

11.03.1941-28.09.1944   *Reichsstatthalter of the Westmark*. Succeeded by Willi Stöhr.

06.04.1941-28.09.1944   *Bevollmächtigter für den gesamten Arbeitseinsatz* (Plenipotentiary for the entire Labor Allocation) in *Gau Westmark*. Succeeded by Willi Stöhr.

16.11.1942-28.09.1944   *Reichsverteidigungskommissar in Reichsverteidigungsbezirk Westmark*. Succeeded by Willi Stöhr.

16.11.1942-29.09.1944	*Beauftragter des RKFDV* in Westmark and Lothringen. Succeeded by Willi Stöhr.
07.02.1943	At *Führer* HQ "Wolffschanze" in Rastenburg / Ostpreußen, during which Hitler addressed most of the *Reichs-* and *Gauleiter* on the subject of the recent German defeat at Stalingrad.
	(c) *Gauleiter* Bürckel will be responsible for the line of the Moselle from the boundary of *Gau* Westmark via the arsenal of Metz-Diedenhofen-south of St. Avoid (part of the Maginot Line) to Saaralben.
08.09.1944	In a letter to Martin Bormann (with whom he did not get along), Bürckel opined that the lack of combat-ready troops to occupy this defensive line made such construction efforts useless. Bormann responded by dispatching Willi Stöhr (who was to succeed Bürckel after his death at the end of the month) to oversee the construction work.

Published Works

Kampf um die Saar (Bohnenberger, 1934)

Wirtschaftspolitik ist Sozialpolitk (Eher-Verlag, 1939)

Volk und Kultur (Wien: Generalreferat für staatlicher Kunstverwaltung und Volksbildung des Reichsgaues Wien, 1940)

"Lothringen!" (Article in the *Metzer Zeitung*, 1940)

Wir sind der Zukunft verpflichtet! (Gauleitung Westmark der *NSDAP*, 1940)

Die Metzer Rede (Gauleitung Westmark der *NSDAP*, 1940)

Decorations and Awards

01.10.1944	*1. Klasse Goldenes Kreuz des Deutschen Ordens, höchste Stufe mit Lorben und Schwertern am Hals zu tragen* (posthumous award)
00.00.194_	*Kriegsverdienstkreuz I.Klasse mit Schwertern*
02.10.1940	*Kriegsverdienstkreuz I.Klasse ohne Schwertern*
00.00.1940	*Kriegsverdienstkreuz II.Klasse mit Schwertern*
ca. 1939	*Spange "Prager Burg" zur Medaille zur Errinerung an den 1. Oktober 1938*
ca. 1939	*Medaille zur Errinerung an den 1. Oktober 1938*
ca. 1939	*Medaille zur Errinerung an den 13. März 1938*
ca. 1934	*Ehrenkreuz des Weltkrieges 1914-1918 mit Schwertern*
00.00.193_	*Goldenes Ehrenzeichen der NSDAP*
00.00.194_	*Dienstauszeichnung der NSDAP in Silber*
00.00.194_	*Dienstauszeichnung der NSDAP in Bronze*
23.12.1939	*Deutsches Schutzwall-Ehrenzeichen*
[01.12.1937]	*Ehrendegen des Reichsführer-SS*
00.00.193_	*Totenkopfring der SS*
00.12.1937	*Julleuchter der SS*
00.00.193_	*Ehrenwinkel für alte Kämpfer*
10.08.1940	*Ehrenbürgerrecht der Gemeinde Wien*

February 1943: *Führer* HQ "Wolfsschanze" in Rastenburg, East Prussia.
From left to right: Fritz Wächtler, Ernst Wilhelm Bohle, Artur Görlitzer,
Josef Bürckel, Paul Wegener, and Arthur Greiser. (NARA photo)

27.04.1933   *Ehrenbürgerrecht der Stadt Neustadt an der Weinstraße*

Notes

- Youngest of four children of the master baker Michael Bürckel and his wife Magdalena, née Zoller, who had married in Lingenfeld on 09.08.1886.
- Religion: Roman Catholic.
- Married 11.07.1920 in Lingenfeld to Hilde Spies (born in Landau, 29.12.1899). The couple had two sons: Josef Artur (born 22.08.1921 in Landau / Pfalz; died of unknown causes as an *Unteroffizier* 20 km north of Calais, 20.01.1943), and Hermann-Jakob (born 10.03.1925)

Sources

Bytwerk, Prof. Randall (CAS Department-Calvin College, Grand Rapids, Michigan): German Propaganda Archive, at www.calvin.edu/academic/cas/gpa/ww2era.htm.

Domarus, Dr. Max (Editor): *Hitler. Speeches and Proclamations, 1932-1945: The Years 1935-1938.* Bolchazy-Carducci Publishers, 1990.

Grunberger, Richard: *The 12-Year Reich.* Holt, Rinehart and Winston, 1971.

Guenther, Irene: *Nazi Chic?: Fashioning Women in the Third Reich.* Berg, 2004.

Haar, Ingo & Fahlbusch, Michael: *German Scholars and Ethnic Cleansing, 1920-1945.* Berghahn Books, 2005.

Hamilton, Charles: *Leaders and Personalities of the Third Reich, Volume I.* R. James Bender Publishing, 1984.

Führer HQ "Wolfsschanze", Rastenburg, East Prussia, 07.10.1943. From left to right: *Gauleiters* Dr. Hugo Jury, Bürckel, Franz Schwede, and August Eigruber. (NARA photo)

Höffkes, Karl: *Hitlers politische Generale: Die Gauleiter des Dritten Reiches*. Grabert-Verlag-Tübingen, 1986.

Hüttenberger, Peter: *Die Gauleiter: Studie zum Wandel des Machtgefüges in der NSDAP*. Deutsche Verlags-Anstalt, 1969.

Janowsky, Oscar I. & Fagen, Melvin M.: *International Aspects of German Racial Policies*. Oxford University Press, 1937.

Lang, Jochen von: *The Secretary. Martin Bormann: The Man Who Manipulated Hitler*. Random House, 1979.

Lilla, Joachim; Döring, Martin; & Schulz, Andreas: *Statisten in Uniform. Die Mitglieder des Reichstags 1933-1945*. Droste Verlag, 2004.

Miller, Michael D.: – *Leaders of the SS & German Police, Volume I (Reichsführer-SS-SS-Gruppenführer, Georg Ahrens to Karl Gutenberger)*. R. James Bender Publishing, 2006.

- *Gauleiter: The Regional Leaders of the Nazi Party and Their Deputies, Volume I* (with Andreas Schulz). R. James Bender Publishing, 2012.

Namier, L. B.: *Diplomatic Prelude, 1938-1939*. Macmillan, 1948.

National Archives and Records Administration, College Park, Maryland: *SS-Personalakte* of Josef Bürckel. Microfilm document collection A3343SS.

Gauleiter and *Reichskommissar* Bürckel on the cover of *Wiener Bilder*, 01.05.1938.

- "Interrogation Records Prepared for War Crimes Proceedings at Nuernberg, 1945-1947/OCCPAC Interrogation Transcripts And Related Records: Speer, Albert"; Publication Number M1270, Record Group RG238.

Noakes, Jeremy: "Federalism in the Nazi State", in Maiken Umbach, ed, *German Federalism*. Macmillan, 2002.

Orlow, Dietrich: *History of the Nazi Party: 1933-1945*. University of Pittsburgh Press, 1973.

Schirach, Baldur von: *Die Pioniere des Dritten Reiches*. Zentralstelle fur der deutschen Freiheitskampf, 1933.

Schuschnigg, Kurt von: *Austrian Requiem*. G. P. Putnam's Sons, 1946.

Sydnor, Charles W., Jr.: *Soldiers of Destruction, The SS Death's Head Division, 1933-1945* (8th edition). Princeton University, 1999.

Trevor-Roper, H. R., ed.: *The Bormann Letters: The Private Correspondence between Martin Bormann and His Wife from January 1943 to April 1945*. Weidenfeld and Nicholson, 1954.

Wettstein, Dr. Lothar: *Josef Bürckel: Gauleiter Reichsstatthalter Krisenmanager Adolf Hitlers*. Books on Demand, 2009.

Friedrich *Leopold* Damian
SA-Obergruppenführer

Born:	04.04.1895 in Böbingen / Landesamt Landau / Pfalz / Bayern.
Died:	19.04.1971 in Neustadt.
NSDAP-Nr.:	133 642 (Joined 01.05.1929)

Promotions

24.08.1915	*überzähliger Gefreiter*
05.08.1916	*überzähliger Unteroffizier*
19.08.1916	*Offizier-Aspirant*
31.07.1917	*Leutnant d. R.*
09.09.1932	*SA-Standartenführer* (mit Wirkung vom 01.05.1931)
15.04.1936	*SA-Oberführer*
20.04.1936	*SA-Brigadeführer*
01.05.1937	*Oberleutnant d. R.*
09.11.1938	*SA-Gruppenführer*
00.00.19__	*Hauptmann d. R.*
09.11.1944	*SA-Obergruppenführer*

Career

00.00.1901-00.00.1908	Attended *Volksschule*.
00.00.1908-00.00.1911	Attended a *Präparandenanstalt*
00.00.1911-00.00.1913	Attended a *Lehrerseminar*.
00.00.1913-ca. 1914	Entered the public school service, eventually employed as an *Oberlehrer* (senior teacher) in Neustadt an der Weinstraße.
23.01.1915-00.05.1915	Entered military service as a *Rekrut*, assigned to *Rekruten-Depot 2 / Ersatz-Bataillon / Bayerische Infanterie-Leibregiment*.
00.05.1915-00.11.1915	Frontline service with *Bayerische Feld-Infanterie-Bataillon Nr. 1*.
00.11.1915-00.03.1916	Assigned to *3. Kgl. Bayerisches Feldartillerie-Regiment*.
20.03.1916	Wounded in action near Malancourt in the area of Verdun (shrapnel in the left forearm).
23.03.1916-28.03.1916	Hospitalized in *Reserve Lazarett Langensalza*.
00.04.1916-00.04.1917	Assigned to *Ersatz-Bataillon Augsburg*; attended an *Aspirantenkurs* (officer candidate course) in Grafenwöhr during this period.
00.04.1917-00.10.1918	Assigned as a *Zugführer*, then *Kompanieführer* to 3. *Bayerische Reserve-Infanterie-Regiment*.
13.09.1917-30.09.1917	Convalescent leave in Böbingen.
07.06.1918-24.06.1918	Convalescent leave in Böbingen.
29.06.1918-03.07.1918	In sick bay due to influenza.
14.10.1918-00.02.1920	In Belgian captivity.
00.02.1920	Repatriated to Germany, then discharged from military service at the *Durchgangslager* in Griesheim / Hessen.

SA-Gruppenführer Damian, ca. 1938.

01.05.1929	Joined the *NSDAP / Ortsgruppe Neustadt an der Weinstraße*.
00.00.1929-00.00.1931	*Kassenverwalter* (treasurer) of the *Ortsgruppe Neustadt an der Weinstraße der NSDAP*.
01.01.1930	Joined the *SA*, assigned to *SA-Standarte V* in Neustadt / Haardt
01.01.1930-01.05.1931(?)	*Geldverwalter* (m.d.W.d.G.b.) of *SA-Standarte V*. He was formally sworn in as an *SA-Führer* on 01.04.1930 by the *Führer* of *SA-Standarte V, SA-Standartenführer* Fritz Schwitzgebel.
01.05.1931-30.04.1935	*Führer* of *SA-Standarte 18 "Emil Müller"* (Neustadt / Haardt).
00.10.1933-00.08.1934	*Sonderbeauftragter der Obersten SA-Führung* to the *Bezirksämtern* of *Neustadt/Haardt, Landau,* and *Germersheim*.
25.02.1934-24.03.1934	Attended the *23.Lehrgang* at the *Reichsführerschule der SA* in München.
01.05.1935	Assigned as a *hauptamtlicher SA-Führer*.
29.03.1936	Unsuccessful candidate for election to the *Reichstag* (*Wahlkreis 27, Rheinpfalz-Saar*).
01.05.1935-14.09.1936	*Führer* of *SA-Brigade 51* (m.d.F.b., 01.05.1935-19.04.1936, then permanent) of *SA-Brigade 51 "Ostpfalz"* (Neustadt/Haardt).
00.06.1936-00.06.1936	Participated in training exercises with *III. Bataillon / Infanterie-Regiment* 21.
15.07.1936-14.09.1936	Attached to the *Stab der Obersten SA-Führung* with assignment to the *Gerichts- und Rechtsamt* (Courts and Legal Office).
15.10.1936-31.10.1937	*Abteilungschef* in the *Gerichts- und Rechtsamt* of the *Obersten SA-Führung* in München.

01.11.1937-13.04.1941	*Chef* of the *Gerichts- und Rechtsamt* of the *Obersten SA-Führung* and Chef of the *Obersten SA-Gericht*; also assigned as a *Beisitzer* (associate judge) to the *Obersten Parteigericht der NSDAP*.
10.04.1938	Unsuccessfully proposed for the *Liste des Führers zur Wahl des Großdeutschen Reichstages am 10. April 1938*. (List of the *Führer* for election to the Greater German *Reichstag* on 10. April 1938).
00.07.1938-00.07.1938	Participated in a 3-week training exercise with *Infanterie-Regiment* 63.
00.09.1938-00.09.1938	Participated in a 3-week training exercise with *Infanterie-Regiment* 63.
26.08.1939-00.00.1941	Assigned, initially as *Hauptmann d. R.* and *Kompaniechef* (ultimately as *Bataillonskommandeur*) to *Infanterie-Regiment* 63 (Ingolstadt), seeing action in the Polish and French Campaigns.
14.04.1941-00.05.1945	*Führer* (m.d.F.b., 14.04.1941-31.12.1941, then permanent) of *SA-Gruppe Oberrhein* (Straßburg).
ca. 1941-00.00.1945	Honorary member of the *Volksgerichtshof* (for the duration of the war).
09.05.1941	Discharged from *Wehrmacht* service by *Infanterie-Ersatzbataillon 365* in Köln-Kalk.
09.05.1941-30.06.1943	Unknown assignment.
28.04.1942-08.05.1945	Member of the *Reichstag (Wahlkreis 32, Baden)*. Succeeded the fallen Willy Ziegler.

Decorations and Awards

09.10.1940	*1939 Spange zum 1914 Eisernen Kreuz I. Klasse*
00.00.1939	*1939 Spange zum 1914 Eisernen Kreuz II. Klasse*
20.09.1918	*1914 Eisernes Kreuz I. Klasse*
14.07.1917	*1914 Eisernes Kreuz II. Klasse*
16.01.1918	*Bayerisches Militärverdienstorden IV. Klasse mit Schwertern*
00.00.1918	*Verwundetenabzeichen, 1918 in Schwarz*
05.10.1940	*Medaille zur Erinnerung an den 1. Oktober 1938*
05.10.1940	*Medaille zur Erinnerung an den 13. März 1938*
ca. 1934	*Ehrenkreuz des Weltkrieges 1914-1918 mit Schwertern*
30.01.1941	*Dienstauszeichnung der NSDAP in Silber*
20.04.1940	*Dienstauszeichnung der NSDAP in Bronze*
ca. 1931	*Abzeichen des SA-Treffens Braunschweig 1931*
24.03.1934	*Tyr-Rune*
00.00.1934	*Ehrendolch der SA*
00.02.1934	*Ehrenwinkel für alte Kämpfer*

Notes

- Son of the farmer Jakob Damian (born 11.11.1861 in Böbingen) and his wife Berta, née Reitz (born 09.11.1864 in Böbingen).
- Religion: Declared himself "gottgläubig", 19__.
- Married to Lise Mugler (born 14.11.1905 in Frankenthal / Pfalz). No children.

Sources

Bayerisches Hauptstaatsarchiv, München, Abteilung IV Kriegsarchiv: Excerpts from various Kriegsranglisten containing data on the Royal Bavarian Army service of Leopold Damian.

Bundesarchiv, Berlin-Lichterfelde (former Berlin Document Center): *Personalunterlagen von SA-Angehörigen: SA-Personalakte* of Leopold Damian.

Lilla, Joachim; Döring, Martin; & Schulz Andreas: *Statisten in Uniform. Die Mitglieder des Reichstags 1933-1945*. Droste Verlag, 2004.

August Eigruber
SA-Obergruppenführer

Born:	16.04.1907 in Steyr / Oberösterreich.
Executed:	28.05.1947 at U.S. War Criminals Prison No. 1, Landsberg am Lech / Bayern (hanged at approximately 0900 hours by Master-Sergeant John C. Woods). The prison chaplain at Landsberg, Karl Morgenschweiss, later gave a statement about Eigruber's last words. As he left his cell, Eigruber called out to his fellow prisoners, "Es lebe Deutschland!" (Long live Germany!). His final words were: "Gott schütze Deutschland, Gott schütze meine Familie, Gott schütze meine Kinder. Ich empfinde es als eine Ehre von diesen, den brutalsten Siegern gehängt zu werden." (God protect Germany, God protect my family, God protect my children. I consider it an honor to be hanged by this, the most brutal victor.) As the gallows trapdoor opened, Eigruber cried out "Will of God, preacher!" Unfortunately for Eigruber, the executioner botched the job and his neck was not broken. His mouth and nose were stuffed with cotton and he suffocated.
NSDAP-Nr.:	83 432 (Joined 18.04.1928)
SS-Nr.:	292 778 (Joined 25.07.1938, with seniority from 12.03.1938)

Promotions

00.04.1930-00.00.1933	*Bezirksleiter der NSDAP*
00.00.1933-00.05.1935	*Kreisleiter der NSDAP*
01.08.1935-11.03.1938	*Gauleiter der* [illegal] *NSDAP*
22.05.1938-08.05.1945	*Gauleiter der NSDAP*
01.06.1938	*SA-Oberführer* (mit Wirkung vom 12.03.1938)
25.07.1938	*SS-Mann* (mit Wirkung vom 12.03.1938
25.07.1938	*SS-Standartenführer* (mit Wirkung vom 12.03.1938)
11.09.1938	*SS-Oberführer* (mit Wirkung vom 25.07.1938)
09.11.1937	*SA-Gruppenführer*
30.01.1939	*SS-Brigadeführer* (mit Wirkung vom 01.01.1939)
09.11.1940	*SS-Gruppenführer*
21.06.1943	*SS-Obergruppenführer*
09.11.1943	*SA-Obergruppenführer*

Career

00.00.1913-00.00.1917	Attended *Volksschule*.
00.00.1917-00.00.1921	Attended *Mittelschule*.
00.00.1921-00.00.1925	Trained as survey technician and precision engineer at the *Bundeslehranstalt für Eisen- und Stahlbearbeitung* ([Austrian] Federal Institute of Iron and Steel Work)(graduated).

A portrait of August Eigruber in civilian attire, autographed by the subject on 15.11.1938. (Hermann-Historica, Auctioneers, München)

16.11.1922	Joined the *Steyr Ortsgruppe* of the *Nationalsozialistische Arbeiterjugend Österreichs* (the Austrian Nazi youth organization).
00.00.1923-00.00.192_	Head of the *Steyr Jugendgruppe* (youth group) of the *National-sozialistischen Arbeiterjugend Österreichs*.
00.00.1925-00.08.1927	*Jugendleiter* of the *Nationalsozialistischen Arbeiterjugend Österreichs* for the whole of Austria.
	Employed as mechanic in the replacement parts facility of the *Steyr-Werke* and as a mechanic on calculating machines for the *Gummifabrik Reithofer* (Reithofer Rubber Factory). He also held positions as mechanical assistant in the *Bezirksvermessungsamt* (district surveying office), Steyr and laborer in a crane factory during this period.
00.04.1927	Joined the *Hitler-Jugend*.
19.08.1927-21.08.1927	Attended the *Reichsparteitag der NSDAP in Österreich* (*Reich* Party Day of the Nazi Party in Austria).
21.08.1927-00.04.1930	*Gauführer* of the *Hitler-Jugend* for Oberösterreich.
18.04.1928	Joined the *NSDAP-Österreich*.
00.00.1929-00.00.1934	Warehouse worker in the spare parts facility of the *Steyr-Werke*. He was fired from this position as a result of his political activities.
00.04.1930-00.00.1931	*Bezirksleiter* of *Bezirk Steyr-Land der NSDAP*.
00.00.1931-00.00.1933	*Bezirksleiter* of *Bezirk Steyr-Stadt der NSDAP*.
00.00.1933-00.05.1935	*Kreisleiter* of *Kreis Steyr der* [illegal] *NSDAP*.
13.06.1933	Arrested by Austrian police in Steyr.

Left to right: Dr. Artur Seyss-Inquart, Dr. Otto Gustav Wächter, Dr.-
Ing. Hanns Blaschke, August Eigruber, Dr. Hugo Jury, unknown,
Konrad Henlein, and Karl Gerland. (NARA photo)

Summer 1934	Arrested for illegal political activities following the assassination of Chancellor Engelbert Dollfuss and interned for several months at the *Anhaltelager* [detention camp] Wöllersdorf. He spent a total of 12 months in Austrian government custody during the 1930's and his home was searched by the authorities on 21 occasions.
00.05.1935-00.03.1936	*Gaugeschäftsführer* for the *Gauleitung Oberösterreich der* [illegal] *NSDAP*.
01.08.1935-11.03.1938	*Gauleiter* of *Gau Oberösterreich der* [illegal] *NSDAP* (Seat: Linz). Succeeded Karl Scharizer. The *Gau* was dissolved after the *Anschluß* of March 1938, then reestablished- as *Gau* [after 15.03.1940, *Reichsgau*] *Oberdonau*- under Eigruber on 22.05.1938.
00.00.1934	Founded the National-Socialist newspaper *Österreichischer Beobachter*.
21.02.1938-11.03.1938	Acting *Landeshauptmann* for *Oberösterreich*.
12./13.03.1938	Telephoned by Dr. Wilhelm Keppler, Hitler's representative in Austria, who authorized him to take over power in Oberösterreich from the Austrian official Dr. Heinrich Gleißner. This was to be done without informing the new Chancellor, Dr. Artur Seyss-Inquart.
14.03.1938-05.05.1945	*Landeshauptmann* of *Oberösterreich* (later redesignated *Oberdonau*). A temporary assignment until 00.05.1938, then permanent). Succeeded Dr. Heinrich Gleißner, who had been removed from office on 13.03.1938

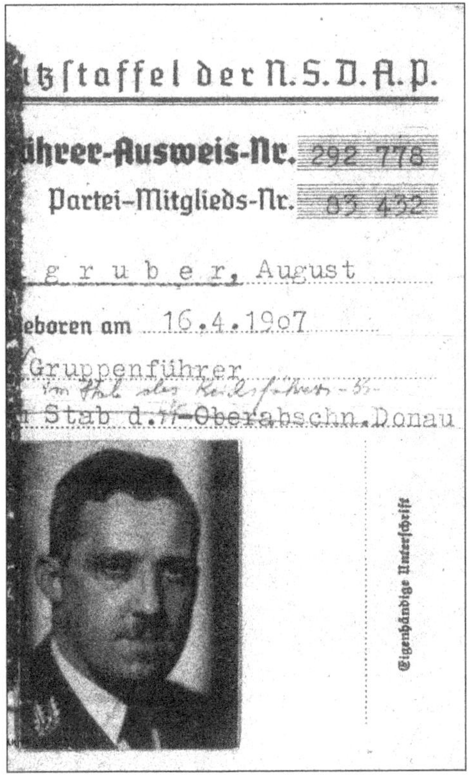

Eigruber's *SS-Führer-Ausweis*. (Source: His *SS* personnel file, on microfilm at NARA)

[and was placed in "Schutzhaft" – protective custody – on 15.03.1938]. Succeeded by Adolf Eigl (who was installed by U.S. authorities on 16.05.1945; Dr. Gleißner was reelected to the post in December 1945).

22.05.1938-08.05.1945	*Gauleiter* of *Gau Oberdonau der NSDAP* (Seat: Linz). Appointed by Dr. Friedrich Rainer in his capacity as political advisor in Austria, he developed a reputation as an independent and forceful administrator. His policies were often at odds with those of the senior Party leadership, and he refused, as he stated, "to implement and follow unconditionally the instructions from Berlin." Among his early accomplishments in this post was the reduction of unemployment in his *Gau*, from 22,932 in October 1937 to 3,195 by October 1938.
23.05.1938-04.05.1945	*Gauwalter der DAF* in *Gau* [after 15.03.1940, *Reichsgau*] *Oberdonau*.
00.03.1938-10.04.1938	*Gauwahlleiter der NSDAP* (regional elections coordinator of the *NSDAP*) in *Oberösterreich*.
10.04.1938-08.05.1945	Member of the *Reichstag (Wahlkreis Österreich)*.
01.06.1938-08.05.1945	Joined the *SA*, assigned as an *Führer z. V.* to *SA-Gruppe Alpenland*.
25.07.1938-09.11.1940	Joined the *SS* (with seniority from 12.03.1938), attached to the staff of *SS-Abschnitt VIII* (Linz).

Gauleiter Eigruber, ca. 1940. (NARA photos)

00.08.1938-27.07.1942	Member of the *Aufsichtsrat* for *Steyr-Daimler-Puch AG*, Wien, Austria's largest manufacturer of automobiles, locomotives, tanks, etc. He was involved in the assimilation of this company into the *Hermann Göring* combine, joining the boards of both companies.
00.00.193_-00.05.1945(?)	Member of the *Aufsichtsrat* for *Wohnungs-AG der Reichswerke "Hermann Göring"* and the *Wolfsegg-Traunthaler Kohlenwerks-AG*, Linz.
00.00.193_-00.05.1945(?)	Member of the *Beirat, Oberösterreichischen Landes-Brandschaden-Versicherungsanstalt* (Upper Austrian Provincial Fire Damage Insurance Institute), Linz.
00.00.1938-00.00.1938	Played a role in the establishment of a large concentration camp near *Mauthausen*, which opened its gates on 08.08.1938. The *Völkischer Beobachter* (Wien edition) of 29.03.1938 printed the following report:

> Bastion of Salzkammergut
> Upper Austria will soon be guarding the country's traitors ... In Gmunden a mass rally of the National Socialists of the Salzkammergut took place on the Adolf Hitlerplatz. At which the Oberbürgermeister of Saarbrücken, Party comrade Schwitzgebel, and the *Gauleiter* and governor of Oberösterreich, Eigruber, addressed the assembled masses from the town hall balcony.... The *Gauleiter* of the *Führer*'s home Gau, Eigruber, mentioned, among other things, and being constantly interrupted by enthusiastic jubilation:

> "We Upper Austrians are now receiving another, special distinction for our part in the struggle ... The concentration camp for traitors from all over Austria will be here."
>
> Thunderous applause almost drowns the announcement, so that the *Gauleiter* can hardly continue with his speech. (Website of the "*Dokumentationsarchiv der österreichischen Widerstandes*," at www.doew.at/thema/thema_alt/wuv/maerz38_2/propaganda. html)

22.09.1939-16.11.1942	*Beauftragter des Reichsverteidigungskommissars für den Wehrkreis XVII in den Gau* [after 15.03.1940, *Reichsgau Oberdonau* and member of the *Verteidigungsausschuß* for *Wehrkreis XVII.*
18.10.1939-15.04.1942	*Stellvertretender Vorsitzender des Aufsichtsrates* of *AG Alpine Montanbetriebe "Hermann Göring"*, Linz.
15.03.1940-04.05.1945	*Reichsstatthalter des Reichsgaues Oberdonau.* On 12.04.1940, he and the other newly appointed *Reichsstatthalter* of the *Ostmark* (former Austria) were officially sworn in by Hitler at the *Reichskanzlei* in Berlin. As in his *Gauleiter* post, he refused to unconditionally obey the commands of the senior *Reich* authorities in Berlin, stating: "As *Reichsstatthalter* I am not here to follow the instructions of some minor official who signed in the name of a *Reichsminister.*"
18.04.1940	Birthday celebration for *Gauleiter* Eigruber (who had turned 33 on 16.04.1940) in Linz, attended by Adolf Hitler, Rudolf Hess, and Hermann Göring.
25.05.1940	Founded the Gusen camp as a subsidiary of *Mauthausen.* It was reported that the two camps had a combined inmate population of 11,500 at that time. As early as 01.07.1940, there were 25,000 foreign laborers from 17 nations laboring in *Reichsgau Oberdonau.*
30.05.1940	Organized a celebration for the 70th birthday of the composer Franz Lehar.
09.11.1940-01.01.1942	Assigned to the staff of *SS-Oberabschnitt Donau.*
15.11.1940-04.05.1945	*Gauwohnungskommissar* for *Reichsgau Oberdonau.*
11.01.1941	Attended a conference of *Ostmark* (former Austria) *Gauleiter* in Salzburg.
06.03.1941	Accompanied *Reichsführer-SS* Heinrich Himmler on a visit to a *KL-Mauthausen*, wearing his black *SS-Gruppenführer* uniform for the occasion. Eigruber visited the camp a total of 26 times and was also a frequent visitor to the "euthanasia" institute, leased to the *Reich* government by Eigruber in 1939, at Hartheim (where some 18,000 people, including 7,000 inmates from *Mauthausen*, were gassed between 1940 and December 1944; many of the personnel assigned by the *T-4* program to Hartheim were selected by the *Gauinspekteur* of *Reichsgau Oberdonau*, Stefan Schachermayr, from Eigruber's *Gauleitung* staff in Linz. According to an affidavit by former inmate Johann Kanduth (who had worked in the *Mauthausen* crematorium), given 30.11.1945 to Lt. Col. David G. Paston in Salzburg, Eigruber visited the camp in 1942 or '43 together with Dr. Ernst Kaltenbrunner, the camp *Kommandant*

Gauleiter Eigruber, ca. 1940.

Franz Ziereis and members of his staff, and several other people. Kanduth stated: "Kaltenbrunner went laughing into the gas chamber. Then the people were brought from the bunker to be executed and then all three kinds of executions- hanging, shooting in the back of the neck, and gassing- were demonstrated. After the dust had disappeared, we had to take away the bodies."

12.03.1941	Of a visit to *Reichsgau Oberdonau*, *Reichsminister* Dr. Goebbels wrote the following in his diary:

> Göring Works: magnificent plan. Plus a comprehensive program
> of home-building, which Eigruber is pursuing with vigor. In
> general he gives an impression of energy and vitality. Will get on
> top of the job.

01.01.1942-00.05.1945	Assigned to the staff of the *Reichsführer-SS*.
08.03.1942	Obtained personal approval from Hitler for the establishment of a *Reich* Musical *Gymnasium* in St. Florien.
25.05.1942-04.05.1945	*Beauftragter des GBA für den Reichsgau Oberdonau* (Representative of the *Generalbevollmächtigter für den Arbeitseinsatz* [Plenipotentiary General for Labor Allocation, *Gauleiter* Fritz Sauckel] for *Reichsgau Oberdonau*). In this capacity, he was often critical of the brutal *SS*

Eigruber accompanies *Reichsführer-SS* Himmler on two inspections of KL-Mauthausen early in 1941. (Above: *Bundesarchiv Bild* 192-308, below: *Bundesarchiv Bild* 192-140)

policies toward the so-called "Ostarbeiter" (foreign laborers from the occupied eastern territories); this resulted in an angry letter of complaint [specifically concerning a speech of 25.11.1942 delivered by Eigruber in Linz] from Himmler to Martin Bormann on 14.01.1943. His treatment of these workers was relatively relaxed and conciliatory; when Italian laborers in *Reichsgau Oberdonau* began to revolt against their German overseers following Italy's September 1943 armistice with the Allies, Eigruber played a role in calming them.

16.11.1942-04.05.1945   *Reichsverteidigungskommissar für den Reichsgau Oberdonau.*

15.06.1943-30.04.1945   Representative of the *Reichskommissar für die Festigung des Deutschen Volkstums* (*RKFDV* [Himmler]) in *Reichsgau Oberdonau.*

25.06.1944   Together with *Reichsminister* Albert Speer, inspected the *Reichswerke Hermann Göring* in Linz. During this visit, the two conversed with inmates from nearby *KL-Mauthausen.*

25.09.1944-04.05.1945   *Führer des Deutschen Volkssturms in Gau Oberdonau.*

15.02.1945   Established a special "*Standgericht*" (court-martial) for Oberdonau. The following is excerpted from the postwar interrogation records of August Eigruber:

> ... [On 30.] March 1945 ... Himmler empowered [Eigruber] to sentence to death all persons who were sabotaging German resistance and the German war effort. Power to impose capital punishment was not previously relegated to *Gauleiter*s. However, Eigruber himself had circumvented this rule before the Himmler order by establishing the 'Standgericht' ... in Upper Austria on 15 February 1945. The following four members of the Standgericht were all immediately appointed by Eigruber: Chief District Attorney (*Generalstaatsanwalt*) Loederer; *Hauptmann* Reichel; Dr. Sirocki; and Franz Edlinger. Edlinger was later replaced by Letner. The function of this court was first and primary, to try offenses of a political nature; and second, to have jurisdiction over criminal cases. It was able to either impose a death sentence or transfer the case to a regular court. Determination of the place, time and manner of execution were left entirely up to Eigruber. (Interrogation Records Prepared for War Crimes Proceedings at Nuernberg, 1945-1947 / OCCPAC Interrogation Transcripts And Related Records: Eigruber, August; Publication Number M1270, Record Group RG238)

28.03.1945   Attended a conference, chaired by *Reichsführer-SS* Heinrich Himmler, along with other Austrian *Gauleiter* (Dr. Siegfried Uiberreither of *Steiermark*, Dr. Hugo Jury of *Niederdonau*, and Baldur von Schirach of *Wien*) and the *Kommandant* of *KL-Mauthausen*, Franz Ziereis. According to postwar testimony by some of the participants, Himmler instructed them to coordinate an orderly evacuation of inmates from forced labor camps in Austria. These camps held tens of thousands of male and female Hungarian Jews. Although the instructions given were to ensure the care and safety of the evacuated Jews, it was also understood that they were not to fall into Allied hands alive. Thousands were murdered in the ensuing "Todesmärsche" (death marches) by their *Volkssturm, Hitler-Jugend, Gendarmerie, Gestapo,* and *SS* guards.

00.04.1945-00.05.1945   Personally took control of defensive measures in the *Reichsgaue Ober-* and *Niederdonau*. He attempted to turn his own *Gau* into a fortress, establishing a draconian regime against "defeatists and backsliders" and granting temporary police powers to his *Kreisleiter*. He made the following radio broadcast on 07.04.1945:

> I am announcing to one and all that we will stand in Oberdonau. Nobody moves, nobody leaves, nobody heads for the rear, there will be no evacuation. We will remain in our homeland. We are determined to defend this homeland... I turn to the soldiers: Now we will stand and if necessary, fight the deciding battle. Not in Siberia, but in our homeland will we fight and if required, die.

In a further radio broadcast on 14.04.1945, he declared: "Oberdonau will not be evacuated." In these harsh measures he had the support of *Generaloberst* Dr. Lothar Rendulic, Commander-in-Chief of *Heeresgruppe Ostmark*, who, on 14.04.1945, ordered the formation of numerous *fliegende Standgerichte* (flying courts-marital) to try and sentence members of the *Wehrmacht* caught out of uniform or absent without leave from their units. Following the fall of Wien to the Soviet Army, Eigruber ordered the arrest, for "desertion" of Viennese people who had fled to Oberdonau; among these was a member of Wien's city council, Karl Mayerzelt, who was arrested and shot. In his *Rundspruch Nr. 64* of 18.04.1945, Eigruber proclaimed:

> I have henceforth decided to intervene violently absconding political leaders, chiefs of offices of State and against upper military echelons and officers who behave arrogantly, carry out their own sequestrations or have a corrosive effect. I authorise the *Kreisleiter* and district administrators to intervene independently in all these cases with the help of the *Gendarmerie* or the German *Volkssturm*. They are either to be handed over to the drumhead court martial immediately or in certain cases simply shot on the spot. (Nikolaus von Preradovich, *Österreichs Höhere SS-Führer*, p. 12)

On 10.04.1945, Eigruber ordered the destruction by explosives of an enormous art collection- earmarked for Hitler's planned art gallery in Linz- stored at the Salzbergbau, a salt mine at Alt Aussee to avoid its capture by the Americans. According to a summary report by Lt. George Stout, USNR, dated 27.05.1945:

February 1943: Conference of *Reichsleiter* and *Gauleiter* at Führer HQ "Wolfsschanze" in Rastenburg, East Prussia. Eigruber shakes the Führer's hand. (NARA)

Records in the hands of German officials indicate that there are in part 6,577 paintings, 230 drawings and water colors, 954 prints, 137 pieces of sculpture, 128 pieces of arms and armor, 79 baskets of objects, 484 cases of unknown objects presumed to be archives in part, 78 pieces of furniture, 122 tapestries, 181 cases of books, 1,200 to 1,700 cases apparently containing books or similar matter... among these holdings are important and famous looted works such as the Van Eyck altar-piece from the church of St. Bavon in Ghent ... the Bouts altar-piece from Louvain, the Michelangelo Madonna and Child from Bruges, and many others ... . According to the fragmentary testimony so far taken, there was a deliberate plan to destroy all holdings at Alt-Aussee. In early April 1945, the provincial *Gauleiter*, Eigruber, ordered the *Kulturreferent*, Stuppäck, to see that measures were taken to insure total destruction of the mine, and its contents. It is probable that this order was transmitted from higher authority. On 10 April, heavy cases marked "Marmor, nicht Stürzen" [Marble, don't drop] were placed in the chambers with the holdings. It was later discovered that these contained 700 Kilo HE [high explosive] bombs and that detonators for them were on the way. Subsequent happenings are not definitely explained. They involve surreptitious actions, threats of flooding, counter-threats, and a quarrel between Eigruber and a man named [*SS-Obergruppenführer* Dr. Ernst]

Eigruber on trial in the "Mauthausen" case at Dachau, held from 29.03. to 13.05.1946.

Kaltenbrunner ... .who was in refuge at Alt-Aussee. In the course of all this, the bombs were taken out and put under a brush pile a short way below the mine, objects were moved about in a curious fashion but did not receive serious damage, some parts of passages were blown, and the electrical wiring knocked out ... (*Report of the American Commission for the Protection and Salvage of Artistic and Historic Monuments in War Areas*)

On 27.04.1945, he issued an order via teletype to Franz Ziereis, *Kommandant* of *KL-Mauthausen*, directing the immediate execution of 42 Austrian anti-Nazis held at the camp, in order to prevent their falling into Allied hands alive. The inmates were murdered in the *Mauthausen* gas chamber (9 on 27.04.1945 and a further 33 the following day) (Winfried R. Garscha & Claudia Kuretsidis-Haider, "Die Räumung der Justizhaftanstalten 1945 als Gegenstand von Nachkriegsprozessen- am Beispiel des Volksgerichtsverfahrens gegen Leo Pilz und 14 weitere Angeklagte", in Gerhard Jagschitz / Wolfgang Neugebauer, eds, *Das Urteil des Volksgerichts Wien (August 1946) gegen die Verantwortlichen des Massakers im Zuchthaus Stein*, Wien 1995, pp. 12-35, at www.doew. at/service/archiv/materialien/april45/stein_einleit2.pdf)

With the U.S. Army pouring into Austria, he ordered a final defense of the River Inn Valley, however his forces were extremely limited-a few *Waffen-SS* units and a *Volkssturm* force numbering less than 30,000 men. Later in April 1945, as the U.S. 11th Armored Division neared his *Gauhauptstadt* of Linz, Eigruber planned to proclaim the city as a "Festung", to be defended to the bitter end. The *Oberbürgermeister*, *SS-Oberführer* Franz Langoth and the local *Kreisleiter*, Franz Danzer, succeeded in changing his mind on this point. Eigruber, in his fanaticism, still ordered destruction of the Nibelungen Bridge in Linz, but was dissuaded in this, too, by Danzer. The American assault on the city began on the morning of 04.05.1945, prompting Danzer to meet with Eigruber at the latter's villa. Eigruber was preparing to flee, and gave Danzer permission to negotiate with the U.S. forces on behalf of Langoth's city administration. After strenuous and frantic efforts and meetings with German military and political authorities, *Kreisleiter* Danzer and Mayor Langoth managed to save the city through a peaceful surrender to the Americans on 05.05.1945.

04.05.1945	Fled from Graz by car, heading south to Kirchdorf/Krems with members of his staff (including his former *Gauinspekteur* and right-hand man, Stefan Schachermayr). On the same day, at 2030 hours, he announced by radio that Linz was an open city and was not to be defended.
05.05.1945	Ordered the surrender of Linz to troops of the U.S. Third Army, then moved on to Windischgarsten.
06.05.1945	From his new base in Kirchdorf an der Krems, 40 km south of Linz, entered into surrender negotiations with Colonel Smythe, a task force commander in the U.S. 80th Infantry Division (under Major-General Horace L. McBride). He met personally with Smythe in Windischgarsten, offering the surrender of his entire *Gau*, but demanded that German troops be allowed to continue fighting the Soviets; this was, naturally, refused. Eigruber was granted four hours to contact *Generaloberst* Dr. Rendulic. On Eigruber's return he informed Colonel Smythe of Rendulic's refusal to surrender. The following day, however, Rendulic did surrender his forces to Major-General Willard G. Wyman's 71st Infantry Division.
07.05.1945	Transferred authority in *Reichsgau Oberdonau* to *Gauhauptmann SS-Oberführer* Prof. Dr. Armin Dadieu.

Postwar Flight, Confinement and Prosecution

Fled from Krems on 08.05.1945, carrying forged papers identifying him as a farmer, Bernhard Gruber, and moved in the direction of Spital/Pyrhn. Traveling off main traffic routes via Vorderstoder up into the "Immerl Gsoll", he hid for several weeks in a small hunting lodge. There he was discovered by a local forestry officer and reported to U.S. CIC officials. The forestry officer escorted the CIC men to the lodge where Eigruber was staying, but found it deserted. Eigruber had concealed himself in bushes mere meters away from the lodge. He then stayed with a woodworker and his family in Gaisriegl / Vorderstoder. From here he apparently met up with two former leaders of the *Hitler-Jugend* from Steyr. These men were supposed to give him identification documents and a vehicle in which to escape. Unfortunately for Eigruber, his two helpers were now in the employ of the U.S. CIC in Kirchdorf. On the morning of 10.08.1945,

Courtroom of the U.S. Military Tribunal at Dachau, 13.05.1946. Stills
from a newsreel from the German news program *Welt im Film* of August
Eigruber, filmed as he was sentenced to death by hanging.

Eigruber and Stefan Schachermayr broke cover to escape to the Mühlviertel area of Austria where the *Gauleiter* still had friends he believed could be trusted. From here he hoped to travel to the Südtirol / Alto Adige (Italy). The aforementioned ex-*HJ* leaders turned over details of his escape plan to the CIC. Acting on this information, Lieutenant-Colonel Snowden of the CIC in Kirchdorf and Major Cameron of the OSS (Co. "A", 2677 Regiment) in Linz set up a barricade near the Vogel Inn in Steyrling (in the St. Pankratz area of the Salzkammergut). This barrier was situated behind a sharp curve in the road where vehicles had to reduce their speed. On 11.08.1945, as *Gauleiter* Eigruber and Schachermeyr approached the barricade in their vehicle, they were captured after brief resistance. Eigruber had a set of dentures confiscated from him as they contained a cyanide capsule. Eigruber later claimed in a statement that, had he been able, he would have shot his way out of the predicament or killed himself. Initially held at the U.S. Army CIC office in Kirchdorf, then moved to Linz where, on 18.08.1945, the OSS conducted a preliminary investigation into Eigruber's role in war crimes, particularly those committed at the *Mauthausen* concentration camp. That same day he was transferred to Gemünden where he was further interrogated at the U.S. Detailed Interrogation Center, then transferred on 22.10.1945 to Nürnberg where he gave further information on the *Mauthausen* camp to his interrogators. Moved to Dachau on 18.02.1946. Upon his arrival there, he was handed over to the U.S. Army interrogator Lieutenant Paul Guth. On 19.02.1946, after his first interrogation by Guth, Eigruber signed a statement admitting to his role in leasing Schloß Hartheim to the *Reich* government in 1939, with the knowledge that it would be used as a facility for exterminating the mentally and physically disabled (as part of the so-called "Euthanasia" [*T-4*] program. He further admitted to having inspected the *Mauthausen* concentration camp and to witnessing the murder of ten inmates in March or April 1945. Eigruber was tried as a war criminal in the "*Mauthausen* Case" (as defendant #13), 29.03.1946-13.05.1946 (defense attorney: Major Ernst Oeding). Sentenced to death by hanging, 13.05.1946. The court declined a plea for clemency from his wife. A further attempt by Dr. Heinrich Gleissner, a senior government official (whom Eigruber had succeeded in Upper Austria in March 1938) to secure a final visit with Eigruber by family members was also declined. In a letter to his sister-in-law, written shortly before his execution, Eigruber wrote:

> How long do I have to sit here until the day of my final judgment … ? I must also suffer the consequences of crimes committed by others who are not present here. But it doesn't matter because they are going to pin *Mauthausen* on me … I only wanted the best for my country … I very much regret not spending more time with Johanna and the children … Oh how I wish I was with you all …

The following is an official report of evidence presented during Eigruber's trial:

13. August EIGRUBER
Nationality: Austrian
Age: 38
Connection with *Mauthausen*:
a) Period: 1938-5 May 1945
b) Status: Honorary Obergruppenfuehrer (Lt. Gen.) in the *SS* and *SA*
c) Position: Gauleier of Upper Austria from June 1938; *Reichsstatthalter*; in 1942 that district's food office, and in 1943 the district's work office came under his supervision; in 1943 he became the district's *Reich* Defense Commissar.

Evidence: One witness testified that it was reputed to have been upon orders of the accused that rations for sick prisoners were cut in the fall of 1944 (R 169, 172), and that he heard the accused say to the Camp Commandant, with reference to some recaptured Russians, 'All these pigs will have to be finished' (R 208, 211-2). Another witness testified that he saw him at the camp three or four times; that he saw him touring the crematory, bunker and blocks; that he was in civilian clothes on that occasion; and that he was there in the fall of 1944 when 40 Austrians arrived, all but one of whom were later executed (R 356-7, 387-8). A third witness heard him in the fall of 1943 address a group of newly arrived prisoners, advising them that if they did not die from work they would be beaten to death. He was dressed in uniform at the time (R 482-4, 502). Another witness testified that as *Gauleiter*, or State Leader, of Upper Danube, *Mauthausen* was under his control and jurisdiction; that he saw him in *Mauthausen* on

12 to 16 occasions, once on 7 September 1944 when the Austrians were brought into the camp, and once in January 1945 on the night of the arrival of some 20 or 22 Anglo-Americans (sic) who, according to talk in the camp, were later liquidated (R 532). This same witness in April 1945 saw an order which he said was a transcription of either a telephone or telegraph message, signed in the name of but not by the accused pursuant to which order a number of Austrian prisoners were killed (R 533, 564-5, 578). A fifth witness, one of a group of Austrians which arrived in *Mauthausen* in September 1944, testified that he was the only survivor of the group, some having been killed by gassing in the after part of April. A few days after their arrival at the camp the accused had addressed them, shouting rebuke and promising rough treatment. The witness said that he had heard both before and after the gassing that this accused gave the order (R 5__, 584, 489, 592, 596) Another witness saw the accused in the Vienna Ditch early in 1942. He gave a 50 to 70 kilogram stone to a Jew to carry and a little later had that replaced by one weighing 80 kilograms. According to rumor, he told a detail leader that he did not want to see 'any one of this gang any more' (R 745). He was wearing some kind of a uniform and struck a few Jews (R 750-1). A seventh witness testified that this accused visited *Mauthausen* and on one ocacasion, at the end of 1942 or beginning of 1943, inspected an execution folding chair, by which four Czechs were put to death as a demonstration for him; the accused himself pushed down on the 'release pedal' to execute one man (R 777-8, 840-2). This witness also testified that boxes of food were delivered to the accused from the camp storehouse (R 779, 802-3). Two other witnesses testified that the accused was present on the night of 5 April 1945 when two Americans, one Englishman, and nine others were executed by shots in the neck (R 1196, 1215-7). One of these witnesses said that just prior to the execution of one, the accused recited a sentence of death in the name of Himmler (R 1216).

Accused was named in the Statements of several co-accused, i.e. No. 1 stated that since February 1945, the camp commander had been under the direct supervision of the accused, that prior to that time the accused probably had tremendous influence because of his high *SS* rank (R 1309-10; No. 1 did not take the stand) ... No. 20 asserted in his Statement that he personally knew the accused 'inspected several killings and brought prisoners to the killings' (R 366; denied by No. 20 on the stand R 3068-9) ... No 45 mentioned the shooting of some twelve people, including two Englishmen, five Poles, one Belgian and one Russian, by the accused and the camp commandant at 0400 on a morning in April 1945, while the two were in an intoxicated condition (R 1322, No 45 did not take the stand) ... No. 54 stated that this accused personally lead the shooting of nine Czechs in the spring of 1943 (R 1370; No. 5 did not take the stand).

In his own Statement the accused said that he came *Gauleiter* of Upper Austria in June 1938, that in 1943 he became the district's *Reich* Defense Commissar; that from 1942, the district food office and from 1943, the district work office were under his supervision; that administration, health and industrial questions concerning *Mauthausen* made necessary frequent discussions with the camp commandant; that he participated in the execution of ten prisoners one night during March or April 1945; that the district security organizations under his order participated int eh search for the Russians who escaped at the end of 1944 or beginning of 1945; and that in 1939 or 1940 he leased his Gau's property, Castle Hartheim, to the *Reich* for use in an euthanasia program (R 301-3).

The accused took the stand in his own defense and testified that he had no control over

Landsberg, 28 May 1947. (Max Williams photos)

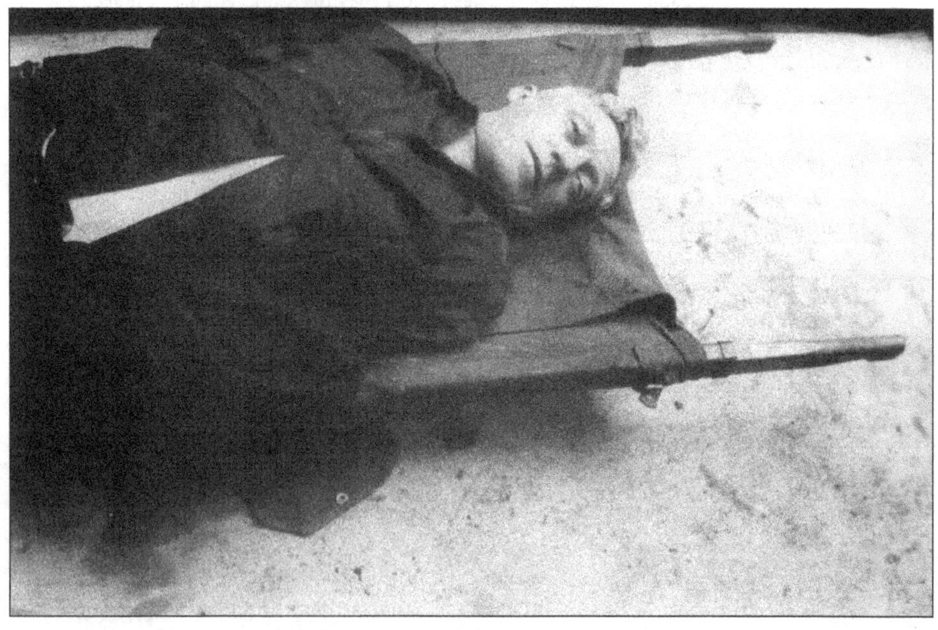

the rations for the camp, which were determined by the *Reich* Food Minister for the entire nation (R 3216-7). He admitted visiting *Mauthausen* ten to fifteen times between 1942 and 1945, but only on business relative to furnishing such services as water, electricity, etc. (R 2223) He always wore civilian clothes on such trips (R 2215). More than one-half the sub-camps were outside his Gau. He denied ever addressing any group in the camp (R 2218-9). He denied ever exercising any authority over the Vienna Ditch, claiming he was here only twice between 1942 and 1945, once on a visit with Himmler and once with Speer ($ 2223-4). He denied ever seeing the execution folding chair (R 2223). He also denied ever receiving prisoners' rations, explaining that he got along on his own family's ration cards (R 2224). He said his official positions of *Gauleiter, Reichsstatthalter* and Defense Commissar (R 2214-15, 2242) gave him no jurisdiction over *Mauthausen* or its out-camps.

Accused admitted he attended an execution at the invitation of the camp commandant (R 2224, 2226-7, 2267-9). He admitted leasing Castle Hartheim, *Gau* property, to the national Government (R 2343-8). He never discussed the internal affairs of the camp with any of the co-accused, nor did he ever give them orders concerning it (R 2228). Neither did he ever request an execution (R 2221). In an emergency he once furnished horse meat for the camp (R 2233). He kept informed relative to the camp epidemics only because of their possible effect on the civilian population (R 2235, 2247-8). The camp was constructed in 1938 without the exercise of any authority on his part (R 2242). He received a monthly food report (R 2245, 2263). He admitted jurisdiction over the *Mauthausen* camp in the event an epidemic therein threatened the civilian population (R 2256). He admitted being an Obergruppenfuehrer (honorary) in the *SS* and *SA* (R 2244, 2258).

Accused No. _2, 38, 42 and 9 denied ever having received orders from this accused (R 1652, 2426, 2501, 2691).

Accused No. 59 said that this accused was in no way responsible for the administration of the camp (R 1584-5); that he always tried to help the food situation (R 1585); and that he certainly had no need to take food from the camp warehouse (R 1629). Accused No. 61 stated that this accused was not in the *Mauthausen* chain of command (R 1_26); that he never saw a written order by the accused concerning the camp (R 1526-7); and that no reports were sent to him (R 1527).

Sentence: Death by hanging.

Petitions: A Petition for Clemency was filed by his wife, Johanna Eigruber, 10 July 1946.

Recommendation: Approval of findings and sentence.

Frau Eigruber was tried by the *Volksgericht* (People's Court) in Graz and sentenced on 13.07.1948 to 21 months' hard labor (with credit for time served in investigative detention). She was immediately released.

Published Work

Ein Gau wächst ins Reich: Das Werden Oberdonaus im Spiegel der Reden des Gauleiters August Eigruber, in *Bücher der Gaupropagandaleitung Oberdonau der NSDAP, Band 1* (Coauthor: Franz J. Huber) (1941)

Decorations and Awards

30.01.1942	*Kriegsverdienstkreuz I. Klasse ohne Schwerter*
ca. 1939	*Spange "Prager Burg" zur Medaille zur Erinnerung an den 1. Oktober 1938*
ca. 1939	*Medaille zur Erinnerung an den 1. Oktober 1938*
ca. 1938	*Medaille zur Erinnerung an den 13. März 1938*
30.01.1939	*Goldenes Ehrenzeichen der NSDAP*
10.06.1941	*Goldenes Hitler-Jugend Ehrenzeichen mit Eichenlaub*
30.01.1939	*Goldenes Hitler-Jugend Abzeichen* (Nr. 3357)
09.02.1942	*Dienstauszeichnung der NSDAP in Gold*
30.01.1940	*Dienstauszeichnung der NSDAP in Silber*
30.01.1940	*Dienstauszeichnung der NSDAP in Bronze*
08.12.1938	*Ehrendegen des Reichsführers-SS*
[30.01.1942]	*Totenkopfring der SS*
00.12.1938 (?)	*Julleuchter der SS*
25.05.1938	*Ehrenwinkel für alte Kämpfer*

[Note: Eigruber claimed in a *Lebenslauf* (*curriculum vitae*) of 01.10.1938 to have been recommended for the *Ehrenzeichen des 9. November 1923 (Blutorden)*, due to his having been imprisoned by Austrian authorities for over a year.]

Notes

- Son of Augustin Eigruber and his wife Aloisia (according to Nikolaus von Preradovich, he was the illegitimate child of Aloisia Eigruber, daughter of Josef Eigruber and his wife, Maria, née Kleestorfer).
- Religion: Roman-Catholic.
- Married on 08.08.1930 (or 25.09.1930) to Johanna Maria Spatzenegger born 18.05.1905 in Steyr; member of the *NS-Frauenschaft*). Three sons (Ad., born 28.01.1929; Hermann, born 02.10.1930; Al., born 22.12.1932) and three daughters (I., born 15.09.1933; G., born 23.11.1934; and ____, born 08.08.1939). Hermann Eigruber, a graduate in electrical engineering from a *Höhere Technische Bundeslehranstalt* was, from 1983 to 1990, a member of the *Nationalrat* (National Council), one of the two houses of the *Österreichisches Parlament*. Previously, he had worked for the firm *Wella und Elektro-Bau AG* until 1955, then was self-employed in the chemical engineering and building material retail industries. He went on to become *Prokurist* (confidential clerk) of the *Asbiton* firm, and after 1975 was *Geschäftsführer* (CEO) of the firm *Compriband-Dichtungen*. His achievements in business resulted in his receiving the honorary title of *Kommerzialrat*. From 1979 to 1988, Eigruber was active as *Bundesobmann* (federal chairman) of the *Ringes Freiheitlicher Wirtschaftstreibender*, and after 1980 he was a *Gemeinderat* in the city of Traun. In 1982, he assumed the position of Vice President of the *Kammer der gewerblichen Wirtschaft für Oberösterreich* (Economic Chamber of Upper Austria). From 19.05.1983 to 18.06.1990, he represented the *Freiheitliche Partei Österreichs* (FPÖ, Freedom Party of Austria) in the *Nationalrat*, continuing to hold a parliamentary seat without party affiliation from 19.06.1990 to 31.08.1990. For his service to his home province, he received the *Silbernes Ehrenzeichen des Landes Oberösterreich* in 2001.

Sources:

American Commission for the Protection and Salvage of Artistic and Historic Monuments in War Areas: *Report of the American Commission for the Protection and Salvage of Artistic and Historic Monuments in War Areas.* U.S. Government Printing Office, 1946.

Bukey, Evan Burr: *Hitler's Hometown: Linz, Austria, 1908-1945.* Indiana University Press, 1986.

Höffkes, Karl: *Hitlers politische Generale: Die Gauleiter des Dritten Reiches.* Grabert-Verlag-Tübingen, 1986.

Hüttenberger, Peter: *Die Gauleiter: Studie zum Wandel des Machtgefüges in der NSDAP.* Deutsche Verlags-Anstalt, 1969.

Lilla, Joachim; Döring, Martin; & Schulz, Andreas: *Statisten in Uniform. Die Mitglieder des Reichstags 1933-1945.* Droste Verlag, 2004.

Miller, Michael D.: *Gauleiter: The Regional Leaders of the Nazi Party and Their Deputies, Volume I* (with Andreas Schulz). R. James Bender Publishing, 2012.

- *Leaders of the SS & German Police, Volume I (Reichsführer-SS-SS-Gruppenführer, Georg Ahrens to Karl Gutenberger).* R. James Bender Publishing, 2006.

National Archives and Records Administration, College Park, Maryland: *SS-Personalakte* of August Eigruber. Microfilm document collection A3343SS.

- "Interrogation Records Prepared for War Crimes Proceedings at Nuernberg, 1945-1947/OCCPAC Interrogation Transcripts and Related Records: Eigruber, August"; Publication Number M1270, Record Group RG238.

Preradovich, Nikolaus von: *Österreichs Höherer SS Führer.* Kurt Vowinckel Verlag KG, 1987.

Slapnicka, Harry: "Um die *Gauleiter*-Posten der 'Ostmark'. Wie Goebbels mitmischt Personalpolitik im Schatten des Krieges", in *Historisches Jahrbuch der Stadt Linz 1996.* Archiv der Stadt Linz, 1997.

Franz Xaver Ritter von Epp
SA-Obergruppenführer

Born:	16.10.1868 in München / Regierungsbezirk Oberbayern / Bayern.
Died:	31.12.1946 in a U.S. military hospital, München. Gravesite: Münchner Waldfriedhof.

NSDAP-Nr.:	85 475 (Joined 18.05.1928)

Promotions

16.08.1887	*Dreijährig-Freiwilliger (Offizier-Aspirant)*
07.12.1887	*Unteroffizier*
09.03.1888	*Portepeéfähnrich*
30.10.1889	*Sekonde-Lieutenant*
13.10.1896	*Premier-Lieutenant* (ohne Patent; 07.11.1896: Granted Patent; 01.01.1899: Redesignated *Oberleutnant*)
27.01.1904	*Oberleutnant* (in the *Kaiserlichen Schutztruppe*, mit Patent vom 07.11.1896)
11.07.1904	*Hauptmann* (in the *Kaiserlichen Schutztruppe*, ohne Patent; 15.09.1904: Granted Patent vom 15.09.1904; 06.03.1906: Granted backdated Patent vom 15.09.1902)
14.12.1906	*Hauptmann* (Reinstated in the Royal Bavarian Army at that rank; mit Patent Nr. 15 vom 28.10.1903)
30.10.1909	*Major* (ohne Patent; 07.03.1910: Granted Patent Nr. 32)
30.11.1914	*Oberstleutnant*
14.12.1917	*Oberst*
02.10.1922	*Generalmajor* (mit RDA vom 01.07.1921)
31.10.1923	*Charakter als Generalleutnant*
18.12.1931	*SA-Gruppenführer (NSKK)*
01.07.1932	*Ehrenführer des NSKK (= Obergruppenführer)*
01.01.1933	*SA-Obergruppenführer*
05.05.1934-30.04.1945	*Reichsleiter* der *NSDAP*
05.07.1935	*Charakter als General der Infanterie*

Career

ca. 1874-ca. 1878	Attended *Volksschule* (presumably)
ca. 1878-00.00.188_	Attended a humanistic *Gymnasium* in Augsburg.
00.00.188_-00.00.1887(?)	Attended the *Ludwig-Gymnasium* in München.
16.08.1887	Entered service as a *Dreijährig-Freiwilliger (Offizier-Aspirant)*, assigned to *Kgl. Bayerisches 9.Infanterie-Regiment* (Würzburg).
01.03.1888-22.01.1889	Attended the *Kriegsschule* in München.
06.02.1888	Received his *Zeugnis der Reife zum Portepeéfähnrich* (diploma for advancement to ensign with sword knot).

A portrait of young Franz Epp, painted by his father in 1893.

06.02.1889	Received his *Zeugnis der Reife zum Offizier* (diploma for commissioning as an officer).
29.01.1894-24.02.1894	Participated in a *Lehrkurs* (instructional course) at the *Gewehrfabrik* (artillery factory) in Amberg.
01.04.1895	Began a four-month visit to the French-speaking area of Switzerland for French language studies.
01.10.1896-30.09.1899	Attended the *Kriegsakademie* in München,
16.07.1900-17.08.1901	Chef of *8. Kompanie / II. Bataillon / Kaiserliche 4. Ostasiatische Infanterie-Regiment* (involved in the "Niederschlagung des Boxeraufstandes" (Suppression of the Boxer Rebellion) in China, 05.08.1900-17.08.1901).
17.08.1901	Returned to Germany.
01.09.1901	Reported for duty with the *Kgl. Bayerisches 19.Infanterie-Regiment* in Erlangen.
11.12.1901-01.02.1904	*Brigadeadjutant* of *Bayerisches 5.Infanterie-Brigade* (Zweibrücken).
26.06.1902-04.07.1902	Participated in a *Generalstabsreise* (general staff study) with *II. Bayerische Armee-Korps*.
04.06.1903-20.06.1903	Participated in another *Generalstabsreise*.
02.02.1904	Transferred from the Bavarian Army to the *Kaiserliche Schutztruppe für Deutsch-Südwestafrika* (Imperial Defense Troops for German Southwest Africa).
25.06.1904	Assigned to *1.Feld-Regiment* of the *Kaiserlichen Schutztruppe für Südwestafrika*.

Hauptmann Epp in the uniform of the *Kaiserlichen Schutztruppe*,
1904. (Ritter von Epp, *Ein Leben für Deutschland, 1938*)

Oberstleutnant Epp as commander of *Kgl. Bayerische Infanterie-Leib-Regiment*
on the Somme, 1915. (Ritter von Epp, *Ein Leben für Deutschland, 1938*)

Oberstleutnant Epp (seated, second from right) with his regimental staff in Romania, 1917. (Ritter von Epp, *Ein Leben für Deutschland, 1938*)

25.06.1904-17.11.1906	Assigned to the *Deutschen Schutztruppe in Deutsch-Südwestafrika*, from 11.07.1904 as *Kompaniechef* in *1.Feld-Regiment*. He took part in the violent suppression of the Herero uprising during this period. Discharged for return to Bavarian Army service, 17.11.1906 (with effect from 30.11.1906).
30.11.1906	Returned to Germany.
03.04.1907	Sustained a sprain to his right knee while off duty.
16.10.1908-21.06.1912	*Adjutant* of the *Bayerische 3. Division* (Landau).
22.06.1912-19.08.1914	*Kommandeur* of II. *Bataillon / Kgl. Bayerisches Infanterie-Leibregiment* (deployed on the Western Front from 07.08.1914).
19.08.1914-16.04.1917	*Kommandeur* (m.d.F.b. until 26.12.1914, then permanent) of *Kgl. Bayerisches Infanterie-Leibregiment* (in France, Belgium, Serbia, Romania, and Italy).
23.09.1914	Wounded in the left foot by shrapnel.
24.09.1914-21.10.1914	Hospitalized in the *Kriegslazaretten Cateau* and *St. Quentin*.
17.02.1916	Ill with sinusitis.
17.02.1916-19.03.1916	Received outpatient treatment in München.
23.06.1916	Granted the personal title of nobility *Ritter von* (in conjunction with the award of the *Militär-Max-Joseph-Orden*.
22.07.1917	Fell ill with a lung infection.

Romania, 07.09.1917: *Oberstleutnant* Ritter von Epp (second from left) poses with *General* Karl Ritter von Wenninger, *Führer* of *XVIII.Reserve-Korps* (to von Epp's left) and his staff; the *General* was killed at Muncelul the following day. (Ritter von Epp, *Ein Leben für Deutschland, 1938)*

01.10.1917-07.10.1917	Hospitalized in the *Reserve-Spital Zavidovic.*
17.11.1917	Granted the title of personal nobility and knighthood (*Ritter von*), with effect from 23.06.1916. He was entered in the *Bayerische Adelsmatrikel* (Bavarian Registry of Nobility, established 1808) on 25.02.1918.
17.04.1917-07.02.1919	*Führer* of *Bayerisches 1.Jägerbrigade des Alpenkorps.*
23.09.1918	Wounded in action.
15.10.1918-18.11.1918	*Kommandeur* of *Kgl. Bayerisches Infanterie-Leibregiment.*
08.02.1919-30.09.1919	*Führer* of *Freikorps Epp* (also known as the *Bayerisches Schützenkorps*). Established in Ohrdruf, his *Freikorps* consisted of the following: *Bayerisches Schützen Regiment 1; Bayerisches Schützen Regiment 2; Jäger-Battalion Bayerisches Schützenkorps* (formerly known as *Freikorps Chiemgau*); *Reiter-Regiment Bayerisches Schützenkorps; leichte Artillerie-Regiment Bayerisches Schützenkorps; schwere Artillerie-Regiment Bayerisches Schützenkorps; Freiwilligen-Flieger-Abteilung Dessloch*; and a *Panzerwagen-Abteilung*. While leading this formation, he played a key role in the destruction of the short-lived *Münchener Räterepublik* in April / May 1919, and was party to the brutal suppression of leftists and "traitors". The following passage concerns the brutal actions of his men during this period, and von Epp's opinion of them:

After the conquest of Munich the Bavarian Free Corps kept its

A formal portrait of *Oberst* Epp, taken in 1917.

Identification card of *Oberst* Franz Ritter von Epp, issued in 1919.

Oberst Ritter von Epp as *Führer* of *Freikorps Epp*, 1919.

name Einwohnerwehr (citizens army), and guarded their illegal arms against "traitors" by committing numerous murders and so-called "Secret Murders" (Fehmemord). When called before the investigation committee of the German *Reichstag*, Franz von Epp defended the reactions as follows:

"It is my opinion that the law protected those traitors who were willing to give away our arms. Therefore, I deem it an act of self-defense and a normal right if patriotic circles act against such traitors, if only to deter others. The murders of traitors (who gave away our arms) had the consent of the patriotic circles, and they have it still. Since the judicial administration remained passive, it seemed to be clear that self-help was necessary. There was no reason to make any distinction between such traitors, who were willing to give the arms away to the Entente or to leftist radical groups, and traitors who wished to betray the arms to the (official) German Disarmament Commission, because in this case also the motives were impure, for the traitors could not know, in any case, whether the arms would be kept for patriotic purposes." (Source: Report, classified "Secret", by the Office of Strategic Services [OSS], Research and Analysis Branch: "Nazi Plans for Dominating Germany and Europe. The Attitude of the *NSDAP* toward Political Terror.", 08.07.1945)

18.10.1919-30.09.1920	*Kommandeur* of *Reichswehr-Schützenbrigade 21*.
00.00.1919-00.00.1928	Member of the *Bayerische Volkspartei* (*BVP*, Bavarian Peoples' Party).
01.10.1920-31.10.1923	*Infanterieführer VII* (München).
00.00.1922-00.00.1933	Member of the *Koloniale Reichsarbeitsgemeinschaft* (*Korag*).
31.10.1923	Retired from the *Reichswehr* as *Char. Generalleutnant*
03.01.1924-00.00.1924	*Führer* of the *Deutscher Notbann* (*DNB*, German Emergency Formation), a paramilitary organization unaffiliated with any political party and established with the unstated approval of the Bavarian government.
00.00.1925-00.05.1945(?)	*Präsident* of the *Deutsche Kolonial-Kriegerbund* (German Colonial War Veterans' League).
18.05.1928	Joined the *NSDAP*.
00.00.1928-00.05.1945	Member of the *Stab der Obersten SA-Führung*.
00.05.1928	Appointed as a member of the *Vorstand* (governing body) of the *Kampfbund für deutsche Kultur* (Fighting League for German Culture).
20.05.1928-18.07.1930	Member of the *Reichstag* (*Wahlkreis 26, Franken*).
14.09.1930-04.06.1932	Member of the *Reichstag* (*Wahlkreis 26, Franken*).
00.00.1931-00.00.1931	*Leiter* of the *Beobachterdelegation der NSDAP* (observers' delegation of the *NSDAP*) to the League of Nations Disarmament Conference in Geneva.
00.00.1931-00.00.1931	Representative of the *Korag* in the *Präsidium* of the *Deutschen Arbeitsgemeinschaft*.
18.12.1931	Granted the rank of *SA-Gruppenführer z. V.* (*NSKK*) (at disposal of the *SA*).
01.07.1932	Named as an *Ehrenführer* of the *NSKK*.
01.07.1932-26.01.1934	Attached to the *Stab der Obersten SA-Führer*s.
31.07.1932-12.09.1932	Member of the *Reichstag* (*Wahlkreis 26, Franken*).
30.08.1932-12.09.1932	Member of the 2. *Ausschuß* (*Auswärtige Angelegenheiten*) (2nd Committee [foreign affairs]) of the *Reichstag*.
08.09.1932-31.12.1935	*Leiter* of the *Wehrpolitische Amt der NSDAP* (Defense Policy Office of the *NSDAP*; dissolved, 31.12.1935) and *Leiter* of the *Wehrpolitische Amt der Obersten SA-Führung*.
08.09.1932-05.05.1934	*Leiter* of the *Kolonialreferat* in the *Wehrpolitische Amt der NSDAP*.
06.11.1932-01.02.1933	Member of the *Reichstag*.
06.12.1932-01.02.1933	Member of the 2. *Ausschuß* (*Auswärtige Angelegenheiten*) of the *Reichstag*.
01.01.1933-27.01.1934(?)	Appointed as an *SA-Führer z.b.V.* to the *Obersten SA-Führung* as consulting *SA-Führer* for defense policy matters.
05.03.1933-14.10.1933	Member of the *Reichstag* (*Wahlkreis 26, Franken*).
09.03.1933-10.04.1933	*Reichskommissar für die polizeilichen Befugnisse* (*Reich* Commissioner for the Police Authority in Bavaria [*Reichs-Polizeikommissar*]) *in Bayern*. On 10.03.1933, he assumed overall command of the Bavarian police.
00.03.1933	Appointed as an honorary member of the *Orden der Bayerischen Tapferkeitsmedaille* (Order of the Bavarian Military Bravery Medal) in Pfalz and the Saar.

Heinrich Hoffmann's formal portrait of *Oberst* Ritter von Epp, 1919. (Max Williams photo)

Another formal portrait of *Oberst* Ritter von Epp, 1919.

16.03.1933-10.04.1933	*(kommissarischer) Bayerischer Ministerpräsident* (acting Bavarian Prime Minister) and *Bayerischer Staatsminister des Äusseren* (Bavarian State Minister for Foreign Affairs). Succeeded by Ludwig Siebert.
10.04.1933-30.04.1945	*Reichsstatthalter in Bayern*. Despite being the highest figure in the Bavarian government and a staunch Bavarian nationalist, von Epp was in fact a figurehead, his name alone commanding respect (he was, after all the savior of Bavaria, having overthrown the short-lived communist regime of 1919). He led an able staff of mostly moderate officials, and the bulk of the organizational planning and work of his office was performed by his *Staatssekretär, SA-Obergruppenführer* Hans Georg Hofmann (until his death on 30.01.1942) and the acting successor to Hofmann, *Oberregierungsrat* (later *Ministerialrat*) Dr. Franz Schachinger (the former deputy *Polizeipräsident* of Nürnberg, who had joined von Epp's staff on 11.09.1933). The real power in the territory was held by *Gauleiter* Adolf Wagner, who held two Bavarian ministerial posts- Education and, most significantly, Interior (which placed him in overall command of the police).
00.00.1933-13.06.1936	*Vizepräsident* of the *Deutsche Kolonialgesellschaft* (German Colonial Society).
06.07.1933	Appointed as *Ehrenpräsident* of the *Verband Deutscher Kunstvereine e.V.* (Association of German Artistic Unions)

A selection of postcard photos showing *Generalleutnant* Ritter von Epp in *Reichswehr* uniform.

14.07.1933	Survived a serious automobile accident without injury in München.
31.08.1933-30.04.1945	*Reichsleiter der NSDAP.*
00.09.1933-00.05.1945	*Ehrenpräsident* of the *Deutschen Gesellschaft für Wehrpolitik und Wehrwissenschaften e.V.*
03.10.1933-00.00.1944	Member of the *Akademie für Deutsches Recht*, München.
13.10.1933-30.04.1945(?)	Senator of the *Deutschen Akademie* in München.
16.10.1933	Appointed as *Ehrenführer* of the *Kyffhäuserbund*. He officially assumed this post on 23.10.1933.
19.10.1933	Appointed as a member of the *Führerrat* of the *Bund Deutscher Osten* (League for the German East).
12.11.1933-08.05.1933	Member of the *Reichstag (Wahlkreis 24, Oberbayern-Schwaben)*.
04.11.1933	Appointed as *Ehrenführer* of the *Bund Deutscher Marinevereine* (League of German Naval Clubs).
12.12.1933-08.05.1945	Member of the *Ältestenrat der Reichstag* (Council of Elders of the *Reichstag*).
27.01.1934-00.05.1945(?)	Assigned as *SA-Führer z. V.* to the *Stab der Obersten SA-Führer*, with the function of *Beratender SA-Führer für wehrpolitische Angelegenheiten* (Consultant *SA* Leader for Matters of Defense Policy).
00.00.1934-00.09.1934	Member of the *Ehrenausschuß im Deutschen Organisations-Ausschuß des VII. Internationalen Straßenkongresses in München vom 03.09.1934 bis 09.09.1934* (Committee of Honor of the German Organizational Committee to the 7th International Roadways Congress in Munich from 03.09.1934 to 09.09.1934).
09.04.1934-15.04.1934	Member of the *Ehrenausschuß für den Reichsberufswettkampf 1934* (Committee of Honor for the *Reich* Professional Competition of 1934).
05.05.1934-17.02.1943	*Reichsleiter* of the *Kolonialpolitische Amt der NSDAP.*
Summer 1934	Protested against the execution of München *SA* leaders (Schneidhuber, Schmid, etc.) during the so-called "Röhm-Putsch" on 30.06.-01.07.1934, consequently losing much of his influence. Hitler's former representative to the foreign press, Ernst "Putzi" Hanfstaengl later recorded:

> I found my old friend General von Epp in despair [at the purge]. He had seriously considered cutting his roots and leaving the country and only the thought that his place as *Reichsstatthalter* would then be taken by the impossible Bavarian *Gauleiter* Wagner had determined him to remain. (Hanfstaengl, *Hitler: The Missing Years*, p. 248)

August 1934 (?)	Unsigned and undated memorandum of complaint by von Epp, sent to Hitler or *Reichsminister* Dr. Frick:

> Abuses in the imposition of protective custody and the resulting disproportionately large number of such internees in Bavaria as compared ... to Prussia have caused me to write a memorandum on matters of principle to the Bavarian government. In that memorandum listed the major abuses and pointed out that such

Ritter von Epp in civilian attire, ca. 1932. (Ritter von Epp, *Ein Leben für Deutschland, 1938)*

Ritter von Epp, ca. 1932, as *Leiter* of the *Wehrpolitische-Abteilung* (Defense Policy Department) in the *Reichsleitung der NSDAP*. The collar patches he wears resemble those of an *SA-Oberführer*, but in fact reflect his status as an *Abteilungsleiter* in the Party's *Reichsleitung* (*Reich* Leadership.) (Hermann-Historica, Auctioneers, München).

The Bavarian Government of 1933. First row: *Ministerpräsident* Dr. Ludwig Siebert, *Reichsstatthalter* Ritter von Epp, Adolf Wagner (Interior), and Hans Schemm (Education). Back row: Heinrich Himmler, Ernst Röhm (Minister without Portfolio), Dr. Hans Frank (Justice), Hermann Esser (Minister without Portfolio, Chief of the Bavarian State Chancellery, and head of the Press Office in the Bavarian Ministry of State), and Georg Luber (Economics).

use of protective custody was undermining that confidence in the law which is the foundation of every commonwealth, and that it was a defiance of the State of Law as demanded and proclaimed by the *Führer*.

I recommended methods for the imposition of protective custody. At the same time I suggested a reexamination of the pending cases of protective custody according to these principles and asked for a list (with ... the grounds for the continuation of custody) of those internees who had not been released by a certain date. Copy ... was ... conveyed to the *Reich* Minister of the Interior. On 18 April 1934, the Bavarian Minister President referred to mae a report from the Bavarian Ministry of the Interior of 14 April. Every sentence in this report could be disproved, it contained distortions, misrepresentations and falsifications in a form which made it impossible for me to deal with it on its merits. This letter of 14 April 1934 points out in this connection: The list asked for by the Herr *Reich* Governor of the persons still held in protective ccustody on 15 April 1934 necessitates a thorough check and examination by officials of my Ministry and of my area of competence in my capacity as supreme authority of Bavaria. In spite of the burden this placed on my staff ... I have had them carry out the work required. But I cannot see my way to forwarding this list, because this would certainly lead to a revision of individual cases by the *Reich* Governorate. This, however, is my own task in my capacity as the supreme responsible authority of Bavaria. I therefore ask that the revision be left to me. I shall make a report about the result ... [no list was ever submitted, and Epp writes of the attitudes of Wagner, Himmler, and Heydrich as follows:] ... a complete misconstruction of the position of the *Reich* Governor, who has the right to call upon every authority ... to report in pursuance of his most distinguished task, namely to ensure compliance with the guidelines of policy laid down by the *Reich* Chancellor ... [Von Epp then cited specific abuses by Wagner and the Bavarian Political Police, including the coercison by physical pressure of communist suspects to make "confessions"] (Shlomo Aronson, *Beginnings of the Gestapo System. The Bavarian Model in 1933*, pp. 46-47)

28.08.1934-30.04.1945	*Bayerischer Landesjägermeister* (Bavarian State Master of the Hunt).
00.01.1935-30.04.1945	*(kommissarischer) Gaujägermeister* of the *Jagdgaue Oberbayern, Oberfranken, Mittelfranken,* and *Oberpfalz.*
00.00.1935	The *Promenadenplatz* of the *Hotel Bayerischer Hof* in München renamed *Ritter-von-Epp-Platz.*
13.06.1936-15.02.1943	*Bundesführer der Reichskolonialbund* (*RKB, Reich* Colonial League), established on 13.06.1936 by the former *Gouverneur von Deutsch-Ostafrika*, Dr. Heinrich Schnee. Dr. Schnee had served, from 1930 to 1936, as the last *Präsident* of the *Deutschen Kolonialgesellschaft* (DFG, German Colonial Society). As Germany had no overseas colonies during the Third *Reich*, it's primary role was to produce agitating propaganda calling for a new German colonial empire. As of 01.01.1943, shortly before it was disbanded by *Führer* decree on 15.02.1943, the RKB numbered some 2,160,000 members- including 50,000 part-time office holders- distributed amongst 41 regional branches known as *Gauverbände* (further divided into approximately 900 *Kreisverbände* and 12,800 *Ortsverbände*).
13.06.1936-08.05.1945	*Vorsitzender* of the *Kolonialrat* (colonial council).
16.08.1937	Telegram of congratulation from Hitler:

Today, on the fiftieth anniversary of your entry into the Army, my thoughts turn to you in appreciation and gratefulness for the great feats you have accomplished in times of war and peace for the benefit of Germany. Let me extend to you my best wishes on your anniversary.

    With kind regards, Adolf Hitler

    (Max Domarus, ed, *Hitler. Speeches and Proclamations, 1932-1945: The Years 1935 to 1938*, pp. 920-921)

16.10.1938	*Chef* (honorary commander) of *Grenadier-Regiment 61.*

A postcard photo of "*General*
von Epp" in civilian attire.

Hoffmann postcard of
Reichsstatthalter Ritter von Epp.

Ritter von Epp with Heinrich Himmler in 1933. The *SS-Oberführer* in the
background is Erasmus Freiherr von Malsen-Ponickau, then head of the
Hilfspolizei (*SS* and *SA* auxiliary police) in München. (Roger Bender photo)

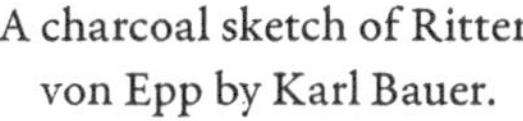

A charcoal sketch of Ritter
von Epp by Karl Bauer.

Franz Ritter von Epp as an
NSKK-Ehrenführer, ca. 1936.

29.10.1938	Honorary member of the *Deutschen Akademie* in München.
22.09.1939	Member of the *Verteidigungsausschuß* (defense committee) of *Wehrkreise VII* and *XIII*.
00.02.1944	*Ehrensenator* (honorary senator) of the *Technische Hochschule* in München.
27.03.1945	Top secret memorandum from the director of the OSS (Office of Strategic Services), General William J. Donovan to the U.S. Joint Chiefs of Staff concerning "Approaches from Austrian and Bavarian Nazis":

> ... Approaches from von Epp
>
> On 23 March [1945], the OSS representative in Bern reported that Heinz Adolf Heintze of the German Foreign Office arrived in Switzerland with a message from Franz Xaver Ritter von Epp, Lieutenant-General, retired, the Reichsstadthalter [sic!] (Governor) of Bavaria. Von Epp declares that, although he has served the Nazis, as an old-school Bavarian officer he wishes to spare Bavaria from becoming a battleground. When central Nazi government controls break down, he intends to assume executive power in Bavaria, aided by several Wehrkreis [army unit] commanders in charge of Bavarian reserve units. The most energetic of these, he says, is General Kriebel, commanding Wehrkreis VII. Von Epp and these Wehrkreis commanders believe that the troops under their command would follow orders

München, 16.10.1938: Birthday celebration for *Reichsstatthalter* Franz Ritter von Epp. Also pictured from left to right: Max Köglmaier, *Ministerpräsident* Ludwig Siebert, Dr. Hans Frank, unknown, Hans Dauser, von Epp, unknown, Reichsleiter Karl Fiehler (*Oberbürgermeister* of München), and *Gauleiter* and *Staatsminister* Adolf Wagner. (NARA photo)

to take action against Himmler and the *SS*.

Von Epp declares that he has acquainted Cardinal Faulhaber and other Bavarian Catholic leaders with his plans and had tried to contact the Vatican through Faulhaber. This contact could not be safely established, he claims, because the *Gestapo* is represented in the Cardinal's entourage. Von Epp apparently asked his emissary, Heintze, to find out whether his appearance at the head of a Bavarian anti-Hitler movement would tend to prejudice the movement in Allied eyes, in view of his own Nazi background and the fact that he had remained in office under Hitler ... (Source: Franklin D. Roosevelt Presidential Library and Museum, administered by the U.S. National Archives and Records Administration)

On 09.04.1945, OSS Assistant Director G. Edward Buxton sent a memorandum to President Roosevelt from which the following is excerpted:

A trusted messenger who is working with Heintze returned to Switzerland on 6 April from Munich, where he made contact with the von Epp group. Source confirms that von Epp and several other

General der Infanterie Franz Ritter von Epp, ca. 1938. (NARA photo)

Hotel Vier Jahreszeiten, München: Ritter von Epp and Adolf Hühnlein
at a party celebrating the 7th anniversary of the *Akademie des Deutschen
Rechts* (Academy of German Law), 22.- 24.11.1940. (NARA photo)

16.10.1938, Ritter von Epp's 70th birthday. Hitler honors the old General, in civilian attire, with a visit. (Roger Bender photo)

high-ranking officers are prepared to do everything within their power to cut short warfare in Bavaria, and, if possible, to prevent unnecessary destruction in Bavaria, and, if possible, to prevent establishment of the "redoubt". Among those working with von Epp are *General* Karl Kriebel, commanding *Wehrkreis VII*, Munich; *Generalleutnant* (Lw) [Lieutenant-General, Air Force] Wolfgang Vorwald, commandant of *Luftgau VII*, Munich, who controls the ground personnel of the large airfields in the Munich area (including Schleissheim and Riehm) and at the Reichenhall airdrome near Salzburg; and *Obergruppenfuehrer* [General] Benno Martin, Higher *SS* and Police Leader for *Wehrkreis XIII*, Nuremberg. The group also includes several younger officers and government officials. (ibid)

Night of 27./28.04.1945   Arrested by *Hauptmann* Dr. Rupprecht Gerngroß, *Chef* of the *Dolmetscher-Kompanie* (Translator Company) of *Wehrkreis VII*, in the course of an uprising in München (known as the *Freiheitsaktion Bayern* [*FAB*, Freedom Action Bavaria] and led by Dr. Gerngroß) On the night of 27./28.04.1945, Dr. Gerngroß and his men arrested *Reichsstatthalter* Ritter von Epp at his headquarters in the Schornerhof. Von Epp's adjutant,

Reichsleiter Franz Ritter von Epp poses for Heinrich Hoffmann, ca. 1941. (NARA photos)

Major von Caracciola, then joined the uprising. A squad of Gerngroß's men occupied the radio stations in Erding and Freimann, broadcasting: "Stop fighting, lay down your arms – destroy the Nazis wherever you meet them. Hoist white flags, the Allied troops are approaching München." At a little after 100 hrs. on the 28th, Gerngross made an announcement over the radio; John Toland writes:

> ... He said *General* von Epp had surrendered all Bavaria and asked the people to help 'the new leaders bring life back to normal as quickly as possible.' Gerngross spoke in good faith, but the uprising had taken another bad turn. Epp had been on the verge of capitulating to *Major* Braun in Freising, when he heard Gerngross broadcast that FAB was pledged to abolish the military. It was more than the old general could stand and he absolutely refused to cooperate. *Major* Braun became so annoyed that he sent 'the old fool' back to his home. (Toland, *The Last 100 Days*, p. 527)

The Putsch was then ruthlessly crushed on the orders of *Gauleiter* Paul Giesler. Forty of Gerngroß's rebels, including *Major* von Caracciola, were quickly murdered by the *SS* and *Gestapo* shortly before the arrival of U.S. troops in the city. Giesler also ordered the arrest of von Epp by the *Gestapo*, but he was released due to the intervention of *Generalfeldmarschall* Albert Kesseling

Ritter von Epp in full regalia wearing the white summer tunic of the *NSDAP*.

A wartime photo of *Reichsleiter* Ritter von Epp. (NARA photo)

28.04.1945	Arrested by the *Gestapo* on orders of *Gauleiter* Giesler, based on the false belief that he had been the leader of the *Freiheitsaktion Bayern* revolt. He was then transported to Schloß Mauterndorf in Taunach-Tal, and from there, together with the recently disgraced and arrested *Reichsmarschall* Hermann Göring, to Fischhorn. He was soon released following the intervention of *Generalfeldmarschall* Albert Kesselring.

Postwar Confinement

Arrested by U.S. Army troops, May 1945, and interned in various Allied facilities (including the Central Prisoner of War Enclosure, known as "Ashcan", at the Palace Hotel in Luxembourg. He was ultimately admitted to a U.S. Army hospital in München where he remained until his death. Posthumous de-Nazification proceedings were carried out for von Epp during 1948/49, resulting in his placement in Gruppe I ("Hauptschuldige", or major offenders). On 07.02.1949, he was recategorized in Gruppe II as a "Belasteter" (incriminated person).

Published Work

Ein Leben für Deutschland (1938; edited by Josef H. Krumbach with assistance from Adolf Dresler).

Decorations and Awards

29.05.1918	*Orden Pour le mérite* as *Oberst* and *Führer* of *Bayerischen 1.Jäger-Brigade des Alpenkorps*
17.11.1917	*Ritterkreuz des Kgl. Bayerischen Militär-Max-Joseph-Ordens* (together with title of personal nobility [*Ritter von*] and knighthood, with effect

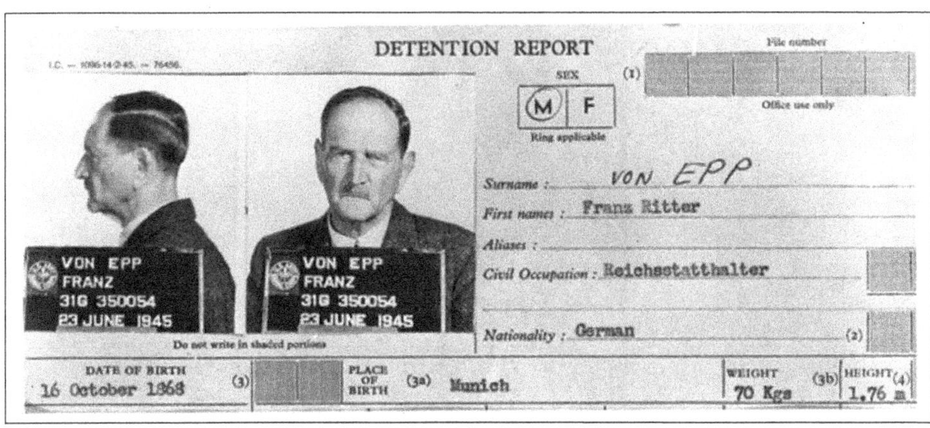

The U.S. Army's detention report on Franz Ritter von Epp.

	from 23.06.1916; Entered in the *Bayerische Adelsmatrikel* [Bavarian Registry of Nobility], 25.02.1918)
20.09.1943	*Ritterkreuz des Kriegsverdienstkreuzes mit Schwertern* for his service as *Reichsleiter* of the *Kolonialpolitische Amt der NSDAP* (05.05.1934-17.02.1943)
05.05.1917	*Königlich Preußische Hausorden von Hohenzollern Ritterkreuz mit Schwertern*
12.03.1915	*1914 Eisernes Kreuz I. Klasse*
22.12.1914	*1914 Eisernes Kreuz II. Klasse*
00.00.194_	*Kriegsverdienstkreuz I. Klasse mit Schwertern*
00.00.194_	*Kriegsverdienstkreuz II. Klasse mit Schwertern*
00.00.19__	*Kgl. Preußischer Roter Adler-Orden IV. Klasse mit Schwertern*
24.10.1901	*Kgl. Preußischer Kronen-Orden IV. Klasse mit Schwertern*
14.02.1916	*Bayerische Offizierskreuz des Militär-Verdienstordens mit Schwertern*
11.09.1914	*Bayerische Militär-Verdienst-Kreuz III. Klasse mit der Krone und mit Schwertern*
06.07.1905	*Bayerische Militär-Verdienstkreuz III.Klasse mit Schwertern*
00.00.19__	*Bayerische Militär-Verdienst-Orden III. Klasse mit Schwertern*
00.00.19__	*Bayerische Militär-Verdienst-Orden IV. Klasse mit Schwertern*
11.05.1918	*Großherzoglich Mecklenburgisches Verdienstkreuz II. Klasse*
00.00.19__	*Offizierskreuz 1903 des Oldenburgische Haus und Verdienstorden von Herzog Peter Friedrich Ludwig*
00.00.19__	*Bayerische Prinz Regent Luitpold Jubiläums Medaille mit der Krone*
14.06.1907	*Südwestafrika-Denkmünze I. Klasse mit 3 Spangen*
00.00.190_	*Chinadenkmünze 1901 mit 2 Spangen*
19.10.1905	*Jubiläums-Medaille*
00.00.189_	*Preußische Kaiser Wilhelm Erinnerungsmedaille 1897*
00.00.1935	*Herzoglich Sachsen-Ernesticher Hausorden Großkreuz mit Schwertern am Ring*
00.00.19__	*Bayerische Dienstauszeichnung 40 jahr.*
28.03.1916	*Dienstauszeichnungskreuz I. Klasse*

A plaque awarded for meritorious service in Germany's former overseas colonies and bearing C. von Ruckteschell's profile of *Reichsleiter* Ritter von Epp.

A medallion featuring Ritter von Epp's profile (in army uniform) and bearing his title of "*Reichsstatthalter von Bayern*", ca. 1935.

15.08.1906	*Dienstauszeichnungskreuz II. Klasse*
19.07.1917	*Österreichischer Militär-Verdienst-Kreuz III. Klasse mit Kriegsdekoration*
26.09.1917	*Österreichischer Orden des Eisernen Krone III. Klasse mit Kriegsdekoration*
29.05.1917	*Österreichischer Orden der Eisernen Krone III. Klasse*
10.10.1915	*Österreichisches Militär-Verdienstkreuz III. Klasse*
ca. 1939	*Medaille zur Erinnerung an den 13 März 1938*
ca. 1938	*Medaille zur Erinnerung an den 1 Oktober 1938*
00.00.1934	*Ehrenkreuz des Weltkrieges 1914-1918 mit Schwertern*
00.00.193_	*Goldenes Ehrenzeichen der NSDAP*
00.00.194_	*Dienstauszeichnung der NSDAP in Silber*
00.00.194_	*Dienstauszeichnung der NSDAP in Bronze*
21.03.1936	*Goldene Medaille des Kolonial-Wirtschaftlichen Komitees*
16.07.1937	*Silberne Ehrenmedaille der Akademie der Bildenden Künste*
29.10.1938	*Großes Ehrenzeichen der Deutschen Akademie*
00.02.1934 (?)	*Ehrendolch der SA*
00.02.1934	*Ehrenwinkel für alte Kämpfer*
00.03.1941	*Ehrenbürgerurkunde der Gemeinde Marktheidenfeld*

00.00.1936	*Ehrenbürgerrecht der Stadt Kolbermoor* (Revoked, 00.00.1946)
20.04.1933	*Ehrenbürgerrecht des Freistaates Bayern*
26.04.1933	*Ehrenbürgerrecht der Stadt Deggendorf*
25.04.1933	*Ehrenbürgerrecht der Stadt Augsburg* (revoked, 00.08.1946).
12.05.1933	*Ehrenbürgerrecht der Stadt Coburg* (revoked, 20.02.1946).
31.03.1933	*Ehrenbürgerrecht der Stadt Bad Reichenhall* (revoked, 04.01.1946)
00.00.1933	*Ehrenbürgerrecht der Stadt Aichach* (Revoked, 00.00.1945)
00.00.1933	*Ehrenbürgerrecht der Stadt Eggenfelden* (Revoked, 00.00.2011)
00.00.1933	*Ehrenbürgerrecht der Stadt Helmbrechts* (Revoked, 00.00.1946)
00.00.1933	*Ehrenbürgerrecht der Gemeinde Kreuth* (Revoked, 00.00.19__)
00.00.1933	*Ehrenbürgerrecht der Stadt Laufen (Salzach)* (Revoked, 00.00.1945)
00.00.1933	*Ehrenbürgerrecht der Stadt Ludwigshafen am Rhein* (Revoked, 00.00.1945)
00.00.1933	*Ehrenbürgerrecht der Stadt Mittenwald* (Revoked, 00.00.19__)
00.00.1933	*Ehrenbürgerrecht der Stadt München* (Revoked, 00.00.1946)
00.00.1933	*Ehrenbürgerrecht der Stadt Neunburg vorm Wald* (Revoked, 00.00.19__)
00.00.1933	*Ehrenbürgerrecht der Stadt Rosenheim* (Revoked, 00.00.1945)
00.00.1936	*Ehrenbürgerrecht der Stadt Kolbermoor* (Revoked, 00.00.1946)
00.00.193_	Commemorative Medal for the World War 1914-1918 with Swords (Hungary)

Notes

- Son of the painter Rudolf Epp (born 30.07.1834 in Eberbach am Neckar, died 08.08.1910 in München; son of the construction business owner Johann *Jakob* Epp) and his wife (whom he married in München in 1867) Katharina, née Steibl (daughter of the master carpenter Rudolf Steibl). Franz had two sisters (Helene and Augusta Anna, born 1870 and 1871, respectively).
- Never married.

Sources

Aronson, Shlomo: *Beginnings of the Gestapo System. The Bavarian Model in 1933*. Israel Universities Press, 1969.

Bosl, Karl (Editor): *Bosls Bayerisches Biographie: 8000 Persönlichkeiten aus 15 Jahrhunderten*, 1983.

Frank, Walter: *Franz Ritter von Epp. Der Weg eines deutschen Soldaten* (1934)

Hamilton, Charles: *Leaders and Personalities of the Third Reich, Volume I*. R. James Bender Publishing, 1984.

Hanfstaengl, Ernst Franz Sedgwick ("Putzi"): *Hitler: The Missing Years*. Arcade Publishing, 1957.

Heideking, Jürgen; Mauch, Christof; and Frey, Marc (Editors): *American Intelligence and the German Resistance to Hitler: A Documentary History*. Westview Press, 1996)

Lilla, Joachim; Döring, Martin; & Schulz Andreas: *Statisten in Uniform. Die Mitglieder des Reichstags 1933-1945*. Droste Verlag, 2004.

Peterson, Edward N.: *The Limits of Hitler's Power*. Princeton University Press, 1969.

Schirach, Baldur von: *Die Pioniere des Dritten Reiches*. Zentralstelle fur der deutschen Freiheitskampf, 1933.

Zorn, Wolfgang: "Epp, Franz Ritter v." In: *Neue Deutsche Biographie (NDB). Band 4*. Duncker & Humblot, Berlin, 1959.

Werner von Fichte
SA-Obergruppenführer

Born:	04.05.1896 in Kassel.
Died:	21.06.1955 in München / Bayern.

NSDAP-Nr.: _______ (Joined 00.00.1928; Expelled from the Party, 00.07.1934; Readmitted 00.00.1937)

Promotions

ca. 1913	*Leutnant*
00.00.19__	*Oberleutnant*
00.00.1928	*SA-Mann*
ca. 1931	*SA-Oberführer*
14.10.1931	*SA-Gruppenführer*
20.07.1933	*SA-Obergruppenführer*
00.00.194_	*Major d. R.*

Career

00.00.1907-ca. 1913	Attended the Kadettenanstalt in Karlsruhe and in Bensburg, and the *Haupt-Kadettenanstalt Groß-Lichterfelde*. Upon commissioning, he was assigned to *Fußartillerie-Regiment Nr. 18*.
00.00.1914-00.00.1918	War service on the Eastern and Western Fronts, in Romania, and in Turkey, ultimately as a member of the *Fliegertruppe*. He sustained minor injuries in an accident on 23.10.1916.
00.00.1919-00.00.19__	*Freikorps* service with the *Brigade Ehrhardt*.
00.00.1923	Discharged from the *Reichswehr* as an *Oberleutnant*.
00.00.1923-00.00.1926	*Führer Westdeutschland* (Leader in western Germany) of the *Bund Wiking*.
00.00.1926-00.00.1928	*Leiter* of the *Nationalklub* (successor organization to the banned *Bund Wiking*).
00.00.1928	Joined the *NSDAP*.
00.00.1928	Joined the *SA*.
08.02.1929	Appointed as a member of the *Stab der Obersten SA-Führer*.
08.02.1929-00.00.1930	Adjutant to the *Oberster SA-Führer West*.
00.00.1930-01.04.1931	*Stellvertreter* (deputy) to the *Obersten SA-Führers* and *Oberster SA-Führer West*.
02.04.1931-13.04.1932	*Führer* (m.d.F.b. until 14.10.1931, then permanent) of *SA-Gruppe Nordwest* (Düsseldorf).
13.04.1932-14.06.1932	*SA* banned by the government of Chancellor Heinrich Brüning.
01.07.1932-31.10.1932	*Führer* of *SA-Gruppe Westfalen* (Münster). Succeeded by Wilhelm Schepmann.
01.11.1932-10.05.1933	*Inspekteur im Stab des Generalinspekteurs der SA und SS* (Inspector on the Staff of the Inspector General of the *SA* and *SS*), München.

Hitler, von Fichte (left), and Wilhelm Friedrich Loeper (right) give the
Party salute during the October 1931 *SA* meeting and march in Braunschweig.

00.03.1933-00.07.1934	*Polizeipräsident* of Erfurt ("vertretungsweise mit der Verwaltung der Stelle beauftragt" [charged with acting administration of the position] until 22.09.1933, then permanent). Removed from this post in connection with the so-called "Röhm-Revolte", 00.07.1934.
00.00.1933-00.07.1934	Landesgruppenleiter of *Landesgruppe XI (Thüringen-Erfurt)* of the *Reichsluftschutzbund*.
02.07.1934	Arrested in connection with the so-called "Röhm-Revolte" and transferred to the *einstweiligen Ruhestand* (temporary retired list).
02.07.1934-16.07.1934	Interned at *KL-Lichtenburg*.
07.07.1934	Expelled from the *SA,* with forfeiture of his *SA* rank and assignments.
00.07.1934	Expelled from the *NSDAP*.
00.00.1934-00.00.1937	Lived in Koblenz.
00.00.1937-00.00.1945	Lived in Berlin-Lichterfelde-West.
00.00.1937	Reaccepted into the *NSDAP*, in accordance with a pardon granted by the *Führer*.
00.00.1939-00.00.194_	*Wehrmacht* service as a reserve artillery officer.

Postwar Prosecution

After 1945, worked as a laborer in München-Solln. On 27.09.1948, the public prosecutor at the *Spruchkammer München VI* requested that he be placed in de-Nazification Gruppe I (as a Hauptschuldiger, or major offender). The *Hauptspruchkammer* in München agreed to this reuest, and on 06.04.1949 classified von Fichte in Gruppe I, sentencing him to two years' detention in a labor camp (which was deemed to have been satisfied by his previous internment). All of his assets except for 500 DM were confiscated. On 25.07.1949, the *Berufungsverfahren* (court of appeals) in München reclassified von Fichte in Gruppe III (Minderbelasteten, lesser

SA-Oberführer Werner von Fichte, ca. 1931 (Baldur von Schirach, *Die Pioniere des Dritten Reiches*, 1933)

SA-Gruppenführer von Fichte in 1933. (Fritz Sauckel, *Kampf und Sieg in Thüringen: Im Geiste des Führers und in treuer Kameradschaft gewidmet den thüringischen Vorkämpfern des nationalsozialistischen Dritten Reiches*, 1934)

offenders), and ordered him to serve half a year on probation. The court waived further punitive measures.

Published Works

Morgentor der Freiheit. Eine Dichtung aus deutschen Schicksalstagen (1934)
Die Sage des Petersklosters zu Erfurt (1934)
Spukflieger (1940)

Decorations and Awards

00.00.191_	*1914 Eisernes Kreuz I. Klasse*
00.00.191_	*1914 Eisernes Kreuz II. Klasse*
ca. 1935	*Ehrenkreuz des Weltkrieges 1914-1918 mit Schwertern*
00.00.1934	*Goldenes Ehrenzeichen der NSDAP*
00.00.1933	*Gauehrenzeichen "Silberner Gauadler" des Gaues Thüringen der NSDAP*
00.00.1934	*Ehrendolch der SA*
00.02.1934	*Ehrenwinkel für alte Kämpfer*

Notes

- Third son of *Generalmajor* Georg Max Eduard von Fichte (born 30.10.1857 in Stuttgart, died 20.05.1933) and his wife (since 05.09.1891), Charlotte Anna, née Rindfleisch (born 19.04.1872).

An official visit to Weimar by Hungarian Prime Minister Gyula Gömbös in 1933. Top, from left to right: Wilhelm Brückner, Prime Minister Gömbös, Alfred Rosenberg, Hitler, Dr. Wilhelm Frick, and *SA-Gruppenführer* Fichte. Bottom, from left to right: Gömbös, an unidentified Hungarian dignitary, Hitler, Frick, Theodor Habicht (*Landesleiter* of the *NSDAP* in Austria), *SA-Obergruppenführer* Curt von Ulrich, Fichte, and *SS-Standartenführer* Paul Hennicke (NARA photos, Heinrich Hoffmann collection)

Weimar, 1933. *SA-Stabschef* Röhm speaks with Hitler. Behind him, from left to right, are August Schneidhuber, Werner von Fichte, Walter Ortlepp, Friedrich Karl Freiherr von Eberstein, Josias Erbprinz zu Waldeck und Pyrmont, and Fritz Sauckel.

In addition to Werner, the couple had the following three sons: Hermann von Fichte (born 22.06.1892 in Posen, died 08.03.1958); Alfred von Fichte (born 03.06.1893 in Posen, died 29.01.1941); and Günther von Fichte (born 18.05.1898 in Kassel, died 16.04.1958). Werner von Fichte was the great-grandson of the German philosopher Johann Gottlieb Fichte (born 19.05.1762 in Rammenau / Sachsen, died 27.01.1814 in Berlin).

Sources

Schirach, Baldur von: *Die Pioniere des Dritten Reiches*. Zentralstelle fur der deutschen Freiheitskampf, 1933.

Heinrich-Georg Wilhelm Werner Graf Fink von Finckenstein
SA-Obergruppenführer

Born:	22.11.1894 in Eisersdorf / Kreis Glatz / Regierungsbezirk Breslau / Provinz Schlesien.
Died:	19.02.1984 in Bielefeld / Regierungsbezirk Detmold / Nordrhein-Westfalen.
NSDAP-Nr.:	19 559 (First joined 00.06.1923; Party banned following the *München-Putsch* of 09.11.1923; Reenrolled 25.09.1925)

Promotions

01.08.1914	*Fahnenjunker*
15.02.1915	*Leutnant* (ohne Patent until 18.06.1915)
00.00.19__	*Oberleutnant*
28.07.1929	*SA-Mann*
21.05.1930	*SA-Sturmführer*
14.06.1931	*SA-Sturmbannführer*
01.07.1932	*SA-Standartenführer*
20.04.1934	*SA-Oberführer*
20.04.1935	*SA-Brigadeführer*
01.05.1937	*SA-Gruppenführer*
30.01.1941	*SA-Obergruppenführer*
01.01.1944 (?)	*Major* (a *Major* Graf Fink von Finckenstein, no first name given, appears in the *Dienstaltersliste T der Offiziere des Deutschen Heeres nach dem Stande vom 1. Mai 1944*; he may or may not be the same man as Heinrich-Georg Graf Fink von Finckenstein)

Career

ca. 1900-ca. 1905	Attended *Volksschule* (presumably).
ca. 1905-00.00.1914	Attended the humanistic *Gymnasium* in Greifenberg / Pommern (passed his *Abitur*, 00.00.1914).
01.08.1914-00.00.1917	Entered service as *Fahnenjunker*, assigned to *Dragoner-Regiment "von Bredow" (1. Schlesische) Nr. 4* (Lüden). Deployed to the front, 01.11.1914; commissioned 15.02.1915 and later assigned as a *Zugführer* in the Regiment.
00.05.1917-00.00.1918	*Führer* of a *Nachrichtenabteilung* (signals detachment) in *5. Reserve-Infanterie-Division*.
00.03.1919-00.07.1919	Service with *Freikorps Freiwilligen Sturmabteilung Schlichtingsheim*.
00.07.1919-00.09.1920	Service with *III. Marine-Brigade von Loewenfeld*, with which he participated in the "Kapp-Putsch" in Kiel during March 1920.
00.09.1920-00.00.192_	Service with the *Organisation Escherich* in Westfalen.

SA-Gruppenführer Fink von Finckenstein, ca. 1938.

00.05.1921-Autumn 1921	Assigned as a *Kompanieführer* to *Regiment Grüner, Bataillon Schlageter,* and *Sturmregiment Heinz* (seeing action in Oberschlesien and the Ruhrgebiet).
00.00.1922	Retired from military service.
00.00.1923-00.00.1933	Employed as a farmer.
20.06.1923	Joined the *NSDAP.*
25.09.1925	Reenrolled in the *NSDAP.*
28.07.1929-20.03.1930	Joined the *SA,* assigned to *Gausturm SC / Bereich Goldberg Haynau.*
20.03.1930-29.03.1931	*Führer* of *SA-Sturm 47/11* (Guhrau).
29.03.1931-01.07.1932	*Führer* of *SA-Sturmbann III/11* (m.d.F.b. until 14.06.1931, then permanent) in Guhrau.
18.10.1931	Participated in the *SA-Aufmarsch in Braunschweig.*
01.07.1932-19.04.1935	*Führer* of *SA-Standarte* 50 (in *SA-Gruppe Schlesien*) (Herrnstadt).
00.00.193_-00.00.19__	Member of the *Kreistag* of *Kreis Guhrau / Schlesien.*
00.02.1934-00.02.1934	Participated in a *Bataillonsführer-Kursus* (instructional course for battalion leaders).
20.04.1935-14.08.1936	*Führer* of *SA-Brigade 21* (m.d.F.b. until 14.09.1935, then permanent from 15.09.1935) in Liegnitz.
29.03.1936	Unsuccessful candidate for election to the *Reichstag.*
15.08.1936-14.06.1939	*Führer* of *SA-Gruppe Schlesien* (m.d.F.b. until 31.03.1937, then permanent from 01.04.1937). Succeeded Otto Herzog.
00.00.1937-00.06.1939	*Preußischer Provinzialrat* for *Provinz Niederschlesien.*
00.00.193_-00.00.19__	Member of the *Arbeitskammer Schlesien der DAF.*

SA-Gruppenführer Fink von Finckenstein, ca. 1938.
(*Bundesarchiv Bild* 146-2004-0013 / CC-BY-SA)

10.04.1938-08.05.1945	Member of the *Reichstag (Wahlkreis 8, Liegnitz)*.
15.06.1939-00.00.194_	Assigned to the *Stab der Obersten SA-Führung* in München.
11.09.1939-00.00.194_	*Wehrmacht* service [assigned to *Füsilier-Regiment 68* as of 01.05.1944?].
00.00.194_-00.00.194_	Assigned as an *Amtschef* to the *Stab der Obersten SA-Führung*.
00.03.1942-00.00.1945	*Ehrenamtlicher Richter am Volksgerichtshof* (honorary judge of the People's Court, a 5-year appointment).
[After 1943]	Assigned as *Major* to *Landesschützen-Ersatz-Bataillon 4* (Glauchau).

Decorations and Awards

00.00.191_	*1914 Eisernes Kreuz I. Klasse*
00.00.191_	*1914 Eisernes Kreuz II. Klasse*
ca. 1934	*Ehrenkreuz des Weltkrieges 1914-1918 mit Schwertern*
00.00.193_	*Goldenes Ehrenzeichen der NSDAP*
ca. 1929	*Nürnberger Parteitagsabzeichen 1929*
ca. 1931	*Abzeichen des SA-Treffens Braunschweig 1931*
30.01.1941	*Dienstauszeichnung der NSDAP in Silber*
30.01.1941	*Dienstauszeichnung der NSDAP in Bronze*
00.00.1934	*Ehrendolch der SA*
00.00.1934	*Ehrenwinkel für alte Kämpfer*

Notes

- Son of the farmer Heinrich Rudolf Albert Graf Fink von Finckenstein (born 29.10.1855 in Reitwein bei Frankfurt am Oder, died 12.02.1939). He was the son of Rudolf Karl Otto Ferdinand,

Graf Fink von Finckenstein, born 03.01.1813, and his wife Elise von Roeder, born 04.07.1824. His first marriage was to Marie-Elisabeth Maximiliane von Haugwitz (born 09.10.1862), with whom he had a son, Rudolf Heinrich Albrecht Otto Graf Fink von Finckenstein (born 29.12.1883). He later married Heinrich-Georg's mother, Sophie, née Freiin von Münchhausen (born 29.12.1858), with whom he had four children: Hilmar Heinrich Konrad Graf Fink von Finckenstein (born 30.09.1889, died 31.05.1934), Marie Elisabeth Gräfin Fink von Finckenstein (born 15.08.1893, died 16.03.1914), *Heinrich-Georg* Wilhelm Werner Fink von Finckenstein (born 22.11.1894), and Hertha Sophie Luise Gräfin Fink von Finckenstein (born 12.08.1899).

Sources

Lilla, Joachim; Döring, Martin; & Schulz Andreas: *Statisten in Uniform. Die Mitglieder des Reichstags 1933-1945*. Droste Verlag, 2004.

Friedrich (Fritz) Karl Florian

SA-Obergruppenführer

Born:	04.02.1894 in Essen.
Died:	24.10.1975 in a hospital in Düsseldorf-Mettman. On 04.11.1975, an urn containing his ashes was buried in Düsseldorf.
NSDAP-Nr.:	16 699 (Joined 18.08.1925)

Promotions

[00.00.1917]	*Unteroffizier*
25.08.1925-00.00.1927	*Ortsgruppenleiter* der *NSDAP*
25.08.1925	*SA-Sturmführer*
00.00.1927-30.09.1929	*Kreisleiter der NSDAP*
01.10.1929-01.08.1930	*Bezirksleiter der NSDAP*
01.08.1930-08.05.1945	*Gauleiter der NSDAP*
25.09.1933	*SA-Gruppenführer*
00.00.1936	*Gauarbeitsführer (ehrenhalber)*
30.01.1937	*SA-Obergruppenführer*

Career

ca. 1900-ca. 1904	Attended *Volksschule* in Wehlau.
ca. 1904-ca. 1909	Attended *Realschule* in Stallupönen.
ca. 1909-ca. 1912	Attended *Realgymnasium* in Stallupönen.
00.00.1912-00.08.1914	*Grubenbeamter* (mining official) with the *Preußischer Berginspektion III* (Prussian Mining Inspectorate 3) in Buer / Westfalen.
00.08.1914-00.00.1916	Entered service as a *Kriegsfreiwilliger*, assigned to *Grenadier-Regiment 1* (Königsberg).
00.00.1916-00.05.1918	Volunteered for the *Fliegertruppe*, assigned after training to *Jagdstaffel (Jasta) 51* in *Jagdgeschwader Richthofen*. Shot down and captured by British troops, 00.05.1918.
00.05.1918-00.11.1919	In British captivity.
00.02.1920-00.00.1929	Resumed employment as a *Grubenbeamter* with the *Preußischer Berginspektion III* in Buer / Westfalen.
00.00.1920-00.00.1922	Member of the *Deutschvölkische Schutz- und Trutzbund*.
ca. 1922	Founder of the *Ortsgruppe Buer* of the *Verband Nationalgesinnter Soldaten*.
00.00.1923	At the direction of the *staatlichen Steinkohlenbergwerk* (state colliery in Buer, banished from the French-occupied territories to Bad Oeynhausen, where he was assigned as a refugee to the management of the State spa.
00.00.1923-00.00.1923	Cofounder and *Unterführer* of the *Westfälische Treuebund*, with which he was active in resistance activities against the French occupation of the Ruhr.
00.00.1924	Returned to Buer.

Gauleiter Florian in 1934. (*Bundesarchiv-Bild* 183-2005-1129-500)

00.00.1924-00.00.1925	*Führer* of the *Völkisch-Sozialen Blocks* and later of the *NS-Freiheitsbewegung* in the French-occupied Ruhrgebiet.
18.08.1925	Joined the *NSDAP*.
25.08.1925-00.00.1927	Founder and *Ortsgruppenleiter* of the *Ortsgruppe Buer* (Gelsenkirchen*)* *der NSDAP*.
25.08.1925	Joined the *SA*.
00.00.1927-30.09.1929	*Kreisleiter* of *Kreis Emscher-Lippe der NSDAP*.
00.00.1927-00.00.1929	*Stadtverordneter* in Gelsenkirchen-Buer (the only National-Socialist to hold that post).
20.05.1928	Unsuccessful candidate of the *NSDAP* for election to the *Preußischer Landtag*.
01.10.1929-01.08.1930	*Bezirksleiter* of the *gauunabhängigen* [non-*Gau*-subordinated, i.e. independent] *Bezirk Bergisch-Land-Niederrhein der NSDAP*. Succeeded Fritz Härtl. The *Bezirk* was upgraded as a *Gau*, with Florian as *Gauleiter*, on 01.08.1930.
14.06.1930	Cofounder of the National-Socialist newspaper *Volksparole*.
01.08.1930-17.04.1945	*Gauleiter* of *Gau Düsseldorf der NSDAP*. He was the first and only holder of this post. *Gau Düsseldorf* was formed from the former *selbständiger Bezirk* [independent district] *Bergisch-Land-Niederrhein* (see 01.10.1929-01.08.1930, above).
14.09.1930-08.05.1945	Member of the *Reichstag* (Wahlkreis 22, Düsseldorf-Ost).
00.00.193_	Founder of the *Völkischer Verlag*, which published the daily *Wuppertaler Zeitung* and *Bergischer Beobachter*, and the Sunday newspaper *Die braune Post*.

Unteroffizier Florian, photographed in Flensburg in 1917.
(Hermann-Historica, Auctioneers, München)

March 1918: *Unteroffizier* Florian poses with comrades in Brügges,
Flanders. (Hermann-Historica, Auctioneers, München)

Gauleiter Florian, ca. 1932, signed and inscribed with "*Heil Hitler*". (Hermann-Historica, Auctioneers, München).

The Golzheimer-Heide near Düsseldorf, 26.05.1933: *Gauleiter* Florian and Hermann Göring carry wreaths during the "*Schlageter-Gedenkfeier*" – a memorial service for the Nazi proto-martyr and saboteur Albert Leo Schlageter (executed ten years earlier by French occupation authorities in Düsseldorf-Golzheim).

SA-Gruppenführer Florian in 1934, wearing the rare cuff title of an *Ehrenführer der SA*.

May 1934: Florian and other former World War I pilots visit England at the invitation of the Royal Aero Club. The photo is autographed by Bruno Loerzer who led the delegation and stands before Florian. (Hermann-Historica, Auctioneers, München)

An autographed photo of Florian as an honorary *SA-Gruppenführer*,
1933. (Hermann-Historica, Auctioneers, München)

24.04.1932	Elected to the *Preußische Landtag* (*Wahlkreis 22 [Düsseldorf-Ost]*).
30.08.1932-14.09.1932	*Stellvertretender Mitglied* (alternate member) of the *1. Ausschuß* (*Wahrung der Rechte der Volksvertreter*) *der Reichstag* (1.Committee [Respect for the Rights of the Peoples' Representatives] of the *Reichstag*).
00.03.1933-End of 1933	Member of the *Arbeitsausschuß der Deutschen Städtetag*.
00.00.1933-00.00.1934	*Stadtverordneter* and *Fraktionsführer der NSDAP* to the *Düsseldorfer Stadtrat*.
12.03.1933	Appointed as a member of the *Rheinische Provinziallandtag*.
30.03.1933	Appointed as *Staatskommissar für die Stadt Düsseldorf*.
00.00.1933-00.00.1945	*Vorsitzender* of the *Rheinische Gemeindetag* and member of the *Reichsvorstand der Deutschen Gemeindetag*.
30.08.1932-14.09.1932	*Stellvertretender Mitglied* (alternate member) of the *1. Ausschuß* (*Wahrung der Rechte der Volksvertreter*) *der Reichstag*.
14.09.1933-08.05.1945	*Preußischer Staatsrat*.
25.09.1933-26.01.1934	Assigned to the staff of *SA-Gruppe Niederrhein*.
27.01.1934-30.06.1936	Assigned as an *SA-Führer z. E.* ("zur Ehrendienstleistung") to *SA-Standarte 39 "Schlageter"* (Düsseldorf).
00.00.1934	Appointed as *Preußische Provinzialrat* for the *Rheinprovinz*.
12.05.1934-17.05.1934	Together with ten other former World War I pilots, including Bruno Loerzer who led the delegation, visited England at the invitation of the Royal Aero Club. The proposed purpose of the visit was to ensure that "the wonderful experience of camaraderie be stronger than dozens of

Friedrich Karl Florian (2nd from left, in *SA* uniform) at the *Berlin Fliegertreffen*, 1934. Also present, from left to right: Fritz Weitzel, unknown DLV officer, and Ernst Udet. (Hermann-Historica, Auctioneers, München)

	diplomatic speeches, thereby uniting two countries separated by the seas." (Hermann-Historica, Auctioneers, München, *66. Auktion Katalog*, 08.-09.05.2013, description of a photo album commemorating the event)
04.02.1935	Appointed by *Reichsforst- und Reichsjägermeister* Hermann Göring as an *Ehrenmitglied* (honorary member) of the the *Reichsbund Deutsche Jägerschaft*.
13.09.1935-00.00.1944(?)	Member of the *Akademie für deutsches Recht*, München.
01.05.1936	Employed in the *Reichsleitung der NSDAP*.
00.00.1936	Appointed as a *Gauarbeitsführer (ehrenhalber)*.
01.07.1936-00.05.1945	Assigned as an *SA-Führer z. V.* to *SA-Gruppe Niederrhein*.
00.00.193_-00.05.1945(?)	Member of the *Aufsichtsrat* of the *Rheinische Heimstätte GmbH, provinzielle Treuhandstelle für Wohnungs- und Kleinsiedlungswesen*, Düsseldorf.
17.11.1938	Attended the state funeral for the German diplomat Ernst vom Rath (whose assassination sparked the pogrom of 09./10.11.1938) in Düsseldorf's Rheinhalle.
07.06.1939	Appointed as *Beauftragter des Führers für die städtebauliche Neugestaltung Düsseldorfs und Wuppertals* (Representative of the *Führer* for the Urban Renewal of Düsseldorf and Wuppertal).

An official portrait of *Gauleiter*
Florian, ca. 1935. (NARA photo)

Florian the hunter, ca. 1935. (Hermann-
Historica, Auctioneers, München)

22.09.1939-16.11.1942	*Beauftragter des Reichsverteidigungskommissars für den Wehrkreis VI in den Gauen Düsseldorf und Essen.*
22.09.1939-17.04.1945	Member of the *Verteidigungsausschuß* of *Wehrkreis VI.*
15.11.1940-17.04.1945	*Gauwohnungskommissar* for *Gau Düsseldorf.*
06.04.1942-17.04.1945	*Beauftragter des GBA für den Gau Düsseldorf.*
01.09.1942	Meeting with *Reichsführer-SS* Himmler and the local *Polizeipräsident*, August Korreng, in Düsseldorf.
16.11.1942-17.04.1945	*Reichsverteidigungskommissar für den Gau Düsseldorf.*
25.09.1944-17.04.1945	*Führer des Deutschen Volkssturms im Gau Düsseldorf.* Subordinated to him as *Gaustabsführer* of the *Volkssturm* was *SA-Obergruppenführer* Franz Bock.
04.03.1945	Issued the following proclamation to the people of *Gau Düsseldorf*:

We are planting the standards of resistance on the Rhein!

The time for the hardest test has now come to our war torn *Gau*. The enemy is on the Rhein. We, my national comrades, must henceforth keep 'Watch on the Rhein' in the literal sense of the words. Germany, our Fatherland bleeding from its severe wounds and fighting with all its strength for its existence, looks to us. We will not let ourselves be shaken in our resolution, our steadfastness and our discipline and our trust in the *Führer* by the newly growing threat and burden of these times. The most German of

Gauleiter Florian, ca. 1938.

Official portrait of *Gauleiter* Florian, ca. 1942.

all rivers shall help us to protect Germany. We want to plant the standards of resistance on its banks and unaffected by the idle chatter and tittle tattle of the weak and continue our fanatic fight for the preservation of the *Reich* and the security of our national and socialist ethnic community (*Volksgemeinschaft*).

Florian, *Gauleiter*.

29.03.1945-06.04.1945	Attended the last *Gauleitertagung* (conference of *Gauleiter*) in Marksuhl bei Eisenach / Thüringen.
06.04.1945	While returning from Marksuhl to Düsseldorf, captured by U.S. troops together with his *Adjutant*, Thiel, but both men managed to flee. They made their way to the headquarters of *Gauleiter* Albert Hoffmann, and from here, Florian spoke via police radio to the commanding general of *III.Flak-Korps*, *Generalleutnant* Heino von Rantzau, who apprised him of the military situation.
16.04.1945	From his temporary headquarters in Neandertal, went with *Adjutant* Thiel to the headquarters of *Generalfeldmarschall* Walter Model (Commander-in-Chief of Heeresgruppe West).
16./17.04.1945	Spent the night in a tunnel in the Grafenberg Wald, together with *Kreisleiter* Karl Walter and other Party members.

Rastenburg / Ostpreußen, 07.10.1943: *Gauleiters* Florian and Robert Wagner during the conference of *Reichs-* and *Gauleiter* at *Führer* HQ *"Wolfsschanze"*. (NARA photo)

Postwar Prosecution and Activities

Recaptured by U.S. Army troops on the outskirts of Düsseldorf at approximately 1830 hours on 17.04.1945, then interned in the Lager Esterwegen where he made two suicide attempts (using poison and attempting to leap from a third-floor window of a barracks in Göttingen; in the latter attempt, his fall was broken by telephone wires and resulted in severe leg and foot injuries as well as a fractured vertebra). In total, he was held for six years in 16 different prisons and internment camps, being transferred to the custody of Dutch authorities for a brief period. Held in investigative detention by German authorities, 02.03.1948 to 17.09.1948, he was eventually charged by the *Schwurgericht* in Düsseldorf for his role in the death sentence, by a summary court, against *Oberstleutnant der Schutzpolizei* Franz Jürgens (*Kommandeur der Schutzpolizei* in Düsseldorf) and four others. The defendants had carried out a plan- code-named "Aktion Rheinland"- which resulted in the City of Düsseldorf being turned over to U.S. Army troops without a fight by the anti-Nazis Dr. August Wiedenhofen (an attorney) and Aloys Odenthal (an architect), and the arrest of the local *Polizeipräsident, SS-Brigadeführer* August Korreng. Jürgens and the others had been tried by a *Standgericht* (court-martial), convened on the orders of *Gauleiter* Florian and under the chairmanship of *Oberstleutnant der Schutzpolizei* Brumshagen on the night of 16./17.04.1945. Sentenced to death along with Jürgens were the German resistance fighters Theodor Andresen, Karl Kleppe, Hermann Weill, and Josef Knab). Jürgens and his co-conspirators were shot in the courtyard of the *Berufschule* on Färberstraße in Düsseldorf; to the head of the execution squad, *Revieroberleutnant der Schutzpolizei* Heinrich Gesell, Jürgens' last words were "Gesell, grüssen Sie mir meine Frau. Es lebe Deutschland!"

Gauleiter Florian with *Reichsjugendführer* Artur Axmann and
Generalfeldmarschall Walter Model, 30.11.1944. (NARA photo)

(Gesell, send my greetings to my wife. Long live Germany!") On 05.03.1949, Florian and his two codefendants were acquitted by the Düsseldorf *Schwurgericht*, under *Landgerichtsdirektor* Dr. Schwieren. On 14.06.1949, the *Spruchgericht* (de-Nazification court) in Bielefeld, citing Florian's "... membership in the *NS-Führungskorps* after 1. September 1939 and... knowledge of the criminal character of this organization", sentenced him to 6 years' imprisonment (with credit for 3 years', 6 months' time already spent in internment camps) and a 20.000 DM fine. Released on 01.05.1951, he was then employed as an industrial representative, and became a member of the anti-communist and anti-Semitic *Deutsches Kulturwerk europäischen Geistes*. After his release, Florian attempted on the one hand to play down his own deeds and the inhuman goals of National Socialism and on the other hand to glorify them; in the Sixties he was thus the object of investigation by the State prosecutor's office. In 1955, he applied for annulment of the remaining fine, because this had been imposed under "non-German law"; in fact, he was let off the remaining DM 10.000. In 1964 and 1973, the former *Gauleiter* brought actions against certain accusations made against him in a number of books. Both cases were dismissed.

Decorations and Awards

00.00.191_	*1914 Eisernes Kreuz II. Klasse*
30.01.1941	*Kriegsverdienstkreuz I. Klasse ohne Schwerter*
00.00.194_	*Kriegsverdienstkreuz II. Klasse ohne Schwerter*
00.00.191_	*Kgl. Preußische Flugzeugführerabzeichen*
ca. 1939	*Medaille zur Erinnerung an den 1. Oktober 1938*

Friedrich Karl Florian, ca. 1960. (Hermann-Historica, Auctioneers, München)

ca. 1938	*Medaille zur Erinnerung an den 13. März 1938*
ca. 1934	*Ehrenkreuz des Weltkrieges 1914-1918 mit Schwertern*
ca. 1933	*Goldenes Ehrenzeichen der NSDAP* ·
00.00.19__	*Goldenes Hitler-Jugend Ehrenzeichen mit Eichenlaub*
ca. 1942	*Dienstauszeichnung der NSDAP in Gold*
00.00.194_	*Dienstauszeichnung der NSDAP in Silber*
00.00.194_	*Dienstauszeichnung der NSDAP in Bronze*
00.00.193_	*Goldene Fliegernadel des NSFK*
04.02.1935	*Goldene Ehrennadel des Reichsbundes "Deutsche Jägerschaft"*
00.00.1936	*Ehrenbürgerbrief der Stadt Buer*
00.02.1934	*Ehrendolch der SA*
00.02.1934	*Ehrenwinkel für alte Kämpfer*
00.10.1937	Large silver and lapis lazuli ring bearing the inscription "Dem Kampfer zur Ehrung", presented to him by Hitler during the *Führer*'s visit to Düsseldorf (Item appears in the catalog for the 64th auction by Hermann-Historica, Auctioneers, München, where it is described as a *Persönlicher Gauleiterring* [personal *Gauleiter* ring])

Notes

- Son of the *Kgl. preußische Landmesser* (Royal Prussian land surveyor) and *Oberbahnmeister* Franz Gottfried Florian and his wife Wilhelmine, née Acktun. The family lived in Essen until 1900, when they relocated to Ostpreußen.
- Religion: Left the church and declared himself and his family "gottgläubig", 00.00.1934.

- Married on 31.03.1923 to Editha ("Ditha") Dorothea Krage (born 13.04.1904 in Neuwied, died 00.00.1992, daughter of a teacher for the blind). The marriage produced one daughter (Gisela) and one son (Armin, born 00.00.1924; *Unteroffizier* and pilot with *6.Staffel / Jagdgeschwader 26 "Schlageter"* [which he joined in August 1944]; he was shot down and killed on 01.03.1945 in combat with a P-47 of 366 Fighter Group while flying a *Focke Wulf Fw 190D-9* near Mönchen-Gladbach).

Sources

Geschichtswerkstatt Düsseldorf e.V.: "Die Ereignisse des 16. und 17. April 1945 in Düsseldorf 'Aktion Rheinland'", at www.geschichtswerkstatt-duesseldorf.de/ downloads/rheinland.pdf.

Hamilton, Charles: *Leaders and Personalities of the Third Reich, Volume I*. R. James Bender Publishing, 1984.

Höffkes, Karl: *Hitlers politische Generale: Die Gauleiter des Dritten Reiches*. Grabert-Verlag-Tübingen, 1986.

Hüttenberger, Peter: *Die Gauleiter: Studie zum Wandel des Machtgefüges in der NSDAP*. Deutsche Verlags-Anstalt, Stuttgart, 1969.

Johsi, Vandana: *Verhaltensmuster von Frauen im NS Alltag (1933-1945): am Beispiel Denunziantinnen*. Doctoral dissertation at the Technischen Universität Berlin, at edocs.tu-berlin.de/diss/2001/joshi_vandana.pdf, 2002.

Lilla, Joachim; Döring, Martin; & Schulz Andreas: *Statisten in Uniform. Die Mitglieder des Reichstags 1933-1945*. Droste Verlag, 2004.

Miller, Michael D. & Schulz, Andreas: *Gauleiter: The Regional Leaders of the Nazi Party and Their Deputies, Volume I*. R. James Bender Publishing, 2012.

Orlow, Dietrich: *History of the Nazi Party: 1933-1945*. University of Pittsburgh Press, 1973.

Schirach, Baldur von: *Die Pioniere des Dritten Reiches*. Zentralstelle fur der deutschen Freiheitskampf, 1933.

Hans Michael Frank

Dr. jur.
SA-Obergruppenführer

Born:	23.05.1900 in Karlsruhe / Baden.
Executed:	16.10.1946 at Nürnberg Prison (hanged by Master-Sergeant John C. Woods, U.S. Army). The journalist Howard K. Smith was a witness to the executions at Nürnberg Prison on 16.10.1946, and wrote:

> Hans Frank was ... the only one of the condemned to enter the chamber with a smile on his countenance. Although nervous and swallowing frequently, this man, who was converted to Roman Catholicism after his arrest, gave the appearance of being relieved at the prospect of atoning for his evil deeds. He answered to his name quietly and when asked for any last statement, he replied in a low voice that was almost a whisper, "I am thankful for the kind treatment during my captivity and I ask God to accept me with mercy."

NSDAP-Nr.:	40 079 (first joined the DAP, 00.00.1919, but did not remain a member when it was redesignated as the *NSDAP* in Spring 1920; Joined *NSDAP*, 03.10.1923; Party banned following the *München-Putsch* of 09.11.1923; Reenrolled with Nr. 14, 15.06.1926; Left again, 10.08.1926; Again reenrolled, with Nr. 40 079, 02.09.1927)

Promotions

18.12.1931	*SA-Gruppenführer*
18.04.1933-31.12.1934	*Staatsminister*
00.06.1933-20.08.1942	*Reichsleiter* der *NSDAP*
19.12.1934-20.08.1942	*Reichsminister*
09.11.1937	*SA-Obergruppenführer*
24.08.1939	*Leutnant d. R.*
26.10.1939-17.01.1945	*Generalgouverneur für die besetzten polnischen Gebiete* (Governor General of the Occupied Polish Territory)

Career

00.00.1906-00.00.1908	Attended the 1st and 2nd classes of *Volksschule* in Rotthalmünster.
00.00.1908	Moved with his family to München-Schwabing, Schnorrstr. 8.
00.00.1908-00.00.1910	Attended his 3rd and 4th classes of elementary school at the *Herrnschule* in München.
00.09.1910-00.00.1916	Attended the *Maximilian-Gymnasium* in München.
00.00.1912	Joined the *Wehrkraftverein*, a military-oriented youth organization.

SA-Gruppenführer Dr. Frank in 1933.

00.09.1916-00.06.1917	In protest against what he regard as antiquated teaching methods at the *Max-Gymnasium*, and following a conflict with his Latin teacher, traveled with his mother to Prague, where he was privately schooled.
00.00.1917-00.00.1918	Resumed his schooling at the *Max-Gymnasium*, but left to report for military service.
00.03.1918-25.06.1918	Assigned to *Kriegsarbeitshilfsdienst* (war work auxiliary service), working in the office of a brown coal (lignite) company.
25.06.1918-28.08.1918	Entered service as a *Kriegsfreiwilliger*, assigned as a *Rekrut* to *1. Ersatz-Kompanie / Ersatz-Bataillon / 1. Kgl. Bayerisches Feldartillerie-Regiment "König"* in the *Marsfeldkaserne*, München.
28.08.1918-26.09.1918	Assigned to *3. Ersatz-Kompanie / Ersatz-Bataillon / 1. Kgl. Bayerisches Feldartillerie-Regiment*.
15.09.1918	Resumed schooling at the *Max-Gymnasium*, where, on 16.12.1918, he began attending a *Kriegssonderkurs* (wartime special course) toward completion of his *Abitur*. He passed with the following grades: German, Latin, and Greek: Good; Mathematics, Physics, and History: Very good. On 15.07.1919, he received his Reifezeugnis (certificate of eligibility for university entrance).
26.09.1918-26.11.1918	Returned to *1. Ersatz-Kompanie / Ersatz-Bataillon / 1. Kgl. Bayerisches Feldartillerie-Regiment*.
07.11.1918-10.11.1918	Hospitalized in the *Reservelazarett München 1*.
09.11.1918-26.11.1918	Member of a *Soldatenrat* (Soldiers' Council).
26.11.1918	Following demobilization, discharged from army service in München, having seen no frontline service during World War I.
00.00.1919-00.00.*1919*	Member of the *Vorstand* of the *Thule-Gesellschaft* in München.
00.04.1919-00.10.1919	Service as a *freiwilliger Reiter* (volunteer cavalry trooper) in *Freikorps Epp*.
00.00.1919	Joined the *Deutsche Arbeiterpartei (DAP)*. He did not, however, join the newly formed *NSDAP*, which evolved from the *DAP* in the Spring of 1920.
00.10.1919	Visited his mother in Prague.
Winter 1919-00.00.1921	Studied law and economics at *Ludwig-Maximilian-Universität*, München.
00.06.1920-00.10.1920	Continued service with *Freikorps Epp*.
00.07.1920-00.09.1920	Briefly served with *Reiter-Regiment 17* (Straubing).
17.03.1921	Joined the *Kriegsteilnehmer-Verband* (war veterans' association) of *Ludwig-Maximilian-Universität*.
27.10.1921-00.00.1923	Studied law and political science at *Christian-Albrechts-Universität*, Kiel.
00.00.1921-00.00.1923	Speaker of an *NS-Studentengruppe* (National-Socialist Students' Group) at the University of München and research assistant to Prof. Karl Haushofer.
21.07.1923	Underwent testing for the *Höheren Justiz- und Verwaltungsdienst* (higher judicial and administrative service), qualifying as a *Referendar*.
00.09.1923-00.08.1924	Performed *Vorbereitungsdienst* (preparatory service) before the *Amtsgericht* of München.
28.09.1923-09.11.1923	Joined the *SA,* assigned to *II. SA-Bataillon / SA-Regiment München*.

03.10.1923	Joined the *NSDAP / Ortsgruppe München*.
08./09.11.1923	Participated in the *München-Putsch*.
00.04.1924	Fled to Italy, briefly staying in Naples and Palermo (Sicily).
00.00.1924	Returned to München.
03.10.1924	Passed his examinations toward his doctorate (Dr. jur.) at the University of Kiel (doctoral thesis: *Die öffentliche juristische Person – Ein Beitrag zur Lehre des Merkmals der öffentlichen Rechtspersönlichkeit* [1924])
00.00.1925	Received his doctorate (Dr. jur.).
20.11.1926	Passed his second state legal examinations (*Assessor-Examen*).
16.06.1925	Reenrolled in the *NSDAP*.
10.08.1926	Left the *NSDAP*.
00.00.1926-00.00.1927	*Assessor* and assistant to Prof. F, van Calker at the *Juristische Seminar* of the *Technische Hochschule* in München.
14.05.1927	Began practicing as an attorney before the *Landgerichtsbezirken München I* and *II*.
02.09.1927	Reenrolled in the *NSDAP*.
00.00.1926	Appointed as *Leiter* of the *Referat für Rechtsangelegenheiten* (Office for Legal Matters) in the *Reichsleitung der NSDAP*.
00.09.1927	Appointed as *2.Beisitzer* (associate judge) of the *Untersuchungs- und Schlichtungsausschuß der NSDAP*.
	Served as an attorney for the *NSDAP* in over 2,400 legal proceedings. Under interrogation at Nürnberg, he stated:

[O]ne day I saw in the newspaper the following ad: "We seek one lawyer to defend members of the party without means before a German court to make it possible to give them a legal defense." This was a trial held in Berlin. And I told Hitler I would like to defend those young people. And therefore the first trial ever held was this trial in Berlin. It happened this way that Hitler learned that here is a young lawyer ready and willing to defend members of his party. In this way Hitler took up connections with me in Munich and when he met me one day at the party office which was at the Schilling Straße he asked me if it was a good idea for me to work for the party. I told him at the time that my object was not to become a lawyer, but rather to pursue an academic career, but I told him, "If you need me, I'm willing to do it." Officially I remained with the School of Technology in Munich until 1929, but at [that] time the trials became so numerous that I had to make a new decision. I joined the Hitler movement as a laywer and I worked in the party as a lawyer ... Since I was willing to do it without money, I did it. ("Interrogation Records Prepared for War Crimes Proceedings at Nuernberg, 1945-1947/OCCPAC Interrogation Transcripts And Related Records: Frank, Hans"; Publication Number M1270, Record Group RG238)

Also under interrogation, on 06.09.1945, he stated:

SA-Gruppenführer Dr. Frank officiates at a meeting of Party officials late in 1933. (NARA, Heinrich Hoffmann collection)

Dr. Frank delivers a speech in 1933. (NARA, Heinrich Hoffmann collection)

... I joined the Party as a lawyer. Hitler asked me to represent him in cases and processes as a lawyer ... I was the personal law counsel of Hitler. That was with the lawyer Roder in Munich. There were some cases of insult against Hitler where I defended him ... (ibid)

11.10.1928-11.05.1936	Founder and *Reichsführer* of the *Bund Nationalsozialistischer Deutscher Juristen* (*BNSDJ*, League of National Socialist German Lawyers). Initially numbering 29 lawyers, it had a membership, by the end of 1929, of 90. This had increased to 701 by the end of 1931, 701, and at year's end, 1932, 1,374 German jurists had enrolled. On 11.05.1936, the organization was rechristened *Nationalsozialistische Rechtswahrerbund* (*NSRB*, National Socialist Legaue of Legal Professionals).
01.08.1929-04.08.1929	Participated in the *4. Reichsparteitag der NSDAP* in Nürnberg.
14.10.1930-08.05.1945	Member of the *Reichstag (Wahlkreis 8, Liegnitz)*.
00.10.1930-00.00.1933	*Sachreferent der NSDAP für Rechtsfragen* (Confidential Advisor of the *NSDAP* for Legal Questions) in the *Reichstag*.
00.00.1930-00.00.1933	*Vorsitzender* of the *Reichstagsausschuß für Rechtspflege* (*Reich* Committee for the Administration of Justice).
01.11.1930-28.10.1931	*Leiter* of the *Rechtsabteilung* (Legal Department) in the *Reichsleitung der NSDAP*, München.
00.00.1931-00.00.1931	*Leiter* of *Unterabteilung III* in the *Quartiermeisterstab der OSAF*.

18.12.1931-13.04.1932	Granted *SA* rank as an *SA-Führer z. V.* and assigned as *Chef* of the *Rechtsabteilung* in the *Obersten SA-Führung*.
28.10.1931-31.12.1934	*Leiter* of the *Rechtspolitische Abteilung* (*RPA*, later redesignated *Rechtsabteilung der Reichsleitung der NSDAP* (until June 1932 a component of the *Reichsorganisationsabteilung II*).
00.08.1932	Appointed as *Sachberater "Rechtsfragen" der NSDAP-Reichstagsfraktion* (Specialist in Legal Questions to the *NSDAP* parliamentary group).
30.08.1932-12.09.1932	Member (as *Obmann der NSDAP-Fraktion*) of the *13.Ausschuß (Rechtspflege) des Reichstages* (13th Committee [legal aid] of the *Reichstag*).
30.08.1932-12.09.1932	Member (as *Obmann der NSDAP-Fraktion"*) of the *18.Ausschuß (Strafgesetzbuch) des Reichstages"* (18th Committee [Penal Code] of the *Reichstag*).
30.08.1932-12.09.1932	Member of the *1.Ausschuß (Wahrung der Rechte der Volksvertreter) des Reichstages* (1st Committee [Protection of Parliamentary Rights] of the *Reichstag*).
06.12.1932-01.02.1933	Member of the *1. Ausschuß (Wahrung der Rechte der Volksvertreter) des Reichstages*.
06.12.1932-01.02.1933	*Obmann der NSDAP* and *stellvertretender Vorsitzender* (deputy chairman) of the *1.Ausschuß (Wahrung der Rechte der Volksvertreter) des Reichstages*. The following is excerpted from an Associated Press article ("Hitlerite Shoves Marxist Out of Committee Chair") of 14.02.1933:

> The *Reichstag's* committee for safeguarding parliamentary rights broke up in disorder today when Chancellor Adolf Hitler's close friend, Hans Frank, pushed the Socialist chairman, Dr. Paul Loebe, bodily off the presidential seat and occupied it himself. 'The Nazi party no longer can tolerate a slanderous Marxist in the chari,' Frank declared ... The Social Democratic newspapers indignantly protested at Herr Frank's "high handed usurpation." The *Vorwaerts* called it to the attention of President von Hindenburg and Chancellor Hitler as a "flagrant breach of constitutional rights punishable under the criminal code."

10.03.1933-18.04.1933	*Staatskommissar für das Bayerischen Staatsministerium der Justiz* (State Commissioner for the Bavarian State Ministry of Justice).
23.03.1933	Appointed as *Vorsitzender* of the *1.Ausschuß (Wahrung der Rechte der Volksvertreter) des Reichstages"*.
01.04.1933	Appointed as a member of *Gauleiter* Julius Streicher's *Zentralkomitee zur Abwehr der jüdischen Greuel- und Boykotthetze* (Central Committee for Defense Against Jewish Atrocity and Boycott Agitation).
18.04.1933-04.12.1934	*Bayerischer Staatsminister der Justiz* (Bavarian State Minister of Justice) and co-editor of the periodical *Deutsche Justiz. Rechtspflege und Rechtspolitik*.
20.04.1933-14.01.1934	Assigned as an *SA-Führer z. V.* to the *Stab der Obersten SA-Führer*.

Dr. Frank in civilian attire, ca. 1933.

Reichsleiter Dr. Frank, ca. 1935.

22.04.1933-19.12.1934	*Reichskommissar für die Gleichschaltung der Justiz in den Ländern und für die Erneuerung der Rechtsordnung (Reichsjustizkommissar) (Reich* Commissioner for the Standardization of Justice in the Länder and for the Renewal of the Legal Order [*Reich* Justice Commissioner]), appointed by *Reich* President Paul von Hindenburg.
13.05.1933	Arrived by plane at Wien's Asper Airport, but was soon expelled from Austria on the orders of Chancellor Engelbert Dollfuss. In an article of 15.05.1933 ("Hitler Man Told to Leave Austria"), the United Press reported:

> Austro-German diplomatic relations became badly strained tonight when the Austrian government informed Dr. Hans Frank, Bavarian minister of justice, and a personal envoy of Chancellor Adolf Hitler's, that it desired his immediate departure from Austria.
>
> Dr. Frank, who with Hans Kerrl, Prussian minister of interior [sic! Minister of Justice], was center of anti-Nazi demonstration on arrival by airplane Saturday, altered his plans for proceeding to Salzburg, and started at once to the Bavarian border.
>
> Police at the town of Gnigl informed Dr. Frank of the Austrian government's desire that he depart. He was motoring from Graz to Salzburg as part of his itinerary of inspection of Nazi groups in Austria when halted.
>
> The incident came on the heels of a formal protest by the

Official portrait of Dr. Hans Frank, ca. 1936. (*Bundesarchiv Bild* 146-1989-011-13)

Reichsleiter Dr. Frank, ca. 1938. (NARA photo)

German minister to Austria against the "inhospitable reception" accorded Dr. Frank and Minister Kerrl at Vienna's airport, when they arrived from Germany last Saturday.

The United Press understood the protest was extremely sharp, and delivered at Chancellor Hitler's personal instigation.

Premier Engelbert Dollfuss, however, at once called a special meeting of the cabinet council, and not only refused to apologize but retaliated by expelling Dr. Frank from the country- the interpretation on the action in permitting him to know he was no longer a desirable alien in Austria.

Dollfuss originally threatened to expel him outright as an undesirable alien, if Dr. Frank attempted Nazi political agitation in Austria.

It also was learned that the Austrian legation in Berlin protested today to the German foreign office against Frank's presence in this country, particularly after his speech at Graz, which the minister plainly told the German authorities was regarded as insulting to Premeier Dollfuss.

An Associated Press article of 16.05.1933 ("Austria Head Gets Rebuke from Hitler") added:

... The Nazi officials [Frank and Kerrl] were cheered ... by thousands of Austrian Nazis, strong opponents of the Dollfuss regime.

Chancellor Dollfuss assured the [German] minister that no affront was intended against the German government, and that the incident was directed only against Frank as an individual.

Frank, it was pointed out in semi-official circles, recently threatened in a speech at Munich that there might be a German invasion of Austria.

01.06.1933	Founded the *Deutschen Rechtsfront* (German Legal Front).
00.06.1933-20.08.1942	*Reichsleiter* der *NSDAP*.
00.06.1933	Permanent employee of the periodical *Gasschutz und Luftschutz* (Gas and Air Raid Protection), Berlin.
26.06.1933	Founded the *Akademie für Deutsches Recht*, München. The following is the statute defining the functions of the Academy:

The Academy's task shall cover primarily:
1. The composition, the initiation, judging and preparing of drafts of law.
2. The collaboration in rejuvenating and unifying the training in jurisprudence and political science.
3. The editing and supporting of scientific publications.
4. The financial assistance for research and work in specific fields of Law and Political Economy.
5. The organization of scientific meetings and the organization of courses.
6. The cultivation of connections to similar institutions in foreign countries.

(Translation of Document 1391-PS, in *Nazi Conspiracy and Aggression, Volume II*)

30.08.1933-03.09.1933	Participated in the 5. *Reichsparteitag der NSDAP* in Nürnberg.
02.10.1933-20.08.1942	*Leiter* (redesignated *Präsident*, 09.08.1934) of the *Akademie für Deutsches Recht* (Academy of German Law) and *Vorsitzender* of the *Akademie-Ausschuß für Rechtsphilosophie* (Academic Committee for Legal Philosophy).
00.00.1933-20.08.1942	*Reichsrechtsführer* and editor of the periodicals *Deutsches Recht. Zeitschrift der Akademie für Deutsches Recht* and *Juristischen Wochenschrift*.
00.00.1933-00.00.1942(?)	*Leiter* of the *Amt für NS-Juristen* in the *Rechtsabteilung* of the *Reichsleitung der NSDAP*.
00.10.1933	Member of the *Führerrat der RDB*.
17.11.1933	Appointed as *Vorsitzender* of the *Rechtsausschuß für die Strafrechtsreform der Akademie für Deutsches Recht* (Law Reform Committee for Criminal Law Reform of the Academy for German Law), together with *Reichsjustizminister* Dr. Franz Gürtner.
12.12.1933-08.05.1945	Member of the *Ältestenrat der Reichstag*.

04.01.1934	Appointed as a *Beisitzer* of the *I.Kammer des Obersten Parteigerichts der NSDAP* (1st Chamber of the Supreme Party Court of the *NSDAP*).
15.01.1934	Appointed as *Referent* of *Abteilung VI* in the *Stab der Obersten SA-Führer*.
00.01.1934	Member of the *Hochschulkommission der NSDAP*.
12.03.1934	*Vorsitzender* of the *Deutschen Landesgruppe* of the International Law Association (assumed office on 26.04.1934).
06.04.1934	Appointed as temporary *Leiter* of *Abteilung 4 (Bund Nationalsozialistischer Deutscher Juristen)* in the *Rechtsabteilung* of the *Reichsleitung der NSDAP*.
30.06.1934	Frank recalls his actions during the *SA* purge of that date:

> [During] the Roehm Putsch I had to sit ... as the Secretary [Minister] of Justice,] I had to sit with the arrested men to avoid to have them shot by the *SS*. This made Hitler very furious. Hitler called me in personally and said, "I demand that you hand these men over immediately for execution." I was sitting in the room of the prison director. I told Hitler, "I am the Secretary of Justice in Bavaria , and without legal authority I cannot hand these men over." Then he said, Hitler said, "I am the Fuehrer of all constitutional and political matters in Germany, and I take the responsibility that these men be handed over." Then I said under reservations of *Reich* law which Hitler had promised to proclaim, I told him to hand me a list of the men who had to be surrendered. While originally there was a list of two hundred names , after long hours of consultation, the list was reduced to eighteen names, so that my interfering in this matter on that day saved the lives of over two hundred men. That was the beginning of the end for me as a justice official and the law was proclaimed and all the steps taken by Hitler were legalized. Hitler had said, "I believe we have made an entirely wrong man the Bavarian Secretary of Justice."
>
> [Question from interrogator: "Did he remove you from that position?"]
>
> Yes, on the 31st of December 1934. Therefore it happened that as far as legal matters were concerned I was the red cloth to the bull. And there never again was any intimate interchange of thoughts between him and me. ("Interrogation Records Prepared for War Crimes Proceedings at Nuernberg, 1945-1947/OCCPAC Interrogation Transcripts and Related Records: Frank, Hans"; Publication Number M1270, Record Group RG238)

27.07.1934-30.11.1936(?)	Appointed as an *SA-Ehrenführer* with the rank of *SA-Gruppenführer* and attached to the staff of *SA-Gruppe Hochland* (München).
04.09.1934-10.09.1934	Participated in the *6. Reichsparteitag der NSDAP* in Nürnberg. On 05.09.1934, he declared:

As chief of the German justice system, I can only say that since the National Socialist legal system is the foundation of the National Socialist State, for us, our supreme *Führer* is also supreme judge. And since we know how sacred the principles of justice are to our *Führer*, we can assure you, fellow citizens, that your life and existence is secure in this National Socialist State of order, freedom, and law!

14.09.1934   Declared in a speech:

It is unbearable to us to permit Jews to play any role whatsoever in the German administration of justice ... It will therefore be our firm aim to exclude Jews increasingly from the administration of the law as time goes on.

04.12.1934   Appointed as *Bayerischer Staatsminister ohne Geschäftsbereich* (Bavarian State Minister without Portfolio).

19.12.1934-30.04.1945   *Reichsminister ohne Geschäftsbereich* (*Reich* Minister without Portfolio). In a handwritten letter of 19.12.1934 concerning Frank's appointment as *Reichsminister*, Hitler wrote:

Dear Herr Minister,
Now that the Ministry of Justice of the *Reich* has been united with that of Prussia and the *Reich* has assumed the direct supervision of the administration of justice by virtue of the law of 5. December 1934, the task of coordinating the judiciary in the Länder has now been accomplished.

In the Academy of German Law, you have created an exemplary and lasting institution for the work of revising the legal system which has enabled you to aid in enforcing the National Socialist Weltanschauung in every sector of the revision of the law without limitation to the judiciary in a narrower sense. May I extend my warmest thanks and my special appreciation for your untiring and successful efforts as *Reich* Commisar for the Coordination of the Judiciary in the Länder and for the revision of the legal system and, at the same time, hereby declare that the task which was assigned to you by the immortal *Reich* President von Hindenburg on 22. April 1933 is now completed. At this time I also appoint you to the position of *Reich* Minister without portfolio in the *Reich* Government.
Mit deutschen Gruß, Adolf Hitler
(Max Domarus, ed, *Hitler. Speeches and Proclamations, 1932-1945: The Years 1932-1934*, p. 547)

01.01.1935-20.08.1942   *Leiter* of the *Reichsrechtsamt der NSDAP* (dissolved, 20.08.1942).

01.03.1935	Appointed as *Leiter* of the *Amt für Rechtsschrifttum* (Office for Legal Literature).
08.03.1935	Appointed as *Führer* of the *Deutsche Landesgruppe* (German National Group) of the *Internationale Kriminalistische Vereinigung* (*IKV*).
08.03.1935	Appointed as *Führer* of the *Deutsche Strafrechtlichen Vereinigung* (German Association for Penal Law).
10.09.1935-16.09.1935	Participated in the *7. Reichsparteitag der NSDAP* in Nürnberg.
25.10.1935	Appointed as a member of the *Ehrenvorstand* of the *Deutsch-Französische Gesellschaft* (*DFG*, German-French Society).
09.11.1935	Appointed as *Führer* of the *Arbeitsgemeinschaft für gewerbliche Rechtsschutz und Urheberrecht* (Association for Legal Protection of Commercial Property and Copyright).
29.01.1936	Delivered a speech to the economic council of the *Deutsche Akademie* in München, from which the following is excerpted:

> We do not care what the world says about our Jewish legislation ... [The *Führer* has settled the Jewish question] in such a dignified and objectively justified form [and the regime would destroy the] Jewish pack of Bolshevistic preachers ... Foreign critics haven't the slightest idea about the vital necessities of the German people. Such Jews as came to live among us showed their gratitude by introducing pestilence into our spiritual life ... We are putting the idea of a people in place of the idea of an institution as our conception of the State. We are placing the commonwealth ahead of private gain. We are substituting the idea of a community built on blood affinity for the idea of a State as the sum total of its subjects. And we are replacing the principle of majority rule by a conception of authority leadership ... The Jew may do business in Germany unhampered but he must under no circumstances deposit his intellectual excrements with us. (Associated Press article "Nazis to Continue Anti-Jewish Drive", Berlin, 29.01.1936)

00.04.1936	Appointed as *Leiter* of the *Forschungsabteilung "Judenfrage"* (Department for Research of the "Jewish Question") in the *Reichsinstitut für die Geschichte des neuen Deutschland* (*Reich* Institute for the History of the New Germany).
11.05.1936-20.08.1942	*Reichsführer* of the *Nationalsozialistische Rechtswahrerbund* (*NSRB*, National Socialist Organization of Legal Professionals, the former *BNSDJ*) (see entry of 11.10.1928-11.05.1936, above). Succeeded by Otto Thierack.
01.06.1936-06.06.1936	*Ehrenpräsident* of the *Internationale Kongress für Gewerblichen Rechtsschutz* (International Congress for Legal Protection of Commercial Property) in Berlin.
Autumn 1936	Participated in an *Offizierslehrgang* (officers instructional course).
08.09.1936-14.09.1936	Participated in the *8. Reichsparteitag der NSDAP* in Nürnberg.

30.11.1936-00.05.1945(?)   Transferred to the active *SA-Führer*korps and assigned as an *SA-Führer z. V.* to *SA-Gruppe Hochland* (with effect from 01.11.1936). Of his *SA* membership, he stated the following during interrogation at Nürnberg on 01.09.1945:

> I had no [*SA*] duties. I had the right to wear the uniform ... Because in former years I had defended many trials for members of the *SA* ... I only wore the uniform at special occasions during marches of the *SA*. If I have worn it often, I have worn it ten times. ("Interrogation Records Prepared for War Crimes Proceedings at Nuernberg, 1945-1947/OCCPAC Interrogation Transcripts And Related Records: Frank, Hans"; Publication Number M1270, Record Group RG238)

00.00.1936   In his article "Legislation and Judiciary in the Third *Reich*" in the *Zeitschrift der Akademie für Deutsches Recht*), wrote:

> To the world we are blamed again and again because of the concentration camps. We are asked, "Why do you arrest without a warrant of arrest?" I say, put yourselves into the position of our nation. Don't forget that the very great and still untouched world of Bolshevism cannot forget that we have made final victory for them impossible in Europe, right here on German soil.

11.05.1937-00.09.1939   One of two *Ehrenpräsidenten* (honorary presidents, the other being *Reichsjustizminister* Dr. jur. Franz Gürtner) of the *Deutschen Gruppe* to the *Arbeitsgemeinschaft für die deutsch-polnischen Rechtsbeziehungen* (Working Association for German-Polish Legal Relations).

06.09.1937-13.09.1937   Participated in the *9. Reichsparteitag der NSDAP* in Nürnberg.

08.05.1938   Accompanied Hitler on his state visit to Italy.

05.09.1938-12.09.1938   Participated in the *10. Reichsparteitag der NSDAP* in Nürnberg.

20.04.1939   Appointed as an *Ehrensenator* of the University of Leipzig.

20.06.1939   Received an honorary doctorate by the University of Sofia (Bulgaria).

24.08.1939-17.09.1939   Military training exercises as a *Leutnant d. R.* in *4.MG (E) Kompanie / Infanterie-Regiment 9* (Potsdam).

08.09.1939-25.10.1939   *Chef der Zivilverwaltung des Militärbezirks Mitte* (Chief of Civil Administration for Military District Center) in Łódź (Ger.: Lodsch) and *Oberster Verwaltungschef der Militärbezirke von Westpreußen, Posen, Lodsch und Krakau* (Supreme Administrative Chief of the Military Districts of West Prussia, Posen, Łódź, and Kraków).

03.10.1939-25.10.1939   *Oberverwaltungschef für die gesamte zivile Verwaltung der besetzten ehemals polnischen Gebiete beim Oberbefehlshaber Ost* (Senior Administrative Chief for all Civil Administration of the Occupied Former Polish Territories attached to Commander-in-Chief East).

26.10.1939-17.01.1945   *Generalgouverneur für die besetzten polnischen Gebiete* (General Governor for the Occupied Polish Territories), with offices at Wawel Castle,

Autographed portrait of Dr. Frank, ca. 1939. (Hermann-Historica, Auctioneers, München)

Kraków. Evacuated and moved his HQ to Neuhaus am Schliersee / Bayern, 17./ 18.01.1945. Of his appointment to this post, Frank testified on 18.04.1946:

> On 24 August: 1939, as an officer in the reserve, I had to join my regiment in Potsdam. I was busy training my company; and on 17 September, or it may have been 16, I was making my final preparations before going to the front when a telephone call came from the *Führer*'s special train ordering me to go to the *Führer* at once.
>
> The following day I traveled to Upper Silesia where the *Führer*'s special train was stationed at that time; and in a very short conversation, which lasted less than ten minutes, he gave me the mission, as he put it, to take over the functions of Civil Governor for the occupied Polish territories." (*Trial of the Major War Criminals Before the International Military Tribunal, Nuremberg, 14 November 1945-1 October 1946, Volume XII*)

Subordinated to Frank in his capacity as *Generalgouverneur* were the following:

Stellvertrender Generalgouverneur
25.10.1939-22.05.1940: *Reichsminister* / *SS-Gruppenführer* Dr.

jur. Artur Seyss-Inquart

22.05.1940-17.01.1945: Dr. jur. Josef Bühler

Amt des Generalgouverneurs (redesignated "Regierung GG", 01.12.1940)

03.10.1939-17.01.1945: Dr. jur. Josef Bühler (Appointed *Staatssekretär*, 07.03.1940)

Deputy *Staatssekretär*

01.09.1941-17.01.1945: *SS-Oberführer* Dr. jur. Ernst Boepple

Chefs of *Hauptabteilungen* (Main Divisions)

Hauptabteilung Innere Verwaltung (Internal Administration)

26.10.1939-01.09.1940: *Ministerialrat* Dr. jur. Friedrich Wilhelm Siebert

01.09.1940-00.11.1940: *Unterstaatssekretar* Ernst Kundt (*kommissarisch*)

06.11.1940-31.01.1942: Eberhard Westerkamp

01.02.1942-01.01.1943: *Ministerialrat* Dr. jur. Friedrich Wilhelm Siebert

01.01.1943-10.10.1943: *SS-Obersturmbannführer* Dr. jur. Ludwig Losacker

18.11.1943-17.01.1945: *SS-Brigadeführer* Dr. jur. Harry von Craushaar

Hauptabteilung Arbeit (Labor; Designated an *Abteilung* until 26.09.1942)

18.11.1939-00.11.1942: *Reichshauptamtsleiter der NSDAP / SS-Obersturmbannführer* Dr. Max Frauendorfer

00.01.1943-17.01.1945: Wilhelm Struve

Hauptabteilung Bauwesen (Construction)

28.02.1940-17.01.1945: *Präsident* Theodor Bauder

Hauptabteilung Ernährung und Landwirtschaft (Food & Agriculture; Designated an *Abteilung* until 06.06.1940)

00.10.1939-00.07.1941: *SS-Brigadeführer* Hellmut Körner

10.07.1941-17.01.1945: Karl Naumann

Hauptabteilung Finanzen (Finance)

02.03.1940-01.01.1942: *Präsident* Dr. Alfons Spindler

01.01.1942-17.01.1945: *SS-Standartenführer* Dr. Hermann Senkowsky

Hauptabteilung Forsten (Forests)

29.10.1939-17.01.1945: *Oberlandforstmeister* Dr. Kurd Eissfeldt

Hauptabteilung Gesundheit (Health)

06.05.1940-20.01.1943: *Obermedizinalrat / SA-Oberführer* Dr. med. Jost Walbaum

20.01.1943-17.01.1945: *SS-Brigadeführer* Prof. Dr. med. Heinrich Teitge

Hauptabteilung Justiz (Justice)

00.10.1939-17.01.1945: *Ministerialrat* Kurt Friedrich Theodor Wille

Hauptabteilung Propaganda (designated an *Abteilung* until 00.00.1940)
00.10.1939-00.07.1940: *SS-Sturmbannführer* Dr. Maximilian Freiherr von du Prel
01.02.1941-17.01.1945: *Oberregierungsrat* Wilhelm Ohlenbusch
Hauptabteilung Unterricht und Wissenschaft (Education & Science)
00.00.1940-00.10.1942: *Hofrat* Dr. Adolf Watzke
05.10.1942-17.01.1945: Prof. Dr. Ludwig Eichholz
Hauptabteilung Wirtschaft (Economy)
00.10.1939-00.06.1940: Dr. Richard Zetzsche
13.06.1940-17.01.1945: *Ministerialdirigent* Dr. Walter Emmerich
Generaldirektor der Ostbahn (Gedob, Eastern Railways)
00.00.1940-17.01.1945: *Präsident* Adolf Gerteis
Hauptabteilung Post (Postal Service)
00.10.1939-17.01.1945: *Präsident* Richard Lauxmann
Emissionsbank (Bank of Issue, i.e. a bank empowered by a government to issue currency)
Reichsbankdirektor a. D. Bankdirigent Dr. Fritz Paersch
Staatssekretär für das Sicherheitswesen (State Secretary for Security Matters)
07.05.1942-09.11.1943: *SS-Obergruppenführer und General der Polizei* Friedrich-Wilhelm Krüger
09.11.1943-17.01.1945: *SS-Obergruppenführer* und General der Polizei Wilhelm Koppe
Chefs des Distrikts (redesignated Gouverneure, 25.04.1941)
Warschau:
26.10.1939-17.01.1945: *SA-Gruppenführer* Dr. jur. Ludwig Fischer
Krakau:
26.10.1939-22.01.1942: *SS-Brigadeführer* Dr. jur. Otto Wächter
31.01.1942-26.05.1943 : *SS-Gruppenführer* Dr. jur. Richard Wendler
26.05.1943-10.10.1943: *SS-Obersturmbannführer* Ludwig Losacker
23.11.1943-18.01.1945: *SA-Gruppenführer* Dr. jur. Curt von Burgsdorff
Lublin:
26.10.1939-01.02.1940: *SS-Brigadeführer* Friedrich Schmidt
01.02.1940-10.04.1943: *Oberbürgermeister a. D.* Ernst Zörner
10.04.1943-27.05.1943: *SA-Gruppenführer* Dr. jur. Ludwig Fischer (kommissarisch)
27.05.1943-22.07.1944: *SS-Gruppenführer* Dr. jur. Richard Wendler Radom
28.12.1939-01.08.1941: Dr. rer. pol. Karl Lasch
01.08.1941-00.01.1945: *Unterstaatssekretär* Ernst Kundt

Reichspropagandaministerium, Berlin: The state visit (25.-27.11.1940) of Dino Grandi,
President of the Chamber of Fasci and Corporations. From left to right: *Reichsamtsleiter*
Wilhelm Heuber (*Bevollmächtigter des Generalgouverneurs* – Frank's representative
in Berlin); Frank; Grandi; and *Reichsminister* Dr. Goebbels. (NARA photo)

Lemberg
01.08.1941-24.01.1942: Dr. rer. pol. Karl Lasch
00.02.1942-00.07.1944: *SS-Brigadeführer* (later *Gruppenführer*)
Dr. jur. Otto Wächter

On 08.04.1946, Dr. Hans Heinrich Lammers testified at length
regarding the numerous difficulties faced by Frank in Poland:

> ... Frank's powers as Governor General were considerably limited
> through the police, since Himmler as Chief of the German Police
> had direct police powers which he was, to be sure, to cooperate
> with those of the Governor General but which he did not always
> do. The Governor General suffered a further loss of power
> through the fact that Himmler was *Reich* Commissioner for the
> Preservation of German Nationality and as such could undertake
> resettlements and did do so without consulting Governor General
> Frank in any way ...
>
> In my opinion Frank always tried to pursue a policy of
> moderation and to create an atmosphere of friendship towards

Germany in Poland. To be sure, he very often was unable to achieve his aim, especially because of the fact that the powers of the police and Himmler's powers were too great in the field of resettlement, so that his measures and his intentions suffered setbacks. He found it difficult to achieve his aims.

In the Government General there had been established a Secretariat of State for the security system. This was under Krüger, then Higher *SS* and Police Chief. This, however, functioned for only 4 to 6 weeks and then differences of opinion in this field broke out once more. The State Secretary for Security, Krüger, stated, "I receive my orders from Himmler." If the Governor General complained about that, then Himmler said, "These are all unimportant matters. I certainly must be able to rule on them directly." The Governor General said, "But for me they are not unimportant; even those things are important to me."

The channels of command and the cooperation with the Governor General were not being observed, and it is therefore perfectly understandable that Herr Frank had a very difficult position with respect to the police system ...

He repeatedly offered his resignation, because of these sharp conflicts which he had, with Himmler in particular, and because Hitler usually decided that he was in the wrong and Himmler in the right. Many statements of his intention or desire to resign were brought to me, some of which I was not even allowed to submit to the *Führer*. But I informed the *Führer* of the Governor General's intentions of resigning and the *Führer* several times refused Frank's offer to resign.

... Herr Frank had often objected to a policy of exploitation and pronounced himself in favor of a policy of reconstruction, in cultural matters as well. He had suggested, for instance, that Polish advisory committees be assigned to the authorities under the Governor General and to the district chiefs, and so forth; that was refused. He spoke in favor of the creation of high schools, theological seminaries, and similar cultural aims, all of which were rejected.

On one occasion he had submitted a long memorandum. This referred to a Polish organization which called itself "The Plough and the Sword." It had offered to cooperate with the Germans, and Frank submitted detailed proposals in a long memorandum, saying that these Poles could be won over to cooperate only if they were met on proper terms. All these suggestions, coming from Frank, were turned down by Hitler ... (*Trial of the Major War Criminals Before the International Military Tribunal, Nuremberg, 14 November 1945 – 1 October 1946, Volume XI*)

04.12.1939-17.01.1945	*Reichsverteidigungskommissar für das Generalgouvernement* (*Reich* Defense Commissioner for the General Government) and *Vorsitzender* of the *Verteidigungsausschuß* (defense committee) in the *Generalgouvernement*.
04.12.1939-31.07.1940	*Generalbevollmächtigter für den Vierjahresplan* (General Plenipotentiary for the Four-Year Plan) in the *Generalgouvernement*.
06.02.1940	In an interview with a *Völkischer Beobachter* correspondent named Kleiss, stated:

> I might, for example, state the following: In Prague, there were put up red posters announcing that seven Czechs had been shot that day. I then said to myself: "If I wished to order that one should hang up posters about every seven Poles shot, there would not be enough forests in Poland with which to make the paper for these posters. Indeed, we must act cruelly." (*Trial of the Major War Criminals Before the International Military Tribunal, Nuremberg, 14 November 1945 – 1 October 1946, Volume VII*)

10.02.1940	Associated Press article concerning the situation in Frank's *Generalgouvernement*:

> Poles Well Off, Nazi Boss Says. Gov. Gen. Frank Scoffs at Charges of Ill Treatment in Occupied Region. Berlin, Germany- (A)- Poland, in the words of her Nazi governor general, is being ruled with more consideration for social and material welfare than ever before, but political opposition "is a luxury which cannot be indulged at present."
>
> Hans Frank, governor of the section occupied by Germany but not annexed to the *Reich*, reviewed charges from abroad against his administration in an interview with 150 correspondents.
>
> Frank denied that the Germans were "shooting priests, sterilizing Polish children, torturing Jews or executing students by the thousands," remarking that "such fantastic accusation merely can make a philosopher smile." [Pamphlets issued by the Polish government in exile in France and by the Polish embassies in Rome and Vatican City have charged Germany with causing the death of thousands of Poles.]
>
> Frank, a ruddy faced lawyer of professorial mien, said he was rigorous in smashing any Polish movement or tendency which would weaken the administration or give enemies of Germany a chance to strike in the east.
>
> "No Concentration Camps"
> "Germany is at war, fighting for its existence," he declared. "So it is not to be supposed that the *Reich* could extend liberties to the Poles greater than those enjoyed by the populations of England,

France or Germany.

"But those who assail me from abroad should know there is
no hunger, there is not even any rationing of bread as in Germany
itself. They should know there is not a single concentration camp
in all Poland."

In reply to charges from abroad that the Poles' health has been
neglected, including reports of typhus outbreaks, he said, "German
medical services in a few months gave free inoculations to 720,000
Jews, Ukrainians and Poles[,]a prodigious accomplishment.

"Never in modern times has there been less typhus than now
in Poland."

Confirming that discipline was strict against political
opposition, Frank said Poles were not permitted to listen to radios
or depart from the prescribed paths of rectitude as laid down by
Germany.

Few Prisoners Returned

"Because it is not military practice to restore prisoners before a war
has ended," Frank said, only a few Polish prisoners of war have
been returned from Germany.

His government is not persecuting Jews, he declared, and has
no special attitude toward the Jewish problem.

Asserting that thousands of Jews from Germany, Hungary,
Slovakia, Rumania and Russia have applied to enter Poland, Frank
said incoming Jews were not being sent to such regions as Lublin,
and that "not a single Jew has left my capital at Cracow."

He said he did not know whether Lublin or some other
area ever would be set aside as a Jewish state, but that he saw no
objection to such a plan.

"Where the newly arrived Jews go is largely decided by the
Jewish communities," which are operating freely in every region,
he said.

At the will of Hitler, Poland is to remain "the home of the
Jews" the governor said, adding that by tradition it seemed more
Jewish than other European states.

Have Own Currency

It must be remembered, however, the area is not a part of Germany
and has not been annexed as has the Poznan district, he said.

"We have our own currency, the value of which we are
holding up by main force despite the fact that the former Polish
government skipped off with the gold reserve," he continued.

"We have our own customs, even against Germany. We are
carrying out the main features of the old Polish trade treaty with
Bulgaria. We hope to make a trade treaty with the *Reich*, and I
guess I know enough about the inner workings of the *Reich*'s

The Anhalter Bahnhof, Berlin: The state visit (25.-27.11.1940) of Dino Grandi.
Right to left: Dr. Frank, Wihelm Heuber, and Grandi. (NARA photo)

economy to wangle a good deal for Poland."

Many government positions, including 19,000 police jobs, are filled by Poles, he reported.

Lack of an army makes a saving whereby the government can pay old Polish pensions and support a social program, Frank said, with the comment that "Poland under German administration isn't going to make war on anyone, so she doesn't need an army."

12.02.1940   Participated in a *Sitzung über Ostfragen* (Meeting on Eastern Questions), chaired by Hermann Göring at his Carinhall estate. The other attendees were Heinrich Himmler, *Reichsminister* Lutz Graf Schwerin von Krosigk; *Staatssekretäre* Paul Körner, Ernst Neumann, Friedrich Landfried, Herbert Backe, Dr. Friedrich Syrup, Wilhelm Kleinmann, Friedrich Alpers; Dr. h.c. Max Winkler (head of *Hauptreuhandstelle Ost*, Main Trusteeship Office East); and the four *Gauleiter* of the eastern *Gaue* (Erich Koch / *Ostpreußen*; Albert Forster / *Danzig-Westpreußen*; Arthur Greiser / *Warthegau;* and Josef Wagner / *Schlesien*). This followed a memorandum from Dr. Frank to Göring's Four-Year Plan Office proposing the use of the resources of Poland to aid Germany's military and economic strength. Göring opened by stating that the primary objective of German policy in the East was to strengthen Germany's war potential. He expressed opposition to the deportation of

Official portraits of *Generalgouverneur* Dr. Frank, ca. 1940 (NARA photos)

economically essential laborers from the eastern *Gaue*. While agreeing that the *Generalgouvernement* should accept Jews deported there, the evacuations should be conducted in an orderly fashion and with Dr. Frank being notified in each case. Himmler responded that his *RKFDV* organization had only deported 300,000 of the 8 million Poles in the incorporated Polish territories and that 240,000 *Volksdeutsche* had come into the freshly vacated areas. But he agreed to work with Dr. Frank in regards to future evacuation measures. However, on 24.03.1940, Göring issued a directive that read:

> The *Generalgouverneur* for the occupied Polish territories has complained that deportations of Jews from the *Reich* into the *Generalgouvernement* are still taking place although the facilities for receiving them are at present not available. I hereby forbid further such deportations without my permission and without proof of the prior agreement of the *Generalgouverneur*...

19.04.1940-00.01.1945   *Präsident* of the *Institut für deutsche Ostarbeit* (Institute for German Work in the East) in Kraków. The Institute was established by Frank's decree of 19.04.1940, the third paragraph of which read:

> It is the task of the Institut für deutsche Ostarbeit scientifically to clear all fundamental questions of the Eastern Space as far as

they concern the *Generalgouvernement* and to disseminate the results of the researches. In accomplishing this task, the *Institut für deutsche Ostarbeit* will collaborate with other institutions of similar aims. (Max Weinreich, *Hitler's Professors: The Part of Scholarship in Germany's Crimes against the Jewish People*, p. 95)

The institute, directed by a legal colleague of Frank, Dr. Wilhelm Coblitz, published books as well as periodicals such as *Die Burg, Das Generalgouvernement,* and *Deutsche Forschung im Osten.* "The first issue of *Die Burg*", writes Max Weinreich

> ... carried a large study by Peter-Heinz Seraphim on "The Jewish Question in the Government General as a Population Problem" and one by Josef Sommerfeldt on "The Development of Jewish Historiography in Poland." The apparent dispassionateness of the writers and the skill they displayed emphasized the viciousness of their purpose. (ibid)

01.05.1940-01.12.1943 — *Präsident* of the *Internationale Rechtskammer* (International Chamber of Law).

15.08.1940 — In a speech to *NSDAP* officials in Kraków, declared:

> Here we are in this land and as Germans we shall never leave this land. The swastika will continue to stream over this land into the farthest future. And the Polish people, as they were once seven hundred years ago, are again under the patronage of the German nation. The Vistula is from now on Germany's river and shall remain Germany's eastern river ... .
>
> We must see to it that particularly this German city of Krakau be completely deprived of its Jewish character and I have decreed, as you know, that until today the Jews have voluntarily [sic] to depart. Whoever, being a Jew, after this midnight remains in the city, tomorrow will be liable to forced labor. It is an impossible situation that the representation of Adolf Hitler's Greater German *Reich* establish itself here in a city which swarms with Jews to the degree that a decent man simply cannot walk the streets.
>
> And it is clear that hereby a serious signal is given: the Jews must disappear from the whole of Europe. ("Die Juden müssen aus Europa verschwinden" [The Jews Must Disappear From Europe] in the *Krakauer Zeitung*, reproduced in *The Jewish Bulletin* [London], no. 45, May 1945)

00.12.1940 — Proposed for the post of *Präsident* of the *Deutschen Akademie*, München.
05.04.1941-00.00.194_ — *Leiter* of the *Internationale Rechtswahrervereinigung.*

23.10.1941	Cofounder of the *Arbeitsgemeinschaft zur Behebung des volksdeutschen Notstands* (Krakau) (Working Group to resolve the Ethnic German Crisis).
18.11.1941	To vigorous applause from his audience, delivered a virulently anti-Semitic speech at the University of Berlin from which the following is an excerpt:

> ... A problem that occupies us in particular is the Jews. These merry little people, which wallows in dirt, and filth, has been gathered together by us in ghettos and special quarters and will probably not remain in the *Generalgouvernement* for very long.
>
> We will get these Jews marching and cause them- as they have already covered the distance from Jerusalem to Poland- to move eastward a few thousand kilometers. But these Jews are not that parasite gang alone, from our point of view, but strangely enough- we only realized it over there [in Poland]- there is another category of Jews, something one would never have thought possible. There are laboring Jews over there who work in transport, in building, in factories, and others are skilled workers such as tailors, shoemakers, etc. We have put together Jewish workshops with the help of these skilled Jewish workers, in which goods will be made which will greatly ease the position of German production, in exchange for the supply of foodstuffs and whatever else the Jews need urgently for their existence. These Jews may well be left to work in this way; in the way in which we are now using them it is something of an achievement for the work-Jews themselves; but for the other Jews we must provide suitable arrangements. It is always dangerous, after all, to leave one's native land. Since the Jews moved away from Jerusalem there has been nothing for them except an existence as parasites: that has now come to an end. If one looks at the Warsaw ghetto *today* in which 480,000 Jews- well, let us say- live, then one must realize that only the determination of the National-Socialist revolution was capable of successfully confronting even this problem. In 1919, at our first meetings in München, we proclaimed the motto: An end must be put to the rule of the Jews in Europe... (Diensttagebuch of Dr. Frank, in Yad Vashem Archives, JM/21)

01.12.1941	In a report of that date sent to Heinrich Himmler, Dr. jur. Günther Reinecke, chief of *SS und Polizeigericht VI* in Kraków, reported on the case of *SS-Untersturmführer* Lorenz Löv, head of Frank's Central Administrative Office in Warsaw. Löv had been tried and sentenced to life imprisonment by Reinecke's court for embezzlement, having operated a black market in furs and other items which were illegally sold from a warehouse he controlled. In the course of this investigation, Reinecke reported, it came to light that "The *Generalgouverneur*'s wife

procured from the warehouse a number of fur coats (at least ten) far in excess of her personal requirements. But even this by no means satisfied her need for furs! From Apfelbaum in Warsaw she ordered a moleskin jacket, a beaver coat, a musquash coat, an ermine coat, two broadtail coats, an ermine jacket, a silver fox cape, a blue fox cape, and other furs. According to *SS-Sturmbannführer* [Albert] Fassbender, the persons making these purchases on behalf of the *Generalgouverneur*'s wife were members of the *Generalgouverneur*'s office and they simply fixed the price at approximately 50 per cent of cost price." From Warsaw's Jewish population, the report went on, the family of *Generalgouverneur* Frank had procured "rings, gold bracelets, gold fountain pens, tinned food, picnic hampers, coffee machines and groceries" ... "at staggeringly low prices." It had also been uncovered that Frank had shipped, from his Kressendorf estate in the *Generalgouvernenement* to his estate at Schobernhof in Germany, "200,000 eggs, a whole year's produce of dried fruit, sheets, blankets, and furniture. In November 1940, two whole convoys of foodstuffs left for Schobernhof. The first contained 150 lb of beef, 50 lb of pork, 20 geese, 50 chickens, 25 lb of salami, 30 lb of ham sausage, 25 lb of ham; the second: 175 lb of butter, 110 lb of cooking oil, 30 lb of cheese, 1,440 eggs, 50 lb of coffee beans and 120 lb of sugar". The report also accused Frank of the theft and transport to Schobernhof of Polish religious artifacts, and ended with "This affair constitutes a case of corruption of the basest sort, all the more deplorable in that it shows that Germans are misusing their positions as senior political leaders of the *Reich* to enrich themselves personally by exploiting the circumstances arising from the war." (all quoted text excerpted from Heinz Höhne, *The Order of the Death's Head*). The situation worsened for Frank when, on 24.01.1942, his close colleague and subordinate, Lemberg Governor Dr. Karl Lasch, was arrested on corruption charges by the *SS* (see entry for that date, below).

24.01.1942	Arrest of Kurt Lasch, *Gouverneur* of Lemberg and a close colleague of Dr. Frank, for massive corruption. The interrogation of Lasch, during which he spoke at length of Frank's own corrupt practices, was devastating to the *Generalgouverneur*. A secret report by Dr. Eberhard Schöngarth (*BdS*, or commander of the Security Police and *SD,* in the *Generalgouvernement*), in which he discusses his personal interview with the imprisoned Dr. Lasch, appears in the "Notes" section, below.
20.04.1942-00.01.1945	*Führer der SA im Generalgouvernement* (Leader of the *SA* in the General-Government).
Summer 1942	Toured a number of universities throughout the *Reich* and spoke to large audiences in protest against what he saw as Germany's descent into an *SS* and Police state. As a result of these and other statements, Hitler forbade Frank to deliver another public speech in the *Reich*. Under interrogation at Nürnberg, Frank stated:

... during the last years I was absolutely stopped from speaking in Germany. It was an expressed order of Hitler that none of my speeches were supposed to be printed in books or newspapers, and I wasn't even allowed to speak. Secretary Lammers gave me this order in a written form in 1942. ("Interrogation Records Prepared for War Crimes Proceedings at Nuernberg, 1945-1947/OCCPAC Interrogation Transcripts And Related Records: Frank, Hans"; Publication Number M1270, Record Group RG238)

In testimony before the International Military Tribunal, Nürnberg on 08.04.1946, witness Dr. Hans Heinrich Lammers (former *Reichsminister* and chief of the *Reichskanzlei*) described the rapid deterioration in Hitler's relationship with Hans Frank:

> The relationship between the two [Hitler and Dr. Frank] was, at the beginning, I should like to say, good and proper, but not particularly close. At any rate, during the whole time he did not belong to those who could be called the closest advisers of the *Führer.*
>
> ... Frank repeatedly made speeches in public in which he stood up for the constitutional state, for right and law, by attacking the 'Police State' and in which- although not in very strong term- she always took a stand against internment in concentration camps, because such internment was without a legal basis. These speeches made by Frank were frequently the cause of severe disapproval on the part of Hitler, so that in the end the *Führer* instructed me to forbid his making speeches and he was forbidden to publish the printed version of these speeches. Finally, Frank's activity in standing up for the constitutional state resulted in his being removed from his office as the *Reich* Chief of the Legal Division of the Party. (*Trial of the Major War Criminals Before the International Military Tribunal, Nuremberg, 14 November 1945 – 1 October 1946, Volume XI*)

24.08.1942 & 00.11.1942	Submitted his resignation as *Generalgouverneur*; rejected by Hitler each time.
00.00.1942-00.01.1945	*Leiter* of the *Arbeitsbereich Generalgouvernement der NSDAP* (Working Sphere of the Nazi Party for the General Government).
19.06.1943	Report submitted by Frank to Hitler summarizing the previous 3½ years of Germany's occupation of Poland:

> In the course of time, a series of measures or of consequences of the German rule have led to a substantial deterioration of the attitude of the entire Polish people in the *Generalgouvernement*. These measures have affected either individual professions or the entire population and frequently also- often with crushing severity- the fate of individuals. Among these are in particular:

Hotel Vier Jahreszeiten, München: Dr. Frank celebrates the 7th anniversary of the *Akademie des Deutschen Rechts*, 22.- 24.11.1940. (NARA photos)

1- The entirely insufficient nourishment of the population, mainly of the working classes in the cities, whose majority is working for German interests. Until the war of 1939, its food supplies, though not varied, were sufficient and generally secure, due to the agrarian surplus of the former Polish state and in spite of the negligence on the part of their former political leadership.

2- The confiscation of a great part of the Polish estates and the expropriation without compensation and resettlement of Polish peasants from manoeuvre areas and from German settlements.

3- Encroachments and confiscations in the industries, in commerce and trade and in the field of private property.

4- Mass arrests and mass shootings by the German police who applied the system of collective responsibility.

5- The rigorous methods of recruiting workers.

6- The extensive paralyzation of cultural life.

7- The closing of high schools, junior colleges, and universities.

8- The limitation, indeed the complete elimination influence from all spheres of State administration. of Polish

9- Curtailment of the influence of the Catholic Church, limiting its extensive influence -an undoubtedly necessary move- and, in addition, until quite recently, the closing and confiscation of monasteries, schools and charitable institutions. (IMT Nuernberg Document 437-PS)

17.01.1945

Fled from Kraków in his Mercedes (license plate "Ost-4") ahead of a column of vehicles which departed the city at 1325 hours. At 2200 hours, he reached Oppeln in Oberschlesien. On the 18th and 19th of the month, Kraków fell to the Soviet Army. On 24.01.1945, he and his entourage reached Bad Aibling, approximately 35 miles southeast of München (where he met with his mistress, Lilly Groh), and on the following day moved on to Fischhausen-Neuhaus am Schliersee where he established the so-called *Aussenstelle des Generalgouvernements Polen* in the Haus "Café Bergfrieden". For the next three months, he lived at his Schobernhof estate (where his family was housed), in "Haus Bergfrieden", and at the residence of Lilly Groh in Bad Aibling.

Postwar Prosecution

On 04.05.1945, Frank was arrested at the Haus "Café Bergfrieden" by 2nd Lieutenant Walter Stein of the U.S. 7th Army and two American soldiers, accompanied by a German police NCO, and taken to the *Stadtgefängnis* (municipal prison) in Miesbach. In his possession were found numerous works of art by Leonardo da Vinci, Rembrandt, Rubens, and others. On the day of his capture, he slashed both sides of his neck, but survived. He was then held in a POW camp of the U.S. 36th Infantry Division at Berchtesgaden where, on 06.05.1945, he again attempted to take his own life, this time slashing his left wrist. In Summer 1945, Frank was transferred along with other high-ranking *NSDAP* and *Wehrmacht* members to Prisoner of War Camp No. 32 (also known as "Ashcan") at Bad Mondorf, Luxembourg, where he remained until

Official portrait of *Generalgouverneur* Dr. Frank, ca. 1941. (NARA photo)

12.08.1945. He was then moved to Nürnberg Prison pending his trial as a major war criminal by the International Military Tribunal. A particular damning piece of evidence against him was his complete *Dienst-Tagebuch* (service diary), amounting to 11,367 pages, which had been found by the U.S. 7th Army at his home in Neuhaus on 18.05.1945. First taken to the 7th Army document center in Heidelberg, it was, on or about 20.09.1945, sent to the Office of U.S. Chief of Counsel in Nürnberg where it was made available to the prosecution team. On trial from 18.10.1945 to 01.10.1946, with Dr. Alfred Seidl as his defense counsel, he was sentenced to death by hanging. In the courtroom on 18.04.1946, he was asked by Dr. Seidl, "Did you ever participate in the annihilation of Jews?" He responded:

I say "yes", and the reason why I say "yes" is because, having lived through the 5 months of this trial, and particularly after having heard the testimony of the witness Hoess, my conscience does not allow me to throw the responsibility solely on these minor people. I myself have never installed an extermination camp for Jews, or promoted the existence of such camps; but if Adolf Hitler personally has laid that dreadful responsibility on his people, then it is mine too, for we have fought against Jewry for years; and we have indulged in the most horrible utterances-my own diary bears witness against me. Therefore, it is no more than my duty to answer your question in this connection with "yes." A thousand years will pass and still this guilt of Germany will not have been erased. (*Trial of the Major War Criminals Before the International Military Tribunal, Nuremberg, 14 November 1945-1 October 1946, Volume XII*)

The following is the text of Frank's judgment by the International Military Tribunal, Nürnberg:

Frank is indicted under Counts One, Three, and Four. Frank joined the Nazi Party in 1927. He became a member of the *Reichstag* in 1930, the Bavarian State Minister of Justice in March, 1933, and when this position was incorporated into the *Reich* Government in 1934, *Reich* Minister without Portfolio. He was made a *Reichsleiter* of the Nazi Party in charge of Legal Affairs in 1933, and in the same year President of the Academy of German Law. Frank was also given the honorary rank of Obergruppenfuehrer in the *SA*. In 1942 Frank became involved in a temporary dispute with Himmler as to the type of legal system which should be in effect in Germany. During the same year he was dismissed as *Reichsleiter* of the Nazi Party and as President of the Academy of German Law.

Crimes against Peace

The evidence has not satisfied the Tribunal that Frank was sufficiently connected with the common plan to wage aggressive war to allow the Tribunal to convict him on count one.

War Crimes and Crimes against Humanity

Frank was appointed Chief Civil Administration Officer for occupied Polish territory and, on 12th October, 1939, was made Governor General of the occupied Polish territory. On 3rd October, 1939, he described the policy which he intended to put into effect by stating: "Poland shall be treated like a colony, the Poles will become the slaves of the Greater German World Empire." The evidence establishes that this occupation policy was based on the complete destruction of Poland as a national entity, and a ruthless exploitation of its human and economic resources for the German war effort. All opposition was crushed with the utmost harshness. A reign of terror was instituted, backed by summary police courts which ordered such actions as the public shootings of groups of twenty to two hundred Poles, and the widespread shootings of hostages. The concentration camp system was introduced in the General Government by the establishment of the notorious Treblinka and Maydanek camps. As early as 6th February, 1940, Frank gave an indication of the extent of this reign of terror by his cynical comment to a newspaper reporter on von Neurath's poster announcing the execution of the Czech students: "[If] I wished to order that one should hang up posters about every seven Poles shot, there would not be enough forests in Poland with which to make the paper for these posters." On 30th May, 1940, Frank told a police conference that he was taking advantage of the offensive in the West which diverted the attention of the world from Poland to liquidate thousands of Poles who would be likely to resist German domination of Poland, including "the leading representatives of the Polish intelligentsia." Pursuant to these instructions the brutal A. B. action was begun under which the Security Police and *SD* carried out these exterminations which were only partially subjected to the restraints of legal procedure. On 2nd October, 1943, Frank issued a decree under which any non-Germans hindering German construction in the General Government were to be tried by summary courts of the Security Police and *SD* and sentenced to death.

Dr. Frank delivers a speech, ca. 1941.

The economic demands made on the General Government were far in excess of the needs of the army of occupation, and were out of all proportion to the resources of the country. The food raised in Poland was shipped to Germany on such a wide scale that the rations of the population of the occupied territories were reduced to the starvation level, and epidemics were widespread. Some steps were taken to provide for the feeding of the agricultural workers who were used to raise the crops, but the requirements of the rest of the population were disregarded. It is undoubtedly true, as argued by counsel for the defence, that some suffering in the General Government was inevitable as a result of the ravages of war and the economic confusion resulting there from. But the suffering was increased by a planned policy of economic exploitation.

Frank introduced the deportation of slave labourers to Germany in the very early stages of his administration. On 25th January, 1940, he indicated his intention of deporting one million labourers to Germany, suggesting on 10th May, 1940, the use of police raids to meet this quota. On 18th August, 1942, Frank reported that he had already supplied 800,000 workers for the *Reich*, and expected to be able to supply 140,000 more before the end of the year.

The persecution of the Jews was immediately begun in the General Government. The area originally contained from 2,500,000 to 3,500,000 Jews. They were forged into ghettos, subjected to discriminatory laws, deprived of the food necessary to avoid starvation, and finally systematically and brutally exterminated. On 16th December, 1941, Frank told the Cabinet of the Governor General: "We must annihilate the Jews wherever we find them and

SA-Obergruppenführer Dr. Frank with *SA* officers assigned to the
Generalgouvernement, December 1943. (NARA photo)

wherever it is possible, in order to maintain there the structure of *Reich* as a whole." By 25th
January, 1944, Frank estimated that there were only 103,000 Jews left.

At the beginning of his testimony, Frank stated that he had a feeling of "terrible guilt"
for the atrocities committed in the occupied territories. But his defence was largely devoted
to an attempt to prove that he was not in fact responsible; that he ordered only the necessary
pacification measures; that the excesses were due to the activities of the police which were
not under his control; and that he never even knew of the activities of the concentration
camps. It has also been argued that the starvation was due to the aftermath of the war and
policies carried out under the Four Year Plan; that the forced labour programme was under
the direction of Sauckel; and that the extermination of the Jews was by the police and *SS*
under direct orders from Himmler.

It is undoubtedly true that most of the criminal programme charged against Frank was
put into effect through the police, that Frank had jurisdictional difficulties with Himmler
over the control of the police, and that Hitler resolved many of these disputes in favour of
Himmler. It therefore may well be true that some of the crimes committed in the General
Government were committed without the knowledge of Frank, and even occasionally
despite his opposition. It may also be true that some of the criminal policies put into effect
in the General Government did not originate with Frank but were carried out pursuant to
orders from Germany. But it is also true that Frank was a willing and knowing participant in
the use of terrorism in Poland; in the economic exploitation of Poland in a way which led to
the death by starvation of a large number of people; in the deportation to Germany as slave

labourers of over a million Poles; and in a programme involving the murder of at least three million Jews.

Conclusion
The Tribunal finds that Frank is not guilty on Count One but guilty under Counts Three and Four.

Published Works
Books
Reden, gehalten auf der ersten Kundgebung der Berufsgruppe Verwaltungsbeamte im Bund nationalsoz. dt. Juristen am 14. Sept. 1933 in Berlin (speeches, 1933)

Neues deutsches Recht: Rede vor d. Diplomatischen Korps u. d. ausländischen Presse am 30. Januar 1934 bei e. Empfangsabend d. Aussenpolit. Amtes d. N.S.D.A.P. (speech, 1934)

Kampfziele der Wirtschaftsrechtler im B[und] N[ational-] S[ozialistischer] D[eutscher] J[uristen] (1934)

Nationalsozialistisches Handbuch für Recht und Gesetzgebung (Editor, 1935)

Schriften der Akademie für deutsches Recht: Rectsfindung im neuen deutschen Staat (1935; with Dr. jur. Heinz Hildebrandt)

Die Jugend und das Recht (1936; with Gottfried Neeße and Hans Schwarz van Berk),

Grundfragen der deutschen Polizei. Bericht über die konstituierende Sitzung des Ausschußes für Polizeirecht der Akademie für Deutsches Recht am 11, Oktober 1936 (coauthor, 1937)

Deutsches Verwaltungsrecht (1937)

Rechtsgrundlagung des nationalsozialistischen Führerstaates (1938)

Rede des Reichsministers Dr. Hans Frank anlässlich seines Besuches der Ausstellung Recht und Rechtswahrer im Spiegel der Kunst in Leipzig am 4. November 1938 (speech, 1939)

Recht und Kunst (1939)

Recht und Verwaltung (1939)

Die Technik des Staates (1942)

Im Angesicht des Galgens. Deutung Hitlers und seiner Zeit auf Grund eigener Erlebnisse und Erkenntnisse (1946)

Articles:
"Deutsche Juristen" (in *Deutsches Recht 4, Nr. 1*, 01.1934)

"Der deutsche Rechtsstaat Adolf Hitlers" (in *Deutsches Recht IV*, 1934)

"Recht- eine Angelegenheit der Volksgemeinschaft" (in *Völkischer Beobachter* [VB], 02.07.1935)

"Revolution im Strafrecht" (in *VB*, 05.07.1935) "Die richtlinien Grundlagen der heutigen Polizeiarbeit" (in *VB*, 20.01.1937)

"Das *Generalgouvernement* in der Neuordnung Europas" (in *Europäische Revue 18*, May 1942)

Decorations and Awards
04.09.1940	*Kriegsverdienstkreuz I. Klasse ohne Schwerter*
00.00.1940	*Kriegsverdienstkreuz II. Klasse ohne Schwerter*
17.11.1933	*Goldenes Ehrenzeichen der NSDAP*
00.00.1934	*Ehrenzeichen des 9. November 1923 (Blutorden)* (Nr. 532; with effect from 09.11.1933)
00.00.1929	*Nürnberger Parteitagsabzeichen 1929*
00.00.194_	*Dienstauszeichnung der NSDAP in Gold*
00.00.194_	*Dienstauszeichnung der NSDAP in Silber*

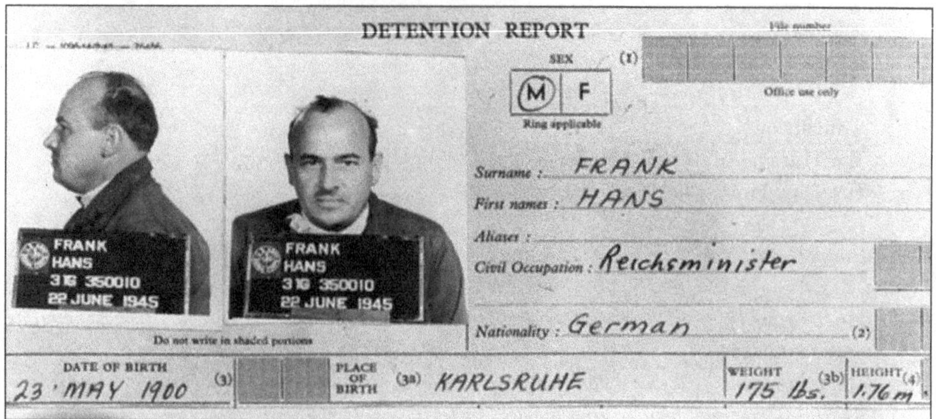

Mug shots of Hans Frank affixed to his U.S. Army detention
report, taken shortly after his arrest in May 1945.

00.00.194_	*Dienstauszeichnung der NSDAP in Bronze*
19.05.1940	*Kreuz von Danzig I. Klasse*
01.09.1940	*Goldene Plakette des Deutschen Auslands-Instituts*
00.01.1934	*Ehrendolch der SA*
21.04.1934	*Ehrenbürgerbrief der Stadt Karlsruhe*
26.09.1936	Order of Saints Maurice and Lazarus, Grand Cross (*Gran Croce dell'Ordine dei Santi Maurizio e Lazzaro*) (Italy)

Notes:

- Son of the attorney Karl Frank (born 22.04.1869 in Edenkoben / Rheinpfalz-Bayern as son of that towns *Bürgermeister*, died 15.01.1945 in Gernlinden bei München) and his wife Magdalena, née Buchmaier (born 16.06.1874 in München as daughter of a businessman and Ratsbürger of München; she was still living as of January 1945). He had two siblings:
  ◊ Karl Frank, born 17.09.1892, a dentist by profession, married with no children; he was wounded in an enemy gas attack and died in München on 29.06.1916.
  ◊ Elisabeth Frank, born 27.09.1903, married on 29.09.1942 to *Regierungsrat* Dr. Schultze of the *Reichspatentamt* in Berlin.
- Religion: Roman Catholic (his mother was Catholic, his father Protestant).
- Married in München on 02.04.1925 to Brigitte Herbst (born 29.12.1895 in Eitorf, died 09.03.1959 in München). She was one of five children born to the marriage of the textile factory director Heinrich Herbst (suicide, 1908, due to indebtedness) and his wife, Maria, née Langer (died 09.03.1959); her siblings were named Elsie, Martha, Otto, and Henry. In the early 1920's she found employment as a stenographer in the bayerischen Landtag, and later worked as a secretary at the University of München where she would often type the doctoral theses of students. In this way she became acquainted with Hans Frank in May 1924. The newly married couple honeymooned in Venice. Three sons and two daughters, all born in München, resulted from the marriage of Hans and Brigitte Frank:
  ◊ Sigrid Leonore (born 17.03.1927)
  ◊ Norman Karl (born 03.06.1928)
  ◊ Brigitte Maria (born 13.01.1935, died 00.00.1981)

Defendant Frank reading in his cell at Nürnberg, 1946. (U.S. Army Signal Corps photo)

◊ Michael Hans (born 15.02.1937, died 00.00.1990)
◊ Nicklas (born 09.03.1939)

• After the war, Brigitte Frank continued to live at Schobernhof where, on 29.05.1947, she was arrested. Two days later, she was admitted, with internee number to the internment camp at Augsburg-Göggingen, from which she was released in 1948.

• While in Poland, Frank had as his mistress the married Lilly Groh (1898-1977). In a lengthy letter of 09.07.1942 to her husband, Dr. Groh, he pleaded with the man to release her from the marriage, and himself sought a divorce from Brigitte in the same year. Although the marriage had already collapsed, however, she did not wish to give up her position as wife to a man of Frank's stature. When Frank traveled to Berlin to request Hitler's permission to divorce, the *Führer* refused to allow it until after the war. He wrote frequently and quite candidly to Lilly, and a number of his letters were offered for auction by Hermann-Historica, Auctioneers, München; the following are excerpts:

13.08.1942: My Lilly! I've said goodbye to it all, the appointments and the ministries. This time, I am proscribed by the all-powerful. The greatest disgrace has been declared (underlined) to me.

20.02.1943: How despairing one becomes, as we now are, when one sees the last miserable formation of the few thousand wounded and sick brought here from Stalingrad, they're completely down and out, unhinged and apathetic shambling.

18.07.1944: The duty of war is now to fulfill our only destiny. That is our lot, as it was the lot of thirty generations who before us endured the same hardships. But because I know from the horrifying reports I hear day after day that our splendid München has sunk to

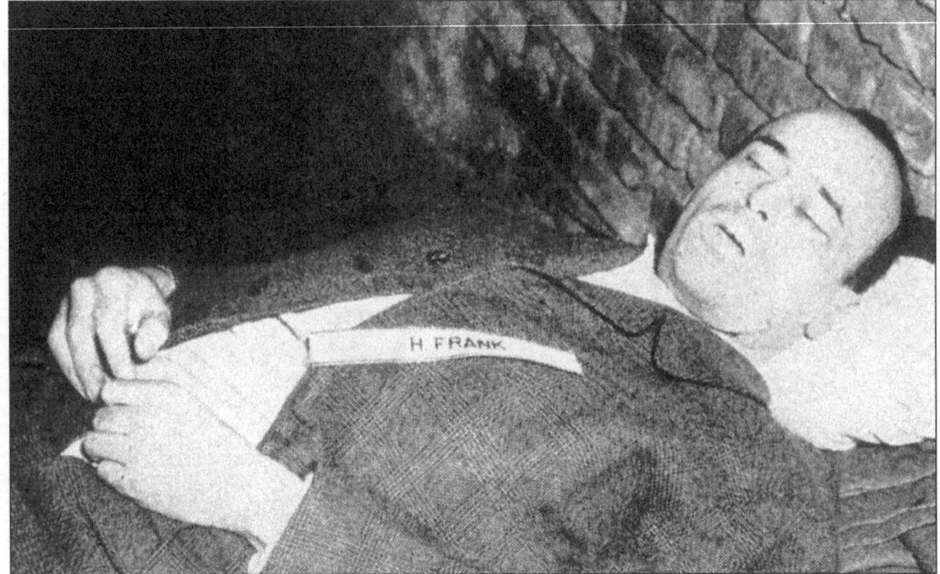

16.10.1946: Convicted war criminal Hans Frank after his execution.

debris and ashes, people and buildings reduced to the depths, I am now homeless.

19.07. 1944: The hardest weeks in the history of our people have begun, the news from the East is very serious ...

28.07.1944: The most serious crisis of our war has arrived, and it must be met with the utmost strength: Germany wants to live!

01.08.1944: Just as I was about to board the train, the very serious news reached me that an uprising has broken out in Warsaw.

22.01.1945: I've just come out of Cracow in the last hour.

28.03.1945: I think now only of the general situation, of my inexcusable role and the ugly guilt that rises in me daily in an agonizing way.

12.12.1945 (Christmas postcard): Dear Lilly! I cordially greet you and your husband. My life is now only the walk to the Pieta! Gentle creature – I've often troubled your golden silence, and through me you've learned much about life's deep, secret anguish. Oh, forget – forgive! Soon like the clouds in front of the peaceful moon up there I will vanish – and you rest and shine again in your beauty, sweet light! Eternal – eternal! Your Hans.

14.10.1946 (two days before his execution): Dear Lilly! If you are holding these lines, then I am dead and part of the final ranks of countless victims of this war marching into eternity. If I hold out to the end, I'll die with a strong inner and outer composure ... . I die after the *Reich*, whom I served, and with this the savage circle now closes, my soul folds its wide wings...and what suffering you had to bear yourself! It is this terrible awareness: What ruins of human fortune we created and have left behind. Live well, live well forever, you, my Lilly, you. Your Hans. (Website of Hermann-Historica, at http://www.hermann-historica.com)

• The following are excerpts from the minutes of a number of *Abteilungsleitersitzungen*" (Conferences of Department Heads in the *Generalgouvernement*) and other meetings presided over by Dr. Frank, and entered in his *Diensttagebuch* (service diary). The source for all of this translated material is *Nazi Conspiracy and Aggression, Volume IV*):

02.12.1939 – First *Abteilungsleitersitzung*: The condemning to death of an archbishop and bishop gives cause to the fundamental observation that a total war against any kind of resistance is being waged in the *Generalgouvernement*. The two bishops have been condemned quite rightly, because arms were found in their possession. If, despite that, they were pardoned to hard labor, then certain other considerations were the cause for that. Reports by the press concerning the shooting of Jews are not desirable because such reports would intimidate the Jews.

08.12.1939 – Second *Abteilungsleitersitzung*: The question of forced labor for the Jews could not be solved satisfactorily from one day to the other. Prerequisite for this would be the card indexing of the male Jews from 14 to 50 years of age. In this it had to be ascertained which trade the Jews had so far carried on, because just in those territories the Jews had had various skilled trades, and it would be a loss if this manpower would not be usefully exploited. To do this, sweeping planning is necessary. For the time being the Jews had to be gathered in columns and had to be employed wherever there was a pressing need. It is the task of the chief of the district to determine these needs.

The police was being reinforced by 4 police battalions. The commitment would be such that each government section was to receive one battalion. Moreover, it was ordered that the police battalions in the *Generalgouvernement* would be relieved and interchanged from time to time with battalions from home. Already before Xmas the relief of 4 battalions would take place. No insecurity would result.... *Gouverneur* Dr. Lasch called attention to the fact that the order of the *Generalgouverneur* regarding the institution of compulsory work provided, that Poles from their 18th year upward were to be conscripted for compulsory labor. A regulation of the age groups from 14 to 18 would also be desirable. One should not neglect the danger that particularly the youth of this age group in high schools could become a source of national resistance.

Generalgouverneur Reichsminister Dr. Frank orders the preparation of a supplementary decree, according to which the compulsory labor was to be extended to the age groups from 14 to 18.

11.01.1940 – Meeting with *Reichshauptamtsleiter* Dr. Fraundorfer: ... Dr. Frauendorfer reports to the *Generalgouverneur* on the possibilities and the extent of the transport of workers into the *Reich*. The German State Railway sends daily 10 transports, each carrying 1000 workers. Beginning with 15 January, these transports, which were stopped temporarily- will be taken up again. In regard to the project of resettlement, it should be avoided, that it be carried out further in the brutal manner which has prevailed up till now, because otherwise it is to be feared that people will not volunteer any more to go into the *Reich*, and also because of the way in which the resettlement is carried out, the possibility of using these people in the *Reich* will be impaired. As an extra allowance for the unemployment pay the *Reich* Ministry for Labor has held out the prospect of an additional sum of 3 million Reichmarks a month.

19.01.1940 – *Abteilungsleitersitzung*: Dr. Frank: My relationship with the Poles is like the relationship between ant and plant louse. When I treat the Poles in a helpful way, so to speak tickle them in a friendly manner, then I do it in the expectation that their work performance redounds to my benefit. This is not a political but a purely tactical- technical problem In cases where in spite of all these measures the performance does not increase, or where the slightest act gives me occasion to step in, I would not even hesitate to take the most draconic action.

08.03.1940 – *Abteilungsleitersitzung*: Frank: One thing is certain. The authority of *Generalgouvernement* as the representative of the *Führer* and the will of the *Reich* in this territory is certainly strong, and I have always emphasized that I would not tolerate the misuse of this authority. I have allowed this to be known anew at every office in Berlin, especially after Herr Field Marshal Göring on 12.2.1940 from Karin Hall had forbidden all Administrative Offices of the *Reich*, including the Police and even the *Wehrmacht*, to interfere in administrative matters of the *Generalgouvernement* ...

There is no authority here in *Generalgouvernement* which is higher as to rank, influence, and authority than that of the *Generalgouverneur*. Even the *Wehrmacht* has no governmental or official functions of any kind in this connection; it has only security functions and general military duties- it has no political power whatsoever. The same applies here to the Police and *SS*. There is here no state within a state, but we are the representatives of the *Führer* and of the *Reich*. In final conclusion, this applies also to the Party, which has here no far- reaching influence, except for the fact that very old members of the National Socialist Party and loyal veterans of the *Führer* take care of the general matters ...

Wherever there is the least attempt by the Poles to start anything, an enormous campaign of destruction directed against the Poles will follow. Then I would not hesitate to set up a regime of terror with all its consequences. I have issued the order to place under arrest for three months several hundred members of such secret organizations, so that nothing can happen in the immediate future. The last word of the *Führer* at my departure was: See to it that there is absolute peace over there, I cannot allow anything to disturb peace in the East. I will see to it.

06.03.1940 – Conference with *Reichshauptamtsleiter* Dr. Frauendorfer: *Reichshauptamtsleiter* Dr. Frauendorfer reports, that 73,000 Polish workers have been sent now into the *Reich*. At present 4,000 men are transferred daily. Further, Dr. Frauendorfer reports on the organization in the plants. It is as simple as possible. On Monday, he visited a few plants in Warsaw, and he could only say that they all were running excellently, this applies also to the machinery. Kitchens were also installed everywhere now and furthermore, the necessary quantities of soap had been provided. 5,000 sets of underwear and linen, 32,000 Kg. flour, 4,000 Kg. of beef and of pork and other merchandise had been distributed. The *Generalgouverneur* ordered that the exact figures should be published in the press. Now, as before, he thinks that the soup given out daily was one of the best solutions, for the problem of the welfare for the working men.

Reichshauptamtsleiter Dr. Frauendorfer remarks that this soup is being served daily now with one pound of bread.

The *Generalgouverneur* will not out of basic considerations give his consent to the proposal that an introductory note should be added to the 2nd decree for the regulation of social insurance, but he thinks it better to point out in a suitable way, that the chief of the labor section has given an explanation about it to the press chief.

In this conference the question of the sending of Polish agricultural workers into the *Reich* was then discussed. Here, Dr. Frauendorfer points out that workmen from the Lublin area had never gone to the *Reich* before, from which naturally certain difficulties arise.

Besides numerous letters from the agricultural workers to their families had already come from the *Reich*, in which the Polish agricultural workers expressed themselves very gratefully concerning their treatment in Germany. The *Generalgouverneur* orders that some of these letters should be published in the press ...

17.03.1940 – Conference with *Generalmajor* Buehrmann, *Stabsleiter* Reichert, *Reichshauptamtsleiter* Dr. Frauendorfer, and two special advisors: Subject of the discussion is the question of the shipment of Polish agricultural workers into the *Reich*.

Generalmajor Buehrmann started by saying that Stabsleiter Reichert has been appointed by the *Reich* Food Ministry, to make sure, that Polish workers would be sent into the *Reich* under all circumstances, and- if necessary- a compulsory service should be enforced upon them. *Generalmajor* Buehrmann would like to recommend, that the *Generalgouverneur*- for the time being does not make any decision for the introduction of a compulsory service or employment of force against the Polish agricultural workers.

Stabsleiter Reichert emphasizes that Berlin insists that the one million agricultural workers be sent into the *Reich*.

Reichshauptamsleiter Dr. Frauendorfer replies to a question of the *Generalgouverneur*, that so far 81,477 Polish agricultural workers were sent into the *Reich*- of which 56,721 were men, and 24,756 women.

Since 12 February, 154 special trains have been sent off, that was the utmost that could have been accomplished in this time. To these figures- just mentioned- are to be added 42,000 workers, who had been already in the *Reich* so that the amount is increased to 130,000.

The *Generalgouverneur* has the opinion, that the 480,000 prisoners of war should also be included in the sum of one million. On account of the railroad and the highway conditions, it is not at present possible to do anything by force, also there were not sufficient police forces there at disposal, to carry such measures out. If any force were exercised, then it would affect workers who were employed as specialized workers for the plants ...

10.05.1940 – Conference held at the *Bergakademie* [Mining Academy]): Then the *Generalgouverneur* deals with the problem of the Compulsory Labor Service of the Poles. Upon the demands from the *Reich* it has now been decreed that compulsion may be exercised in view of the fact that sufficient manpower was not voluntarily available for service inside the German *Reich*. This compulsion means the possibility of arrest of male and female Poles. Because of these measures a certain disquietude had developed which according to individual reports, was spreading very much, and which might produce difficulties everywhere. *Generalfeldmarschall* Göring some time ago pointed out in his long speech the necessity to deport into the *Reich* a million workers. The supply so far was 160,000. However, great difficulties had to be overcome. Therefore it would be advisable to consult the district and town chiefs in the execution of the compulsion, so that one could be sure from the start that this action would be reasonably successful. The arrest of young Poles when leaving church service or the cinema would bring about an ever- increasing nervousness of the Poles. Generally speaking, he had no objections at all if the rubbish, capable of work yet often loitering about, would be snatched from the streets. The best method for this, however, would be the organization of a raid, and it would be absolutely justifiable to stop a Pole in the street and to question him what he was doing, where he was working, etc.

10.07.1940 – *Abteilungsleitersitzung*: The *Generalgouverneur* then addresses the assembly with the following words: My dear Comrades!

There are so few of us here that no one can actually really conceal himself. Everybody has to fear that the spotlight will now and then rest on him.

It is clear that education will perhaps still be necessary here and there; furthermore, it is clear that this open- minded comradeship, this common spirit of close contact finds its counter- part in the unstinted observation of authority in inner office relations. We cannot permit the offices to become 5 o'clock tea rooms. But, of course, our position as Germans here must be such that the lowest of us is still far above the highest Pole in this room.

And another thing was told me by the *Führer* in all seriousness, a few days ago: that the old Japanese proverb:- after the war tighten your helmet strap- should retain its validity. Comrades, never again shall we be a weak *Reich*. The *Wehrmacht* will represent the crown of community education. Just as the *NSDAP* is the crown of social, political and ideological leadership, so the *Wehrmacht* will be the essence of military training, of the proud and immaculate bearing of our people. And you can say: you took part in it as soldiers. I am very happy about this hour of the *Wehrmacht*, for it joins us all together. Some of you left your mothers, your parents at home, others their wives, their brides, their brothers, their children. In all these weeks, they will be thinking of you, saying to themselves: my God, there he sits in Poland where there are so many lice and Jews, perhaps he is hungry and cold, perhaps he is afraid to write. It would not be a bad idea then to send our dear ones back home a picture, and tell them: well now, there are not so many lice and Jews any more, and conditions here in the *Generalgouvernement* have changed and improved somewhat already. Of course, I could not eliminate all lice and Jews in only one year's time. (public amused) But in the course of time, and above all, if you help me, this end will be attained. After all, it is not necessary for us to accomplish everything within a year and right away, for what would otherwise be left for those who follow us to do?

19.12.1940 – Conference in Warsaw: Dr. Frank: In this country the force of a determined leadership must rule. The Pole must feel here that we are not building him a legal state, but that for him there is only one duty, namely, to work and to behave himself. It is clear that this leads sometimes to difficulties, but you must in your own interest see, that all measures are ruthlessly carried out in order to become master of the situation. You can rely on me absolutely in this.

09.09.1941 – *Abteilungsleitersitzung: Obermedizinalrat* Dr. Walbaum expresses his opinion of the health condition of the Polish population. Investigations which were carried out by his department proved that the majority of Poles eat only about 600 calories, whereas the normal requirement for a human being is 2,200 calories. The Polish population was enfeebled to such an extent that it would fall an easy prey to spotted fever. The number of diseased Poles amounted today already to 40%. During the last week alone 1,000 new spotted fever cases have been officially recorded. That represented so far the maximum number. This health situation represented a serious danger for the *Reich* and for the soldiers who were coming into the *Generalgouvernement*. A spreading of the pestilence into the *Reich* is absolutely feasible. The increase in tuberculosis, too, was causing anxiety. If the food rations were to be diminished again, an enormous increase of the number of illnesses could be predicted.

16.12.1941 – *Abteilungsleitersitzung:* Dr. Frank: Severe measures must and will be adopted against Jews leaving the ghettos. Death sentences pending against Jews for this reason must be executed as quickly as possible. This order according to which every Jew found outside the ghetto is to be executed, must be carried out without fail.

[...]

Amtschef in Warsaw, Dr. Hummel: In Warsaw, in spite of the setting up of a third court chamber, we have been able to decree only 45 death sentences, only 8 of which have been carried out since in each individual case, the *Gnadenkommission* [Clemency Commission] in Kraków has to make the final decision. A further 600 sentences were demanded and are under consideration. An effective isolation of the ghetto is not possible by way of the Special Court Procedure. The procedure to be followed up to liquidation takes too much time. It is burdened with too many formalities and must be simplified.

16.12.1941 – Speech of *Generalgouverneur* Dr. Frank in the Government Building at Kraków: ... As far as the Jews are concerned, I want to tell you quite frankly, that they must be done away with in one way or another. The *Führer* said once: should united Jewry again succeed in provoking a world war, the blood of not only the nations, which have been forced into the war by them, will be shed, but the Jew will have found his end in Europe. I know, that many of the measures carried out against the Jews in the *Reich*, at present, are being criticized. It is being tried intentionally, as is obvious from the reports on the morale, to talk about cruelty, harshness, etc. Before I continue, I want to beg you to agree with me on the following formula: We will principally have pity on the German people only, and nobody else in the whole world. The others, too had no pity on us. As an old National- Socialist, I must say: This war would only be a partial success, if the whole lot of Jewry would survive it, while we would have shed our best blood in order to save Europe. My attitude towards the Jews will, therefore, be based only on the expectation that they must disappear. They must be done away with. I have entered negotiations to have them deported to the East. A great discussion concerning that question will take place in Berlin in January, to which I am going to delegate the *Staatssekretär* Dr. Bühler. That discussion is to take place in the Reichssicherheitshauptamt with *SS- Obergruppenführer* Heydrich. A great Jewish migration will begin, in any case.

But what should be done with the Jews? Do you think they will be settled down in the "*Ostland*", in villages [*Siedlungdörer*]? This is what we were told in Berlin: Why all this bother? We can do nothing with them either in the "*Ostland*" nor in the "*Reichkommissariat*;". So, liquidate them yourself.

Gentlemen, I must ask you to rid yourself of all feeling of pity. We must annihilate the Jews, wherever we find them and wherever it is possible, in order to maintain there the structure of the *Reich* as a whole. This will, naturally, be achieved by other methods, than those pointed out by Bureau Chief Dr. Hummel. Nor can the judges of the Special Courts be made responsible for it, because of the limitations of the frame work of the legal procedure. Such outdated views cannot be applied to such gigantic and unique events. We must find at any rate, a way which leads to the goal, and my thoughts are working in that direction.

The Jews represent for us also extraordinarily malignant gluttons. We have now approximately 2,500,000 of them in the *Generalgouvernement*, perhaps with the Jewish mixtures and everything that goes with it, 3,500,000 Jews. We cannot shoot or poison those 3,500,000 Jews, but we shall nevertheless be able to take measures, which will lead, somehow, to their annihilation, and this in connection with the gigantic measures to be determined in discussions from the *Reich*. The *Generalgouvernement* must become free of Jews, the same as the *Reich*. Where and how this is to be achieved is a matter for the offices which we must appoint and create here. Their activities will be brought to your attention in due course.

18.03.1942 – Conference of the Distriktstandortführer der *NSDAP* in Kraków: Dr. Frank: As you know, I am a fanatic as to unity in administration It is therefore clear that the

Higher *SS* and Police Leader is subordinated to me, that the Police is a component of the Government, that the *SS* and Police Leader in the district is subordinated to the Governor, and that the Kreis chief has the authority of command over the gendarmerie in his Kreis. This the *Reichsführer-SS* has recognized; in the written agreement all these points are mentioned word for word and signed. It is also self- evident that we cannot set up a closed shop here which can be treated in the traditional manner of small states. It would, for instance, be ridiculous if we would build up here a security policy of our own against our Poles in the country, while knowing that the Polacks in West Prussia, in Posen, in Wartheland and in Silesia have one and the same movement of resistance. The Reichsführer- *SS* and Chief of the German Police thus must be able to carry out with the aid of his agencies his police measures concerning the interests of the *Reich* as a whole. This, however, will be done in such a way that the measures to be adopted will first be submitted to me and carried out only when I give my consent. In the *Generalgouvernement*, the Police is the *Wehrmacht*. As a result of this, the Leader of this Police system will be called by me into the Government of the *Generalgouvernement*; he is subordinate to me, or to my deputy, as a *Staatssekretär* for the Security System.

[Pages 195- 196]

Incidentally, the struggle for the achievement of our aims will be pursued cold bloodedly. You see how the state agencies work. You see that we do not hesitate before anything, and stand whole dozens of people up against the wall. This is necessary because here simple consideration says that it cannot be our task at this period when the best German blood is being sacrificed to show regard for the blood of another race. For out of this one of the greatest dangers may arise. One already hears today in Germany that prisoners- of-war for instance with us in Bavaria or in Thuringia are administering large estates entirely independently, while all the men in a village fit for service are at the front. If this state of affairs continues then a gradual retrogression of Germanism will show itself. One should not underestimate this danger. Therefore, everything revealing itself as a Polish power of leadership must be destroyed again and again with ruthless energy. This does not have to be shouted abroad, it will happen silently.

05.08.1942 – Dr. Frank's statements at an official meeting of *NSDAP* leaders in Kraków: "Dr. Frank: The situation in regard to Poland is unique insofar as on the one hand- I speak quite openly- we must expand Germanism in such a manner that the area of the *Generalgouvernement* becomes pure German colonized land at some decades to come; and, on the other hand, under the present war conditions we have to allow foreign racial groups to perform here the work which must be carried out in the service of greater Germany.

[...]

What a dirty people made up of Jews swaggered around here before 1939! And where are the Jews today? You scarcely see them. If you see them then they are working.

18.08.1942 – Discussion with *Gauleiter* Sauckel, Kraków: Dr. Frank: I am pleased to report to you officially, Party Comrade Sauckel, that we have up to now supplied 800,000 workers for the *Reich*.

[...]

Dr. Frank: Recently, they have requested us to supply them with a further 140,000. I have pleasure in informing you officially that in accordance with our agreement of yesterday 60% of the newly requested workers will be supplied to the *Reich* by the end of October, and the balance of 40% by the end of the year.

[…]

Dr. Frank: Beyond the present figure of 140,000 you can, however, next year reckon upon a higher number of workers from the *Generalgouvernement*. For we shall employ the Police to conscript them.

24.08.1942 – Cabinet session in Kraków (Subject: "A new Plan for seizure and for food of the *Generalgouvernement*"): Gentlemen, I have called you together today with special speed and emphasis in order to acquaint you with a measure which is unusually important and decisive for all the work in the *Generalgouvernement* in the year to come. What I tell you, I tell you in strictest confidence. I call your attention to the fact that every word which leaks out of this meeting, unofficially, might mean a tremendous damage to our country.

A few days ago a meeting with the *Reichsmarschall* took place in Berlin. The *Reichsmarschall* had the reports concerning the almost catastrophic developments in the food situation in Germany. According to all confidential reports of the police, as well as of the *Gauleiter*, which, as he expressed himself, also confirmed by his own experiences, the situation is as follows: unless a considerable improvement in the food situation in Germany can be achieved in a short time, serious consequences as to the health of the people, especially the German working people, would result. In hundreds of thousands of sick cases, one can already see the tragic consequences not only of this food shortage but also a deterioration of foodstuffs which endangers health …

A serious situation, therefore, has arisen since Germany not only has to feed herself but also a large proportion of other European people. We must also take care that in the months to come and during the coming winter sufficient food will be distributed to the German people that they will be able to withstand the great nervous strain of the coming months in every case.

Under these circumstances you probably will not be surprised that the saying now has become true: Before the German people are to experience starvation, the occupied territories and their people shall be exposed to starvation. In this movement, therefore, we here in the *Generalgouvernement* must also have the iron determination to help the Great German people, our fatherland.

Germany had almost sufficient rye to tide them over until the new harvest, but not sufficient wheat. In large parts of Germany, therefore, no more wheat can be distributed in the near future. We therefore must aid the fatherland until the beginning of the new wheat harvest.

The *Generalgouvernement* therefore must do the following: The *Generalgouvernement* has taken on the obligation to send 500,000 tons bread grains to the Fatherland in addition to the foodstuffs already being delivered for the relief of Germany or consumed here by troops of the *Wehrmacht*, Police or *SS*. If you compare this with our contributions of last year you can see that this means a six- fold increase over that of last year's contribution of the *Generalgouvernement*.

The new demand will be fulfilled exclusively at the expense of the foreign population. It must be done cold- bloodedly and without pity; for this contribution of the *Generalgouvernement* is still more important this year since the occupied Eastern territories- Ukraine and Ostland- will not yet be able to make an important contribution toward the relief of Germany's food problem. Even if a million tons of bread grains could be delivered from Ostland and Ukraine, it would in the face of Germany's food situation be only a 'drop in the bucket'.

For this reason I wanted to acquaint you, Gentlemen, here in this governmental session with the decisions which I have made known today to Party member Naumann. You will essentially find an additional increase of the quota of foodstuffs to be shipped to Germany and new regulations for the feeding of the population; especially of the Jews and of the Polish population, whereby, if possible, the provisioning of the working people, especially of those working for German interests, shall be maintained.

The step which we are taking together today, is one of the most decisive ones, because it will surely have certain consequences as to the internal order of this country in January or February of next year. These consequences have to be accepted, because before the German people be starved, others, as a matter of course, must undergo the same.

I first of all give the word to Party member Naumann, who will give you a general report about this problem.

[Naumann, President of the Main Department for Food and Agriculture, then spoke at length regarding the food crisis in the *Generalgouvernement*, with the following comments on the feeding of Jews and Poles]:

... The feeding of a Jewish population, estimated heretofore at 1.5 million, drops off to an estimated total of 300,000 Jews, who still work for German interests as craftsmen or otherwise. For these the Jewish rations, including certain special allotments which have proved necessary for the maintenance of working capacity, will be retained. The other Jews, a total of 1.2 million, will no longer be provided with foodstuffs.

Non-German normal consumers will receive, from 1 January 1943 to 1 March 1943, instead of 4.2 kg bread per month, 2.8 kg; from 1 March 1943 to 30 July 1943 the total bread ration for these non- German normal consumers will be cancelled ...

[Frank concluded the meeting:] ... In whatever difficulties you observe some place here, in the form of the sicknesses of your workers, the breakdown of your associations, etc., you must always think of the fact that it is still much better when a Pole breaks down than that a German succumb. That we sentence 1.2 million Jews to die of hunger should be noted only marginally. It is a matter, of course, that should the Jews not starve to death it would, we hope, result in a speeding up of anti- Jewish measures ...

Not unimportant manpower has been taken from us in form of our old proven Jewish communities. It is clear that the working program is made difficult when in the middle of this program, during the war, the order for complete annihilation of the Jews is given. The responsibility for this cannot be placed upon the government of the *Generalgouvernement*. The directive for the annihilation of the Jews comes from higher quarters. We have to be content with the consequences and can only report that the Jew has caused tremendous difficulties with regard to the work-program. I was able to prove, the other day, to Staatssekretaer Ganzenmüller, who was complaining that a big building project in the *Generalgouvernement* came to a halt, that this would not have happened if the many thousands of Jews working at it had not been deported. Now the order is given that the Jews will have to be removed from the armament projects. I hope that this order, if not already cancelled, will soon be cancelled, for then the situation will be still worse.

14.12.1942 – Official Meeting of Political Leaders of the *NSDAP Arbeitsbereich* in the *Generalgouvernement*: Dr. Frank: You know that we have delivered over 940,000 Polish workers to the *Reich*. Thereby the *Generalgouvernement* absolutely and relatively stands at the head of all the European countries. This achievement is enormous: it has also been recognized as such by *Gauleiter* Sauckel.

[...]

Dr. Frank: I will endeavor to get out of the reservoir of this territory everything that is yet to be got out of it. When you consider that it was possible for me to deliver to the *Reich* 600,000 tons of bread grain, and in addition 180,000 tons to the *Wehrmacht* stationed here; further an abundance amounting to many thousands of tons of other commodities such as seed, fats, vegetables, besides the delivery to the *Reich* of 300 million eggs, etc.- you can estimate the significance this territory possesses for the *Reich*. In order to make clear to you the significance of the consignment from the *Generalgouvernement* of 600,000 tons of bread grain, you are referred to the fact that the *Generalgouvernement* by this achievement alone covers the raising of the bread ration in the Greater German *Reich* by two- thirds during the present rationing period. This enormous achievement can rightfully be claimed by us.

25.01.1943 – Meeting in Warsaw: *Staatssekretär* Krüger: When we settled about the first 4,000 in Kreis Zamosc shortly before Christmas I had an opportunity to speak to these people.

(*Staatssekretär* Krüger) It is understandable that in resettling this area we did not make friends of the Poles.

[...]

(*Staatssekretär* Krüger) In colonizing this territory with racial Germans, we are forced to chase out the Poles.

(*Staatssekretär* Krüger) We are removing those who constitute a burden in this new colonization territory. Actually, they are the asocial and inferior elements. They are being deported, first brought to a concentration camp, and then sent as labor to the *Reich*. From a Polish propaganda standpoint this entire first action has had an unfavorable effect. For the Poles say: After the Jews have been destroyed then they will employ the same methods to get the Poles out of this territory and liquidate them just like the Jews.

[...]

(*Staatssekretär* Krüger) As I have mentioned a great deal of unrest in Polish territory has resulted because of this resettlement.

Dr. Frank: We will discuss each individual case of resettlement in the future exactly in the same manner as in the case of Zamosc, so that you will, Mr. *Staatssekretär*, appear before me and render a report.

[...]

Dr. Frank: Gentlemen, be assured that this composite structure of *Generalgouvernement*, on which all who are gathered around this table have worked so splendidly, really has the power to endure over this period. The great task which is given us will grow more difficult. No one will help us; we are fully and entirely dependent on ourselves. The *Führer* can only help us as a kind of administrative island or administrative pill- box. We must defend ourselves from all sides. To all criticism of methods which we have heard- you know my basic principle, I don't need to say it in this circle- I would like to stress one thing: we must not be squeamish when we learn that a total of 17,000 people have been shot. These persons who were shot were nothing more than war victims. If we compare this number against the irreplaceable blood sacrifices which the German people uninterruptedly day by day and every hour are making, then it weighs as nothing in the balance. We are now duty bound to hold together. Each must bring with him understanding for the other, he must be convinced that he is doing his best. The main thing is that we do not allow any personal slackness to arise. We must remember that we, who are gathered together here, figure on Mr. Roosevelt's list of war criminals. I have the honor of being Number One. We have, so to speak, become accomplices in the world historic sense. Just because of this we must hold together, and be in agreement with one another, and it would be ridiculous if we were to let ourselves get involved in any squabbles over methods.

14.01.1944 – Speech to Members of District *Standortführung* to Political Leaders, Conference Room of District *Standortführung* Krakau: Frank: Once we have won the war, then, for all I care, mincemeat can be made of the Poles and the Ukrainians and all the others who run around here- it doesn't matter what happens.

15.01.1944 – Meeting of the Political Leaders of the *NSDAP* in *Arbeitsbereich Generalgouvernement* Area, Kraków: Frank: I have not been hesitant in declaring that when a German is shot, up to 100 Poles shall be shot too.

- The following is Dr. Eberhard Schöngarth's secret report detailing his interrogation of *Gouverneur* Dr. Lasch (see entry of 24.01.1942 in the "Career" section, above:

On 26.4.1942 I personally interviewed Dr. Lasch at his own request. On this occasion Dr. Lasch learnt for the first time, according to him that the *Reichsführer-SS* had duly given him the opportunity of proving himself in the front line. He was furious that *Staatssekretär* Bühler had not only said nothing of this offer of the *RFSS* but, in reply to his question as to whether he could not make up his lapses by employment at the front, had said that it was now too late. After his memory had been further assisted Dr. Lasch declared he would drop his reserve with regard to his former friends, and speak the truth about them frankly. He asked for my support in approaching the *RFSS*, as he proposed to send a plea to the *Reichsführer-SS* requesting him to give him once more the opportunity of proving himself at the front. He said he had recognized the true feelings of Secretary of State Bühler and the clique at the Castle and no longer had any intention of sparing them. He recognized that they had dropped him and tried to put all the blame on him. He could only state in his defense that his lapses were not the only ones, but that nearly all leaders in the administration of the *Generalgouvernement* had enriched themselves in this way and even more extensively.

I requested Dr. Lasch to write down everything he knew himself. Up to now Dr. Lasch has written as follows:

'Dr. Frank's attitude to the Party, the *SS* and the State is, to a very great extent, influenced by his Secretary of State Bühler who is, again, agreeable to Dr. Frank simply because he is entirely lacking in any creative power. Dr. Bühler owes his position to the fact that he can do nothing, has no initiative and in emergencies is presented with a fait accompli. The assertion is continually being made that there is a struggle between Bormann and Dr. Frank. It can be assumed that *Reichsleiter* Bormann knows nothing about it. Dr. Frank is a pronounced adversary of Bormann.

Between the Reichsführer and Dr. Frank and also between *SS-Obergruppenführer* Heydrich and Dr. Frank lies, in the opinion of Frank/Bühler, a whole world of injustice, police authority, oppression of the people, concentration camps, cruelty! Dr. Frank preaches a State controlled by a legal code, and by that he means a Legal State. *RFSS* Himmler desires a police state! By means of speeches and writings, propaganda is said to be made for a Ministry of Justice which excludes police influence which has become so strong. According to Dr. Frank's interpretation, its 'justice' is opposed to the *RFSS* Himmler's 'injustice.' The dangerous influence of the *SS* in the judicial sphere was considered proven by the example of the Prague Trials. An *SS Führer* is said to have acted as Public Prosecutor at these. (Trial of the former Prime Minister of the Protectorate). The dangerous influence of the *SS* in the State is considered proven by the fact that *SS* rank was conferred on all leading men in the *Reich-* Ribbentrop, Lammers, Bormann, Seyss-Inquart, etc. Herein lies the danger of the leading position of the *SS* in the State. Dr. Frank wishes, after the war, when the *Führer* is no longer in such great need of the *RFSS*, to inaugurate the greatest fight for justice, with the assistance of the *Wehrmacht* and the SA! A fight against the *SS* in the *Generalgouvernement* will take this course! Krüger and Schöngarth must be dismissed. Perhaps they know too much. Bühler's ambition plays a great part here. Moreover these efforts will be vigorously furthered by Dr. Keith, the 'Golen' Gutbred (*SS-Unterscharführer*) and Wächter. Since Wächter has been in Galicia he may have changed his attitude. He always wanted *Obergruppenführer* Krüger's post and tried to get it. Wächter is *SS-Brigadeführer* but was always less friendly disposed towards the *SS* than all other Governors. In token of his power, Dr. Frank, in October charged Bühler and Boepple with the duty of assembling the Special Service and of settling the quarrel between Zörner and Globocnik in Lublin by banning the 'Globus.' Later on exactly similar things were to be done in Cracow. Bühler and Boepple were to be in command and were charged with making strategic plans. After much discussion, it was possible to impress upon Bühler the craziness of such action. I suggested telephoning Dr. Wendler, inviting him to this conference. Bühler saw the advantage of this. Wendler came to Krakow after my arrest. He is said to be completely under Bühler's thumb now! At first against Krüger-Schöngarth and then against others. In the meantime Bühler quickly dismissed his *SS* Adjutant, he suddenly became a private-such things take place at the Castle.'

In this connection a statement by Minister Ruemelin (retired) regarding *SS-Obergruppenführer* Krüger, made on the occasion of a visit on 22.4.42, is interesting. Ruemelin happened to say that people had been struck by the fact that, on the occasion of the establishment of the *SA* in the *Generalgouvernement* on 20.4.42 and the taking over of the leadership of the *SA* by the *Generalgouverneur* as *SA-Obergruppenführer*, no leader of the *SS* was represented. Ruemelin declared that there was already talk of a Röhm situation. Since 3 Police Officers had replaced the 3 *Wehrmacht* officers who retired from the Special Service.

The only person who is welcome is the man who brings something. Woe to him who possesses something more beautiful or who owns more. When Dr. Ley furnished his estate, there was tremendous gossip, also regarding Dr. Goebbels, Funk, etc.

He was no example to us, for his day consists of running round from castle to castle in a magnificent carriage with guards of honor, books, music, plays, and banquets. The best value is to be got out of every reception. There is nothing natural, no simplicity, all is pose, playacting and serves to satisfy his intoxication, brought about by ambition and lust for power and, at the same time, his likeness to Mussolini, of which he is convinced by his flatterers, is interpreted as fate and destiny.

Astrology, palmistry, telling fortunes by cards, lead melting, all methods of looking into the future are employed. Frau Frank has often gone to two women in one day, in order to see if what has been said tallies. Fortune smiles on him. Jupiter is his lucky star. That being so, nothing can happen to him. Others will fall, he will always remain on top. Frau Frank has often said in large circles that she intends to die a Minister's wife. It is a Prince's Court in which we, his closest colleagues, live as strangers. It is cold magnificence in which, as at the Castle, one is smothered in the Gothic style ...

His attitude to the other *Reich* Ministers is determined by his exaggerated ambition, by his overweening opinion of himself and his autocracy. He waves everybody aside, some more than others. Thus he often says that Dr. Lammers owes his position to him. He would like to have been Foreign Minister in succession to Baron von Neurath, after having realized for some time that there was no prospect of becoming Minister of Justice during Dr. Guertner's life-time and so he tried hard to obtain a post as Ambassador. He wished to be Ambassador in Rome.

His spiritual opposition to Himmler and Heydrich is deep-rooted. Before he became *Generalgouverneur* and as only Reichsminister, he was always criticizing the fact that the *Führer* did not summon the *Reich* Cabinet. He even did this publicly in the sessions of the Council of the Academy for German Law and referred to it as being the kind of Cabinet in which everyone could speak openly. He also did this as *Generalgouverneur*. His Cabinet meetings do not consist of deliberations but of lectures, with much affectation, about fictitious conditions, and conclude with a long monologue by Frank, praising the achievements of the *Generalgouverneur* in all strains... Particularly interesting are his efforts to create, after the war, a Reichsrechtsministerium [*Reich* Law Ministry] which is to house the Ministries of the Interior, Justice, Church and Education, and Public Worship, as well as a number of subsidiary authorities. This is to become the principal Command Post for all transactions of domestic policy in the *Reich*, which are all to be based on 'justice.' For justice is the most valuable possession of the nation. True, everything is a conglomeration of legislation, jurisdiction and administration, but it must all be given a fine name, It is his desire to amalgamate everything-over and above judicial law-under the title of justice, everything which can be understood by the moral conception of justice. The saying: "Everything is right which is of use to the nation" is cleverly utilized here. Accordingly, everything was to be under the control of the Reichsrechtsführer [*Reich* Law Leader]. In his modesty he does not call himself Rechtswahrerführer [Leader of the Legal Profession] but Reichsrechtsführer. Legislation, that will of the *Führer* or the people which is made standard, would have to go through him. In this sphere too, he hopes to be given special assignments by the *Führer* after the war. Frau Frank often says: "You see, if my husband had the opportunity of having long talks with the *Führer*, it would be a good thing for the *Führer*, for there is no one with whom he could speak so well as with my husband."

Every little token given him by the *Führer* is immediately magnified a thousand fold and utilized for his ambitious plans. If he ever has the opportunity to talk for example to Schaub or Prof. Hoffmann or anyone else in the *Führer*'s entourage, he repeats what he has been told as if it originated from the *Führer* himself.

His conversations with foreign statesmen are also on this plane, and have often made our blood boil. If, in his capacity as President of the Academy for German Law, he travels abroad and is received by the Duce, Count Ciano, King Boris of Bulgaria, or the Bulgarian Prime Minister, he always acted as if, in his capacity as Reichsminister, he had received instructions from the *Führer*; he discussed foreign policy but was always able to turn the conversation when it became dangerous. He did this also in Hungary and in France before the war, where he met the former Premier, Chautemps, in Paris.

After the shooting of Röhm he violently criticized Dr. Gürtner's statement in the *Reich* Cabinet, according to which, the execution of the sentences passed was legal ... Lasch gives the following details regarding the personal relations between Dr. Frank and his wife:

Through Dr. Heuber, Frau Frank received regular monthly donations during 1935/36.

For several years she received RM. 750.-monthly from the Academy for German Law, but the Treasurers of the Academy for German Law put an end to that. Furthermore Frau Frank had, for over a year, a free ticket from Berlin to Munich which had to be paid for by the Academy for German Law ...

When traveling abroad she bought principally in Ghettos and from Jews. First of all clothes-baskets and then their contents. The foreign exchange was taken from the combined quota of her co-travelers. The articles were brought duty free across the frontier under diplomatic pass.

Otto Herbst (brother of Frau Frank) was called up at the beginning of the Polish campaign. Applications for his release were immediately made, although he was only a clerk in a Veterinary Company which collected horses far behind the front line. He was reserved for the *Generalgouvernement*, but so far has not done a day's service in the *Generalgouvernement*. He keeps his post. In the meantime he has become the director of the Academy for German Law. He had only been reserved for this post. Richard Schneider-Edenkoben, cousin of Frau Frank, was called up as N.C.O. and immediately requested a reserved job. He was given that of film expert for the *Generalgouvernement*. He once drew up a short memorandum (four pages), then disappeared and was never seen again.

Above all, Frau Frank's relations, as well as her women friends, are given posts everywhere.

Pictures, as well as great quantities of food of all kinds, were despatched regularly to the *Reich* in the *Generalgouverneur*'s Pullman car. The whole family often traveled in the coach without tickets. The guard often had trouble in this way. At the frontier it is known as the smuggling coach!

Frau Frank received a large diamond ring of 5 carats from the Governor, Dr. Fischer. It is said to have cost RM. 5,000 (Note: it is rumored in Warsaw that Dr. Fischer has so far tried in vain to get back from Frau Frank the money said to have been spent on this ring).

Dr. Fischer also gave the *Generalgouverneur* a Rembrandt which was brought by Landgerichtsrat Dorn to Schliersee where it was hung. It is, however, no longer there. Dorn is said to have objected and the picture was then returned.

As soon as Dr. Lasch had written down, without prejudice, everything he knows, I intend to examine him exhaustively on his statements regarding some points. I have had to allow a few days' grace as, at present, Dr. Lasch is not feeling well. I wish to avoid any objection by Dr. Lasch that when in ill health he revealed facts, the effects of which he could not perceive in his condition. After Lasch has been interrogated on these lines I will submit a final copy of Lasch's statements.

2. Submitted to *SS-Obergruppenführer* Krüger requesting him to note contents.

[Signed] Schöngarth

SS-Oberführer (Translation of Document 3815-PS, in *Nazi Conspiracy and Aggression, Volume VI*)

Sources

Domarus, Dr. Max (ed): *Hitler. Speeches and Proclamations, 1932-1945: The Years 1932-1934*. Bolchazy-Carducci Publishers, 1990.

Gilbert, G.[ustav] M.[ahler]: *Nuremberg Diary*. Farrar, Straus, 1947.

Lilla, Joachim; Döring, Martin; & Schulz Andreas: *Statisten in Uniform. Die Mitglieder des Reichstags 1933-1945*. Droste Verlag, 2004.

National Archives & Records Administration, College Park, Maryland: "Interrogation Records Prepared for War Crimes Proceedings at Nuernberg, 1945-1947/OCCPAC Interrogation Transcripts and Related Records: Buehler, Josef and Frank, Hans; Publication Number M1270, Record Group RG238.

Office of United States Chief Counsel for Prosecution of Axis Criminality: *Nazi Conspiracy and Aggression (11 volumes). U.S. Government Printing Office, District of Columbia, 1946.*

Schenk, Dieter: – *Hans Frank. Hitlers Kronjurist und Generalgouverneur.* S. Fischer, 2006.

- *Krakauer Burg: Die Machtzentrale des Generalgouverneurs Hans Frank 1939-1945.* Ch. Links Verlag, 2010.

Schirach, Baldur von: *Die Pioniere des Dritten Reiches.* Zentralstelle fur der deutschen Freiheitskampf, 1933.

Steiner, John Michael: *Power Politics and Social Change in National Socialist Germany: A Process of Escalation into Mass Destruction.* Walter de Gruyter, 1976.

Weinreich, Max: *Hitler's Professors: The Part of Scholarship in Germany's Crimes against the Jewish People.* Yale University Press, 1999.

Johann ("Hans") Baptist Fuchs
SA-Obergruppenführer

Born:	08.06.1877 in Regensburg / Regierungsbezirk Oberpfalz / Bayern.
Died:	18.11.1938 in München / Regierungsbezirk Oberbayern / Bayern.

NSDAP-Nr.:   411 570 (Joined 01.02.1931)

Promotions

11.01.1897	*Fahnenjunker*
10.11.1897	*Gefreiter*
01.12.1897	*Unteroffizier*
25.01.1898	*Portepeéfähnrich*
01.03.1899	*Leutnant*
07.03.1910	*Oberleutnant*
10.09.1914	*Hauptmann*
17.11.1919 (?)	*Major a. D.*
01.10.1922	*Polizei-Major (Landespolizei)*
24.02.1931	*SA-Gruppenführer*
00.00.1933	*Polizeioberst a. D.*
09.11.1933	*SA-Obergruppenführer*
01.04.1933	*Polizeioberstleutnant*
01.04.1933	*Polizeioberst*

Career

00.00.1884-00.00.1888	Attended *Volksschule*.
00.00.1888-00.00.1897	Attended a humanistic *Gymnasium* (passed his *Abitur*).
11.07.1897-00.00.1899	Entered military service as a *Fahnenjunker*, assigned to the *17. Kgl. Bayerisches Feldartillerie-Regiment* (Germersheim).
00.00.1898-00.00.1899	Attended the *Kriegsschule* in München.
18.03.1899	Commissioned and returned to service as an officer of *17. Kgl. Bayerisches Feldartillerie-Regiment*.
[00.00.1906]	Adjutant of *II. Bataillon / 17. Kgl. Bayerisches Feldartillerie-Regiment*.
02.08.1914-18.08.1914	*1. Adjutant* of *Generalkommando III. Armee-Korps / königlich-Bayerisches Festung* (Royal Bavarian Fortress) Germersheim / Pfalz.
19.08.1914-03.09.1916	*1. Brigadeadjutant* of *13. Kgl. Bayerisches Landwehr-Infanterie-Brigade*.
03.09.1916-28.09.1916	*1. Adjutant* of the *Infanterie-Division* of *VIII. Preußische Armee-Korps*.
28.09.1916-28.11.1917	*Brigadeadjutant* of *17. Kgl. Bayerisches Reserve-Infanterie-Brigade*.
18.06.1917-23.06.1917	Participated in the *32. Lehrgang* at the *Heeresgasschule Berlin*.
23.10.1917-19.11.1917	Participated in the *4. Generalstabskurs* (4th General Staff training course) in Sedan.
20.11.1917-28.11.1917	Participated in the *22. Kurs für höhere Truppenführer*.
29.11.1917-20.02.1918	*Führer* of *III. Bataillon / Kgl. Bayerisches Reserve-Infanterie-Regiment 23*.

SA-Gruppenführer Johann Fuchs in 1933. (*Reichstagshandbuch* photo, 1933)

21.02.1918	Assigned as a *Generalstabsanwärter* (general staff officer candidate) to the staff of *8. Kgl. Bayerisches Reserve-Division.*
22.02.1918-23.02.1918	Attached to the staff of *8. Kgl. Bayerisches Reserve-Division.*
23.02.1918-02.03.1918	Attached to *7.Batterie / Kgl. Bayerisches Reserve-Feldartillerie-Regiment 9.*
02.03.1918-25.03.1918	Returned to staff, *8. Kgl. Bayerisches Reserve-Division.*
25.03.1918-17.04.1918	Attached to the staff of *Kgl. Bayerisches Reserve-Feldartillerie-Regiment 9.*
18.04.1918-29.04.1918	On leave.
02.05.1918-17.05.1918	Returned to service with the staff of *Kgl. Bayerisches Reserve-Feldartillerie-Regiment 9.*
17.05.1918-15.06.1918(?)	Assigned as *2. Generalstabsoffizier (Ib)* to the staff of *15. Kgl. Bayerisches Infanterie-Division.*
30.05.1918	Fell ill with a severe heart condition.
15.06.1918-30.06.1918	Attached to *Maasgruppe Ost* (Meuse [River] Group East), under *General der Kavallerie* von Garnier.
22.11.1918	Granted 4 weeks' leave in Nürnberg and München to regain his health.
10.12.1918	Admitted to the *Reservelazarett* at the *Alte Frauenklinik* in München due to heart and intestinal complaints.
03.01.1919-10.04.1919	Assigned to *II. Bataillon / 17.Infanterie-Regiment* (as *Kommandeur* of that unit from 05.02.1919).
08.01.1919-04.02.1919	Underwent a cure at the *Pflegestätte* [nursing facility] *Hainzenbad* in Partenkirchen.
05.02.1919-26.05.1919	Service with *Freikorps Bamberg.*
29.03.1919-27.04.1919	On leave in München.

SA-Obergruppenführer Fuchs, ca. 1934.

11.04.1919-19.04.1919	*Chef des Stabes beim Kommando der Regierungstruppen Bamberg* (Chief of Staff in the Headquarters of the Government Troops in Bamberg).
20.04.1919-26.05.1919	*1.Offizier im Stab* of *Freikorps Bamberg.*
26.05.1919-03.07.1919	Attached to the *Versorgungsamt* of *I.Armee-Korps.*
04.07.1919	Appointed to serve as a *Bataillon-Kommandeur* with *Wehrregiment München.*
05.07.1919-30.09.1919	*Kommandeur* of *III. Bataillon / Wehrregiment München.*
01.10.1919-06.02.1923	*Chef des Stabes* of *1. Gruppe / Staatliche Polizeiwehr Bayern* (later redesiganted *Kommando der Landespolizei München*).
07.02.1923-01.03.1923	Assigned as a *Referent* (advisor) to *Abschnittskommando III* of the *Landespolizei* in München.
02.03.1923-15.11.1923	Assigned as a *Referent* to *Abschnittskommando II* of the *Landespolizei* in München.
16.11.1923-31.10.1925	Attached to the *Kommando* (headquarters) of the *Landespolizei* in Ansbach. Discharged from police service due to "Dienstuntauglichkeit" (unfitness for duty), 31.10.1925.
00.00.1923-00.00.1925	Studied law and economics (4 semesters) at the University of München.
09.11.1923	Participated in the *München-Putsch.*
00.00.1924-00.00.1925	Underwent training in business.
00.10.1925	Discharged from the *Landespolizei* as a *Polizei-Major.*
00.11.1925-00.00.1931	Self-employed businessman, working in the insurance field.
01.02.1931	Joined the *NSDAP / Ortsgruppe Braunes Haus.*

24.02.1931-15.03.1933	Joined the *SA* as a *hauptamtlicher SA-Führer*, assigned as *Stellvertreter des Stabschefs der SA* (Deputy to the Chief of Staff of the *SA* (Ernst Röhm) and *Leiter* of the *Quartiermeisterstab (Abteilung QU) beim Stab der Obersten SA-Führung*".
01.07.1932-00.00.193_	Assigned to the *Stab der Obersten SA-Führer*.
20.02.1933-15.03.1933	*Führer* (m.d.F.b.) of *SA-Gruppe Hochland* (München).
12.03.1933-31.12.1933	*Sonderkommissar der Sicherheits-Hilfspolizei* (Special Commissioner for the Auxiliary Security Police) in the *Bayerische Staatsministerium des Innern*.
15.03.1933-15.11.1934	Attached, as *SA-Führer z. V. der Obersten SA-Führung*, to the *Bayerische Staatsministerium des Innern*.
01.04.1933-08.05.1933	Reentered police service, with attachment to *Schutzpolizei Abschnitt II* in München.
01.05.1933-29.08.1933	*Sonderkommissar der Obersten SA-Führung für die Sicherheitshilfpolizei in Bayern* (Special Commissioner of the Supreme *SA* Leadership for the State of Bavaria) attached to the *Bayerischen Staatsministerium des Innern*.
01.05.1933-00.00.193_	*Verbindungsführer der Obersten SA-Führung zu allen (bayerischen) Staatsministerien und staatlichen Behörden* (Chief of Liaison between the Supreme *SA* Leadership and all [Bavarian] State Ministries and Government Authorities).
08.05.1933-01.09.1933	Active police service in München. Placed in retirement, 01.09.1933.
29.08.1933-14.11.1934	*Sonderbevollmächtigter der Obersten SA-Führung für das Land Bayern* (Special Plenipotentiary of the Supreme *SA* Leadership for the State of Bavaria) attached to the *Bayerischen Staatsministerium des Innern*.
00.00.1934-00.00.193_	Full-time employment with the *Bayerische Staatsministerium des Innern*.
15.11.1934-30.04.1937	Assigned as an *SA-Führer z. V.* to the *Obersten SA-Führung*.
01.05.1935-00.00.193_	*Verbindungsführer der Obersten SA-Führung zum Beauftragten für das Siedlungswesen im Stab des Stellvertreters des Führers* (Chief of Liaison between the Supreme *SA* Leaderhip and the Representative for Settlement Matters on the Staff of the Deputy to the *Führer* [Rudolf Hess]).
01.05.1937-18.11.1938	Assigned to the active *SA-Führer*korps.

Decorations and Awards

15.07.1916	*1914 Eisernes Kreuz I. Klasse*
17.09.1914	*1914 Eisernes Kreuz II. Klasse*
10.02.1918	*Kgl. Bayerischer Militärverdienstorden IV. Klasse mit Schwertern und Krone*
17.12.1914	*Kgl. Bayerischer Militärverdienstorden IV. Klasse mit Schwertern*
00.00.19__	*Dienstauszeichnungskreuz II. Klasse*
00.00.191_	*Prinzregent-Luitpold-Medaille*
ca. 1934	*Ehrenkreuz des Weltkrieges 1914-1918 mit Schwertern*
00.00.1934 (?)	*Ehrenzeichen des 9. November 1923 (Blutorden)*
00.00.194_	*Dienstauszeichnung der NSDAP in Bronze*
00.00.1934	*Ehrendolch der SA* (mit Wirkung vom 03.02.1934)

00.02.1934   *Ehrenwinkel für alte Kämpfer*

Notes
- Son of a railway official (*Bahnkonduktor*) Franz Xaver Fuchs (born 03.10.1836 in Lappersdorf bei Regensburg, died 11.08.1883) and his wife Creszentia, née Schmidt (born 24.04.1844 in Donaustauf bei Regensburg, died 02.11.1909).
- Married on 15.10.1903 to Luise Christenn (born 21.09.1860 in Kulmbach). One son (born 13.02.1905) and one daughter (born 21.11.1906).

Sources:

Bundesarchiv, Berlin-Lichterfelde (former Berlin Document Center): *Personalunterlagen von SA-Angehörigen: SA-Personalakte* of Johann Fuchs.

Herbert Robert Gerhard Heinrich Fust
SA-Obergruppenführer

Born:	01.06.1899 in Langenfelde / Kreis Grimmen / Vorpommern.
Died:	11.11.1974 in Bucholz / Nordheide.

NSDAP-Nr.:	385 173 (Joined 01.12.1930)

Promotions

00.00.1918	*Unteroffizier*
00.00.1919	*Vizefeldwebel*
15.11.1930	*SA-Mann*
02.01.1931	*SA-Scharführer*
25.01.1931	*SA-Truppführer*
20.02.1931	*SA-Sturmführer*
24.06.1931	*SA-Sturmbannführer*
28.09.1931	*SA-Standartenführer*
15.04.1933	*SA-Oberführer*
24.12.1933	*SA-Brigadeführer*
20.04.1936	*SA-Gruppenführer*
00.00.1940	*Leutnant d. R.*
00.00.194_	*Oberleutnant d. R.*
30.01.1941	*SA-Obergruppenführer*

Career

00.00.1905-00.00.1906	Attended *Volksschule* in Glewitz / Kreis Grimmen.
00.00.1906-00.00.1907	Privately schooled.
00.00.1907-00.00.1909	Attended *Volksschule* in Demmin.
00.00.1909-00.00.1911	Attended *Gymnasium* in Demmin.
00.00.1911-00.00.1915	Attended *Realschule* in Eldena / Pommern (graduated *Obersekunda*).
00.00.1915-00.00.1917	Attended the *Landwirtschaftsschule* (agricultural school) in Eldena / Pommern.
00.00.1917-00.00.1918	Entered service as *Kriegsfreiwilliger*, assigned to *1.MG-Kompanie / Jäger-Bataillon 9*.
00.12.1918-00.05.1919	Service with *Freikorps Graf Kanitz* and *MG-A* in Kurland.
00.05.1919	Returned to Germany.
00.05.1919	Discharged from military service.
00.01.1920-00.01.1920	*Freikorps* service in Berlin.
00.00.1920-00.00.1921	Served an agricultural apprenticeship.
00.00.1921-00.00.1922	Studied agriculture (2 semesters) in München. Simultaneously employed as an administrator and inspector of his father's and others' estates..
00.00.1923-00.00.1924	Member of the *"Stahlhelm"-Bund*.
00.00.1923-00.00.1927	Member of the *Deutsch-Völkische Freiheitspartei* (*DVFP*).
00.00.1924-00.00.1927	Member of *Frontbann*.

SA-Brigadeführer Fust, 1934.

00.00.1927-00.00.1930	Member of the *Freikorps Roßbach*.
00.02.1927-00.00.19__	Self-employed as a farmer in Warrenzin bei Dargun.
15.11.1930	Applied for *NSDAP* membership.
15.11.1930-01.01.1931	Joined the *SA*, assigned as to *SA-Sturm 5* (Gnoien).
01.12.1930	Joined the *NSDAP / Ortsgruppe Levin*.
02.01.1931-24.01.1931	*Führer* of a *Schar* in *SA-Sturm 5* (Gnoien).
25.01.1931-14.02.1931	*Führer* of a *Trupp* of *SA-Sturm 5* (Gnoien).
15.02.1931-23.06.1931	*Führer* of *SA-Sturm 245* (Dargun).
Summer 1931	Sentenced to pay a 75 RM fine due to an "offense against the State and justice".
24.06.1931-27.09.1931	*Führer* of *SA-Sturmbann II/90*.
28.09.1931-16.02.1933	*Führer* of *SA-Standarte 90*.
01.01.1932-00.00.1933	Member of the *Amtsversammlung* (official assembly, also known as *Amtsausschuß* [official committee]) and *Amtsvertreter* of the *Amt Malchin*.
01.01.1932-00.00.1933	*Stellvertretender Gemeindevorsteher* (deputy mayor) of Warrenzin bei Dargun.
05.06.1932-14.10.1933	Member of the *Landtag* of Mecklenburg-Schwerin.
17.02.1933-31.08.1933	*Führer* (m.d.F.b.) of the independent *SA-Untergruppe* Mecklenburg (Schwerin).
01.03.1933	Entered full-time *SA* service.
05.03.1933-14.10.1933	Member of the *Reichstag (Wahlkreis 35, Mecklenburg)*.
00.05.1933-00.07.1934	*Sonderkommissar* [later redesignated *Sonderbevollmächtigter] der OSAF für Mecklenburg-*

SA-Brigadeführer Fust, 1934.

	Strelitz und Mecklenburg Schwerin sowie [as well as] *Lübeck.*
01.09.1933-14.09.1933	*Führer* of *SA-Brigade 11 "Mecklenburg"* (Schwerin).
15.09.1933-31.10.1937	*Führer* (m.d.F.b. until 23.12.1933, then permanent from 24.12.1933) of *SA-Gruppe Hansa* (Hamburg).
	Succeeded Arthur Böckenhauer.
12.11.1933-08.05.1945	Member of the *Reichstag (Wahlkreis 35, Mecklenburg*; after 29.03.1936: *Wahlkreis 34, Hamburg*; after 10.04.1938, *Wahlkreis 27, Rheinpfalz-Saar*).
11.10.1934-00.05.1945	*Hamburgischer Staatsrat.*
00.02.1936	Received an official reprimand from the *Obersten Parteigericht der NSDAP* for drunkenness and sexual affairs.
01.11.1937-31.01.1942	*Führer* of *SA-Gruppe Kurpfalz* (Mannheim). Succeeded Friedrich Fenz. Succeeded by Carl Caspary.
10.11.1938	Issued an order to the *Führer* of *SA-Brigade 50* which read:

> On the orders of the *Gruppenführer* all Jewish synagogues in the area of *Brigade 50* are immediately to be blown up or set on fire. Neighboring houses which are inhabited by the Aryan population must not be damaged. The action is to be carried out in civilian clothing. Mutinies and acts of looting are to be forbidden.

	As a result, *SA-Brigade 50* carried out the destruction and partial destruction of 35 synagogues during "Reichskristallnacht".
26.08.1939	Entered *Wehrmacht* service.

A painting of *SA-Brigadeführer* Fust from 1934.

00.09.1939-00.00.1942	War service, with action in France and North Africa. He was ultimately assigned as an *Oberleutnant d. R.* with a *Panzer-Pionierbataillon* (armored engineer battalion). Wounded in action in North Africa, 1941.
21.01.1942-00.05.1945	*Führer* of *SA-Gruppe Hansa* ("vertretungsweise mit der Führung beauftragt" [charged with acting leadership] until 31.01.1942, then permanent from 01.02.1942). Succeeded Siegfried Kasche.
01.02.1942-00.05.1945	*Hamburgischer Staatsrat.*

Postwar Prosecution
Interned after 1945. He was tried by the *Landgericht* in Wiesbaden for his role in "Reichskristallnacht", but acquitted on 25.07.1952.

Decorations and Awards

00.00.1940	1939 Eisernes Kreuz I. Klasse
00.00.19__	*1939 Spange zum 1914 Eisernes Kreuz II. Klasse*
00.00.191_	*1914 Eisernes Kreuz II. Klasse*
00.00.194_	*Allgemeines-Sturmabzeichen*
ca. 1941	*Verwundetenabzeichen, 1939 [in Schwarz?]*
ca. 1919	*Baltenkreuz*
00.00.1943	*Ärmelband "Afrika"* (presumably)
ca. 1934	*Ehrenkreuz des Weltkrieges 1914-1918 mit Schwertern*
30.01.1938	*Goldenes Ehrenzeichen der NSDAP*
00.00.194_	*Dienstauszeichnung der NSDAP in Silber*
00.00.194_	*Dienstauszeichnung der NSDAP in Bronze*

00.00.1931	*Abzeichen des SA-Treffens Braunschweig 1931*
00.00.193_	*SA-Sportabzeichen in Bronze*
00.00.1934	*Ehrendolch der SA* (mit Wirkung vom 03.02.1934)
00.02.1934	*Ehrenwinkel für alte Kämpfer*

Notes

• * Son of the farmer Adolf Fust (born 01.02.1863 in Remlin of Barsdorf / Mecklenburg, died 19.02.1927) and his wife Martha, née Wegener (born 21.11.1872 in Jördensdorf).

• * Religion: Protestant until 00.00.1938, then left the church and declared himself "gottgläubig".

• * Married to Thea Marsmann (born 16.11.1901 in Pnesdorf / Vorpommen).

Sources

Bundesarchiv, Berlin-Lichterfelde (former Berlin Document Center): *Personalunterlagen von SA-Angehörigen: SA-Personalakte* of Herbert Fust.

Lilla, Joachim; Döring, Martin; & Schulz Andreas: *Statisten in Uniform. Die Mitglieder des Reichstags 1933-1945*. Droste Verlag, 2004.

Paul Giesler

SA-Obergruppenführer

Born:	15.06.1895 in Siegen / Westfalen.
Suicide:	08.05.1945 near Berchtesgaden (?). Several accounts exist concerning the death of Giesler. According to *General der Flieger* Karl Koller, the last Chief of the *Luftwaffe* General Staff, the *Gauleiter* and his wife unsuccessfully attempted to poison themselves on 01.05.1945. However another account states both were found on 04.05.1945 in the Märchenwald Forest near Ramsau [in the area of Berchtesgaden] together with their chauffeur in their car and had been shot (they likely committed suicide, but it is possible that they could have been killed by persons unknown). The *Gauleiter* biographer Karl Höffkes presents the following variation: "Giesler went together with his wife, mother-in-law and his brother Hermann over to Rosenheim from Berchtesgaden. A reconstruction of further events is extraordinarily difficult. Different circumstances point thereupon that the opinion put forwards until now that Paul Giesler killed himself together with his wife in the park of a hotel at the Hintersee in Berchtesgaden on the 02.05.1945 by a shot into the temple is not correct." The following course of events seems to be the more probable: Paul Giesler together with his wife unsuccessfully attempted suicide with soporific drugs on 01.05.1945. On the same day both were brought into the district hospital in Berchtesgaden. The local *Landrat*, Theodor Jacob Giesler (probably no relation!), later claimed to have disarmed them upon their arrival at the hospital, and a police officer placed in front of the sickroom of *Gauleiter* and Frau Giesler, in order to prevent their escape and to facilitate their being turned over to U.S. authorities. On the night of 02.05.1945 the local *Kreisleiter*, Bernhard Stredele, freed the *Gauleiter* with members of the local Gauleitung by force. Later on 02.05.1945, *Gauleiter* Giesler shot his wife in a wooded area between Hintersee and Hirschbichl, after his mother-in-law had poisoned herself. Paul Giesler then shot himself in the temple, likely surviving this suicide attempt. According to the witness testimony of a baker named Ertl, the severely injured Giesler was brought to Ramsau. A *Waffen-SS* lieutenant then took him in his car to the military hospital in Stanggaß. Paul Giesler probably survived for another six days, finally succumbing to his injuries on 08.05.1945. A local doctor practicing in Stanggaß at that time, Dr. Gottschalk, certified his death on that date and Giesler was buried in the cemetery in Berchtesgaden on 10.05.1945.
NSDAP-Nr.:	72,741 (First joined, 00.00.1922; Party banned following the *München-Putsch* of 09.11.1923; Reenrolled 01.01.1928)

Heinrich Hoffmann's formal portrait of *Gauleiter* Paul Giesler, ca. 1943. (NARA photo)

Promotions

ca. 1915	*Vizefeldwebel d. R.*
18.12.1915	*Leutnant d. R.*
00.00.1929-01.01.1931	*Ortsgruppenleiter* der *NSDAP*
15.09.1931	*SA-Sturmbannführer*
09.09.1932	*SA-Standartenführer* (mit Wirkung vom 01.07.1932)
15.11.1933	*SA-Oberführer*
20.04.1934	*SA-Brigadeführer*
09.11.1937	*SA-Gruppenführer*
00.00.1940	*Hauptmann d. R.*
00.08.1941-08.11.1941	*Stellvertretender Gauleiter der NSDAP*
00.09.1941	*Hauptdienstleiter der NSDAP*
09.11.1941-22.04.1944	*Gauleiter (m.d.W.d.G.b.) der NSDAP*
23.06.1942-22.04.1944	*Staatsminister (m.d.W.d.G.b.)*
30.01.1943	*SA-Obergruppenführer*
22.04.1944-08.05.1945	*Staatsminister* (permanent)
22.04.1944-08.05.1945	*Gauleiter der NSDAP* (permanent)

Career

00.00.1902-00.08.1914	Attended *Volksschule* and *Realgymnasium* in Siegen
00.08.1914-00.00.1918	Entered service as a *Kriegsfreiwilliger,* assigned to the *Garde-Pionier-Bataillon.*
00.00.191_-00.12.1915	Assigned to *1.Garde-Infanterie-Division.*

00.12.1915-00.01.1918(?)	*Zugführer* in the *Garde-Minenwerfer-Kompanie / Garde-Pionier-Bataillon.*
00.01.1918-00.11.1918	*Führer* of the *Minenwerferkompanie* of *2.Garde-Regiment zu Fuß* (wounded several times).
00.00.1919-00.00.1921	Studied architecture at the *Höheren Landesbauschule* in Darmstadt. He also attended the *Technischen Hochschule* in Darmstadt during this period.
00.00.1919	Joined the *"Stahlhelm"-Bund.*
00.00.1922-00.00.1933	Self-employed as an architect in Siegen.
00.00.1922	Joined the *NSDAP / Ortsgruppe Siegen.*
00.00.1922	Cofounder of the *SA* in Siegerland.
00.00.1924	Appointed as a *Parteiredner der* [illegal] *NSDAP.*
01.01.1928	Reenrolled in the *NSDAP.*
00.00.1929-01.01.1931	*Ortsgruppenleiter* of the *Ortsgruppe Siegen der NSDAP.*
00.01.1931-09.09.1932(?)	Reentered the *SA,* assigned as *Führer* (m.d.F.b. until 15.09.1931, then permanent) of *Sturmbann IV/ SA-Standarte 132* (Hagen / Westfalen). His command presumably ended with the government ban on the *SA* of 13.04.1932- 01.07.1932, and may have resumed when the ban was lifted on 01.07.1932.
00.02.1931	Formally sworn in as an *SA-Führer* by *SA-Standartenführer* Friedrich Esscher.
18.10.1931	Participated in the *SA-Aufmarsch in Braunschweig.*
24.04.1932	Unsuccessful candidate for election to the *Preußische Landtag (Wahlkreis 18, Westfalen-Süd).*
09.09.1932-31.08.1933	*Führer* of *SA-Standarte 130* (Siegen). In this assignment, he organized and personally participated in brutal attacks on enemies of the Party and *SA.* Among the units under his command was the *Rollkommando Odendahl* which instigated a number of violent riots before and after the Nazi seizure of power. When the *SA, SS,* and *Stahlhelm* were designated as *Hilfspolizei* (auxiliary police) in March 1933, Giesler ensured that members of his *Rollkommando* were at the top of the list for *Hilfspolizei* appointments. On 02.05.1933, Giesler and the *Führer* of the *Rollkommando* led an *SA* unit in attacking the *Haus der Arbeit* in Siegen (which housed local offices of the *Allgemeiner Deutscher Gewerkschaftsbundes [ADGB],* the *SPD,* and the Social-Democratic newspaper *Siegener Volks-Zeitung*). Giesler's thugs devastated the building, and subjected the leftists therein to severe physical abuse before arresting them. On 19.07.1933, Giesler led an *SA-Kommando* in the arrest of Gotthold Reinhardt, director of the *fürstlich-berleburgischen Rentkammer* and a strong critic of Nazi economic ideas who had refused to make payments for the support of the *SA.* According to a statement by Reinhardt dated 06.07.1946 (in the *Stadtarchiv Bad Berleburg*), Giesler personally flogged him with such force that he sustained a fractured skull. He then ordered his subordinates: "If he leaves the cell, shoot!" On 07.09.1933, Reinhardt's attorney was dragged out of his home by *SA* men- again led by Giesler- and forced to wear a sign bearing the words

	"Ich habe die *SA* beleidigt, ich bin ein gemeingefährlicher Volksschädling" (I have insulted the *SA* and am a publicly dangerous pest harmful to the people).
00.04.1933-00.04.1934	Member of the *Stadtrat* in Siegen.
01.09.1933-14.03.1934	*Führer* (m.d.F.b. until 14.11.1933, then permanent from 15.11.1933) of *SA-Brigade 68* (Siegen).
12.11.1933-08.05.1945	Member of the *Reichstag* (representing *Wahlkreis 18, Westfalen-Süd*; after 29.03.1936, *Wahlkreis 14, Weser-Ems*; and from 12.03.1942, Wahlkreis 18, Westfalen-Süd [traded *Reichstag* seats with Gustav Bertram]).
15.03.1934-30.06.1934	*Führer* (m.d.F.b.) of *SA-Gruppe Westfalen* (Dortmund).
30.06.1934	Narrowly escaped arrest and/or murder during the "suppression" of the so-called "Röhm-Putsch".
00.07.1934	Discharged from his posts in *Gau Westfalen* in connection with the "Röhm-Putsch", at the instigation of *Gauleiter* Josef Wagner. Together with his superior, Wilhelm Schepmann (the Dortmund-based *Führer* of *SA-Obergruppe X* comprising *SA-Gruppen Westfalen* and *Niederrhein*), he was accused of planning an armed *SA* rebellion on 30.06.1934 and of acting against the authority of *Gauleiter* Wagner and the *SS*.
00.07.1934-14.05.1935	Attached as an *SA-Führer z. V.* to the *Obersten SA-Führung.*
30.07.1934	Disciplinary proceedings before the *Obersten Parteigericht* initiated by *Gauleiter* Josef Wagner. In a letter from the *SA-Sondergericht* (Special Court, under the chairmanship of *SA-Gruppenführer* Arthur Böckenhauer and tasked with investigating the so-called "Röhm-Affäre") dated 27.08.1934, Giesler was informed: "Your case is by no means unusual. A large number of high *SA* leaders find themselves in the same situation." Edward N. Peterson writes: "The 21 charges brought against him in October 1934 read like the total indictment against the SA: corruption, violence against citizens, threats of violence even against Goering." (Peterson, *The Limits of Hitler's Power*, pp. 165-166) On 10.04.1935, he was finally acquitted by the *II.Kammer des Obersten Parteigerichts* (2nd Chamber of the Supreme Party Court) due to lack of evidence, having been charged with violating Section 4, paragraph 2 b of the *Satzung der NSDAP* (Statute of the *NSDAP*).
15.05.1935-31.08.1936	*Führer* (m.d.F.b. until 14.09.1935, then permanent from 15.09.1935) of *SA-Brigade 63 "Oldenburg-Ostfriesland"* (in *SA-Gruppe Nordsee*).
19.02.1936	Appointed as a member of the *Kulturkreis der SA.*
00.06.1936	Appointed as *Verbindungsführer* (chief of liaison) between the *Gauleitung Weser-Ems der NSDAP* and *Bezirksgruppe 3* (Bremen) of the *Soldatenbund.*
01.09.1936-31.05.1938	*Stabsführer* (m.d.W.d.G.b. until 28.02.1937, then permanent from 01.03.1937) of *SA-Gruppe Hochland* (München).
01.06.1938-31.10.1938	*"Mit dem Aufbau der SA in Österreich beauftragt"* (Charged with the establishment of the *SA* in Austria) as *Führer* (m.d.F.b.) of *SA-Gruppe Alpenland* (Linz). He assumed the duties of this post on 02.07.1938.
00.06.1938-00.06.1938	Participated in military training exercises.

SA-Gruppenführer Giesler (2nd from left) in 1936. Also present, from left to right, are Fritz Wächtler, Adolf Wagner, and Friedrich Christiansen.

02.07.1938-06.09.1941	*Führer* (m.d.F.b. until 31.10.1938, then permanent from 01.11.1938) of *SA-Gruppe Alpenland* (Linz an der Donau). Acting leadership of the *Gruppe* passed to *SA-Oberführer* Vincenz Pohl until 01.02.1942, when *SA-Gruppenführer* Wilhelm Dittler took command.
11.09.1938	Attended the *10. Reichsparteitag der NSDAP* in Nürnberg.
00.09.1939-00.00.1940	War service as a *Kompanieführer* in *Infanterie-Regiment 131 / 44.Infanterie-Division* in Poland and France. During the Polish Campaign, he participated in the advance from Kraków to Jaroslaw (despite being wounded in the early days of the invasion, sustaining two broken ribs), the attack on Teschen, the attack on a three-tier line of bunkers near Skotschau, in the liberation of a Polish camp being constructed to hold German POWs at Lubaczów, and finally, in the battle for Lwów (Lemberg). The following year, during the invasion of France, he was again wounded in action.
00.08.1941-08.11.1941	Attached with the *NSDAP* rank of *stellvertretender Gauleiter* to the *Parteikanzlei* in München for the purpose of assessing his suitability for assignment as a *Gauleiter*.
06.09.1941-00.05.1945	Due to his transfer to full-time *NSDAP* duties, released from active *SA* service and assigned as an *SA-Reserveführer* to the staff of *SA-Gruppe Alpenland*.
09.11.1941-18.06.1943	*Gauleiter* (m.d.W.d.G.b.) of *Gau Westfalen-Süd der NSDAP*. Succeeded his old nemesis, the disgraced Josef Wagner (who had been personally

expelled by Hitler). On 26.01.1943, Albert Hoffmann, still holding the rank of *stellvertretender Gauleiter*, assumed the duties of *Gauleiter* (m.d.W.d.G.b.) in *Westfalen-Süd*, while Giesler was acting *Gauleiter* in *München-Oberbayern*. Hoffmann remained as acting *Gauleiter* until 17.04.1944, then took over as permanent *Gauleiter* (he was confirmed in this post by *Reichsleiter* Dr. Ley on 03.05.1944).

09.11.1941-08.05.1945	*Preußischer Staatsrat.*
15.11.1941-18.06.1943	*Gauwohnungskommissar* for *Gau Westfalen-Süd*. Succeeded by Albert Hoffmann.
06.04.1942-18.06.1943	*Beauftragter des GBA für den Gau Westfalen-Süd*. From 26.01.1943, acting *Gauleiter* Albert Hoffmann performed the duties of this post.
23.06.1942-08.05.1945	*Gauleiter* of the *Traditionsgau München-Oberbayern der NSDAP* (m.d.W.d.G.b. [deputizing for *Gauleiter* Adolf Wagner, on leave due to illness] until 22.04.1944, then permanently appointed upon the death of Wagner). Of the new *Gauleiter*, the local *HSSPF*, Dr. Benno Martin, later commented: "Giesler was more dangerous [than Wagner] because he was less insane." (Edward N. Peterson, *The Limits of Hitler's Power*, p. 165)
23.06.1942-08.05.1945	*Bayerische Staatsminister des Innern* (Bavarian State Minister of the Interior; m.d.W.d.G.b. until 12.04.1944, then permanent). Succeeded Adolf Wagner.
23.06.1942-08.05.1945	*Bayerische Staatsminister für Unterricht und Kultur* (Bavarian State Minister of Education and Culture; m.d.W.d.G.b.). Succeeded Adolf Wagner.
23.06.1942-08.05.1945	*Gauwohnungskommissar* for *Gau München-Oberbayern*.
23.06.1942-08.05.1945	*Beauftragter des GBA für den Gau München-Oberbayern*.
30.06.1942	Attended a meeting with *Reichsführer-SS* Himmler, Martin Bormann, and *NSKK-Korpsführer* Erwin Kraus.
09.07.1942	Together with *Reichsminister* Dr. Goebbels and Artur Görlitzer (deputy *Gauleiter*, Berlin) visited the *NS-Deutsche Oberschule* in Feldafing am Starnberger See / Bayern.
09.11.1942	Participated in the remembrance ceremony for victims of the *München-Putsch*.
16.11.1942-26.01.1943	*Reichsverteidigungskommissar für den Gau Westfalen-Süd*.
16.11.1942-08.05.1945	*Reichsverteidigungskommissar für den Gau München-Oberbayern (Wehrkreis VII)*.
02.11.1942-08.05.1945	*Bayerische Ministerpräsident* (Bavarian Minister President [Prime Minister]; m.d.W.d.G.b. until 12.04.1944, then permanent). Succeeded the late Ludwig Siebert, who died 01.11.1942. Officially appointed to this post by Adolf Hitler at the *Deutsches Museum*, München, on 17.04.1944.
02.11.1942-08.05.1945	*Bayerische Staatsminister der Finanzen und Wirtschaft* (Bavarian State Minister of Finance and Economics; kommissarisch m.d.W.d.G.b.). Succeeded Ludwig Siebert.
13.01.1943	In a speech before the student body of the University of München, commemorating the 470th anniversary of the founding of that

Haus der Technik, Wien, 20.01.1939: *SA-Gruppenführer* Giesler, prior to his lecture *Der Sinn des neuen Baues* (The Purpose of Modern Architecture) in conversation with *Gauamtsleiter* Benno Gürke (right) and the *Fachgruppenwalter für Bauwesen*, Heinrich Goldemund. (Weltbild photo, in *Wissenschaft und Bildung*)

institution, Giesler spoke of the duty of his listeners to serve the *Reich*. John W. Wheeler-Bennett writes:

> The men, [Giesler] said, would be combed out lest there should be found, even among these physical weaklings, military cripples, and effete intellectuals, some who might be more usefully employed in shouldering a gun in the defence of the Fatherland. As for the girls, continued *Gauleiter* Giesler with a leer, "They have healthy bodies, let them bear children. That is an automatic process which, once started, continues without requiring the least attention. There is no reason why every girl student should not for each of her years at the University present an annual testimonial in the form of a son. I realize that a certain amount of co-operation is required and, if some of the girls lack sufficient charm to find a mate, I will assign to each of them one of my adjutants, whose antecedents I can vouch for, and I can promise her a thoroughly enjoyable experience." (Wheeler-Bennett, *The Nemesis of Power: The German Army in Politics, 1918-1945*, p. 540)

Several of the women in the audience who indignantly attempted to leave were taken into *Gestapo* custody; the rioting that ensued lasted several weeks. The protests were orchestrated by the student resistance group "White Rose", led by Hans and Sophie Scholl (aged 25 and 21, respectively), which Giesler subsequently played a key role in destroying. In February 1943, Giesler informed the authorities in Berlin that he had dismissed male "White Rose" members (Hans Scholl, Christoph Probst, Alexander Schmorrel, and Willi Graf) from the *Wehrmacht*, thus placing them under civil rather than military law. On 19.02.1943, the *Gauleiter* advised a special commission of the *Gestapo* that "*Generalfeldmarschall* Keitel has dismissed the soldiers in question from the *Wehrmacht* and agrees that they be sentenced by the *Volksgerichtshof* [under Dr. Roland Freisler]." Giesler pressed for a quick sentencing and execution of the first group of defendants, who were tried by Freisler's court on 22.02.1943. The Scholl siblings and Christoph Probst were found guilty of distributing anti-Nazi leaflets, sentenced to death, and went to the guillotine the same day. In the second trial, on 19.04.1943, Professor Kurt Huber, Alexander Schmorrel, and Willi Graf were likewise sentenced to death.

25.09.1944-08.05.1945   *Führer des Deutschen Volkssturms im Gau München-Oberbayern.* Subordinated to him as *Gaustabsführer* was *SA-Gruppenführer* Bernhard Hofmann. In November 1944, Giesler delivered a speech in München, probably during the swearing-in of a *Volkssturm* unit, in which he declared:

> We do not fall under the spell of the enemy's current material superiority, rather we will make all the enemy's hopes turn to dust through the longed-for revival of our strength in the form of the German *Volkssturm*. In it rests the last chance, one which will never recur, to turn the spirit of our Volk into a zest for war to defend the National Socialist nation with utter fanaticism. (*Münchener Neueste Nachrichten*, issue of 25.11.1944; Translation courtesy of Richard Hargreaves)

13.02.1945   Excerpt from the diary of *Reichsminister* Dr. Goebbels:

> The *Führer* talks about *Gauleiter* Giesler very positively. People suggested he appoint [Hermann] Esser instead of Giesler. You couldn't imagine what would have happened in München after the heavy air raids if Esser had been the leader ... . (Elke Fröhlich, ed, *Die Tagebücher von Joseph Goebbels, Teil II, Bd. 15*, translation courtesy of Richard Hargreaves)

24.02.1945   Attended the last meeting of *Reichs-* and *Gauleiter* with Hitler in the *Reichskanzlei*, Berlin, to mark the 25th anniversary of the establishment of the *NSDAP*.

00.03.1945-00.04.1945   On the authority of Dr. Ernst Kaltenbrunner, chief of the *RSHA*, planned to exterminate the surviving inmates of *KL–Dachau* and several of its satellite camps. The following is excerpted from the 20.11.1945 interrogation of Giesler's *Gaustabsamtsleiter,* Hubertus "Bertus" Gerdes- in which he spoke of his role in sabotaging the plans for this mass murder- conducted by Special Agent Johannes Imhoff of the Counter Intelligence Corps (CIC), Nürnberg Sub-Regional Office:

... In December 1944 or January 1945, I had an opportunity to see a secret decree issued by Kaltenbrunner in the office of *Gauleiter* Giesler from a courier in my presence and after I had been permitted to read it was destroyed in accordance with the classification to be destroyed after reading. The order which was signed by Kaltenbrunner read approximately as follows: "In agreement with the *Reichsführer-SS* I have brought about and directed all higher police officers that all Germans shall go unpunished who in the future participate in the persecution and annihilation of enemy aircrews who parachute down". Giesler told me that Kaltenbrunner was in constant touch with him because he was greatly worried about the attitude of the foreign workers and especially inmates of concentration camps Dachau, Mühldorf and Landsberg which were in the path of the approaching Allied armies. On a Tuesday in the middle of April 1945, I received a telephone call from *Gauleiter* Giesler asking me to be available for a conversation that night. In the course of our personal conversation that night, I was told by Giesler that he had received a directive from Kaltenbrunner by order of the *Führer* to work out a plan without delay for the liquidation of the concentration camp at Dachau and the two Jewish labor camps in Landsberg and Mühldorf. The Directive proposed to liquidate the two Jewish labor camps at Landsberg and Mühldorf by use of the German *Luftwaffe,* since the construction area of these camps had previously been the targets of repeated enemy air attacks. This action received the code name of "Wolke A 1" [Cloud A 1]. I was directed by Giesler to take up connections with *General[leutnant* Adolf] Galland regarding the execution of this plan. I had not met General Galland up to the time he stayed at the airport of Riem, therefore we invited him to dinner the next day at Soehaus together with a very small party of people. During the meal only questions of a general nature were discussed and the use of the new jet propelled fighter plane. On this occasion I was to agree on a date with Galland so that we could discuss the above mentioned operation. The conversation however never took place.

I attempted repeatedly to get in touch with men at the *Führer*'s Headquarters such as [Heinrich] Walkenhorst, chief of the personal staff of Bormann, and Treitsch, go-between officer for Himmler and Bormann. I was finally able to make these calls although both were very brief. Walkenhorst informed me that he

Münchener Hauptbahnhof, 18.10.1942: *Reichsminister* Dr. Goebbels is greeted on his arrival in the city by acting *Gauleiter* Giesler. (NARA photo)

knew nothing of such an order by the *Führer*, however, that he was going to make detailed inquiries and would call back. This second call did not materialize. The connection with Treitsch was very faint but he informed me that Kaltenbrunner was Himmler's deputy and that we had to comply absolutely with his directives.

Since I was not able to obtain the desired information from either of these two men, I tried to dissuade the *Gauleiter* from execution of this abominable plan. Finally *Gauleiter* Giesler informed me that I would have to let my conscience be my guide in the execution of this directive.

I was certain that I would never let this directive be carried out. As the action Wolke A 1 should have become operational already for some time, I was literally swamped by couriers from Kaltenbrunner and moreover I was supposed to have discussed the details of the Mühldorf and Landsberg actions in detail with the two *Kreisleiter*s concerned. The couriers who were in most cases *SS* officers usually *SS-Untersturmführer*s, gave me terse and strict orders to read and initial. The orders threatened me with the most terrible punishment including execution if I did not comply with them. However, I could always excuse my failure to execute the plan because of bad flying weather and lack of gasoline and bombs. Therefore, Kaltenbrunner ordered to have Jews in Landsberg marched to Dachau in order to include them in the Dachau extermination operations, and that the Mühldorf action was to be carried out by the *Gestapo*. Kaltenbrunner also ordered an operation -Wolkenbrand -for the concentration camp Dachau which provided that the inmates of the concentration camp at Dachau were to be liquidated by poison with the exception of Aryan nationals of the Western Powers. *Gauleiter* Giesler received this order direct from Kaltenbrunner and discussed in my presence the procurement of the required amounts of poison with Dr. Harrfeld, the *Gau* Health Chief. Dr. Harrfeld promised to procure these quantities when ordered and was advised to await my further directions. As I was determined to prevent the execution of this plan in any event, I gave no further instructions to Dr. Harrfeld.

The inmates of Landsberg had hardly been delivered at Dachau when Kaltenbrunner sent a courier declaring the action Wolkenbrand was operational. I prevented the execution of the Wolke A 1 and Wolkenbrand by giving Giesler the reason that the front was too close and asked him to transmit this on to Kaltenbrunner. Kaltenbrunner therefore issued directives in writing to Dachau to transport all Western European prisoners by truck to Switzerland and to march the remaining inmates into Oetztal [Tirol], where the final liquidation of these prisoners was to take place without fail.

The *Gauleiter* informed me that Kaltenbrunner's office was literally enraged when I did not set upon the release of the code word Wolkenbrand and he informed me in the strictest confidence that I would have to be extremely careful, since the *Gestapo* was after me. It was clear to me from Kaltenbrunner's threats that my failure to execute his orders in regard to Operations Wolke A 1 and Wolkenbrand would result in not only my personal extermination but also that of my wife and four children. Giesler insisted that I leave my home in München at once and stay in the underground command posts of the *Gauleiter* under his personal protection.

On April 27, the *Gauleiter* made it possible for me to leave Munich under the pretext of official business in order to look after my family whom I unfortunately did not find again. I did not return to Munich. (Document 3462-PS, in *Nazi Conspiracy and Aggression, Volume VI*; used in evidence against Kaltenbrunner)

According to the testimony, given before the International Military Tribunal at Nürnberg in August 1946, of the former *HSSPF* in München, Karl Friedrich Freiherr von Eberstein:

At the beginning of March 1945, the *Gauleiter* and *Reich* Defense

Commissar Giesler in München ordered me to come to him, and made the monstrous request to me that I should use my influence with the commandant of Dachau [*SS-Obersturmbannführer* Eduard Weiter] so that when the American troops approached, the prisoners-there were 25,000 people there at the time-were to be shot. I refused this demand with indignation, and I pointed out that I could not give any orders to the commandant, whereupon Giesler said to me that he, as *Reich* Defense Commissar, would see to it that the camp would be bombed to bits by our own forces. I told him that I considered it impossible that any *Luftwaffe* commander would be willing to do this. Then Giesler said he would see to it that something would be put into the soup of the prisoners. That is, he threatened to poison them. On my own initiative I sent a teletype inquiry to the Inspector of Concentration Camps [Richard Glücks] and asked for a speedy decision from Himmler as to what was to be done with the prisoners in case the American troops approached. Shortly afterwards the news came that the camps were to be surrendered as a whole to the enemy. I showed that to Giesler. He was very indignant because I had frustrated his plans and because I was of a different opinion. Shortly after we had another clash regarding the defense of München, which was completely hopeless. The *Wehrmacht* Commander was thrown out eight days before me, and on the 20th of April I was also dismissed and all my offices were taken away from me, and I was without power. (*Trial of the Major War Criminals Before the International Military Tribunal, Nuremberg, 14 November 1945-1 October 1946, Volume XX*)

00.04.1945-08.05.1945	*Reichsverteidigungskommissar Süd* (*Reich* Defense Commissioner South, with his jurisdiction encompassing *Gaue München-Oberbayern, Schwaben, Tirol, Oberdonau,* and *Salzburg*).
27./28.04.1945	Oversaw the brutal suppression by *SS* units of an uprising in München (known as the *Freiheitsaktion Bayern* [*FAB*, Freedom Action Bavaria]) led by *Hauptmann* Dr. Rupprecht Gerngroß (1915-1996), Chef of the *Dolmetscher-Kompanie* (translator company) of *Wehrkreis VII.* On the night of 27./28.04.1945, Dr. Gerngroß and his men arrested *Reichsstatthalter* Franz Ritter von Epp at his headquarters in the Schornerhof. Von Epp's adjutant, *Major* von Caracciola, then joined the uprising. A squad of Gerngroß's men occupied the radio stations in Erding and Freimann, broadcasting: "Stop fighting, lay down your arms-destroy the Nazis wherever you meet them. Hoist white flags, the Allied troops are approaching München." The putsch failed owing to the refusal of *Reichsstatthalter* von Epp to lend his support, and the *SS* and *Gestapo* quickly murdered 40 of Gerngroß's rebels, including Von Caracciola, shortly before the arrival of U.S. troops in the city. Despite the failure of the revolt, it helped to plunge the Nazi administration into

The Deutsches Museum, München, 17.04.1944: The *Führer* officially appoints Paul Giesler as Bavarian *Ministerpräsident*, as *Reichsleiters* Dr. Ley and Bormann look on, 1944. (NARA photo)

chaos, and resulted in Giesler's abandoning his headquarters as the U.S. 3rd, 42nd, and 45th Infantry Divisions neared the city. On the night of 28./29.04.1945, Giesler ordered *SA-Brigadeführer* Johann "Hans" Zöberlein, commanding *Gruppe Hans* of the *Freikorps Adolf Hitler* (a component of the so-called *Werwolf-Oberbayern*), to suppress an anti-Nazi uprising by citizens of Penzberg / Oberbayern, 50 km south of München, which was led by the former *Bürgermeister* Hans Rummer (a Social Democrat). In what became known as the "Penzberger-Mordnacht", 16 men and women (one of whom was pregnant) were hanged without trial for their part in the revolt.

29.04.1945	Appointed as *Reichsminister des Innern*, as successor to the disgraced Heinrich Himmler, in Hitler's final political testament.
30.04.1945	Fled with his wife to Berchtesgaden.

Decorations and Awards

00.00.191_	*1914 Eisernes Kreuz I. Klasse*
00.10.1939	*1939 Spange zum 1914 Eisernes Kreuz II. Klasse*
00.00.191_	*1914 Eisernes Kreuz II. Klasse*
00.00.194_	*Kriegsverdienstkreuz I. Klasse ohne Schwerter*
00.00.194_	*Kriegsverdienstkreuz II. Klasse ohne Schwerter*
00.00.1940	*Verwundetenabzeichen, 1939 in Silber*
00.00.1918 (?)	*Verwundetenabzeichen, 1918*

ca. 1934	*Ehrenkreuz des Weltkrieges 1914-1918 mit Schwertern*
00.00.1934	*Goldenes Ehrenzeichen der NSDAP*
00.00.19__	*Goldenes Hitler-Jugend Ehrenzeichen mit Eichenlaub*
00.00.194_	*Dienstauszeichnung der NSDAP in Silber*
00.00.194_	*Dienstauszeichnung der NSDAP in Bronze*
ca. 1931	*Abzeichen des SA-Treffens Braunschweig 1931*
00.00.1934	*Ehrendolch der SA* (mit Wirkung vom 03.02.1934)
00.02.1934	*Ehrenwinkel für alte Kämpfer*

Notes:
* Son of an architect and older brother of Prof. Hermann Giesler (born 02.08.1898 in Siegen, died 20.01.1987 in Düsseldorf). A veteran of World War I, Hermann Giesler joined the *NSDAP* (with *NSDAP*-Nr. 622 515) and the *SA* on 01.10.1931, and was promoted to *SA-Obersturmbannführer* (ca. 1939), *SA-Standartenführer* (30.01.1941), *SA-Oberführer* (30.01.1942), and *SA-Brigadeführer* (20.04.1945). Appointed as a Professor at Weimar's *Hochschule für Baukunst* on 30.01.1938, he held a variety of important architectural posts in the Third *Reich*, and from 24.12.1938 to 08.05.1945 served as *Generalbaurat für die Hauptstadt der Bewegung* (München) (he was also appointed, ca. 1939, as *Generalbevollmächtigter für das Bauwesen in Bayern und die Ostmark*). Also on 24.12.1938, he was granted the title of *Reichskultursenator*. Awarded the Grand Prize and Gold Medal at the World Exposition in Paris, 00.00.1937, he received an honorary award of the *Goldenes Ehrenzeichen der NSDAP* on 30.01.1939. In December 1941 he was appointed as *Leiter* of the *OT* [*Organisation Todt*] *Eisenbahneinsatz Riga* (also known as *Baugruppe Giesler*), and from 00.00.1942 to 00.00.1944 served as *Leiter* of the *OT-Einsatzgruppe Rußland-Nord*. In his capacity as *Reichsbaurat für die Stadt Linz an der Donau*, a post he held from 1941 to May 1945, he was charged by Hitler with reconstruction of the city of Linz (a program near and dear to the Führer's heart, as he hoped to make his hometown of Linz- under the name "Germania"- the future capital city of the *Reich*. Hermann Giesler also served as a member of the *Reichstag* from 04.08.1943 to 08.05.1945 (representing *Wahlkreis 3, Berlin-West*). Assigned as *Leiter* of *OT-Einsatzgruppe Deutschland VI* (Bayern), 00.00.1944-00.05.1945, he was arrested by U.S. Army troops at Zell am See in 1945 and interned at Dachau. Tried by the U.S. Military Tribunal, Dachau (in the "Mühldorf Case"), he was sentenced to life imprisonment in 1947. Released from U.S. War Criminals Prison No. 1 at Landsberg am Lech / Bayern, 00.10.1951, he was then employed as a freelance architect in northern Germany. His memoirs, *Ein anderer Hitler* (Another Hitler), were published in 1977.
* Religion: Declared himself "gottgläubig", 19__.
* Married to Margret Patt (born 13.08.1896 in Weidenau an der Sieg, died with her husband, 00.05.1945).
* In a posthumous de-Nazification hearing on 01.06.1949 before the *Hauptkammer* in München, declared a "Hauptschuldiger" (major offender), resulting in the confiscation of his entire estate.

Sources
Goebbels, Dr. phil. Paul Joseph: *The Goebbels Diaries 1942-1943* (edited by Louis P. Lochner). Doubleday, 1948.
Final Entries 1945: The Diaries of Joseph Goebbels (edited by Prof. Hugh Trevor-Roper, Translated from the German by Richard Barry). Putnam, 1978.

Hamilton, Charles: *Leaders and Personalities of the Third Reich, Volume I.* R. James Bender Publishing, 1984.

Helmes, Dieter: *Aufbau und Entwicklung der NSDAP im Siegerland vor der Machtübernahme.* 1974.

Höffkes, Karl: *Hitlers politische Generale: Die Gauleiter des Dritten Reiches.* Grabert-Verlag-Tübingen, 1986.

Lilla, Joachim; Döring, Martin; & Schulz Andreas: *Statisten in Uniform. Die Mitglieder des Reichstags 1933-1945.* Droste Verlag, 2004.

Miller, Michael D. & Schulz, Andreas: *Gauleiter: The Regional Leaders of the Nazi Party and Their Deputies, Volume I.* R. James Bender Publishing, 2012.

Office of United States Chief Counsel for Prosecution of Axis Criminality: *Nazi Conspiracy and Aggression (11 volumes). U.S. Government Printing Office, District of Columbia, 1946.*

Orlow, Dietrich: *History of the Nazi Party: 1933-1945.* University of Pittsburgh Press, 1973.

Peterson, Edward N.: *The Limits of Hitler's Power.* Princeton University Press, 1969.

Pfau, Dieter: *2. Mai 1933-Zerschlagung von Arbeiterbewegung und Gewerkschaften. (= Beiträge zur Geschichte der Siegerländer Arbeiterbewegung. Band 4).* 2003.

Seidler, Franz: *Deutscher Volkssturm: Das letzte Aufgebot 1944-1945.* Bechtermunz Verlag, 1999.

Toland, John: *The Last 100 Days.* Random House, 1966.

Wheeler-Bennett, John W.: *The Nemesis of Power: The German Army in Politics, 1918-1945.* St. Martin's Press, 1954.

Günther Gräntz
SA-Obergruppenführer

Born:	26.07.1905 in Frankfurt-am-Main / Hessen.
Killed in Action:	30.04.1945 in Berlin-Zehlendorf as *Gaustabsführer* of the *Volkssturm* in *Gau Berlin*.
NSDAP-Nr.:	5 274 (Joined 06.02.1925)

Promotions

00.11.1922	*SA-Mann*
00.05.1931	*SA-Sturmbannführer*
15.09.1932	*SA-Standartenführer* (mit Wirkung vom 01.07.1932)
15.11.1933	*SA-Oberführer*
09.11.1934	*SA-Brigadeführer*
01.05.1937	*SA-Gruppenführer*
01.05.1940	*Leutnant d. R.*
30.01.1941	*SA-Obergruppenführer*
00.00.194_	*Oberleutnant d. R.*
00.00.194_	*Hauptmann d. R.*
01.05.1944	*Major d. R.*

Career

00.00.1912-00.00.1914	Attended *Mittelschule* in Frankfurt-am-Main.
00.00.1914-00.00.1923	Attended the *Reform-Realgymnasium* in Frankfurt-am-Main.
00.00.1921-mid-1922	Member of the Frankfurt-am-Main *Ortsgruppe* of the *Deutschnationale Jugendbund* (German Nationalist Youth League).
00.11.1922	Joined the *SA*.
01.12.1922-00.11.1923	Assigned to the *SA-Hundertschaft Frankfurt-am-Main*.
00.00.1923	Passed his *Reifeprüfung* (matriculation examination).
00.00.1923-00.00.1925	Served an apprenticeship in the chemical trade.
00.11.1923-00.00.1924	Member of of the *Deutsch-Völkische Freiheitsbewegung (DVFB)*, the *Kampfbund zur Brechung der Zinsknechtschaft*, the *Deutschen Partei*, the *NS-Freiheitsbewegung (NSFB)*, and the *Großdeutschen Volksgemeinschaft (GVG)*.
00.00.1925	Joined the *NSDAP*.
00.00.1924	Joined the *Frontbann*.
00.00.1925-00.00.1930	Studied economics and law at *Johann Wolfgang Goethe-Universität* in Frankfurt-am-Main and, after 01.04.1927, at the University of München.
00.02.1925-31.03.1927	Reentered the *SA* in Frankfurt-am-Main, with successive assignments as *Schriftwart*, *Kassenwart*, and *Adjutant*.
02.06.1925	Joined the *NSDAP / Ortsgruppe Frankfurt-am-Main*.
00.00.1926-00.00.1930	Member of the *NSDStB*.
01.04.1927-30.04.1928	Assigned to *SA-Sturm 1* (München).

Official portrait of *SA-Gruppenführer* Gräntz in 1934.

01.05.1928-30.04.1931	*Adjutant* (m.d.W.d.G.b. until 30.04.1930, then permanent) of *SA-Standarte II* (Frankfurt-am-Main).
01.08.1928-20.08.1928	Assigned as an assistant to the *Stab der Obersten SA-Führer*.
00.00.1930-00.00.193_	Employed as a commercial clerk.
01.05.1931-31.12.1931	*Führer* of *SA-Sturmbann I/81* and *Standartenadjutant* of *SA-Standarte 81*.
18.10.1931	Participated in the *SA-Aufmarsch in Braunschweig*.
01.01.1932-13.04.1932	*Führer* of *SA-Sturmbann IV/88*.
01.07.1932-14.08.1933	*Adjutant* of *SA-Untergruppe Hessen-Nassau-Süd*.
11.09.1932-15.10.1932	Participated in the *12.Lehrgang* at the *Reichsführerschule der SA*, München.
01.12.1932	Appointed as a *hauptamtlicher SA-Führer*.
00.03.1933-00.05.1933	Adjutant to the *Polizeipräsident* in Frankfurt-am-Main.
05.03.1933-14.10.1933	Member of the *Preußischer Landtag (Wahlkreis 19, Hessen-Nassau)*.
12.03.1933-00.11.1933	Member of the *Kommunallandtag* in Wiesbaden and the *Provinziallandtag* in Hessen-Nassau (Kassel).
06.06.1933-14.08.1933	*Leiter* of the *SA-Führervorschule der SA-Gruppe Hessen*.
15.08.1933-31.03.1934	*Führer* (m.d.F.b.) of *SA-Brigade 49* (m.d.F.b. until 14.11.1933, then permanent from 15.11.1933) in Frankfurt-am-Main.
15.09.1933-00.04.1934	*Stabsführer* of *SA-Gruppe Hessen* (Frankfurt-am-Main).
12.11.1933	Unsuccessful candidate for election to the *Reichstag (Wahlkreis 19, Hessen-Nassau)*.

SA-Gruppenführer Gräntz congratulating an *SA* athlete, 1934.

Berlin, November 1944: *SA-Obergruppenführer* Gräntz reviews new recruits to the city's *Volkssturm*, of which he was *Gaustabsführer* under *Gauleiter* Dr. Goebbels. (NARA)

01.02.1934-01.03.1934	*Führer* ("vertretungsweise beauftragt mit der Führung") of *SA-Brigade 47* (Kassel).
00.02.1934-00.04.1934	*Ratsherr der Stadt Frankfurt-am-Main* (Councilman of the City of Frankfurt-am-Main).
01.04.1934-08.06.1934	Attached to the *Stab der Obersten SA-Führer*.
09.06.1934-30.11.1935	*Abteilungschef* in the *Personalamt der Obersten SA-Führung* (München).
01.11.1935	Attached to *SA-Gruppe Sachsen*.
01.12.1935-30.09.1936	*Führer* of *SA-Brigade 162 "Minden-Nienburg"* in *SA-Gruppe Sachsen* (Minden / Westfalen).
29.03.1936	Unsuccessful candidate for election to the *Reichstag*.
00.06.1936-00.06.1936	Attended a *Lehrgang* at the *Gruppenschule der SA-Gruppe Nordsee*.
01.10.1936-31.01.1942	*Führer* of *SA-Gruppe Westmark* (m.d.F.b. until 31.03.1937, then permanent from 01.04.1937; redesignated *SA-Gruppe Mittelrhein*, 01.07.1941). Succeeded Emil Steinhoff. Succeeded by Karl Lucke.
10.04.1938-30.04.1945	Member of the *Reichstag (Wahlkreis 21, Koblenz-Trier)*.
06.02.1939-06.05.1939	Military training exercises with *Infanterie-Regiment 19*.
00.00.1940-18.09.1943 (?)	War service as a reserve officer in *Infanterie-Regiment 468*. Wounded in action during the Western Campaign of May/June 1940; assigned as a *Leutnant d. R.* to the Regiment's *4.(MG.)Kompanie* as of 23.02.1942.
01.02.1942-31.12.1943	*Führer* of *SA-Gruppe Niedersachsen*.
18.09.1943	Granted 6 months' leave (through 31.03.1944) from military service by the *Oberkommando des Heeres* to exercise leadership over *SA-Gruppe Niedersachsen*.
01.01.1944-30.04.1945	*Führer* of *SA-Gruppe Berlin-Brandenburg* (redesignated *SA-Gruppe Berlin*, 01.09.1944). Succeeded Willi Künemund.
00.00.1944	Discharged from military service as a *Major d. R.*
00.10.1944-30.04.1945	*Gaustabsführer* of the *Volkssturm* in *Gau* Berlin.

Decorations and Awards

23.02.1942	*Kriegsorden des Deutschen Kreuzes in Gold* as *Leutnant d. R.* and *Zugführer* (?) in *4.(MG.)Kompanie / Infanterie-Regiment 468 / 268. Infanterie-Division*, Eastern Front
00.00.194_	*1939 Eisernes Kreuz I. Klasse*
31.05.1940	*1939 Eisernes Kreuz II. Klasse*
00.00.1940	*Verwundetenabzeichen, 1939 in Schwarz*
07.05.1934	*Goldenes Ehrenzeichen der NSDAP*
00.00.193_	*Abzeichen des SA-Treffens Braunschweig 1931*
30.01.1942	*Dienstauszeichnung der NSDAP in Gold*
20.04.1940	*Dienstauszeichnung der NSDAP in Silber*
20.04.1940	*Dienstauszeichnung der NSDAP in Bronze*
30.09.1942	*Anerkennungsschreiben des Stabschefs der SA*
00.00.193_	*SA-Sportabzeichen in Silber*
24.09.1935	*Deutsches Reiterabzeichen in Bronze*
15.10.1932	*Tyr-Rune*
00.00.1934	*Ehrendolch der SA* (mit Wirkung vom 03.02.1934)
00.02.1934	*Ehrenwinkel für alte Kämpfer*

Notes

* Son of the Studienrat Fritz Gräntz (born 28.12.1875 in Chemnitz) and his wife Susanne, née Weidner (born 01.09.1880 in Dresden).
* Unmarried.

Sources

Bundesarchiv, Berlin-Lichterfelde (former Berlin Document Center): *Personalunterlagen von SA-Angehörigen: SA-Personalakte* of Günther Gräntz.

Lilla, Joachim; Döring, Martin; & Schulz Andreas: *Statisten in Uniform. Die Mitglieder des Reichstags 1933-1945*. Droste Verlag, 2004.

Kurt Wilhelm Günther
SA-Obergruppenführer

Born:	31.10.1896 in Gera / Thüringen.
Executed:	03.04.1947 in Weimar (shot).

NSDAP-Nr.:	30,179 (First joined 00.09.1922; Party banned following the *München-Putsch* of 09.11.1923; Reenrolled, 15.02.1926) ·

Promotions

27.01.1918	*Gefreiter*
00.09.1922	*SA-Mann*
01.08.1927	*SA-Truppführer*
30.03.1929	*SA-Sturmführer*
01.10.1929	*SA-Standartenführer* (01.07.1932: Reinstated at that rank upon the lifting of the government ban on the SA)
29.11.1932	*SA-Oberführer* (mit Wirkung vom 01.07.1932)
10.11.1933	*SA-Brigadeführer*
09.11.1935	*SA-Gruppenführer*
09.11.1938	*SA-Obergruppenführer*
15.11.1939	*Gefreiter d. R.*
01.01.1940	*Unteroffizier d. R.*
01.03.1940	*Reserve-Offizier-Anwärter*
01.04.1940	*Feldwebel d. R.*
01.07.1940	*Leutnant d. R.*
01.05.1941	*Oberleutnant d. R.*

Career

00.00.1903-00.03.1912	Attended *Mittelschule* in Gera.
00.04.1912-00.10.1915	Training as a *Vermessungs- und Kulturtechniker* (measurement and culture technician).
23.10.1915-21.03.1916	Drafted (?) and assigned to *Pionier-Ersatz-Bataillon 11* (Hannoversch-Münden).
22.03.1916-01.09.1916	Assigned to *6. Kompanie / Pionier-Bataillons 11*.
01.09.1916-31.01.1918	Assigned to *Minenwerferkompanie 408* in France and Russia.
00.11.1916-00.00.191_	Hospitalized in the *Feldlazarett Asfeld-Vieux* (due to rheumatism).
01.02.1918-15.04.1918	Detached for *Spezial-Ausbildung* (specialized training) with the *Kgl. Preußischen Landesaufnahme* (Royal Prussian Land Survey) in Berlin.
15.04.1918-12.07.1918	Detached for *Spezial-Ausbildung* (special training in topography and trigonometry) to the *Versuchsstelle für Kriegsvermessungswesen* at the *Technischen Hochschule* in Stuttgart.
13.07.1918-27.11.1918	Assigned as a *Topograph und Trigonometer* (topography and triganometry technician) to *Vermessungsabteilung 8* in the Vosges (France).
28.11.1918-09.12.1918	Assigned to *1. Kompanie / Ersatz-Infanterie-Regiment 96*.

SA-Brigadeführer Günther in 1934. (Fritz Sauckel, *Kampf und Sieg in Thüringen: Im Geiste des Führers und in treuer Kameradschaft gewidmet den thüringischen Vorkämpfern des nationalsozialistischen Dritten Reiches,* 1934)

09.12.1918-10.12.1918	Assigned to *2. Kompanie / Ersatz-Bataillon / Infanterie-Regiment 172.*
10.12.1918	Discharged from military service.
00.01.1919-00.07.1919	*Vermessungstechniker* (measurement technician) with the *Gemeinnützigen Baugenossenschaft* (non-profit building society) in Gera (project engineering for settlement plans).
00.08.1919-00.12.1933	Topological surveyor with the *Bauabteilung* (construction department) of the leather factory *Hirschberg vorm. Heinrich Knoch & Co.* in Hirschberg/Saale.
00.09.1922	Cofounder of the *NSDAP Ortsgruppe Hirschberg/Saale.*
00.09.1922	Joined the *NSDAP / Ortsgruppe Hirschberg/Saale.*
00.09.1922-00.11.1923	*SA* service in Hirschberg/Saale.
00.09.1922-00.00.1929	*Schriftführer* (secretary) and *Schatzmeister* (treasurer) of the *Ortsgruppe Hirschberg/Saale der NSDAP* (with several breaks in service due to bans on the *NSDAP*).
00.11.1923	Briefly arrested and detained for supporting the *München-Putsch.*
03.06.1924	Ordered by the *Amtsgericht* (district court) in Hirschberg/Saale to pay a 25 *Reichsmark* fine due to his support for the *NSDAP* and unlawful possession of a weapon.
15.02.1926	Reenrolled in the *NSDAP.*
01.08.1927-01.10.1929	*SA-Führer* in Schleiz / Thüringen.
01.10.1929-13.04.1932	*Führer* of *I. Sturmbann / SA-Standarte 153* (Gera).

SA-Gruppenführer Günther chats with boys of the *Hitler-Jugend*, ca. 1937.

08.12.1929-31.12.1933	Member of the *Gemeinderat* in Tiefengrün.
31.07.1932-12.09.1932	Member of the *Reichstag (Wahlkreis 12, Thüringen)*.
29.11.1932-14.09.1933	*Führer* of *SA-Untergruppe Thüringen-Ost* (Gera)(with effect from 01.07.1932). This command was known as *SA-Brigade 41"Thüringen-Ost"* from 15.09.1933.
00.01.1933 00.01.1933	Participated in a five-day training course with the *Reichskuratorium für Jugendertüchtigung* (*Reich* Committee for Youth Training) in Ohrdruf.
05.03.1933-14.10.1933	Member of the *Reichstag (Wahlkreis 12, Thüringen)*.
15.09.1933-28.02.1935	*Führer* of *SA-Brigade 41 "Thüringen-Ost"* (Gera).
12.11.1933-08.05.1945	Member of the *Reichstag (Wahlkreis 12, Thüringen)*.
11.02.1934-14.02.1934	Participated in an *Informationskurs* (informational course) in Döberitz.
30.06.1934-02.07.1934	Arrested and detained in connection with the "Röhm-Putsch".
01.03.1935-00.05.1945	*Führer* (m.d.F.b. until 14.09.1935, then permanent from 15.09.1935) of *SA-Gruppe Thüringen* (Weimar). Assumed the post on 13.03.1935, and officially sworn in on 22.03.1935. Succeeded Gustav Zunkel, who had been killed in an auto accident on 08.12.1934.
25.01.1936-00.05.1945	Member of the *Arbeitskammer Mitteldeutschland*.
27.02.1936-12.04.1945	*Thüringischer Staatsrat*.
02.03.1937-04.03.1937	Completed a *P-Schein-Informationskurs* in Dresden.
01.06.1938-00.05.1945	*Präsident* of the *Mitteldeutsche Rennverein* (Central German Racing Club) in Gotha.
15.11.1939-09.12.1939	Entered *Wehrmacht* service as a *Gefreiter d. R.*, assigned to *3. Kompanie / Infanterie-Ersatz-Bataillon 8* (Frankfurt am Oder).

A postcard photo of *SA-Gruppenführer* Günther, signed by him in 1938. (Igor Karpov photo)

SA-Obergruppenführer Günther, ca. 1939.

10.12.1939-18.01.1940(?)	Assigned to *1. Kompanie / Infanterie-Regiment 307*.
18.01.1940-25.05.1941(?)	Assigned to *1. Kompanie / Infanterie-Regiment 324*, with which he participated in a *Luftlande* (air landing) operation in Norway on 09.04.1940).
25.05.1941-28.11.1941	*Führer* of *8.(MG-)Kompanie / Infanterie-Regiment 324* (operating in northern Finland from 07.07.1941).
01.09.1941	Wounded in a Soviet bombing raid.
04.11.1941-14.11.1941	Hospitalized in the Feldlazarett Alakurtti, Finland (for grippe and chronic kidney infection).
29.11.1941-11.12.1941(?)	Assigned to *5. Kompanie / Infanterie-Ersatz-Bataillon 203* (Güterfelde).
11.12.1941-23.01.1942	Hospitalized in the *Reservelazarett Weimar*.
23.02.1942-08.11.1943(?)	Attached to the staff of *Division 153* (Potsdam).
08.11.1943	Discharged from military service as an *Oberleutnant d. R.*
04.08.1944-00.05.1945	*Ehrenamtlicher Richter* (honorary judge) of the *Volksgerichtshof*.

Postwar Prosecution

Arrested by the Soviet NKVD, 17.05.1946, Günther was sentenced to death by the *Militärgericht des Bundeslandes Thüringen* (Military Court of the Federal State of Thuringia) on 20.02.1947.

Decorations and Awards

01.06.1940	*1939 Spange zum 1914 Eisernen Kreuz II. Klasse*
18.11.1917	*1914 Eisernes Kreuz II. Klasse*

25.05.1918	*Reußische Silberne Verdienstmedaille mit Schwertern*
15.11.1941	*Infanterie-Sturmabzeichen in Silber*
28.09.1941	*Verwundetenabzeichen, 1939 in Silber*
05.10.1940	*Medaille zur Erinnerung an den 1. Oktober 1938*
05.10.1940	*Medaille zur Erinnerung an den 13. März 1938*
18.12.1934	*Ehrenkreuz des Weltkrieges 1914-1918 mit Schwertern*
00.00.1934	*Goldenes Ehrenzeichen der NSDAP*
00.00.194_	*Dienstauszeichnung der NSDAP in Silber*
00.00.194_	*Dienstauszeichnung der NSDAP in Bronze*
01.12.1934	*SA-Sportabzeichen in Gold* (Nr. 1 822)
00.00.1937	*Deutsches Reitersportabzeichen in Silber*
07.11.1934	Authorized to wear the *Armelstreifen* for 8 years' active *SA* service (1925-1932)
00.00.1934	*Ehrendolch der SA* (mit Wirkung vom 03.02.1934)
00.02.1934	*Ehrenwinkel für alte Kämpfer*
29.12.1927	Portrait photo of Adolf Hitler (personally dedicated by the *Führer*)

Notes

- Son of the weaving mill foreman Wilhelm Günther (born 12.11.1857 in Lössnitz / Erzgebirge) and his wife Ida Marie (born 13.09.1859 in Meerane / Sachsen; died 21.03.1902).
- Married on 28.02.1925 to Marianne Schmidt (born 16.02.1902 in Schleiz / Thüringen, daughter of the *Geheime Archivrat* Dr. Berthold Schmidt). He fathered at least three children, including one son from a previous relationship (Hans Rudolf König, born 20.01.1920), as well as a son (Ewald) and daughter (Brigitte, born 11.02.1926) from his marriage with Marianne.
- Religion: Protestant until 00.00.1934, then left the church and declared himself "gottgläubig".

Sources

Bundesarchiv, Berlin-Lichterfelde (former Berlin Document Center): *Personalunterlagen von SA-Angehörigen: SA-Personalakte* of Kurt Günther

Lilla, Joachim; Döring, Martin; & Schulz Andreas: *Statisten in Uniform. Die Mitglieder des Reichstags 1933-1945*. Droste Verlag, 2004.

Heinrich ("Heinz") Josef Haake
SA-Obergruppenführer

Born:	24.01.1892 in Köln.
Died:	17.09.1945 in a British internment camp at Recklinghausen (or, according to Ernst Klee and Horst Romeyk, in the military hospital of the internment camp at Velen).
NSDAP-Nr.:	13 328 (First joined, 00.00.1922; Party banned following the *München-Putsch* of 09.11.1923; Reenrolled, 14.04.1925)

Promotions

00.00.191_	*Unteroffizier*
22.02.1925-01.06.1925	*Gauführer (= Gauleiter) der NSDAP*
00.07.1925-00.00.1927	*Ortsgruppenleiter der NSDAP*
00.00.193_	*Reichshauptamtsleiter der NSDAP*
20.04.1936	*SA-Brigadeführer*
09.11.1938	*SA-Gruppenführer*
24.01.1942	*SA-Obergruppenführer*

Career

ca. 1898-Winter 1905	Attended *Volksschule* in Köln.
Winter 1905-ca. 1910	Attended the *Schillergymnasium* in Köln-Ehrenfeld. He also served as a leader of the *Wandervogel* movement in Köln during this period.
ca. 1910-00.00.1914	Employed as a bank clerk in Köln.
00.08.1914-00.00.1918	Entered service as a *Kriegsfreiwilliger*, assigned to various *Infanterie-Regimenter* on the Western Front. Wounded in action four times, he was a participant in the assault on Langemarck during which he was wounded and subsequently classified as "schwerkriegsbeschädigt" (severely war-disabled).
00.00.1918-00.00.1919	In British captivity.
00.00.1919-00.00.1924	Returned to Köln, initially working in an insurance office and later in a branch of the *Deutsche Bank*.
00.00.1919-00.00.1922	Member of the *Deutschvölkischen Schutz- und Trutzbund*.
00.00.1922	Joined the *NSDAP* in Köln.
00.11.1923-00.00.1925	Member of the *NS-Freiheitsbewegung (NSFB)*.
00.09.1924-00.02.1925	*Gauleiter* of *Gau Rheinland-Süd* of the *Völkische Block*.
07.12.1924-14.10.1933	Member of the *Preußischer Landtag* (initially *Landeswahlvorschlag NSDAP*; from 31.03.1925-20.05.1928, he was the first and only delegate of the *NSDAP* to that body; after 05.03.1933, he represented *Wahlkreis 20, Köln-Aachen*). The *Nationalsozialistische Jahrbuch* of 1925 features a treatise, entitled "Parliamentarismus oder Diktatur?" (Parliamentarianism or Dictatorship?) by Haake which summarizes his political philosophy during this period. The following is an excerpt:

Heinz Haake, ca. 1933. (Igor Karpov photo)

Under a different form of government it will still be possible to control the great masses of the people who uphold the state, in order to lead them towards a unifying goal in times of necessity. The character of the parliament, however, not only obstructs this unifying leadership, but it also is responsible for the fact that there is flaming up in Germany a fight of all against all, and the individual against the individual for their daily bread … .

… So the first act of annihilation must be directed against the Jewish-democratic parliamentarianism. (Robert M. W. Kempner, *"Blueprint of the Nazi Underground- Past and Future Subversive Activities"*, in *Research Studies of the State College of Washington, Volume XIII, Number 2, June 1945*)

22.02.1925-27.03.1925	*Gauführer der NSDAP* in Köln.
27.03.1925-01.06.1925	*Gauführer* of *Gau Rheinland-Süd der NSDAP*. Soon after his appointment, Haake developed a reputation for stubbornly refusing to cooperate with other Party members, and particularly with his assigned *Gaugeschäftsführer* (business manager), Josef Grohé. He eventually asked Dr. Robert Ley to take over leadership of the *Gau*, which the latter did at the start of June.
14.04.1925	Officially reenrolled in the *NSDAP*.
00.07.1925-00.00.1927	*Ortsgruppenleiter* of the *Ortsgruppe Köln der NSDAP*.

An autographed portrait of Heinz Haake, ca. 1933. (Igor Karpov photo)

10.09.1925-00.00.1926	Member of the *Arbeitsgemeinschaft der Nord- und Nordwestdeutschen Gaue der NSDAP.*
19.08.1927-21.08.1927	Participated in the *3. Reichsparteitag der NSDAP* in Nürnberg.
00.06.1928-00.04.1932	Geschäftsführer of the *NSDAP-Fraktion* in the *Preußischer Landtag.*
01.10.1928-18.05.1929	*Führer der SS* in *Gau Rheinland-Süd.*
00.00.1930	Appointed as a *Sonderberater* (special advisor) to the *Stab des OSAF-Stellvertreters Ruhr-West.*
18.05.1929	Appointed as *Leiter* of the *Schutzstaffel-Rhein der NSDAP.*
00.00.1932-00.00.1933	*3. Vizepräsident* (3rd Vice-President) and *Kassenbeirat* (cash advisor) of the *Preußischer Landtag.*
00.04.1932-14.10.1933	*2. Stellvertreter* (2nd Deputy) to the *Fraktionsführer der NSDAP* in the *Preußischer Landtag.*
00.05.1932-01.02.1933	*Vorsitzender* of the *Ausschuß für das Vergebungswesen.*
15.07.1932-09.12.1932	*Landesinspekteur West der NSDAP* (responsible for the *Gaue* of *Düsseldorf, Essen, Koblenz-Trier, Köln-Aachen,* and *Saar*).
00.12.1932-00.00.1933	*Kommissar-West der NSDAP* and attached as *z.b.V.* (*zur besondere Verwendung,* for special employment) to the *Stabsleiter der Politischen Organisation der NSDAP* (Dr. Robert Ley).
00.12.1932-00.04.1933	*Leiter* of the *Organisationsamt der NSDAP.*
11.03.1933	Elected to succeeded Johannes Horion, deceased the previous month, as *Landeshauptmann* of the *Rheinprovinz.*
00.03.1933-14.10.1933	*1. Vizepräsident* of the *Preußischer Landtag (5. Periode).*

Köln, 1933: Visit of *Reichsminister* Dr. Goebbels. To Goebbels' right is
Gauleiter Josef Grohé and to his left, *Landeshauptmann* Heinz Haake.

11.04.1933-10.05.1945	*Landeshauptmann* of the *Rheinprovinz* (officially appointed to this post, 16.04.1933).
00.04.1933-00.12.1933	Member of the *Reichsvorstand der Deutsche Gemeindetag.*
00.00.1933-00.00.1934	*Landschaftsführer Rheinland* and member of the *Führerring der Reichsbund "Volkstum und Heimat".*
30.08.1933-03.09.1933	Participated in the 5. *Reichsparteitag der NSDAP* in Nürnberg.
00.10.1933-00.00.1944	Member of the *Akademie für Deutsches Recht*, München.
12.11.1933-08.05.1945	Member of the *Reichstag (Wahlkreis 20, Köln-Aachen).*
00.00.1934-00.05.1945	*Preußischer Provinzialrat* for the *Rheinprovinz.*
00.00.1934-00.05.1945	*Reichsinspekteur der NSDAP* in the *Obersten Leitung der PO.*
11.12.1934-00.05.1945	Member of the *Vorstand der Deutsche Gemeindetag* and *stellvertretender Vorsitzender* (deputy Chairman) of the *Provinzialdienststelle Rheinlandeinschliesslich Hohenzollern der Deutsche Gemeindetag* (Provincial Office for the Rhineland, including Hohenzollern, of the German Municipal Parliament).
00.00.193_-00.00.19__	*Stellvertreter des Leiters der Verwaltung* (Deputy to the Head of Administration) of the *Provinzialverband der Rheinprovinz* (Provincial Federation of the Rhine Province) in Düsseldorf.
00.00.193_-00.00.19__	*Vorsitzender der Landesplanung der Rheinprovinz* (Regional Planning Chairman of the Rhine Province).

An exquisite formal portrait of *SA-Gruppenführer* Haake, ca. 1937. (Laurens Hessels photo)

SA-Obergruppenführer Haake.

00.00.193_-00.00.19__	*Geschäftsführer* of the *Rheinischen landwirtschaftlichen Berufs-genossenschaft.*
00.00.193_-00.00.19__	*Leiter* of the *Landesversicherungsanstalt Rheinprovinz* in Düsseldorf.
00.00.193_-00.00.19__	*Vorsitzender* of the *Landesjugendamt Rheinland.*
00.00.193_-00.00.19__	*Vorsitzender* of the *Gemeindeunfallversicherungsverbandes Rheinprovinz und Hohenzollern.*
00.00.193_-00.00.19__	Member of the *Bundesleitung* of the *Verein für das Deutschtum im Ausland* (VDA).
00.00.193_-00.00.19__	*Beauftragter der Volksdeutschen Mittelstelle für volksdeutsche Fragen* (Representate of the Ethnic German Central Office for Ethnic German Questions).
00.00.193_-00.00.19__	*Leiter* of the *Forschungsstelle "Rheinländer in aller Welt"* (Research Office "Rhinelanders Throughout the World") of the *Deutschen Auslands-Institut*, Düsseldorf.
00.00.193_-00.00.19__	*Vorsitzender* of the *Kommunale Aufnahmegruppe GmbH*, Düsseldorf, and *Rheingas GmbH*, Düsseldorf.
00.00.193_-00.00.19__	*Vorsitzender des Verwaltungsrates* of *Rheinischen Girozentrale und Provinzialbank*, Düsseldorf, and *Provinzial Feuerversicherungsanstalt der Rheinprovinz*, Düsseldorf.
00.00.193_-00.00.19__	*Stellvertretender Vorsitzender des Aufsichtsrates* (Deputy Chairman of the Supervisory Board) of *Rheinisch-Westfälisches Elektrizitätswerk AG*, Essen; *Rheinischen Girozentrale und Provinzialbank*, Düsseldorf, *Rheinische Heim GmbH*, Bonn; and *Rheinische Heimstätte GmbH, Rheinisches Heim*, Bonn.

00.00.193_-00.00.19__	Member of the *Aufsichtsrat* of *Preußische Elektrizitäts AG*, Berlin.
00.00.193_-00.00.19__	Member of the *Verwaltungsrat* of *Preußische Landespfandbriefanstalt*, Berlin.
00.00.193_-00.00.19__	Member of the *Beirat* of *Ruhrgas AG*, Essen.
00.00.193_-00.00.19__	Member of the *Führerbeirat der Reichsverbande der öffentlich rechtlichen Versicherungen* (Lead Advisory Council of the *Reich* Federation of Public Legal Insurance).
00.00.193_-00.00.19__	*Führer* of the *Deutschen Bundes Heimatschutz* (German Federation for the Preservation of Regional Tradition).
00.00.193_-00.00.19__	Editor of the monthly magazine *Die Rheinprovinz.*
10.12.1935	Appointed as a member of the *Archäologischen Instituts des Deutschen Reiches.*
20.04.1936-00.05.1945	Assigned as *SA-Führer z. V.* (with the duties of *Berater für Fragen der Kultur und des Siedlungswesens* [Advisor for Questions of Culture and Settlement Matters]) to the staff of *SA-Gruppe Niederrhein.*
00.00.193_-00.00.19__	*Vorsitzender* of the *Arbeitsgemeinschaft der Landeshauptleute und Reichsgauhauptleute.*
25.06.1938	Appointed as an *Ehrensenator* of the University of Köln.

Postwar Confinement
Arrested by British authorities, 00.05.1945, then held in an internment camp until his death.

Published Work
Das Ehrenbuch des Führers. Der Weg zur Volksgemeinschaft (1934)
Erinnerungen aus der Kampfzeit (three albums of letters, documents, photos, and political leaflets dated 1924-1927, 1928-1930, and 1933)

Decorations and Awards

00.00.191_	*1914 Eisernes Kreuz II. Klasse*
00.00.194_	*Kriegsverdienstkreuz II. Klasse ohne Schwerter*
ca. 1918	*Verwundetenabzeichen, 1918 in Silber*
00.00.191_	*Langemarck-Kreuz*
ca. 1939	*Medaille zur Erinnerung an den 1. Oktober 1938*
ca. 1938	*Medaille zur Erinnerung an den 13. März 1938*
ca. 1934	*Ehrenkreuz des Weltkrieges 1914-1918 mit Schwertern*
00.00.193_	*Goldenes Ehrenzeichen der NSDAP*
00.00.193_	*Goldenes Hitler-Jugend Ehrenzeichen mit Eichenlaub*
00.00.194_	*Dienstauszeichnung der NSDAP in Gold*
00.00.194_	*Dienstauszeichnung der NSDAP in Silber*
00.00.194_	*Dienstauszeichnung der NSDAP in Bronze*
00.00.19__	*Ehrenzeichen für deutsche Volkspflege*
ca. 1939	Deutsches Schutzwall-Ehrenzeichen
00.00.1939	*Goldene Plakette der Universität Bonn*
00.00.193_	*Ehrenwinkel für alte Kämpfer*
00.00.193_	*Ehrenbürgerrecht der Technischen Hochschule Aachen.*

Notes

- First child of the Köln city architect Louis Haake (born 00.00.1853, died 00.00.1953) and his wife Cäcilie, née Antoni.
- Religion: Declared himself "gottgläubig", 00.00.19__.
- Married on 05.07.1921 to Christine Heuser (born 06.12.1893 in Köln, died 09.01.1960 in Köln; daughter of a *Postassistent* and later *Oberpostsekretär*). One child, a son, was born to this marriage.

Sources

Höffkes, Karl: *Hitlers politische Generale: Die Gauleiter des Dritten Reiches.* Grabert-Verlag-Tübingen, 1986.

Kempner, Robert M. W.: "Blueprint of the Nazi Underground- Past and Future Subversive Activities", in *Research Studies of the State College of Washington, Volume XIII.*

Lilla, Joachim; Döring, Martin; & Schulz Andreas: *Statisten in Uniform. Die Mitglieder des Reichstags 1933-1945.* Droste Verlag, 2004.

Miller, Michael D. & Schulz, Andreas: *Gauleiter: The Regional Leaders of the Nazi Party and Their Deputies, Volume I.* R. James Bender Publishing, 2012.

Romeyk, Horst: "Heinrich Haake (1892 – 1945)", in *Rheinische Lebensbilder 17*, 1997.

- *Die leitenden staatlichen und kommunalen Verwaltungsbeamten der Rheinprovinz 1816-1945.* Publikationen der Gesellschaft fur Rheinische Geschichtstkunde, 1994.

Friedrich ("Fritz") Haselmayr

SA-Obergruppenführer

Born:	11.04.1879 in Kirchenlaibach bei Bayreuth / Regierungsbezirk Oberfranken / Bayern.
Died:	18.06.1965 in Locham bei München / Bayern.

NSDAP-Nr.:	85,034 (Joined 01.04.1928)

Promotions

16.07.1897	*Fahnenjunker*
17.12.1897	*Fahnenjunker-Unteroffizier*
25.01.1898	*Portepeefähnrich*
10.03.1899	*Leutnant*
28.10.1909	*Oberleutnant*
01.10.1913	*Hauptmann*
18.08.1918	*Major*
01.02.1924	*Oberstleutnant*
31.01.1928	*Charakter als Oberst a. D.*
01.04.1933	*SA-Gruppenführer*
00.00.193_	*Reichshauptamtsleiter der NSDAP*
25.07.1935	*Charakter als Generalmajor*
00.00.1938	*Hauptdienstleiter der NSDAP*
11.04.1939	*SA-Obergruppenführer*
01.12.1940	*Oberst z. V.*
01.07.1941	*Generalmajor z. V.*
01.01.1943	*Generalleutnant z. V.*

Career

00.00.1885-00.00.1889	Attended *Volksschule.*
00.00.1889-00.00.1897	Attended a humanistic *Gymnasium* in Passau (graduated and passed his *Abitur*, 00.00.1897)
16.07.1897-01.10.1912	Entered service as *Fahnenjunker*, assigned to 16. *Kgl. Bayerisches Feldartillerie-Regiment "Großherzog von Toskana"* (Passau).
01.03.1898-24.01.1899	Attended the *Kriegsschule* in München.
01.06.1900-28.06.1900	Attended a *Pionierkurs* with 3. *Kgl. Bayerisches Pionier-Bataillon.*
22.11.1905-30.09.1907	*Bezirksadjutant* of the *Bezirkskommando* in Passau.
01.10.1907-12.09.1910	Attended the *Kgl. Bayerisches Kriegsakademie* in München.
12.09.1910	Returned to duty with *16. Kgl. Bayerisches Feldartillerie-Regiment.*
15.12.1910-15.04.1911	Attached to the *Linienkommandantur K I* in München.
01.10.1912-30.09.1913	Attached to the *Kriegsschule* in München as a *Hilfslehrer (Taktiklehrer*, tactics instructor) for the duration of the *Lehrkurses 1912/13* (1912/13 instructional courses).

Hauptmann Haselmayr, ca. 1914. (Victor Gross Militärische Antiquitäten photo)

01.10.1913-00.00.1914	Attached to the *Kriegsschule* in München as a *Hilfslehrer (Taktiklehrer)* for the duration of the *Lehrkurses*
01.10.1913	*Hauptmann beim Stabe* of 16. *Kgl. Bayerisches Feldartillerie-Regiment.*
02.08.1914-05.09.1914	*Führer* of 7. *Kompanie / Kgl. Bayerisches Reserve-Infanterie-Regiment 2* on the Western Front.
05.09.1914 – 12.09.1914	*Führer* of III. *Bataillon / Kgl. Bayerisches Reserve-Infanterie-Regiment 2.*
12.09.1914 – 02.10.1914	*Führer* of 7. *Kompanie / Kgl. Bayerisches Reserve-Infanterie-Regiment 2.*
02.10.1914-18.02.1915	*Führer* of II. *Bataillon / Kgl. Bayerisches Reserve-Infanterie-Regiment 2.*
18.02.1915 – 28.02.1915	*Führer* of I. *Bataillon / Kgl. Bayerisches Reserve-Infanterie-Regiment 1.*
01.03.1915-04.06.1915	*Führer* of 9. *Kompanie / Kgl. Bayerisches Reserve-Infanterie-Regiment 1.*
04.06.1915-05.07.1915	*Führer* of I. *Bataillon / Kgl. Bayerisches Reserve-Infanterie-Regiment 12.*
06.07.1915-14.08.1915(?)	*Adjutant* of 1. *Kgl. Bayerisches Landwehr-Division.*
14.08.1915-27.09.1916	*2.Adjutant* of 1. *Kgl. Bayerisches Landwehr-Division.*
27.09.1916-05.04.1917	*1.Adjutant* of 1. *Kgl. Bayerisches Landwehr-Division.*
06.04.1917-15.02.1918	Assigned to the *Kgl. Bayerisches Generalkommando z.b.V. 63.*
15.02.1918-07.03.1918	*Generalstabsoffizier* in 15. *Kgl. Bayerisches Infanterie-Division.*
07.03.1918-28.07.1918	Assigned as a *Generalstabsoffizier* to the *Generalstab* of *Etappen-Inspektion 2.*
28.07.1918-16.08.1918	Assigned to the *Generalstab* of *Etappen-Inspektion 7.*
16.08.1918-06.11.1918	Assigned to the *Generalstab* of *Etappen-Inspektion 2.*
06.11.1918-04.12.1918	*Ia* in the *Generalstab* of *Etappen-Inspektion 2.*
05.12.1918-04.06.1919	*Adjutant* to the *Generalkommando* of I. *Kgl. Bayerisches Armee-Korps.*

SA-Gruppenführer Haselmayr in 1933. (Victor Gross Militärische Antiquitäten photos)

04.06.1919-04.10.1919	*Wehrkommissar* (Defense Commissioner with assignment as *Kommandeur der Einwohnerwehr*) of the government in Niederbayern (Landshut).
04.10.1919-28.08.1920	Assigned as a *Stabsoffizier* to the staff of *II. Bataillon / Reichswehr-Infanterie-Regiment 45* (Würzburg).
01.02.1920-04.05.1920	*Stellvertretender Kommandeur* (m.d.F.b.) of *II. Bataillon / Reichswehr-Infanterie-Regiment 45.* Involved in the suppression of the uprising in the Ruhr.
28.08.1920-01.01.1921	*Kommandeur* (m.d.F.b. until 01.10.1920, then permanent) of *III. Bataillon / Reichswehr-Infanterie-Regiment 45.*
01.01.1921-01.10.1923	*Kommandeur* of *II. Bataillon / Infanterie-Regiment 20* (Ingolstadt).
01.10.1923-30.06.1925	*Generalstabsoffizier*, assigned as Instructor for *Führergehilfen-Ausbildung* to the staff of *7.Division* and *Wehrkreiskommando VII* (München).
09.11.1923	Intermediary between the *Reichswehr* and the Putschists during the *München-Putsch.*
00.00.1925	Founder of the *Arbeitsgemeinschaft für wehrgeistige Forschung e. V.*
01.07.1925-01.10.1926	Temporarily assigned as an instructor of *Führergehilfen-Ausbildung* in *Reiter-Regiment 8* (Öls).
01.10.1926-15.01.1927	Temporarily assigned to *7.Division* (München).
15.01.1927-01.03.1927	Temporarily assigned to the *Kommandantur München.*
01.03.1927-31.01.1928	Assigned to the staff of the *Kommandantur München.*
31.01.1928	Discharged from the *Reichswehr.*

SA-Gruppenführer Haselmayr with *SA-Obergruppenführer* Franz Ritter von Epp during the 5.*Reichsparteitag der NSDAP* in Nürnberg, September 1933. Visible behind von Epp is Franz Pfeffer von Salomon. (Victor Gross Militärische Antiquitäten photo)

01.04.1928	Joined the *NSDAP / Ortsgruppe München*.
00.00.1928	Joined the *NS-Deutsche Studentenbund* (*NSDStB*, National Socialist German Students League).
00.00.1928-00.00.1930	Studied history and law at the University of München.
00.00.1928-00.00.1929	Member (as representative of the *NSDAP*) of the *Allgemeine Studentenausschuß* (*AstA*) at the University of München.
00.00.1927	Cofounder of the National-Socialist student's newspaper *Akademischer Beobachter*.
00.00.1929-00.00.1933	Active as a freelance writer.
00.00.1929-00.00.1933	*Mitarbeiter*, dealing with military topics, for the *Völkischer Beobachter*.
18.01.1930-31.12.1933	*Leiter* of the *Arbeitsgemeinschaft für die deutsche Wehrverstärkung* (Working Group for German Rearmament).
00.00.1930-00.00.193_	*Beauftragter für wehrpolitische Fragen* (Representative for Defense Policy Questions) in the *Reichsleitung der NSDAP*.
00.00.1931-00.00.1931	In Switzerland as member of the *Beobachterdelegation* (observers' delegation) of the *NSDAP* to the Disarmament Conference in Geneva and the Reparations Conference in Lausanne.
08.09.1932-00.10.1935	*Stellvertretender Leiter* of the *Wehrpolitische Amt der Obersten SA-Führung*.

SA-Obergruppenführer Haselmayr poses for formal portraits in 1939. (Victor Gross Militärische Antiquitäten photo)

08.09.1932-31.03.1934	*Hauptgeschäftsleiter* (head business manager) of the *Wehrpolitische Amt der NSDAP.*
21.11.1932	Appointed to the *Stab der Obersten SA-Führung.*
28.04.1933-14.10.1933	Member of the *Kgl. Bayerisches Landtag.*
01.04.1933	Joined the *SA.*
06.05.1933	Appointed as *stellvertretender Vorsitzender* (deputy chairman) of the *Deutschen Liga für Völkerbund.*
01.08.1933-00.01.1936	Honorary member of the *Deutschen Gesellschaft für Wehrpolitik und Wehrwissenschaften.*
12.11.1933-28.03.1936	Member of the *Reichstag (Wahlkreis 24, Oberbayern-Schwaben).*
27.01.1934-00.05.1945	Assigned as an *SA-Führer z. V.* to the *Stab der Obersten SA-Führung.*
01.04.1934-00.10.1935	*Leiter* of *Abteilung I (Politische Abteilung)* of the *Wehrpolitische Amt der NSDAP.*
00.00.1934(?)-00.00.19__	*Vizepräsident* of the *Deutsche Gesellschaft für Völkerbundfragen.*
30.09.1935	Founder and *Leiter* of the *Arbeitsgemeinschaft für Wehrgeistige Forschung.*
05.10.1935	Appointed as a member of the *Sachverständigenrat* (council of experts) of the *Reichsinstitut für Geschichte des neuen Deutschland (Reich* Institute for the History of the New Germany), Berlin.
00.00.193_-00.00.19__	*Präsident* of the *Deutsche Studienkomitee der New Commonwealth* (German Study Committee of the New Commonwealth).
00.00.193_-00.00.19__	*Hauptlektor* (head lecturer) of the *Reichsstelle zur Förderung des deutschen Schrifttums.*

Generalmajor z. V. Haselmayr, ca. 1941. (Victor Gross Militärische Antiquitäten photo)

29.03.1936	Unsuccessful candidate for election to the *Reichstag*.
26.11.1936-00.00.19__	*Ehrenamtlicher 1. Beigeordneter* of the *Gemeinde Gräfelfing*.
10.04.1938	Unsuccessfully proposed for the *Liste des Führers zur Wahl des Großdeutschen Reichstages am 10. April 1938*. (List of the *Führer* for election to the Greater German *Reichstag* on 10. April 1938).
01.07.1938	Placed "at disposal" (*zur Verwundung / z. V.*) of the German Army.
20.06.1940-01.09.1940	Assigned as *Bearbeiter für französische Kriegsgefangene* (officer responsible for matters related to French prisoners of war) to the *Generalstab* of the *Generalkommando z.b.V.* and later to the *Kommandoamt der Kriegsgefangenen in Frankreich*.
01.09.1940-26.12.1941	*Feldkommandant 569 (FK 569)* in Lille.
26.12.1941-01.12.1942	*Oberfeldkommandant 579 (OFK 579)* and *Kommandeur* vom *Sicherungsgebiete "Wolhynien-Podolien"* (Commander of the Security Region for Volhynia-Podolia [Ukraine]).
01.12.1942-31.01.1943	In *Führerreserve OKH*.
31.01.1943	Discharged from military service.

Postwar Prosecution

In Allied captivity, 00.07.1945-28.06.1947. After three de-Nazification hearings (before the *Hauptkammer München-Land* on 09.02.1949, the *Hauptkammer München* on 15.01.1951, and the *Berufskammer München* on 24.07.1951), he was placed in *Gruppe* II (as a *Belasteter*, or incriminated person). On 11.03.1952, at the intercession of the *Minister für Politische Befreiung in Bayern*, those previous court decisions were nullified.

Published Works

Deutschlands Schicksalsstunde bevorstehende Entscheidung in der Wehrfrage (1930)
"Deutschlands Recht auf Wehrverstärkung" (Essay, 1931)
Wehrpolitik und wehrpolitische Erziehung (1934)
Die Wehrmacht (1935)
Wehrgeistiges Schrifttum (1938)
Grundlagen, Aufbau und Wirtschaftsordnung des Nationalsozialistischen Staates: Der Aufbau des nationalsozialistischen Staates. Der verwaltungsrechtliche Aufbau, Bd. 1, 2, & 30 (1939)

Decorations and Awards

17.02.1915	*1914 Eisernes Kreuz I. Klasse*
14.12.1914	*1914 Eisernes Kreuz II. Klasse*
15.09.1915	*Kgl. Bayerischer Militär-Verdienstorden IV. Klasse mit Schwertern mit der Kronee und mit Schwertern*
11.09.1914	*Kgl. Bayerischer Militär-Verdienstorden IV. Klasse mit Schwertern*
ca. 1934	*Ehrenkreuz des Weltkrieges 1914-1918 mit Schwertern*
12.03.1905	*Kgl. Bayerisches Jubiläums-Medaille*
00.00.1917	*Dienstauszeichnung II. Klasse*
00.00.19__	*Wehrmacht-Dienstauszeichnungen*
01.12.1933	*Goldenes Ehrenzeichen der NSDAP*
30.01.1941	*Dienstauszeichnung der NSDAP in Silber*
30.01.1941	*Dienstauszeichnung der NSDAP in Bronze*
05.12.1933	*Goldenes Ehrenzeichen der NSDAP*
30.01.1941	*Dienstauszeichnung der NSDAP in Silber*
30.01.1941	*Dienstauszeichnung der NSDAP in Bronze*
00.00.1934	*Ehrendolch der SA* (mit Wirkung vom 03.02.1934)
00.02.1934	*Ehrenwinkel für alte Kämpfer*

Notes

• Son of the *Bezirksgeometer* (district topological surveyor) Josef Haselmayr (born 18.06.1838 in Dillingen, died 25.02.1907) and his wife Rosa, née Steiner (born 30.05.1856 in Kirchenlaibach, died 29.01.1910).
• Married on 18.04.1911 to Annaluise Ott (born 06.02.1885 in Neu-Ulm, died 07.11.1962). One daughter, Erika (born 10.04.1920, died 20.10.2001 [?]), resulted from this marriage.
• Religion: Catholic.

Sources

Bayerisches Hauptstaatsarchiv, München, Abteilung IV Kriegsarchiv: Excerpts from various *Kriegsranglisten* containing data on the Royal Bavarian Army service of Friedrich Haselmayr.

Bundesarchiv, Berlin-Lichterfelde (former Berlin Document Center): *Personalunterlagen von SA-Angehörigen: SA-Personalakte* of Friedrich Haselmayr.

Lilla, Joachim; Döring, Martin; & Schulz Andreas: *Statisten in Uniform. Die Mitglieder des Reichstags 1933-1945*. Droste Verlag, 2004.

Edmund Heines
SA-Obergruppenführer

Born:	21.07.1897 in München / Bayern.
Executed:	30.06.1934 in the Strafgefängnis München-Stadelheim (shot by a firing squad of the *Leibstandarte "Adolf Hitler"*).
NSDAP-Nr.:	________ (First joined 00.00.1921 or 00.12.1922; Party banned following the *München-Putsch* of 09.11.1923; Reenrolled, 00.00.1925)

Promotions

26.10.1915	*überzähliger Gefreiter*
21.06.1916	*Unteroffizier*
12.02.1917	*Offizier-Aspirant*
24.04.1917	*Vizewachtmeister d. R.*
08.01.1918	*Leutnant d. R.*
00.00.1927 (?)	*SA-Sturmführer*
00.00.1929 (?)	*SA-Standartenführer*
ca. 1929	*SA-Oberführer*
00.00.1930-00.00.1930	*Ortsgruppenleiter* der *NSDAP*
00.06.1930-00.11.1930	*(kommissarischer) Gauleiter* der *NSDAP*
14.10.1931	*SA-Gruppenführer*
00.00.1933-30.06.1934	*Stellvertretender Gauleiter* der *NSDAP*
25.03.1933-30.06.1934	*Polizeipräsident.*
20.04.1933	*SA-Obergruppenführer*

Career

ca. 1903-ca. 1907	Attended *Volksschule* in München (presumably)
ca. 1907-00.00.1915	Attended *Gymnasium* and *Realgymnasium* in München (passed his *Abitur*).
30.01.1915	Mobilized as a *Kriegsfreiwilliger* at Rekruten-Depot 1 of *Ersatz-Abteilung / 9. Kgl. Bayerisches Feldartillerie-Regiment.*
17.03.1915-21.04.1915	Assigned to *2. Ersatz-Batterie / 9. Kgl. Bayerisches Feldartillerie-Regiment.*
21.04.1915-30.04.1916(?)	Assigned to *1.Batterie / 9. Kgl. Bayerisches Feldartillerie-Regiment* and deployed to the front.
05.06.1915-07.06.1915	Received medical treatment for boils on the upper right arm.
25.08.1915-28.08.1915	Treated for a skin abscess behind and below the right ear.
11.10.1915	Severely wounded in the head by an artillery blast, then hospitalized in Krefeld (11.10.1915-19.10.1915) and München (20.10.1915-30.04.1916.
30.04.1916-19.07.1916	Returned to service with assignment to *2. Ersatz-Batterie / Ersatz-Abteilung / 9. Kgl. Bayerisches Artillerie-Regiment.*
19.07.1916-20.07.1916	Assigned for just one day to *Gebirgs-Kanonen-Ersatz-Abteilung 213.*

An autographed photo of Edmund Heines in civilian
attire, ca. 1923. (Snyder's Treasures photo)

20.07.1916-18.09.1918	Assigned to *Bayerische Infanterie-Geschütz-Batterie 8*, with which he deployed to the front on 02.09.1916.
18.09.1918-00.04.1919 (?)	Assigned to *Infanterie-Geschütz-Batterie 24*.
00.04.1919-00.05.1919	Member of *Freikorps Oberland*, with which he participated in the suppression of the *Münchener Räterepublik* (the short-lived communist government in Bavaria).
00.00.1919-00.12.1922	*Führer* of an *MG-Kompanie* in the *Freikorps Roßbach*, (1921). He participated in the "Kapp-Putsch" of March 1920, then in combat on the Baltic (1920) and in Oberschlesien (1921). His unit was eventually based at the Resenfeld estate in Kreis Greifenhagen near Stettin. In July 1920, he committed the "Fememord" of a 26-year-old "Rossbacher" and farm worker named Willi Schmidt, who had attempted to leave the organization and been accused of conspiring to reveal the location of secret *Freikorps* weapons dumps to the Allied Control Commission in Stettin. He personally shot the victim, a crime for which he was later tried and convicted (see entry for 00.00.1929, *below*).
00.01.1920	Became *Geschäftsführer* of the "Tiergarten-Club" in Berlin (which Gerhard Roßbach had taken over that month for use as the headquarters of his *Organisation Roßbach*).
00.00.1922-00.12.1922	*Führer* of the *München Ortsgruppe* of *Freikorps Roßbach*.
00.12.1922	Transferred, together with his *Freikorps Roßbach Ortsgruppe*, to the *SA*.

00.12.1922	Joined the *NSDAP* in München (date per Andreas Schulz; Sten Lundø gives 00.00.1921).
00.12.1922-00.11.1923	*Führer* of *2. Bataillon / SA-Regiment München*, numbering nearly 300 men. In this assignment, on 08./09.11.1923, he participated in the *München-Putsch*, tasked with securing the Hotel Vierjahreszeiten (where Belgian and French officers of the Allied Military Control Commission were quartered); John Dornberg writes: "Whether it was Göring's idea to take the Allied officers into 'protective custody' or Heines's own, as he later admitted to police, 'because I wanted to do something special for the revolution,' has remained unknown" (Dornberg, *Munich 1923*, p. 175) After ordering his troops to surround the building, Heines and 20 men – among them the future *Reichsleiter* Dr. Hans Frank – stormed into the building. With the aid of his pistol, he forced Christian Tauber, the night manager, to provide him with the list of Allied officers' room numbers. At one point, he and a French colonel faced one another with pistols drawn, but the Frenchmen backed down. Hines originally intended to take some of the officers hostage, but instead placed guards at their doors. He then led his men out of the hotel to the Infantry School where they were encamped until being ordered by Göring to report to the Bürgerbräukellar at 0300 hours.
00.00.1924	Sentenced to 15 months' fortress arrest for his involvement in the *München-Putsch*.
00.04.1924-00.09.1924	Imprisoned with Hitler and other Putschists at the Festung Landsberg am Lech.
00.09.1924-00.00.1925	*Führer* of 2. *Bataillon* / Regiment München of the *Frontbann*.
00.00.1925	Reenrolled in the *NSDAP*.
00.00.1925-00.00.1926	*Landesführer Bayern* of the *Abteilung Roßbach* (absorbed by the *SA*, 00.00.1926).
00.00.1925-00.08.1926	*Bundesleiter* of the *Wehrjugendverband Schill* (*Schill-Jugend*) (a National-Socialist youth organization in southern Germany).
06.05.1925	Appointed as *Führer* of the *Jugendbund der NSDAP* (Youth League of the *NSDAP*).
00.00.1926-00.00.192_	Studied law at the University of Erlangen.
00.00.1926	Reentered the *SA*.
00.00.1926-31.05.1927	*Führer* of *SA-Sturm 3*, then *SA-Sturm* 9 (München) and *Leiter* of the *Zeugmeisterei* (quartermaster's branch) of *Sportverband Schill*. He was also active as a *Reichsredner der NSDAP* during this period.
00.05.1927	Initiated a rebellion by the *SA* in München.
31.05.1927	Expelled from the *NSDAP* and *SA* for his role in the aforementioned rebellion.
00.07.1927	*Führer* of the *Treuschaften Neuhausen, Innere Stadt und Rossbach* in München.
00.05.1928	Tried by the *Schwurgericht* (assize court) in Stettin in the so-called "*Stettiner Fememordprozeß*" (Stettin Feme Murder Trial) for the July 1920 murder of Willi Schmidt (see entry for 00.00.1919-00.12.1922, above). Although the prosecution sought the death penalty, he was

An autographed photo of *SA-Sturmführer* Heines in 1926. (*Bundesarchiv Bild*)

SA-Gruppenführer Heines, ca. 1932.

instead sentenced to 15 years' imprisonment for manslaughter. The court sentence stated "Heines thrust his pistol into Schmidt's face and fired twice." On 13.03.1929, the third criminal court of the Reichsgericht suspended his original sentence e and commuted it to just 5 years' imprisonment. After he'd spent just one and one-half years' behind bars, his attorney appealed the sentence once more and he was released on 5.000 RM bail on 14.05.1929.

00.00.1929	Resumed law studies.
00.00.1929	Reenrolled in the *NSDAP* and the *SA*.
00.08.1929-00.06.1930	*Führer* of *SA-Standarte X München-Land*. Succeeded by Hans Kallenbach.
00.00.1930-00.00.1930	*Ortsgruppenleiter* of the *Ortsgruppe München-Haidhausen der NSDAP*.
00.00.1930-00.00.1930	*Adjutant* to *Gauleiter* Adolf Wagner.
00.05.1930-00.08.1930	*Gaupropagandaleiter* of *Gau Gross-München (Oberbayern) der NSDAP*".
00.06.1930-00.11.1930	*Gau-SA-Führer* and *(geschäftsführender) stellvertretender Gauleiter* of *(Unter-)Gau Oberpfalz der NSDAP* (Seat: Regensburg).
00.06.1930-00.11.1930	*(kommissarischer) Gauleiter* of *(Unter-)Gau Oberpfalz der NSDAP*. Succeeded Franz Maierhofer. Succeeded by the same man.
14.09.1930-30.06.1934	Member of the *Reichstag* (after 30.07.1932, representing *Wahlkreis 8, Liegnitz*).

00.11.1930-00.04.1931	Assigned as a *Referent z.b.V.* (responsible for the *Abteilung Nachrichtenwesen und Presse* [News and Press Department]) to the *Begleitstab der Obersten SA-Führung* (Escort Staff of the Supreme *SA* Leadership).
07.04.1931-28.04.1931	*(kommissarisch) Gausturmsturmführer* Berlin. Succeeded Ernst Wetzel. Succeeded by Horst von Petersdorff.
01.(?).05.1931-30.06.1934	*Stellvertreter des Stabschefs der SA* (Deputy to the Chief of Staff of the *SA* [Ernst Röhm]).
31.07.1931-30.06.1934	*Führer* of *SA-Gruppe Schlesien* (Breslau) ("beauftragt mit der kommissarischen Führung" until 13.10.1931, then permanent from 14.10.1931). Succeeded by Otto Herzog.
12.05.1932	Involved in an assault on the journalist Helmuth Klotz in the restaurant of the *Reichstag*. In 1929, Klotz (a distinguished naval aviator in World War I, early member of the *NSDAP*, and participant in the *München-Putsch*) had switched his allegiance from the *NSDAP* to the SPD, and had publicized Ernst Röhm's homosexuality in March 1932. As a consequence of the crime, Heines and the other perpetrators (Wilhelm Stegmann and Fritz Weitzel) were excluded from the *Reichstag* for the 30 days and, on 14.05.1932, sentenced to the 3 months' imprisonment for "gemeinschaftlicher Körperverletzung" (joint assault) by the *Schnellschöfengericht* of Berlin-Mitte. Klotz fled to Prague on 15.03.1933, then moved to Paris on 01.05.1933. He was arrested near the French capital on 08.07.1940 and interned at KL-Sachsenhausen, then sentenced to death for high treason by the *Volksgerichtshof* on 27.11.1942 and executed at Berlin-Plötzensee on 03.02.1943 (aged 48).
00.08.1932	Appointed as *Sachberater* (specialist) for *N.S. Notwehr (Gefangenenhilfe)* (National Socialist Self Defense [Prisoners' Aid]) in the *NSDAP-Reichstagsfraktion*.
06.12.1932-01.02.1933	Member of the *1. Ausschuß (Wahrung der Rechte der Volksvertreter) der Reichstag* (1st Committee [Respect for the Rights of the Peoples' Representatives] of the *Reichstag*).
00.00.1933-30.06.1934	*Stellvertretender Gauleiter* of *Gau Schlesien der NSDAP* (Seat: Breslau). Succeeded by Walter Gottschalk, who assumed the post on 01.09.1934.
00.04.1933-10.07.1933	Member of the *Preußische Staatsrat*.
25.03.1933-30.06.1934	*Polizeipräsident* in Breslau. Succeeded *Regierungsrat* Joachim von Alt-Stutterheim. Succeeded by Albrecht Schmelt.
00.00.1933-30.06.1934	*Sonderkommissar der OSAF* [later redesignated *Sonderbevollmächtigter der OSAF*] in der *Provinz Schlesien*.
28.04.1933	Established an unofficial ("wild") concentration camp at Dürrgoy in the southern part of Breslau. Among the inmates of this camp, where beatings and torture were the norm, were the former Social Democratic *Oberpräsident* of *Niederschlesien* (Hermann Lüdemann, 1880-1959, imprisoned until 1935), the former Social Democratic mayor of Breslau, and the former editor of the Social Democrats' daily newspaper in Breslau.

Heines in 1933.

01.05.1933-30.06.1934	*Führer* of *SA-Obergruppe I* (consisting of *SA-Gruppen Berlin-Brandenburg, Ostmark, Schlesien, Ostland, Pommern,* and *Nordmark*).
00.06.1933	Began work for the periodical *Gasschutz und Luftschutz*, Berlin.
06.08.1933	Participated in the "*SA-Appell*" on Tempelhofer Feld, Berlin, together with *SA-Stabschef* Ernst Röhm.
01.07.1933-14.03.1934	*Führer* of *SA-Obergruppe III* (consisting of *SA-Gruppen Berlin-Brandenburg, Ostmark, and Schlesien*; HQ: Breslau). Succeeded Manfred Freiherr von Killinger. Succeeded by Siegfried Kasche.
14.09.1933-30.06.1934	*Preußischer Staatsrat.*
07.10.1933-08.10.1933	Accompanied by *SA-Stabschef* Röhm, presided over the *Herbstparade* [Autumn Parade] *der SA* in Breslau.
26.11.1933	Witness at the 32nd session of the "*Reichstagsbrandprozeß*" (*Reichstag Fire Trial*) before the *Reichsgericht* in Leipzig.
00.00.1933-30.06.1934	*Landesgruppenleiter* of *Landesgruppe XV (Schlesien)* of the *Reichsluftschutzbund*.
22.01.1934	Participated in the *SA-Führertagung (SA* leadership conference) at the *Reichspropagandaministerium*, Berlin.
15.03.1934-30.06.1934	*Führer* of the newly formed *SA-Obergruppe VIII* (m.d.F.b. until 13.04.1934, then permanent from 14.04.1934). In this post, with headquarters in Breslau, he oversaw *SA-Gruppe Schlesien*.
00.00.1934-30.06.1934	*Preußischer Provinzialrat* for *Provinz Niederschlesien*.
00.00.1934-30.06.1934	*Leiter* of the *Staatspolizeistelle* Breslau.
00.06.1934	Addressed his *SA* troops in Breslau:

SA-Obergruppenführer Heines on horseback in 1933.

Keep a stiff upper lip and be ready for all emergencies. The German people, in their true German way, have been far too kind to their enemies. The Jews are more insolent than ever. The virus of Judaism is a plague infecting the whole world .... Germany will be eternal because the storm-troop organization is eternal. (*New York Times*, 26.06.1934)

30.06.1934   Arrested together with Ernst Röhm and other *SA* leaders at the Pension Hanslbauer in Bad Wiessee. The following is the postwar account by Hitler's chauffeur, Erich Kempka:

I run quickly up the stairs to the first floor where Hitler is just coming out of Röhm's bedroom. Two detectives come out of the room opposite. One of them reports to Hitler: "My *Führer*... the Police-President of Breslau is refusing to get dressed!"

Taking no notice of me, Hitler enters the room where *Obergruppenführer* Heines is remaining. I hear him shout: "Heines, if you are not dressed in five minutes I'll have you shot on the spot!" I withdraw a few steps and a police officer whispers to me that Heines had been in bed with an 18-year-old *SA-Obertruppführer*. At last Heines comes out of the room with an 18-year-old fair-haired boy mincing in front of him. "Into the laundry room with

Breslau, 1933: *SA-Obergruppenführer* Edmund Heines with the *Gauleiter* of Schlesien (Silesia), Helmuth Brückner (in his uniform as an honorary *SA-Gruppenführer*). (Laurens Hessels photos)

them!" cries [Julius] Schreck.... Now the bus arrives which had been fetched by Schreck. Quickly, the *SA* leaders are collected from the laundry room and walk past Röhm under police guard. (ibid, p. 179)

Another account, appearing in Heinz Höhne's *The Order of the Death's Head*, states that shortly after 0630 hours, Hitler personally hammered on the door of Heines' room, whereupon Heines answered; a male companion could be seen in the room behind him. Viktor Lutze then entered the room, searching it for weapons, whereupon Heines implored, "Lutze, I've done nothing. Can't you help me?" Lutze replied, "I can do nothing ... I can do nothing." Heines and his fellow *SA* "conspirators" were then transported to the prison at München-Stadelheim. Heines was confined to Cell 483 until evening, when he and five other senior *SA* leaders were shot by a firing squad of the *Leibstandarte "Adolf Hitler"*. In 1938, the Australian author, historian, and astute observer of 1930's Germany, Stephen R. Roberts, wrote a scathing epitaph of Heines:

Murderer ... parasite, [and] sadist ... , he was a pervert if ever there was one. But even so, the blame must all go to Hitler for allowing such a man to attain a position of prominence in the Brown Army and for giving him executive positions long after he knew his real nature. For eighteen months Heines had terrorized tens of thousands of people with all the irresponsibility and whims of an Oriental despot, yet Hitler, knowing this, had done nothing. He had promoted him Brownshirt commander of all Silesia and had given him control of the Breslau police-this self-confessed murderer who, a few short years before, had been scrounging a living in the cafés of Munich, working intermittently as a waiter and between times begging or bullying people for drinks. Handy with his fists, he naturally became a platoon commander of the local Nazis and ultimately leader of 100 men, with a liking for fighting and plenty of time on his hands for political activities and discreet blackmailing. "When Edmund came along," one of his Munich intimates told me, "the tradesmen paid up." Caught in an act of flagrant moral delinquency in Röhm's villa, Heines, infamous Feme murderer, worst of all the Nazi types, was ignominiously shot, regretted by nobody. The very fact of his existence and power must remain an eternal reproach to Hitler. (Roberts, *The House That Hitler Built*, p. 118)

30.06.1934	Expelled from the *Preußische Staatsrat* by order of *Ministerpräsident* Göring.
01.07.1934	Posthumously relieved of all posts, stripped of his rank, and expelled from the *SA*.

SA-Obergruppenführer Heines and *SA-Stabschef* Röhm
inspecting *SA* troops from horseback, ca. 1934.

Published Work

Der Heines-Prozess. Ein Kapitel deutscher Notzeit. Verlag F. Eher Nachf.,G.m.b.H. (Editor, 1929)

Schlesisches SA-Liederbuch. Völkischer Verlag W. Uttikal, Breslau (1933)

Vom Kampf und Sieg der schlesischen SA: Ein Ehrenbuch. SA-Gruppe Schlesien, Wilhelm Gottl. Korn, Breslau (ca. 1933)

Luftschutz, die deutsche Schicksalsfrage. Plesken, Stuttgart (ca. 1934)

Decorations and Awards

00.00.1916	*1914 Eisernes Kreuz I. Klasse*
11.03.1916	*1914 Eisernes Kreuz II. Klasse*
05.07.1918	*Kgl. Bayerischer Militär-Verdienstorden IV.Klasse mit Schwertern*
11.05.1917	*Kgl. Bayerischer Militär-Verdienstorden III.Klasse mit Krone und Schwertern*
ca. 1918	*Verwundetenabzeichen, 1918 in Schwarz*
00.00.19__	*St.-Annen-Orden*
04.01.1934	*Ehrendolch der SA*
00.02.1934	*Ehrenwinkel für alte Kämpfer*
ca. 1932	*Tyr-Rune*

Notes

• Illegitimate son of the housemaid Martha Heines (daughter of the Esslingen mechanic Johann Baptist Heines) and her employer, *Oberleutnant* Edmund von Parish, who hailed from a Hamburg businessman's family. Edmund's half-sister was the costume artist Hermine von Parish (born

10.04.1907, died 00.00.1998), founder, in 1970, of the Parish-Kostümbibliothek in München.

- Religion: Protestant.
- Heines' younger brother, Oskar Heines (born 03.02.1903 in München) was also an *SA* officer; promoted to *SA-Sturmbannführer* on 14.06.1931, he was assigned on the same date as Adjutant to the *Gruppenstaffelführer* of *SA-Gruppe Schlesien*; promotion to *SA-Obersturmbannführer* followed in late 1933 or early '34. On the morning of 01.07.1934, he heard a radio report concerning the execution of his brother. Soon after, he and *SA-Obersturmbannführer* Werner Engels reported to the *Polizeipräsidium* in Breslau where they were immediately placed under arrest by *SS* men. From there, they were driven that night to a forested area near Deutsch-Lissa. At dawn on 02.07.1934, the two were shot on orders of *SS-Obergruppenführer* Udo von Woyrsch. In his memoirs, the early *Gestapo* official and anti-Nazi conspirator Hans Bernd Gisevius describes the fate of Oskar Heines:

> [Edmund] Heines ... had a tubercular brother who was convalescing in the Silesian mountains. Apparently it would have been too much of an effort to take him all the way to [Berlin-] Lichterfelde. The executioners found it more convenient to have him attempt to escape in the mountains; thus the Heines brothers were wiped out. Some weeks later [*Reich* Interior Minister Dr. Wilhelm] Frick received a letter from Heines's aged mother, written in a clumsy hand. The mother pleaded that this last of her sons be released from "protective custody." He was, she wrote, her sole remaining bread-winner; moreover, he was sick. An investigation was made. It developed that by some oversight the mother had been sent the ashes of only one of her sons. The second cardboard box was promptly dispatched. (Gisevius, *To the Bitter End*, p. 164)

Sources

Bleuel, Hans Peter: *Sex and Society in Nazi Germany*. J. B. Lippincott Company, 1973.

Gisevius, Hans Bernd: *To the Bitter End*. Houghton Mifflin Company, 1947.

Höffkes, Karl: *Hitlers politische Generale: Die Gauleiter des Dritten Reiches*. Grabert-Verlag-Tübingen, 1986.

Höhne, Heinz: *The Order of the Death's Head*. Martin Secker and Warburg, 1969.

Lilla, Joachim; Schulz, Andreas; & Döring, Martin: *Statisten in Uniform. Die Mitglieder des Reichstags 1933-1945*. Droste Verlag, 2004.

Miller, Michael D. & Schulz, Andreas: *Gauleiter: The Regional Leaders of the Nazi Party and Their Deputies, Volume I*. R. James Bender Publishing, 2012.

Noakes, Jeremy & Pridham, Geoffrey (Editors): *Nazism 1919-1945: A History in Documents and Eyewitness Accounts, Volume I. The Nazi Party, State and Society, 1919-1939*. University of Exeter, 1983.

Office of Strategic Services [OSS], Research and Analysis Branch: *Nazi Plans for Dominating Germany and Europe. The Attitude of the NSDAP toward Political Terror*, 08.07.1945.

Roberts, Stephen H.: *The House That Hitler Built*. Harper & Brothers, 1938.

Schirach, Baldur von: *Die Pioniere des Dritten Reiches*. Zentralstelle fur der deutschen Freiheitskampf, 1933.

Schuster, Martin, M.A.: *Die SA in der nationalsozialistischen »Machtergreifung« in Berlin und Brandenburg 1926-1934*. Doctoral Dissertation at "Fakultät I-Geisteswissenschaften der Technischen Universität Berlin", 2004.

Wilhelm Helfer
SA-Obergruppenführer

Born:	26.12.1886 in Kaiserslautern / Rheinland-Pfalz.
Died:	05.08.1954 in Laufen / Regierungsbezirk Oberbayern / Bayern.

NSDAP-Nr.: 305 (First joined, 00.06.1922; Party banned following the *München-Putsch* of 09.11.1923; Reenrolled, 28.03.1925)

Promotions

ca. 1910	*Unteroffizier d. R.*
00.06.1922	*SA-Mann*
09.09.1923	*SA-Kompanieführer* (an early *SA* rank)
01.04.1929	*SA-Brigadeführer* (an early version of the rank)
01.07.1932	*SA-Oberführer*
01.03.1933	*SA-Gruppenführer*
09.11.1937	*SA-Obergruppenführer*

Career

00.00.1893-00.00.1897	Attended a *Seminarschule*.
00.00.1897-00.00.1904	Attended *Realschule*.
00.00.1904-00.00.1907	Served a commercial apprentice in spinning and weaving mills.
00.00.1904-00.00.1905	Attended a *kaufmännische Bürgerschule*, then employed as a textile buyer (with positions in Germany and abroad).
00.00.1907-00.00.1908	Attended the *Textiltechnikum* (textile technical school) in Reutlingen (2 semesters).
00.10.1908-00.10.1910	Military service with *1. Kompanie / Bayerische Infanterie-Leibregiment* (München).
00.10.1910(?)-00.00.1912(?)	Employed as a cashier.
00.00.1911-00.00.1911	Participated in a 6-week reserve training exercise with the *Bayerische Infanterie-Leibregiment*.
00.00.1912-00.00.1912	Participated in a 4-week reserve training exercise.
00.00.1912-00.00.1922	*Minenbeamter* (mininig official) and *Zahlmeister* (paymaster) of the *Vereinigten Diamantenminen AG* in Lüderitzbuch, Deutsch-Südwest-Afrika, and with Consul. Diamond Mines, Ltd. in Cape Town, South Africa.
00.08.1914-00.07.1915	War service as an *Unteroffizier*, assigned to *1.Gebirgsbatterie Narubis* and *1.reitenden Batterie / Regiment Franke* (Naulila) in the *Kaiserliche Schutztruppe in Deutsch-Südwest-Afrika*.
00.07.1915-00.00.1918	Captured and interned by the British in Transvaal. In 1915, he was punished by his captors for violation of censorship laws with 8 days' hard labor in a camp at Windhoek, and later the same year, with 14

SA-Gruppenführer Helfer in 1933.

days' imprisonment for horse thievery at the Karibib camp in Southwest Africa.

00.00.1922	Returned to Germany.
00.00.1922-00.00.1923	Assigned as a *Zugführer* and *Stoßtruppführer* with the *Einwohnerwehr* and in the *Kampfbund München*.
00.00.1922	Joined the *Deutschvölkischen Schutz- und Trutzbund*.
00.06.1922	Joined the *NSDAP*.
00.06.1922	Joined the *SA*.
00.06.1922-00.03.1923	Assigned as an *SA-Mann* (*z.b.V. des Kommandeurs*) to the *Stab Göring*.
00.04.1923-08.09.1923	*Zugführer* and *Kompanieführer*stellvertreter in 5. *Kompanie / SA-Regiment München*.
09.09.1923-00.00.1927	*Kompanieführer* of 5. *Kompanie / SA-Regiment München*. He later wrote: "The 5th Company continued to exist despite the prohibition [by the Bavarian Government following the *München-Putsch* of 09.11.1923]. An interruption of service never took place." (Source: Document in his *SA* personnel file at the *Bundesarchiv*, Berlin).
09.11.1923	Participated in the *München-Putsch*, München.
00.00.1924-00.00.1930	Member of the *Frontkriegerbund München*. He was also assigned as *Gauführer Oberbayern* of that veterans' organization, and as a member of its *Bundesleitung*.
28.03.1925	Reenrolled in the *NSDAP*.
01.04.1929-00.02.1931	*Brigadeführer* of *SA-Brigade I* (München).

SA-Gruppenführer **Helfer** in 1933.   *SA-Gruppenführer* **Helfer**, ca. 1936.

00.02.1931-00.04.1931	*Gausturmführer München-Oberbayern*. Voluntarily resigned from this position.
18.10.1931	Participated in the *SA-Aufmarsch in Braunschweig*.
01.03.1932-01.07.1932	Tasked with the organization of *SA-Sturm "Reichsleitung"*.
01.07.1932-22.05.1933	*Führer* of *SA-Untergruppe München-Oberbayern*.
05.03.1933-14.10.1933	Member of the *Bayerische Landtag*.
12.03.1933-00.00.1934(?)	*Sonderkommissar* [later redesignated *Sonderbevollmächtigter*] *der Obersten SA-Führung bei der Regierung von Oberbayern*.
15.03.1933-14.09.1933	*Führer* of *SA-Gruppe Hochland* (München; m.d.F.b. until 22.05.1933, then permanent from 23.05.1933).
15.09.1933-31.10.1933	Attached to the *Stab der Obersten SA-Führer*.
01.11.1933-19.11.1933	Assigned to the *Stab der Obersten SA-Führung*.
12.11.1933-08.05.1945	Member of the *Reichstag (Wahlkreis 24, Oberbayern-Schwaben)*.
20.11.1933-09.07.1934	*Standortführer der SA (SA* Garrison Leader) in München.
10.07.1934-30.01.1942	*Führer* of *SA-Gruppe Hochland* (München; m.d.F.b.until 14.09.1935, permanent from 15.09.1935). After 01.08.1941, the duties of this post were carried out by *SA-Brigadeführer* Albert Heinz.
[00.00.1938]	*III. Vorsitzender* of the *Bayerische Landesverband für Wanderdienst*.
00.00.193_-00.00.19__	*Beisitzer* of the *Ordenskommission der Reichsleitung der NSDAP* (Awards Commission of the *Reich* Leadership of the *NSDAP*).
01.05.1941-00.05.1945	Assigned as *SA-Führer z. V.* to *SA-Gruppe Hochland*.
01.05.1941-00.05.1945	*Reichszeugmeister* and *Leiter* of the *Reichszeugmeisterei der NSDAP* (*Hauptamt VIII* in the office of the *Reichsschatzmeister der NSDAP*

SA-Gruppenführer Helfer, ca. 1936. (NARA photo)

	[National Treasurer of the Nazi Party, *Reichsleiter* Franz Xaver Schwarz]). Succeeded Richard Buchner.
22.09.1942	Charged with *Durchführung der Luftschutzmaßnahmen* (Implementation of Air-Raid Protection Measures) in the *Gebäudekomplex der Reichsleitung der NSDAP* (Building Complex of the *Reich* Leadership of the Nazi Party) on the Tegernseelandstraße, München.
12.11.1944	Participated in the swearing-in ceremony, led by the *Reichsführer-SS*, of the *Deutschen Volkssturm* in the Zirkusgebäude auf dem Marsfeld, München.

Decorations and Awards

00.00.191_	*1914 Eisernes Kreuz II. Klasse*
00.00.191_	*Kolonialauszeichnung für Verdienste in den deutschen Kolonien (Löwenorden)*
00.00.191_	*Preußisches Kreuz für Hilfsdienste im Kriege*
ca. 1918	*Bayerische Goldene Hochzeitmedaille*
ca. 1934	*Ehrenkreuz des Weltkrieges 1914-1918 mit Schwertern*
30.01.1939	*Goldenes Ehrenzeichen der NSDAP*
00.00.1934	*Ehrenzeichen des 9. November 1923 (Blutorden)* (Nr. 14, with effect from 09.11.1933)
00.00.194_	*Dienstauszeichnung der NSDAP in Gold*
00.00.194_	*Dienstauszeichnung der NSDAP in Silber*
00.00.194_	*Dienstauszeichnung der NSDAP in Bronze*

An autographed portrait of *SA-Obergruppenführer* Helfer, ca. 1937. (Hermann-Historica, Auctioneers, München)

| 00.00.193_ | *Abzeichen des SA-Treffens Braunschweig 1931* |
| 00.00.1933 | *Große Goldene Adolf-Hitler-Medaille der NSV* |

Notes

- Son of the lawyer Heinrich Helfer and his wife Julie, née Berg.
- Religion: Protestant.

Sources

Bundesarchiv, Berlin-Lichterfelde (former Berlin Document Center): *Personalunterlagen von SA-Angehörigen: SA-Personalakte* of Wilhelm Helfer.

Lilla, Joachim; Döring, Martin; & Schulz Andreas: *Statisten in Uniform. Die Mitglieder des Reichstags 1933-1945*. Droste Verlag, 2004.

Wolf-Heinrich Julius Otto Bernhard Fritz Hermann Ferdinand, Graf von Helldorf

SA-Obergruppenführer

Born:	14.10.1896 in Merseburg / Sachsen.
Executed:	15.08.1944 at Plötzensee Prison, Berlin (hanged for his involvement in the conspiracy against Hitler).
NSDAP-Nr.:	325 408 (Joined 01.08.1930; Expelled from the Party, 05.08.1944)

Promotions

02.08.1914	*Fahnenjunker*
22.03.1915	*Leutnant* (ohne Patent)
31.07.1931	*SA-Oberführer*
10.09.1931	*SA-Gruppenführer*
25.03.1933-24.07.1944	*Polizeipräsident*
09.11.1938	*SA-Obergruppenführer*
00.00.19__	*Rittmeister d. R.*
00.00.194_	*Berechtigung zum Tragen der Uniform eines Generalleutnants der Polizei* (authorized to wear the uniform of a Lieutenant General of the Police)
00.00.1944	*Berechtigung zum Tragen der Uniform eines Generals der Polizei*

Career

00.00.1902-00.07.1914	Privately instructed by a *Hauslehrer* in his parents' home; upon reaching "Quarta", he went on to attend the humanistic *Gymnasium* in Wernigerode, and the *Klosterschule* in Rossleben a. d. Unstrut.
02.08.1914-23.10.1916	Entered service as a *Fahnenjunker*, assigned to *Thüringische Husaren-Regiment Nr. 12* (Torgau). Deployed to the Western Front from 28.09.1914. He participated in the first Battle of Ypres and was later transferred to the Eastern Front where he remained until being sent to France in August 1918. Also serving in the Regiment during this period was future *Reich* Foreign Minister Joachim von Ribbentrop.
Late-1915	Under medical care for venereal disease for three months.
24.10.1916	Transferred to infantry service.
20.11.1916-Spring 1918	Assigned as *Führer* of the *MG-Kompanie* of the *Südarmee* on the Eastern Front
Spring 1918-00.00.1918	Returned to service with *Husaren-Regiment 12*.
00.00.1919-00.00.1919	Service with *Freikorps Lützow*, involved in fighting against communist uprisings in Braunschweig, Jena, and München.
00.00.1919-00.12.1919	Service with *Dragoner-Regiment 8*.
00.12.1919	Discharged from active military service.
00.00.1919-00.00.1920	*Führer* of the *Offiziers-Stoßtrupp* in *Freikorps Roßbach*, with which he participated in the "Kapp-Putsch" on 13.03.1920.
00.00.1919-00.00.1924	Member of the *"Stahlhelm"-Bund*.

SA-Oberführer Graf Helldorf in 1931.

00.00.1921-00.00.1928	*Fideikomissherr auf Wohlmirstedt* (Unstruttal).
00.00.1923-00.00.1924	*Führer* of the *"Stahlhelm"-Bund* in the Unstrut-Gebiet and *Adjutant* to the *2. Bundesführer* of the *"Stahlhelm"*, Theodor Duesterberg.
00.00.1924-00.00.1924	Fled to Italy where he lived in exile.
00.00.1924	Amnestied and returned to Germany.
00.08.1924-00.00.1925	*Führer* of *Gruppe Mitte* of the *Frontbann* in Halle an der Saale.
07.12.1924-03.03.1928	Member of the *Preußische Landtag (Wahlkreis 11, Halle-Merseburg; Liste der NSFB)*.
01.05.1925-22.09.1925	*Kommandeur* of the *Frontbann*.
00.00.1927-00.00.1930	*Präsident* of the *Landwirtschaftskammer* (Chamber of Agriculture) for *Provinz Sachsen*.
00.00.1928-00.00.1930	*Leiter* of the Gestüt Harzburg.
00.00.1930	Resettled in Berlin.
01.08.1930	Joined the *NSDAP*.
00.01.1931	Joined the *SA*.
00.04.1931-00.07.1931	*Führer* of *SA-Standarte 2* (Berlin).
31.07.1931-09.09.1931	*Führer* of the independent *SA-Untergruppe Groß-Berlin (Oberführer Berlin)* and *Gausturmführer Berlin*.
00.00.1931	*Vorsitzender* of the *Deutsche Volkssportverein (DVV)* in Berlin.
10.09.1931-13.04.1932	*Führer* of *SA-Gruppe Berlin-Brandenburg* (Berlin) and *Führer der SS* in *Gau Brandenburg*.
12.09.1931	On the evening of Yom Kippur, anti-Semitic actions organized and directed by Graf von Helldorf (*Führer* of *SA-Gruppe Berlin-*

SA-Gruppenführer Graf Helldorf as *Führer* of *SA-Gruppe Berlin-Brandenburg* with *Gauleiter* Dr. Goebbels, 1931.

Brandenburg) and his *Adjutant*, Karl Ernst took place on Berlin's Kurfürstendamm. Von Helldorf and Ernst were given their instructions by Dr. Goebbels earlier that month. For about two hours, plainclothes *SA* men yelled anti-Semitic slogans, harassing and assaulting any passerby who appeared to be Jewish. Upon the arrival of the police, the *SA* men promptly stood down. In his article in the *London Times* of 14.09.1931, the journalist and eyewitness Douglas Reed wrote:

> Several hundreds of Nazis gathered last evening in the crowded Kurfürstendamm and the adjoining streets and attacked Jews or persons of Jewish appearance, smashed café chairs and windows and generally created pandemonium in that part of the city where diners-out and pleasure-seekers mostly meet. Foreign obsearvers who, like your Correspondent, arrived by chance during the disorders, saw a display of organizaed hooliganism almost incredible in a city like Berlin. The occasion was the Jewish New Year and the disorders occurred mainly near the Jewish synagogue in the Fasenstraße, where the customary service was being held. As I left the tramcar near the Kaiser Friedrich Memoiral Church about nine o'clock I noticed tha that the usual Saturday evening crowds had been reinforced by large numbers of young men recognizable as members of the Nazi Storm Detachments. Within

SA-Gruppenführer **Graf Helldorf** and an unnamed *SA-Oberführer*, ca. 1932.

minutes of my arrival, two passers-by who looked like Jews were set upon, knocked down and savagely beaten. Later I saw other instances, and altogether a large number of persons were attacked. [...] One has the evidence of one's eyes that this was all done to plan. The Nazis had a system of messengers and connecting links of superiors who were distinguished by black arm bands and then men who appeared at street corners inciting their followers to disorders. Groups seem to have been allotted to certain stretches of the Kurfürstendamm and at given signals these came together or dispersed and mingled with the crowd.

In a second *Times* article dated 24.09.1931, Reed wrote:

Twenty-seven persons, nearly all Nazis, who were charged with complicity in the anti-Jewish disorders in the Kurfürstendamm on the evening of Saturday September 12, were sentenced today to severe terms of imprisonment, varying between nine and twenty-one months. The trial, which has lasted seome days, began with the hearing of witnesses attacked on the evening in question, when a number of persons were set upon and a café wrecked. These included two German dentists, neither of them Jews, a Rumanian, an Armenian student and others.

September 1932: The inauguration of the *Gaugeschäftsstelle der NSDAP* on Voßstrasse, Berlin Left to right: Johann Meinshausen, Alfred Ernst, Albert Speer, Graf von Helldorf, *Gauleiter* Dr. Goebbels, Dagobert Dürr, and Karl Hanke. (NARA photo)

During this "*1.Kurfürstendamm-Prozeß*" (1st Kurfürstendamm Trial), von Helldorf fled from Berlin, and a warrant was issued for his arrest. He was soon discovered in Nürnberg and sent back to Berlin, to be charged with "schweren Landesfriedenbruchs" (Severe Breach of Peace). He was tried together with Karl Ernst in the "*2.Kurfürstendamm-Prozeß*", 00.10.1931-00.11.1931. In evidence against them was a statement of one of the lower-ranking arrested *SA* men, summarized in the 24.09.1931 article by Reed as follows:

A Nazi arrested in a car observed by many witnesses said that he had earlier been accompanied by the "commanding officer" of the "Berlin Nazi Storm Detachments" [von Helldorf], and his "chief of staff" [Ernst], who was driving up and down the Kurfürstendamm several times in order to "persuade their friends to go home". The Public Prosecutor here intervened to say that the Nazi leader in question, Count Helldorf, and his companion, Herr Ernst, had been sought by the police, but had disappeared. After the publication of this statement in the Press they have presented themselves at the Court and gave evidence. They denied having instructed the rioters, but are to be brought to trial Friday.

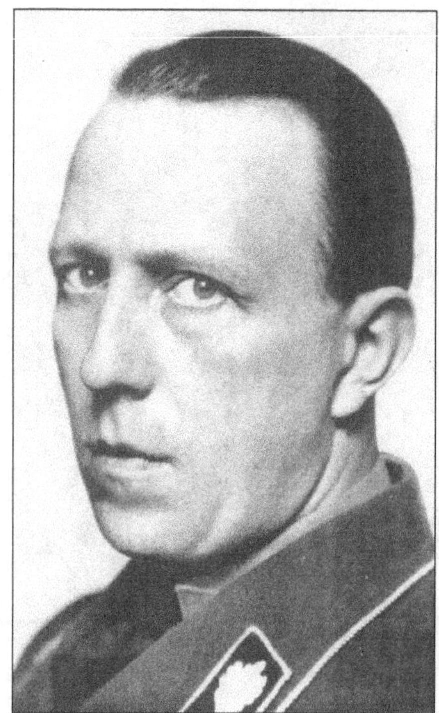

Formal portrait of *SA-Gruppenführer* Graf
Helldorf, ca. 1932. (Baldur von Schirach,
Die Pioniere des Dritten Reiches, 1933)

Formal portrait of *SA-Gruppenführer*
Graf Helldorf, ca. 1933.

On 07.11.1931, both defendants were sentenced by the Charlottenburg *Schöffengericht* (district court consisting of two lay judges and one professional judge) to 6 months' imprisonment. Thanks to the efforts of von Helldorf's lawyer, Dr. Roland Freisler (who, as president of the Nazi People's Court, was later to condemn him to death), his sentence was reduced from 6 months' imprisonment to a fine (his subordinates, on the other hand, received prison sentences of as long as two years). Dr. Goebbels also lent his support, making a great spectacle in court during von Helldorf's appeal proceeding in January 1932. As a result, Helldorf's sentence was reduced to a 100 RM fine.

00.12.1931-00.02.1932	Imprisoned.
13.04.1932-14.06.1932	Official government ban on the *SA* and *SS*.
24.04.1932-14.10.1933	Member of the *Preußische Landtag (Wahlkreis 3, Potsdam II)*.
01.07.1932-31.03.1933	*Führer* of *SA-Gruppe Berlin-Brandenburg* (Berlin).
15.09.1932-31.03.1933	*Führer* of *SA-Obergruppe I* (consisting of *SA-Gruppen Berlin-Brandenburg, Ostmark, Schlesien, Pommern*, and *Nordmark*; HQ: Berlin).
00.02.1933-00.03.1933	*SA-Standortführer von Groß-Berlin* (*SA* Garrison Leader for Greater Berlin).

00.03.1933	On orders from the acting Prussian Interior Minister, Hermann Göring, occupied the *Karl-Liebknecht-Haus* (headquarters of the KPD in Berlin) and appropriated it for use by the *SA*.
00.03.1933	Appointed as *Führer der deutschen Pferderennsportes* (Leader of German Equestrian Sports) and *Vizepräsident* of the *Deutschen Sportbehörde für Boxsport* (German Sports Authority for Boxing).
25.03.1933-18.07.1935	*Polizeipräsident* in Potsdam. Succeeded by Wilhelm Ernst Graf von Wedel.
01.04.1933-31.01.1937	Assigned as an *SA-Führer z. V.* to the *Obersten SA-Führung.*
00.05.1933-18.07.1935	*Leiter der Staatspolizeistelle* in Potsdam.
03.06.1933	Appointed as *Berater für den Pferdesport im Preußischen Ministerium des Innern* (Advisor for Equestrian Sports in the Prussian Ministry of Interior).
00.00.1933-00.00.1933	Participant in the *Kongress der Internationalen Kriminalpolizeilichen Kommission* in Wien.
02.11.1933-10.08.1944	Member of the *Reichstag* (after 10.04.1938, representing *Wahlkreis 2, Berlin West*).
19.07.1935-24.07.1944	*Polizeipräsident* in Berlin ("mit der vorläufigen Wahrnehmung der Dienstgeschäfte beauftragt" [with provisional oversight of the post until 01.12.1935, then permanent; presented with Ernennungsurkunde [certificate of appointment], 26.11.1935). Succeeded *Konteradmiral a. D.* Magnus von Levetzow. After von Helldorf's arrest for involvement in the 20.07.1944 bomb plot against Hitler, he was succeeded by *SS-Gruppenführer und Generalleutnant der Polizei* Kurt Göhrum.
01.12.1935-24.07.1944	*Örtliche Luftschutzleiter der Luftschutzort Berlin* (in his capacity as *Polizeipräsident*).
01.12.1935-24.07.1944	Member of the *Provinzialrat* for Berlin.
01.12.1935-24.07.1944	Member of the *Internationale Kriminalpolizeiliche Kommission.*
20.07.1935	Issued orders for the closure of Jewish businesses on the Kurfürstendamm in Berlin.
Spring 1936	Together with *Gauleiter* Dr. Goebbels, organized a terror campaign against the Jews of Berlin. In his diary entry for 20.04.1936, Goebbels wrote of his orders to von Helldorf's police with regard to the Jews: "Police act with appearance of legality, Party stands by as observer … " Targeted were, in Goebbels' words, "Work-shy and asocial Jews" (arbeitsscheu und asoziale Juden), however all Jews were ultimately open to officially-sanctioned harassment. In conference with von Helldorf, Goebbels stated his goal was "to drive the Jews out of Berlin … without any sentimentality. 300 police officers were mobilized for the purpose." Of a meeting with police officials, Goebbels recorded in his diary [16.06.1936]: "Legality is not the watchword, but harassment. On 19.06.1936, he wrote: "Helldorf is now proceeding radically on the Jewish question. The Party is helping him. Many arrests … . The police understand my instructions. We will free Berlin of Jews. I will not let up. Our path is the right one." (all quotes from Ralf-Georg Reuth, *Goebbels*, p. 234)

SA-Gruppenführer Graf Helldorf at his desk in Berlin, ca. 1935.

12.05.1936	Appointed as *Leiter* of the *Polizei-Befehlsstab für die XI. Olympiade Berlin 1936* (Police Command Staff for the 11th Summer Olympic Games in Berlin).
12.05.1936-31.05.1936	Member of the German delegation (led by Kurt Daluege) to the 12th Conference of the International Criminal Police Commission in Belgrade.
01.02.1937	Transferred to the active *SA-Führerkorps* as *Referent für Polizeifragen* (Advisor for Police Questions) in the *Stab der Obersten SA-Führung*.
01.02.1937-00.07.1944	*Verbindungsführer der SA zur Deutschen Polizei* (Liaison Chief of the *SA* to the German Police).
01.02.1937-00.07.1944	Member of the *Polizeiausschuß* (Police Committee) of the *Akademie für Deutsches Recht*.
21.01.1938	Informed *Generaloberst* Wilhelm Keitel, Chef of the *Wehrmacht*samt, that the new wife, Erna Gruhn, of *Reich* War Minister *Generalfeldmarschall* Werner von Blomberg had once been a prostitute. Keitel passed this news on to Hermann Göring who informed Hitler of the details on 24.01.1938. This, coupled with false accusations of homosexuality leveled against the Army's Commander-in-Chief, *Generaloberst* Werner Freiherr von Fritsch, resulted in the forced retirement of both officers and Hitler's assuming overall control of the *Wehrmacht*.

00.04.1938

Accompanied *Gauleiter* Dr. Goebbels in a meeting with Hitler regarding the intensification of anti-Jewish actions in Berlin; the goal discussed was the "clean[ing] up" the *Reich* capital. In his diary entry of 23.04.1938, Goebbels wrote:

> Then we put it to the *Führer*. He agrees, but only after his trip to Italy. Jewish establishments will be combed out. Jews will then get a swimming pool, a few cinemas, and restaurants allocated to them. Otherwise entry forbidden. We'll remove the character of a Jew-paradise from Berlin. Jewish businesses will be marked as such. At any rate, we're now proceeding more radically. The *Führer* wants gradually to push them all out. Negotiate with Poland and Romania. Madagascar would be the most suitable for them. (Ian Kershaw, *Hitler 1936-1945: Nemesis*, pp. 133-134)

In early June, von Helldorf arrested 300 Jews, using various pretexts. The police president went on leave soon after, and all but a few (Jews with criminal records) were released. Goebbels recalled von Helldorf from leave and informed him that it was his objective to clear the Jews from the *Reich* capital; he then reiterated this point to an assembly of police officers. On 13.06.1938, von Helldorf established a special Judendezernat (Jewish section) to work with the Berlin Gauleitung on Jewish matters. Within a few days of this, his police had taken into custody 1,644 Jews (1,122 criminals, 445 "asocials", and 77 foreigners); 66 of these were incarcerated in prisons, 1,029 sent to concentration camps, and the remainder held in police custody for a number of days. Goebbels wrote approvingly in his diary entry of 19.06.1938: "Helldorf is now taking radical steps on the Jewish Problem. The Party is helping him in this ... The police have understood my instructions."

00.11.1938

Following the "Reichskristallnacht" pogrom of 09./10.11.1938, von Helldorf, though notorious for his outspoken anti-Semitism, reprimanded his officers for following orders to not interfere with the violence. His subordinate, the *Kriminalpolizei* official Hans Bernd Gisevius, writes:

> During the riots he had been absent from Berlin. Immediately after his return he called a conference of all police officers and berated them for their passivity – even though under orders. To the dismay of all the Nazis he announced that if he had been present he would have ordered his police to shoot the rioters and looters. It was a remarkably courageous statement for a chief of police and high officer of the *SA* to make. Precisely because of Helldorf's position it was particularly dangerous for him to condemn the official Party line. (Gisevius, *To the Bitter End*, p. 335)

SA-Gruppenführer Graf Helldorf in 1938. (Max Williams photo)

Autumn 1938

Established contacts with the German military and civilian resistance movement. He was joined in this effort by his deputy *Polizeipräsident*, Fritz-Dietlof Graf von der Schulenburg (born 05.09.1902, hanged at Berlin-Plötzensee on 10.11.1944). Hans Bernd Gisevius, himself an active member of the anti-Hitler conspiracy from its earliest years, writes of his role in initiating the Police President:

> We [in the German resistance] felt so sure of ourselves that, after years of feeling out, I at last talked plainly with Count Helldorf. And Helldorf declared himself ready to go along with us. The speed with which we took control of the regular police was of the greatest moment for our action in Berlin. The technical knowledge and commanding position of the Berlin police chief would be of invaluable aid to us in this respect. Moreover, the collaboration of so highly placed an *SA* leader would serve to confuse the situation further. It would lend a great deal more credibility to our story of an *SS* Putsch headed by Himmler. Nevertheless, I was somewhat cautious with Helldorf and did not lay all our cards on the table. (ibid, p. 319)

Of Helldorf's reckless willingness to join in the conspiracy, Gisevius writes:

Graf Helldorf skiing with friends at Lech, 1940. (Andreas Schulz archives)

The difference between Stauffenberg, Helldorf, and Schulenberg – all three of them counts – was that Helldorf had come to the Nazi Movement as a primitive, I might almost say an unpolitical revolutionary. The other two had been attracted primarily by a political ideology. Therefore, it was possible for Helldorf to throw everything overboard at once: Hitler, the Party, the entire system. Stauffenberg, Schulenberg, and their clique wanted to drop no more ballast than was absolutely necessary; then they would paint the ship of state a military gray and set it afloat again. (ibid, pp. 513-514)

01.07.1938	Appointed as *stellvertretender Leiter* of the *Obersten Behörde für Vollblut-Zucht und -Rennen.*
01.07.1938	Appointed as *Vorsitzender* of the *Reichsverbans für Vollblut-Zucht und -Rennen.*
02.07.1938	In his diary entry for that date, *Gauleiter* Dr. Goebbels wrote: "Helldorf wants to establish a Judenghetto [Jewish ghetto] in Berlin. The rich Jews themselves are to pay [for the creation] of that."
04.12.1938	Issued the following decree:

On the basis of the Police Decree Regarding the Appearance of the Jews in Public of 28. November 1938, the following is decreed

A formal portrait of Wolf Heinrich Graf von Helldorf, taken shortly after he was awarded the *Ritterkreuz des Kriegsverdienstkreuzes mit Schwertern.*

for the police district of Berlin.

Paragraph 1. Streets, squares, parks, and buildings, from which the Jews are to be banned, are to be closed to Jewish subjects of the State and stateless Jews, both pedestrians and drivers.

Paragraph 2. Jewish subjects of the State and stateless Jews who at the time when this decree goes into effect still live within a district banned to the Jews, must have a local police permit for crossing the banned area.

By 1. July 1939, permits for Jews living within the banned area will no longer be issued.

Paragraph 3. Jewish subjects of the State and stateless Jews who are summoned by an office within the banned area, must obtain a local police permit for twelve hours.

Paragraph 4. The ban of Jews in Berlin comprises the following districts: (1) All theatres, cinemas, cabarets, public concert and lecture halls, museums, amusement places, the halls of the Fair, including the Fair grounds and broadcasting station on the Messedamm, the Deutschlandhalle and the Sport Palace, the *Reich* Sport Field, all athletic fields including ice skating rinks; (2) All public and private bathing places... (Bernard Dov Weinryb, *Jewish Emancipation Under Attack*, p. 56)

04.07.1940   Issued a *Polizeiverordnung* (police ordinance) concerning shopping hour restrictions for Jews in Berlin, as follows:

> 1. The shopping hours are fixed for Jews as 16.00-17.00 [4:00 p.m.-5:00 p.m.]. This applies to all public places of sale, all public and private markets, including market halls and street sales.
> 2. Jews within the meaning of this police ordinance are those persons whose ration cards are marked with a 'J' or the word 'Jude'.
> 3. When requested, shopkeepers must place a notice which refers to the restriction of shopping hours for Jews to 16.00-17.00 in all public places of sale, on all public and private market stalls, and on street stalls.

00.00.1942-00.00.1942   *Kommandeur* of a *Panzer-Abteilung* on the Eastern Front (unconfirmed).

09.06.1942   Attended the state funeral of Reinhard Heydrich in the Mosaiksaal of the *Reichskanzlei*, Berlin.

00.00.1943-00.07.1944   *Höherer Polizeiführer von Groß-Berlin* (Higher Police Leader of Greater Berlin). Succeeded by *SS-Gruppenführer und Generalleutnant der Polizei* Kurt Göhrum.

00.06.1944   Informed his colleague and fellow conspirator Arthur Nebe (*Chef* of *Amt V* [*Kriminalpolizei*] / *RSHA*) of the details for the assassination of Hitler and coup d'etat (code named "Valkyrie") planned for the following month. Nebe promised to do his part in the plot by arresting *Reichsminister* Dr. Goebbels and *Reichsleiter* Dr. Ley, dissolving the *Gestapo*, and subordinating the *SS* to the *Wehrmacht*.

20.07.1944   Early on the day of "Valkyrie", Graf Helldorf was in his office with fellow conspirator Hans-Bernd Gisevius, waiting for the telephone call of General Olbricht. Olbricht was to inform him that the assassination of Hitler had succeeded, and give him the order to proceed with the arrest of senior *SS* and Police leaders in Berlin (Kaltenbrunner, "*Gestapo*" Müller, and Walter Schellenberg). He was then to take control of the *Reich* capital with his *Orpo* and *Kripo* forces. The following is Gisevius' account:

> At four o'clock Helldorf at last rushed excitedly into our room. "It's starting!" he exclaimed. Olbricht had just telephoned him to hold himself in readiness; there would be an important message within half an hour.
>
> ... Before four-thirty Helldorf reappeared. "Gentlemen, we're off!" he exclaimed triumphantly. "Olbricht has just given me an official order to report at the Bendlerstraße: he says the *Führer* is dead, a state of siege has been proclaimed, and he has urgent orders to deliver to me in the name of [*Generaloberst*] Fromm."
>
> Helldorf said this with a completely credible imitation of surprise. His energetic manner bore witness to his full appreciation of the significance of this moment. Obviously he was already

> imagining the scene in which he would persuade his police officers
> that they must remain inactive for the time being ... (Gisevius, *To
> the Bitter End*, p. 540)

Gisevius then accompanied Helldorf to the War Office in the Bendlerstraße, where the two met with the leaders of the conspiracy, including the retired General Ludwig Beck and Claus von Stauffenberg (who had recently arrived by air from Rastenburg, erroneously believing he had killed the *Führer*). Gisevius continues:

> Olbricht was the first to speak ... [H]e informed Helldorf in the tone of a military command that the *Führer* had been the victim of assassination that afternoon. The *Wehrmacht* had taken over the direction of the government; a state of siege was being proclaimed. The Berlin police were hereby subordinated directly to the *Oberkommando der Wehrmacht* and he, Helldorf, was to carry out at once the necessary measures ... .
>
> Helldorf played along. He made a brief military bow and started to leave in order to convoke the meeting of his officers, but before he was out the door, Beck's quiet, firm voice reached out to him.
>
> The former chief of the general staff spoke more loudly than usual: "One moment, Olbricht. In all loyalty we must inform the chief of police that according to certain reports from headquarters Hitler may not be dead ..." (ibid)

General Olbricht then declared that the main source of such reports, *Generalfeldmarschall* Keitel, was "lying!"; Beck stated, " ... What is important is that Helldorf must know what the other side has asserted about the failure of the assassination ... "; and Helldorf, Gisevius, and Gottfried Graf von Bismarck-Schönhausen were, continues Gisevius, "suddenly ... confronted with the brutal reality of the Putsch ... " (ibid, p. 541-543)

23.07.1944   Morning report of the *Gestapo*:

> A large number of connections and relationships will be investigated immediately. For example, the behavior of *Polizeipräsident* Graf Helldorf, who had verifiable personal contacts with the clique of conspirators [Verschwörerclique], places him in a peculiar light.

24.07.1944   Arrested in connection with the 20.07.1944 assassination attempt on Hitler. He confessed his part in the conspiracy while under interrogation.

05.08.1944   Expelled from the *NSDAP*.

10.08.1944   Expelled from the *Reichstag*.

15.08.1944   Tried for "Landesverrat" (high treason) by the *1. Senat des Volksgerichtshofes*. At one point during the proceedings, the raving court

Graf Helldorf on trial before Roland Freisler's *Volksgerichtshof*, August 1944.

President, Dr. Roland Freisler, leveled a series of wild accusations and insults at von Helldorf, and pronounced him a traitor to the *Führer*. "In a courageously candid manner, von Helldorf retorted: 'Why the theatrics? Every single one of us [old Party comrades] must try to jump from the sinking ship somehow. And you yourself, you know this only too well.' Stunned, Freisler was at a loss for words and refrained from further cross examination." (Dr. Max Domarus, ed, *Hitler. Speeches and Proclamations, 1932-1945: The Years 1941-1945*, p. 3137) In his closing statement, von Helldorf declared:

> During my interrogation I have confessed my guilt, which consisted of disloyalty towards the *Führer* and the movement. I am aware of the consequences of my mode of action and I have to carry them ... that I affirm National Socialism the way it was taught and understood by us during the years of struggle ... What exhibits itself to me today as the realization of National Socialism, I cannot endorse anymore.

Decorations and Awards

10.02.1944	*Ritterkreuz des Kriegsverdienstkreuzes mit Schwertern* as *Polizeipräsident* of Berlin (presented by *Gauleiter* Dr. Goebbels)
00.00.191_	*1914 Eisernes Kreuz I. Klasse*
00.00.191_	*1914 Eisernes Kreuz II. Klasse*
00.00.194_	*Kriegsverdienstkreuz I. Klasse mit Schwertern*
00.00.194_	*Kriegsverdienstkreuz II. Klasse mit Schwertern*
00.00.191_	*Orden vom Weissen Falken, Ritterkreuz II. Klasse mit Schwertern, Sachsen-Weimar*
ca. 1934	*Ehrenkreuz des Weltkrieges 1914-1918 mit Schwertern*
30.01.1938	*Goldenes Ehrenzeichen der NSDAP*
00.00.194_	*Dienstauszeichnung der NSDAP in Silber*
00.00.194_	*Dienstauszeichnung der NSDAP in Bronze*
03.02.1934	*Ehrendolch der SA*
00.02.1934	*Ehrenwinkel für alte Kämpfer*

Notes

- Parents:
  - ◊ Father: *Preuß. Rittmeister der Landwehr-Kavallerie a. D.* Wolf Heinrich Karl Julius Ferdinand Graf Helldorf (born 04.06.1869 in Runstädt, died ca. 1951; son of Julius Heinrich Graf Helldorf (owner of the estate Jethe bei Simmersdof; born 19.11.1827 in Merseburg, died 13.01.1908 in Merseburg) and his wife Anna, née von Bose (born 22.02.1839, died 24.08.1907 in Runstädt).
  - ◊ Mother: Dorothea Gräfin Helldorf, née von Holy und von Poniecitz (24.06.1874 in Metz, died 29.08.1948 in Berlin; daughter of *Generalmajor* Heinrich von Holy und Poniecitz [1848-1935] and his wife Elisabeth, née von Petersdorff.
- Siblings:
  - ◊ Hans Joachim Otto Ferdinand Heinrich Julius Graf von Helldorf (born 21.08.1898, killed in action as a *Leutnant* of the 12th Hussars near Equancourt, 05.09.1918

◊ Anna Dorothea Henriette Ida Hildegard Gräfin von Helldorf (born 24.09.1900)
- Religion: Protestant.
- Married in Rothenburg/Tauber on 11.10.1920 to Ingeborg von Wedel (born 00.00.1894, the daughter of *Oberst* Benno von edel and his wife Elsbeth, née von Dewitz; her brother was the eventual *SS-Brigadeführer*, and successor to Graf Helldorf as *Polizeipräsident* in Potsdam, Wilhelm Graft von Wedel (born 18.11.1891, died 19.10.1939 in Potsdam). Four sons and one daughter were born to the marriage:
  ◊ Wolf-Ingo Ferdinand Julius Heinrich Benno Graf von Helldorf (born 23.10.1921), captured by Allied forces at La Haye as *Fahnenjunker-Feldwebel* in 2. *Kompanie* / Panzer-Regiment 130.
  ◊ Joachim Ferdinand Hans Heinrich Wedego Graf von Helldorf (born 06.03.1923)
  ◊ Oda Carmen Gisela Henriette Gräfin von Helldorf (born 20.01.1927)
  ◊ Hans-Benno Ferdinand Heinrich Graf von Helldorf (born 03.03.1929)
  ◊ Olaf Rüdiger Heinrich Graf von Helldorf (born 15.05.1936)

Sources

Anderson, Alexander W.: *National Socialist Violence and Anti-Semitism as Propaganda in German: 1928-1934*. Masters Thesis submitted to the Department of History, McGill University, Montreal, 07.02.1993.

Domarus, Dr. Max (Editor): *Hitler. Speeches and Proclamations, 1932-1945: The Years 1941-1945*. Bolchazy-Carducci Publishers, 1990.

Gisevius, Hans Bernd: *To the Bitter End*. Houghton Mifflin Company, 1947.

Harrison, Ted: "'Alter Kämpfer' im Widerstand. Graf Helldorff, die NS-Bewegung und die Opposition gegen Hitler", in *Vierteljahreshefte für Zeitgeschichte, 45.Jahrgang, 3.Heft*. Institut für Zeitgeschichte, München, 1987.

Reuth, Ralf Georg: *Goebbels*. Harcourt Brace & Company, 1993.

Schirach, Baldur von: *Die Pioniere des Dritten Reiches*. Zentralstelle fur der deutschen Freiheitskampf, 1933.

Schulz, Andreas & Zinke, Dr. Dieter: *Die Generale der Waffen-SS und der Polizei 1933-1945, Band 2 (Hachtel-Kutschera)*. Biblio-Verlag, 2005.

Weinryb, Bernard Dov: *Jewish Emancipation under Attack*. American Jewish Committee, New York, 1942.

Otto Friedrich Herzog
SA-Obergruppenführer

Born:	30.10.1900 in Ziskam / Regierungsbezirk Pfalz / Bayern (per *SA-Führer-Fragebogen* in his *SA* file. Per another source, born in Germersheim / Rheinpfalz).
Died:	06.05.1945 in Breslau / Gau Niederschlesien. Richard Hargreaves writes: "Herzog's fate is unclear. Some reports say he shot himself, others that he tried to escape the fortress and was killed when his car struck a mine [per Andreas Schulz, on the Schiesswerderbrücke, together with *SA-Gruppenführer* Aster]." (Hargreaves, *Hitler's Final Fortress- Breslau 1945*)
NSDAP-Nr.:	38 960 (Joined 21.06.1926)

Promotions
01.06.1919	*überzähliger Unteroffizier*
21.06.1926	*SA-Mann* (?)
00.00.1927	*SA-Sturmführer*
01.04.1930	*SA-Standartenführer*
01.07.1932	*SA-Oberführer*
01.04.1933	*SA-Gruppenführer*
09.11.1936	*SA-Obergruppenführer*
03.09.1939	*Unteroffizier d. R.*
01.12.1939	*Feldwebel d. R.*
01.02.1940	*Leutnant d. R.*
11.01.1942	*Oberleutnant d. R.*

Career
00.00.1907-00.00.1914	Attended *Volksschule*.
00.00.1914-00.00.1917	Attended *Fortbildungsschule (Handelsschule)*.
01.12.1916-01.05.1917	Worked as a trade apprentice in Landau.
01.06.1917	Entered the *Bayerische Unteroffiziers*schule (Bavarian Noncommissioned Officer's School) in Fürstenfeldbruck as a *Zögling* (cadet).
01.07.1917-01.10.1917	Assigned to *W-Kompanie* at the *Unteroffiziersschule in Fürstenfeldbruck*.
01.10.1917-17.11.1917	Assigned to *2. Kompanie* at the *Unteroffiziersschule*.
13.10.1917	Sworn in by the King of Bavaria.
17.11.1917-22.07.1918	Assigned to *1. Kompanie* at the *Unteroffiziersschule*.
22.07.1918-04.09.1918	Engaged in agricultural work.
21.09.1918-01.10.1918	Assigned to *2. Kompanie* at the *Unteroffiziersschule*.
18.11.1918	Discharged from military service without having seen combat.
01.05.1919-30.09.1919	Served with *Freikorps Epp* in München (severely wounded [bullets in the lung, abdominal wall, and upper left arm], 00.05.1919).

SA-Gruppenführer Herzog in 1934. (Laurens Hessels photo)

Hitler reviews an *SA* parade in Oldenburg, 1931. Standing in front of the car are, from left to right, *Gauleiter* Roever, Viktor Lutze, and *SA-Standartenführer* Herzog. (Roger Bender photo)

04.06.1919-01.10.1919	Assigned to *3. Kompanie / I. Bataillon / 1.Schützen-Regiment (Freikorps Epp)*.
01.10.1919-31.12.1923	Assigned to *Schützen-Regiment 41* in München.
00.00.1919-00.00.1923	Member of the paramilitary *Reichskriegsflagge*, assigned as an *Ausbilder* (training instructor).
09.11.1923	Actively participated in the *München-Putsch*.
31.12.1923	Discharged from the *Reichswehr* as an *Unteroffizier* due to his involvement in the *Putsch*.
00.00.1924-00.00.1929	Employed as a commercial assistant.
21.06.1926	Joined the *NSDAP*.
21.06.1926-30.09.1928	Joined the *SA*, assigned to *SA-Sturm 55* (Oldenburg)(as *Führer* [m.d.F.b.] of the *Sturm* from 00.00.1927).
00.00.1926-00.00.1927	*Ortsgruppenleiter* of the *Ortsgruppe Varel der NSDAP*.
00.00.1927-00.00.1928	*Bezirksleiter* of the *Bezirk Varel der NSDAP*.
00.00.1928-00.00.1929	*Gaugeschäftsführer* of the *Gauleitung Weser-Ems der NSDAP*.
01.10.1928-30.09.1929	*Führer* (m.d.F.b.) of *SA-Standarte 91* (Oldenburg).
01.10.1928-01.04.1931	*Gau-SA-Führer* [later redesignated *Gausturmführer*] *Weser-Ems*.
00.00.1929-00.00.1932	*Gauorganisationsleiter* of *Gau Weser-Ems der NSDAP*.
00.00.1929-00.00.1934	*Stadtrat* in Oldenburg.
01.10.1929-09.08.1933	*Führer* of *SA-Brigade Weser-Ems* (m.d.F.b. until 01.11.1930, then permanent). The *SA* was banned by the German government, 13.04.1932-14.06.1932, and he resumed command on 01.07.1932.
01.04.1930-13.04.1932	*Führer* of *SA-Standarte 91*.
01.10.1930	Entered full-time *NSDAP* service.
01.11.1930-09.08.1933	*Führer* of *SA-Brigade Weser-Ems*.
00.05.1931-00.00.1933	Member and *Vorsitzender* of the *NSDAP* Faction in the *Oldenburgische Landtag*.
18.10.1931	Participated in the *SA-Aufmarsch in Braunschweig*.
00.10.1931	Officially sworn in as an *SA-Führer* by Viktor Lutze.
01.07.1932-09.08.1933	*Führer* of *SA-Untergruppe Weser-Ems* (redesignated *SA-Brigade Weser-Ems*, 00.00.1933).
31.07.1932-12.09.1932	Member of the *Reichstag (Wahlkreis 14, Weser-Ems)*.
05.03.1933-14.10.1933	Member of the *Reichstag (Wahlkreis 14, Weser-Ems)*.
09.08.1933-10.07.1934	Assigned as an *SA-Führer z. V.* to *SA-Gruppe Nordsee*.
00.00.1933-00.00.1934	Editor of the *Oldenburgischer Staatszeitung*.
12.11.1933-06.05.1945	Member of the *Reichstag (Wahlkreis 14, Weser-Ems;* after 29.03.1936, *Wahlkreis 7, Breslau)*.
00.02.1934-10.07.1934	*Gauorganisationsleiter* of *Gau Weser-Ems der NSDAP*.
06.07.1934 – 00.07.1934	*Führer* (m.d.F.b.) of *SA-Obergruppe VIII* (Schlesien). Succeeded Edmund Heines, who had been executed during the purge of 30.06.1934.
10.07.1934	Transferred to the *aktiven SA-Führerkorps* (active *SA* leadership corps).
10.07.1934-31.07.1936	*Führer* (m.d.F.b. until 14.09.1935, then permanent from 15.09.1935) of *SA-Gruppe Schlesien* (Breslau). Succeeded Edmund Heines (executed on 30.06.1934). Succeeded by Heinrich Georg Graf Fink von Finckenstein.

SA-Gruppenführer Herzog, ca. 1934. (*Reichstagshandbuch* photo)

SA-Gruppenführer Herzog (right) with Wilhelm Bruckner at
an SA function, ca. 1935. (Roger Bender photo)

SA-Obergruppenführer Herzog as *Stabsführer* of the *Obersten SA-Führung*, with *SA-Stabschef* Lutze, ca. 1938. (Laurens Hessels photo)

01.05.1936-14.06.1939	*Stabsführer* of the *Obersten SA-Führung* (m.d.W.d.G.b. until 31.07.1936, then permanent from 01.08.1936). Actually assumed the post, 23.05.1936. Succeeded by Max Jüttner.
15.06.1939-31.01.1942	*Führer* of *SA-Gruppe Schlesien* (Breslau). Succeeded Heinrich Georg Graf Fink von Finckenstein. Succeeded by Richard Aster.
03.09.1939-15.05.1940	Enlisted in the German Army as an *Unteroffizier d. R.*, assigned to *Infanterie-Regiment 49*. Following commissioning as a reserve officer on 01.02.1940, assigned as a *Kompanieführer* to *Infanterie-Regiment 49*. Wounded in action on the Western Front. 15.05.1940.
01.02.1942-06.05.1945	*Inspekteur* of the *Inspektion Gebirgsjäger-SA* in the *Stab der Obersten SA-Führung.*
01.07.1943	Placed in *Führerreserve* with attachment to *Wehrbezirkskommando Breslau I.*
04.08.1944-00.00.1945(?)	Assigned as a judge to the *Volksgerichtshof.*
00.04.1945-06.05.1945	*Führer* of *Volkssturm-Einheiten* (German Home Army units) and *Führer* of a *Kampfgruppe* in the defense of "Festung Breslau". On 05.05.1945, the *Festungskommandant, General der Infanterie* Hermann Niehoff, called a meeting of his officers at his underground headquarters. Richard Hargreaves writes that after Niehoff informed them of his decision to surrender the city to the Soviet Army:

> Every man agreed except Otto Herzog, an inveterate Nazi who had been rewarded with the *Ritterkreuz* for his leadership of Breslau's *Volkssturm*. In a few weeks, the Western Allies and the

Soviet Union would be at war and the German Army would once again be marching in, he told Niehoff as he left the command bunker. (Hargreaves, *Hitler's Final Fortress- Breslau 1945*)

Decorations and Awards

15.04.1945	*Ritterkreuz des Eisernen Kreuzes* as *Führer* of *Volkssturm-Einheiten* and of a *Kampfgruppe* for his role in the defense of *Festung* Breslau
00.00.1940	*1939 Eisernes Kreuz I. Klasse*
00.00.1940	*1939 Eisernes Kreuz II. Klasse*
01.09.1942	*Kriegsverdienstkreuz I. Klasse ohne Schwerter*
01.09.1944	*Kriegsverdienstkreuz II. Klasse mit Schwertern*
01.09.1942	*Kriegsverdienstkreuz II. Klasse ohne Schwerter*
00.00.1940	*Verwundetenabzeichen, 1939 in Schwarz*
ca. 1939	*Medaille zur Erinnerung an den 1. Oktober 1938*
ca. 1938	*Medaille zur Erinnerung an den 13. März 1938*
00.00.193_	*Goldenes Ehrenzeichen der NSDAP*
ca. 1934	*Ehrenzeichen des 9. November 1923 (Blutorden)*
00.00.194_	*Dienstauszeichnung der NSDAP in Silber*
00.00.194_	*Dienstauszeichnung der NSDAP in Bronze*
ca. 1931	*Abzeichen des SA-Treffens Braunschweig 1931*
00.00.1934	*Ehrendegen der SA* (mit Wirkung vom 03.02.1934)
00.02.1934	*Ehrenwinkel für alte Kämpfer*

Notes

* Son of the innkeeper Peter Herzog and his wife Pauline, née Bürger.
* Married to Lotte ____ (born 27.09.1897 in Freiberg / Sachsen). Divorced, 07.09.1938, and later remarried. One son was born to his first marriage (on 27.11.1934) and two to his second (on 11.11.1942 and 18.03.1944).

Sources

Bayerisches Hauptstaatsarchiv, München, Abteilung IV Kriegsarchiv: Excerpts from various Kriegsranglisten containing data on the Royal Bavarian Army service of Otto Herzog.

Bundesarchiv, Berlin-Lichterfelde (former Berlin Document Center): *Personalunterlagen von SA-Angehörigen: SA-Personalakte* of Otto Herzog.

Hargreaves, Richard: *Hitler's Final Fortress- Breslau 1945*. Pen & Sword Books, 2011.

Lilla, Joachim; Döring, Martin; & Schulz Andreas: *Statisten in Uniform. Die Mitglieder des Reichstags 1933-1945*. Droste Verlag, 2004.

Ludwig *Arthur* Hess

SA-Obergruppenführer

Born:	18.07.1891 in Plauen / Vogtland / Sachsen.
Died:	20.06.1959 in Ingolstadt / Bayern.

NSDAP-Nr.:	6 840 (first joined 00.03.1923; Party banned following the *München-Putsch* of 09.11.1923; Reenrolled 02.06.1925)

Promotions

00.00.191_	*Vizefeldwebel d. R.*
01.02.1928	*SA-Standartenführer*
29.11.1932	*SA-Oberführer* (mit Wirkung vom 01.07.1932)
01.07.1933	*SA-Brigadeführer*
25.03.1934	*SA-Gruppenführer*
30.01.1941	*SA-Obergruppenführer*

Career

00.00.1898-00.00.1906	Attended *Bürgerschule* in Plauen.
00.00.1906-00.00.1909	Employed as an apprentice orthopedic shoemaker in his father's business in Plauen. He also attended the *Berufsschule* in Plauen during this period.
00.00.1909-00.00.1911	Employed as a shoemaker in his father's business.
00.00.1911-00.00.1913	Military service with *Infanterie-Regiment 133* (Zwickau).
10.08.1914-Ostern 1918	Entered service as *Kriegsfreiwilliger*, assigned to *2. Kompanie / Infanterie-Regiment 134*. Among his assignments were that of *Gefechts-Ordonnanz* (combat orderly) and *Führer* of the *Bataillon-Ordonnannz-Stab*.
00.00.1918-00.00.19__	Employed as a master shoemaker and co-owner of *Wilhelm Hess & Söhne Orthopädie-Schuhmacherei* in Plauen.
00.03.1923	Joined the *NSDAP / Ortsgruppe Plauen*.
00.08.1923-00.00.1923	*Führer* of the *2.SA-Abteilung* in Plauen.
08.11.1923-11.11.1923	Arrested, placed in investigative detention, and sentenced to 3 months' imprisonment. He was later acquitted on appeal.
02.06.1925	Reenrolled in the *NSDAP / Ortsgruppe Plauen*.
00.00.1925	Reentered the *NSDAP*.
00.00.1925-00.00.1928	*Führer* of *II. SA-Abteilung Plauen*.
00.10.1927-00.00.1933	Teacher of a *Schuhmacher-Fachklasse* (professional shoemakers' class) at the *Gewerbeschule* in Plauen.
01.02.1928-30.06.1930	*Führer* of *SA-Standarte II Plauen*.
ca. 1929	Cofounder of the Plauen *Ortsgruppe* of the *NS-Lehrerbund* (*NSLB*, National-Socialist League of Teachers).
01.07.1930-00.00.1931	*Führer* of *SA-Brigade II Plauen*.
00.00.1931-30.06.1932	*Führer* of *SA-Standarte 2* (Plauen).
01.07.1932-01.09.1933	*Führer* of *SA-Untergruppe Plauen*.
31.07.1932-12.09.1932	Member of the *Reichstag* (*Wahlkreis 30, Chemnitz-Zwickau*).

SA-Brigadeführer Arthur Hess as *Führer* of *SA-Untergruppe*
Plauen, 1933. (Laurens Hessels photo)

06.11.1932-01.02.1933	Member of the *Reichstag* (*Wahlkreis 30, Chemnitz-Zwickau*).
05.03.1933-14.10.1933	Member of the *Reichstag* (*Wahlkreis 30, Chemnitz-Zwickau*).
00.04.1933-31.10.1933	(*ehrenamtlicher*) *Kommissar z.b.V.* to the *Kreishauptmannschaft Zwickau*.
01.09.1933-14.04.1935	*Führer* of *SA-Brigade 36* (Plauen).
12.11.1933-28.03.1936	Member of the *Reichstag* (*Wahlkreis 30, Chemnitz-Zwickau*).
00.00.193_-00.00.19__	*Leiter* of the *Reichsfachgruppe der Orthopädie-Schuhmachermeister* (*Reich* Professional Group of Master Orthopedic Shoemakers), Berlin.
12.02.1934-14.02.1934	Participated in the *SA-Sonderlehrgang* at Döberitz.
17.05.1934	Appointed as *Landeshandwerksführer für den Treuhänderbezirk Sachsen*. He was sworn in to this post by *Reichswirtschaftsminister* Dr. Schmitt on 31.05.1934.
02.08.1934	Requested judicial proceedings be initiated against him so that he could defend his own honor (*Ehrenschutz*) from accusations in connection with the so-called "Röhm-Putsch" of June/July 1i34. In a letter of 14.08.1934, *Gauleiter* Mutschmann of Sachsen accused Hess of having spoken of "a coming second revolution", and to have announced a "night of the long knives" prior to 30.06.1934. He also claimed that Hess had secured the position of *Landeshandwerksführer* through surreptitious means.
25.09.1934	Acquitted by the *Sondergericht der Obersten SA-Führung* (Special Court of the Supreme SA Leadership), which determined that "No accusations can be brought forward which would bring him (Hess) into connection with the intentions of the former chief of staff (Röhm), or with intentions

	against the P.O. [*Politische Organisation* of the *NSDAP*] or against the *Führer*."
09.01.1935	Appointed as *Reichsinnungsmeister beim Reichsinnungsverband des Schuhmacherhandwerks* (*Reich* Guild Master of the *Reich* Guild Association for the Master Shoemaking Craft). Officially appointed to this post by *Reichshandwerksmeister* Wilhelm Schmidt on 10.01.1935. With this post came the following additional assignments: • *Präsident* of the *Verband der Internationale Schuh- und Leder-Wirtschaft* (Association of the International Shoe and Leather Industry) • Member of the *Engeren Beirat der Reichsgruppe Handwerk*, Berlin • Member of the *Beirat der Reichsstelle für Lederwirtschaft* (Advisory Council of the *Reich* Office for the Leather Industry), Berlin • Member of the *Beirat der Reichswirtschaftskammer* (Advisory Council of the *Reich* Chamber of Economics), Berlin • *Ehrenamtlicher Richter* (honorary judge) of the *Reichsarbeitsgericht* (*Reich* Labor Court), Leipzig
08.02.1935	Found guilty by the *II.Kammer* (2nd Chamber) of the *Obersten Parteigerichts der NSDAP* of having violated *§ 4 Abs. 2 b der Satzung der NSDAP* (Section 4, paragraph 2 b of the Statute of the *NSDAP*). The court requested he be censured for this violation, but acquitted him of surreptitiously obtaining his post as *Landeshandwerksführer* in *Gau Sachsen*.
15.04.1935-00.05.1945	Assigned as an *SA-Führer z. V.* to the *Stab der Obersten SA-Führung*.
14.05.1935-00.05.1945	*Verbindungsführer der Stab der Obersten SA-Führung zum Reichshandwerksmeister* (Chief liaison between the Staff of the Supreme *SA* Leadership and the *Reich* Artisan Master).
29.03.1936	Unsuccessful candidate for election to the *Reichstag*.
00.00.193_-00.00.19__	*Direktor* of the *Zentralverband Deutscher Schuhmacher-Rohstoff-Genossenschaften eGmbH* (Central Association of German Shoemaking Commodity Cooperatives).
00.00.193_-00.00.19__	*Direktor* of the *Meisterschule der Schuhmacherhandwerk* (Master Shoemaking School).
00.00.193_-00.00.19__	*Vorstand* (Chairman) of the *Landeslieferungs-Genossenschaft der Schuhmacherhandwerk für die Bezirke Brandenburg, Pommern und Nordmark eGmbH*, Berlin.
07.02.1939-00.05.1945	*Wehrwirtschaftsführer*.
00.00.19__-00.00.19__	*Leiter* of the *Reichsfachgruppe der Orthopädie-Schuhmachermeister* (*Reich* Professional Group of Master Orthopedic Shoemakers) in Berlin.
00.05.1940-00.00.1945	*(ehrenamtlicher) Richter* and *Beisitzer* of the *Volksgerichtshof* (5-year appointment).
00.00.1942-00.05.1945	*Beauftragter der Reichshandwerkmeister für Erziehung und Unterricht im Handwerk* (Representative for Education and Instruction of Artisans to the *Reich* Artisan Master).

Published Works

Wahrheiten über die gegenwärtige Methodik der Beschuhung kranker Füsse (1938)

SA-Gruppenführer Hess leading a parade of the officers of *SA-Untergruppe Plauen* past the *Führer* and *Gauleiter* Mutschmann, probably early in 1934. (NARA, Heinrich Hoffmann collection)

Grundsätzliche Richtlinien für die Berufserziehung im Schuhmacherhandwerk (1938)

Decorations and Awards

00.00.191_	*1914 Eisernes Kreuz II. Klasse*
30.01.1944	*Kriegsverdienstkreuz I. Klasse ohne Schwerter*
00.00.194_	*Kriegsverdienstkreuz II. Klasse ohne Schwerter*
00.00.191_	*Sächsische Friedrich-August-Medaille*
00.00.191_	*Bronzene Friedrich-August-Medaille*
ca. 1939	*Medaille zur Erinnerung an den 1. Oktober 1938*
ca. 1934	*Ehrenkreuz des Weltkrieges 1914-1918 mit Schwertern*
00.00.193_	*Goldenes Ehrenzeichen der NSDAP*
00.00.194_	*Dienstauszeichnung der NSDAP in Gold*
30.01.1940	*Dienstauszeichnung der NSDAP in Silber*
30.01.1940	*Dienstauszeichnung der NSDAP in Bronze*
00.00.1929	*Nürnberger Parteitagsabzeichen 1929*
00.00.193_	*Deutsche Reiterabzeichen in Bronze* (?)
00.00.193_	*Ehrendolch der SA*
00.02.1934	*Ehrenwinkel für alte Kämpfer*

Notes

• Son of the master shoemaker Ludwig Wilhelm Hess (born 25.12.1866 in Oberbreit) and his wife

Hedwig Klara, née Pommer (born 28.01.1866 in Plauen).
- Married to Friederike Wilhelmine Puschert (born 22.12.1891 in Plauen). Four daughters (born 11.01.1911; 08.07.1914; 15.11.1919; 29.08.1924; and 18.04.1926).

Sources

Bundesarchiv, Berlin-Lichterfelde (former Berlin Document Center): *Personalunterlagen von SA-Angehörigen: SA-Personalakte* of Arthur Hess.

Philipp Prinz und Landgraf von Hessen (-Rumpenheim)
SA-Obergruppenführer

Born: 06.11.1896 in Schloß Rumpenheim bei Hanau.

Died: 25.10.1980 in Rome.

NSDAP-Nr.:   418 991 (Joined 01.10.1930)

Promotions

00.00.1914	*Fahnenjunker*
00.00.1914	*Unteroffizier*
18.06.1915	*Leutnant* (ohne Patent; 00.00.1918: Granted Patent vom 19.12.1915)
09.09.1932	*SA-Oberführer* (mit Wirkung vom 01.07.1932)
13.05.1933	*SA-Gruppenführer*
07.06.1933-15.11.1943	*Oberpräsident*
15.10.1936 (?)	*Hauptmann d. R. (Luftwaffe)*
09.11.1938	*SA-Obergruppenführer*

Career

00.00.1905-00.00.1910	Attended *Realgymnasium* in Frankfurt-am-Main.
00.00.1910-00.00.1912	Attended the *deutschen Internatsschule* at Bexhill-on-Sea (East Sussex) in southern England .
00.00.1912-00.00.1914	Attended the *Musterschule* (a progressive school teaching modern languages) in Frankfurt-am-Main.
00.00.1914-00.00.1914	Attended *Realgymnasium* in Potsdam, where he excelled in mathematics and science and qualified for university studies in architecture.
06.08.1914-00.00.1918	Entered service as a *Kriegsfreiwilliger*, assigned first as a *Fahnenjunker* to *Ersatz-Eskadron / Leibdragoner-Regiment (2. Grossherzoglich Hessische) Nr. 24* (Darmstadt). His later attorney before a de-Nazification court wrote: " ... the concerned party did not make use of the then existing right of high nobility to serve in the officer corps, but entered the Dragoon-Regiment Nr. 24 in Darmstadt as a volunteer, where he served as a soldier of the pike." (Jonathan Petropoulos, *Royals and the Reich. The Princes von Hessen in Nazi Germany*, p. 39.) He was soon appointed as a *Fahnenjunker* (officer cadet), however, and was deployed to the front in Flanders on 19.02.1915. In the Spring of 1915, he was transferred to the East Front, servicing mostly in the Regimentsstab and concerned with the supply of weapons and ammunition. On 26.02.1918, he was reassigned to the staff of *25.Kavallerie-Brigade* on the Eastern Front, where he participated in the Kurland offensive (in what is now Latvia). As of 10.06.1917, he was back in Belgium, but returned with his unit to Galicia in 1918. According to his own account, he played a part in fighting near Chilkowa on 21.03.1918, and briefly commanded a

Philipp Prinz von Hessen and his fellow Nazi blueblood, Prinz
August Wilhelm von Preussen ("AuWi") in 1933.

machine gun squadron at Sienkowo in the absence of its commander. He
returned to the family home – Friedrichshof – in the Autumn of 1918.

Autumn 1915	Passed his *Notabitur* (emergency school leaving examination).
16.02.1918-26.03.1920	Assigned as an *Ordonnanzoffizier* to the staff of *25.Kavallerie-Brigade (Großherzoglich Hessische)*.
26.02.1918	Officially transferred to the staff of *25.Kavallerie-Brigade*.
12.11.1918-26.03.1920	*Ordonnanzoffizier* to the *Abwicklungsstelle* (a small staff concerned with the dissolution and disbandment of a unit) of *Dragoner-Regiment 24*.
26.03.1920	Discharged from active military service at the *Abwicklungsstelle* of *Dragoner-Regiment 24*.
00.00.1920	Discharged from military service.
00.04.1920-00.00.1922	Studied architecture and art history at the *Technische Hochschule* Darmstadt (left without graduating).
00.00.1923-00.06.1933	Lived in Rome, where he worked as a freelance interior decorator. He established contacts with the Fascist Party during this period, later claiming to have been a member even before Mussolini's "March on Rome" in October 1922.
00.00.1922-00.00.1923	Employed by the *Kaiser-Friedrich-Museum* in Berlin, where he worked on a catalog of graphic arts.
00.00.1930	Introduced by his cousin, August Wilhelm Prinz von Preußen, to Hermann Göring.
01.10.1930	Joined the *NSDAP*.

SA-Oberführer Philipp Prinz
von Hessen, 1933.

07.06.1933: The Prinz von Hessen stands
beside his wife, Mafalda, Princess of
Savoia, as he is sworn in as *Oberpräsident*.

00.10.1931	Joined the *SA*.
14.10.1930	First meeting with Hitler in Göring's Berlin apartment.
17.10.1931-18.10.1931	Participated in the *SA-Aufmarsch in Braunschweig*.
09.09.1932-26.01.1934	Attached to the *Stab der Obersten SA-Führer*.
00.00.193_-00.00.193_	*Mitarbeiter* of the *Landesgruppe Italien der NSDAP*.
00.05.1933	Nominated by Hermann Göring, in his capacity as Prussian *Ministerpräsident*, as *Oberpräsident* of *Provinz Hessen-Nassau*.
07.06.1933-15.11.1943	*Oberpräsident der Provinz Hessen-Nassau* (kommissarischer until 15.06.1933, then permanent; Seat: Kassel). As such, he also held the following collateral assignments:

• *Leiter* of the *Bezirksverband Hessen;*
• *Leiter* of the *Bezirksverband Nassau*;
• *Vorsitzender* of the *Spruchkammer für Siedlung und Auseinandersetzung Hessen-Nassau.*

Albert Speer writes of the Prince of Hessen:

> He was one of the few followers whom Hitler had always treated
> with deference and respect. Philipp had often been useful to him,
> and especially in the early years of the Third *Reich* had arranged
> contacts with the head of Italian Fascism. In addition he had
> helped Hitler purchase valuable art works. The prince had been

28.09.1937: Philipp, in *SA-Gruppenführer* uniform, speaks with the Italian Foreign Minister, Count Galeazzo Ciano, during Mussolini's visit to Göring's "Carinhall" estate.

> able to arrange their export from Italy through his connections with the Italian royal house, to which he was related. ...: [Hitler] continued to treat him with the greatest outward courtesy and invited him to his meals. But the members of Hitler's entourage, who until then had been so fond of talking with a "real prince," avoided him as if he had a contagious disease. (Speer, *Inside the Third Reich*, p. 399)

Although his relations with Hitler and Göring were very close, such was not the case with the two *Gauleiter* who administered the *NSDAP* within the province: Jakob Sprenger (*Gau Hessen-Nassau*) and Karl Weinrich (*Gau Kurhessen*). A bitter conflict raged between the *Oberpräsident* and the Party bosses. After the war, during an interrogation session on 20.07.1945, the Prince reflected on his earlier difficulties with Sprenger (whom he described as "insufferable"):

> The *Gauleiter* often interfered in matters which were no concern of the party at all, but were purely matters of the state administration; and which the Prince as the *Oberpraesident* of Hessen should have handled alone.
>
> The Prince wanted to maintain the independence of the

SA-Obergruppenführer Philipp Prinz von Hessen, ca. 1939.

State and Provincial administration (*Staatliche* and *Provinziale Selbstverwaltung*). The Prussian province of Hessen had, as a survival from the time of Bismarck, two units (*Bezirksverbande*) each with its own *Selbstverwaltung* (self-administration), viz: the former *Herzogtum Nassau* with the former *Freie Reichsstadt* and the *Kurfurstentum Hessen*. Sprenger was looking toward uniting Nassau with the territory of the former Grossherzogtum Hessen (later Freistaat Hessen) of which he was *Reichsstatthalter* into one *Reichsgau*. The Prince, on the other hand wanted to merge the administration of Nassau and Prussian Hessen and gradually create an administration ... for all Hessen-Nassau. This plan was opposed to Sprenger's intentions. Sprenger was a dishonest and unscrupulous man who appointed dishonest subordinates. He was hated throughout the *Gau*. The prince had his residence in Kassel, and avoided contact with Sprenger ...

It is ... likely that Sprenger and others might have caused his arrest. Sprenger had systematically been collecting evidence against the Prince to build up a case against him, e.g. that the Prince was opposed to the party, that he had sabotaged Sprenger's efforts, that he used to have Jewish friends in Frankfurt, etc. ("Interrogation of the Prince of Hesse", in Institut für Zeitgeschichte, München, ZS-918-44)

00.07.1933-15.11.1943(?)	Member of the *Preußische Staatsrat*.
14.09.1933-15.11.1943(?)	*Preußischer Staatsrat*.
00.00.1933-00.09.1943	*Ehrenpräsident* (honorary President) of the *Deutsch-Italienischen Gesellschaft* (German-Italian Society), dedicated to forging and maintaining social and cultural bonds with Italy (primarily through lectures on science, art, and other subjects, as well as concerts and dramatic performances by guest musicians and actors from Italy). Under postwar interrogation, he stated he had only attended about three meetings of the Gesellschaft. The actual *Präsident* was Hans von Tschammer und Osten (until his death on 25.03.1943). Otto Meissner assumed the office of *Präsident* in September 1943.
27.01.1934	Appointed as an *Ehrenführer der SA* to the staff of *SA-Gruppe Hessen*.
27.01.1934-00.05.1945	Assigned as an *SA-Führer z. V.* to *SA-Gruppe Hessen*.
00.00.1934-15.11.1943	*Präsident* of the *Provinzialrat* of *Provinz Hessen-Nassau*.
00.05.1935-00.05.1935	Accompanied Hermann and Emmy Göring on their honeymoon/goodwill tour of southeastern Europe, with stops in Belgrade, Budapest, and Bucharest.
00.00.1936-08.09.1943	Liaison between Hitler and Mussolini, as well as to King Vittorio Emmanuele III.
00.00.1936-08.09.1943	*Verbindungsführer der SA* (Chief Liaison of the SA) to the Italian "Blackshirts" (also known as *Verbindungsführer "Italien" der SA*).
15.09.1936	Appointed as a *Mitarbeiter* to the *Dienststelle Ribbentrop*.
15.01.1937-24.01.1937	Accompanied Hermann and Emmy Göring on their tour of Italy (Rome and Capri).
15.10.1936-15.11.1936	Participated in reserve training exercises with *Aufklärungsstaffel 124* (Kassel-Rothwesten).
22.09.1939-16.11.1942	*Stellvertretender Reichsverteidigungskommissar* (Deputy *Reich* Defense Commissioner) and member of the *Verteidigungsausschuß* (defense committee) for *Wehrkreis IX*.
00.08.1943	Visit to *Führerhauptquartier* "Wolfsschanze" in Rastenburg / Ostpreußen.
08.09.1943	Arrested together with his wife.
12.09.1943-15.04.1945	Held in solitary confinement at *KL-Flossenbürg* as a so-called "Priviligerter" (privileged inmate).
22.09.1943	Arrest of his wife, who was registered as an inmate at *KL-Buchenwald* in October 1943 under the surname "Weber".
15.11.1943	Redesignated as an *Oberpräsident zur Disposition*.
25.01.1944	Placed in retirement.
15./16.04.1945	Transferred to *KL-Dachau*, then, together with 138 other "Priviligierten" from 17 European nations, marched by an *SS-Kommando* toward the "Alpenfestung" in the Tirol.
28.04.1945-30.04.1945	Held under *SS* guard in Niederdorf / Südtirol.

Postwar Confinement and Activities

Released from *SS* custody at Niederdorf by troops of a retreating *Wehrmacht* unit, 30.04.1945, then remained with 138 other "privileged inmates" at the Hotel "Pragser Wildsee" in the

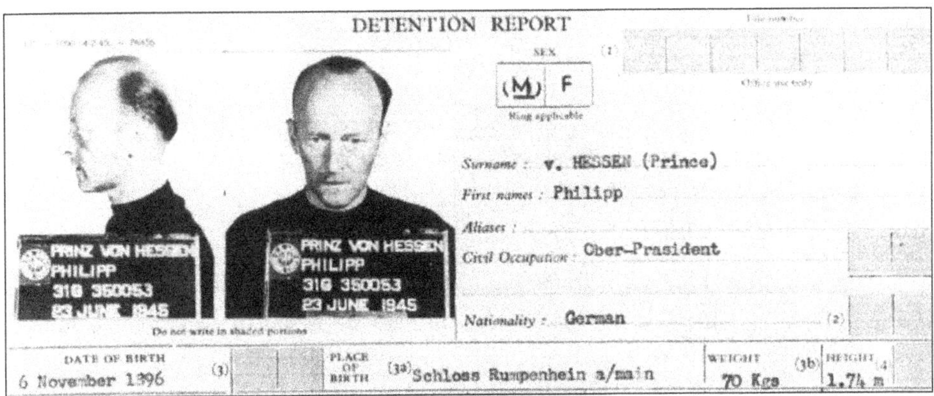

Detention Report of Philipp Prinz von Hessen.

Pragser Dolomiten (Südtirol), 30.04.1945-04.05.1945. There he was arrested by U.S. Army troops on 04.05.1945, and transported to Allied Headquarters in Naples. From there, he was sent with other Nazi luminaries to the Allied internment and interrogation center "Ashcan" at the Palace Hotel in Mondorf-les-Bains, Luxembourg. He was then held at "Haus Alaska" in Oberursel, then briefly confined to several other camps. From 00.01.1946-17.12.1947, he resided in US-Lager CI Camp 91 on the Rheinstraße in Darmstadt, and on 29.04.1947, he was in the Gerichtsgefängnis Nürnberg as a war crimes trial witness. He underwent de-Nazification proceedings before the *Spruchkammer* at Darmstadt-Lager, 15.12.1947-17.12.1947; on the latter date was placed in Kategorie II ("Belasteter"; incriminated person) and sentenced to two years' in a labor camp (with credit granted for time served) as well as forfeiture of 30% of his property and other assets, which had a total value of DM 600,00. An appeals court reclassified him as a "Minderbelasteter" (minor offender, in de-Nazification category III) on 13.08.1948. In February 1949, a further appeal hearing resulted in his final reclassification as a "Mitläufer" (fellow traveler, in Kategorie IV), and his fine was reduced to DM 36,568. On 04.03.1948, the Prince appeared as a witness in the "Wilhelmstraße Trial" before the U.S. Military Tribunal, Nürnberg.Jonathan Petropoulos writes: "In order to pay his bill, Prince Philipp transferred ownership of Schloß Belvedere in Kassel – his official residence when he was *Oberpräsident* of Hessen-Nassau – to the state. In that it was badly damaged by Allied bombs ... relinquishing Schloss Belvedere was a modest sanction." (Petropoulos, *Gray Zones: Ambiguity and Compromise in the Holocaust and Its Aftermath*). He lived in Kronberg, then in Schloß Fasanerie, serving as head of the *Gesamthaus Hessen*, (00.00.1968-25.10.1980) and *Präsident* of the *Deutschen Verkehrswacht* (from 1968).

Decorations and Awards

00.00.191_	*1914 Eisernes Kreuz I. Klasse*
00.00.191_	*1914 Eisernes Kreuz II. Klasse*
00.00.194_	*Kriegsverdienstkreuz I. Klasse ohne Schwerter*
00.00.194_	*Kriegsverdienstkreuz II. Klasse ohne Schwerter*
00.00.191_	*Hessische Tapferkeitsmedaille*
00.00.191_	*Österreichisches Verdienstkreuz __. Klasse*
00.00.19__	*Braunschweigisches Kriegsverdienstkreuz*
00.00.19__	*Ehrenkreuz von Sachsen-Meiningen*

00.00.19__	*Ehrenkreuz von Schaumburg-Lippe*
ca. 1939	*Medaille zur Erinnerung an den 1. Oktober 1938*
ca. 1938	*Medaille zur Erinnerung an den 13. März 1938*
ca. 1934	*Ehrenkreuz des Weltkrieges 1914-1918 mit Schwertern*
30.01.1939	*Goldenes Ehrenzeichen der NSDAP* (personally presented by *SA-Stabschef* Lutze)
00.00.194_	*Dienstauszeichnung der NSDAP in Silber*
00.00.194_	*Dienstauszeichnung der NSDAP in Bronze*
00.00.193_	*Ehrenzeichen I. Klasse des Deutschen Roten Kreuzes*
00.00.19__	Presentation portrait of the *Führer*, personally dedicated to him by Hitler
ca. 1931	*Abzeichen des SA-Treffens Braunschweig 1931*
00.00.193_	*Ehrendolch der SA*
00.00.193_	*Ehrenwinkel für alte Kämpfer*
00.00.193_	*Ordine Supremo della Santissima Annunziata* (the Royal Italian Order of the Most Holy Annunciation)
00.00.193_	*Gran Croce dell'Ordine dei Santi Maurizio e Lazzaro* (Order of Saints Maurice and Lazarus, Grand Cross)
00.00.19__	Mohammed-Ali Order with Sash (Egypt)
00.00.19__	Grand Cross of the Order of the Phoenix (Greece)
00.00.19__	Star of Karajordge 1st Class (Yugoslavia)
00.00.19__	Grand Cross of the Order of St. Alexander (Bulgaria)

Notes

• Third son of Friedrich Karl Landgraf von Hessen (born 01.05.1868 at Schloß Panker / Holstein, died 28.05.1940 in Kassel) and his wife Margarete Beatrice Feodor, née Prinzessin von Preußen (born 22.04.1872 in Potsdam, died 22.01.1954 in Schönberg bei Bronberg / Hessen; daughter of the Deutschen Kaisers und Königs von Preußen Friedrich III.).

• Prince Philipp had six brothers:

◊ Friedrich Wilhelm Prinz von Hessen (born 1893, killed in action 12.09.1915 at Curu Orman, Romania)

◊ Maximilian Prinz von Hessen (born 1894, killed in action at St. Jean-Chappel near Bailleul, Flanders, 12.10.1914)

◊ Wolfgang Prinz von Hessen (his twin, died 12.07.1989 in Kronberg im Taunus / Hessen; *SA-Obersturmführer*)

◊ Ricardo Prinz von Hessen (born 1901, died 1969)

◊ *Christoph* Ernst August Richard Prinz und Landgraf von Hessen (born 14.05.1901 in Frankfurt-am-Main, shot down and killed, 07.10.1943, as a *Major d. R.* while on a reconnaissance flight over Forlì, Emilia-Romagna in the Appenines). He was a *Ministerialdirektor* in the *Reichsluftfahrtministerium*. On 15.12.1930, he married Sophie, Princess of Greece and Denmark (born 26.06.1914), with whom he had five children. He was an honorary *SS* officer (*SS*-Nr. 35 903) attached to the staff of the *Reichsführer-SS*, and received the following SS, Government, and *Luftwaffe* promotions:

04.02.1932	*SS-Anwärter*
12.09.1932	*SS-Mann*
01.02.1933	*SS-Obertruppführer*

A 1938 postcard of Princess Mafalda and her three sons – Moritz, Otto, and Heinrich.

00.05.1933	*Regierungsrat*
19.06.1933	*SS-Sturmführer* (mit Wirkung vom 12.06.1933)
21.04.1934	*SS-Sturmhauptführer* (mit Wirkung vom 20.04.1934)
00.00.1934	*Oberregierungsrat*
00.00.1935	*Unteroffizier d. R.*
20.12.1935	*SS-Obersturmbannführer* (mit Wirkung vom 24.12.1935)
22.04.1936	*SS-Standartenführer* (mit Wirkung vom 20.04.1936)
00.00.193_	*Ministerialdirektor*
01.09.1937	*Feldwebel d. R.*
25.06.1939	*SS-Oberführer* (mit Wirkung vom 01.06.1939)
00.00.1939 (?)	*Leutnant d. R.*
06.07.1940	*Oberleutnant d. R.* (mit Wirkung vom 01.05.1940)
13.09.1940	*Hauptmann d. R.* (mit Wirkung vom 01.09.1940)
00.00.194_	*Major d. R.*

- Religion: Evangelical-Lutheran.
- Married at the Palazzo Racconigi in Torino, Italy on 23.09.1925 to Mafalda, principessa di Savoia (born 19.11.1902 in Rome / killed on 28.08.1944 during an Allied air attack as an inmate of the Buchenwald concentration camp). She was the second daughter of King Vittorio Emanuell III of Italy. Among attendees of the wedding were King George of Greece, King Carol of Romania, Prince Paul of Yugoslavia, and the Italian Prime Minister, Benito Mussolini. The couple spent their honeymoon in Capri, then visited Germany. They had four children:
  ◊ Moritz Prinz von Hessen (born 06.08.1926)
  ◊ Heinrich, Prinz von Hessen (born 30.10.1927, died 00.00.1999)

◦ Otto Adolf Prinz von Hessen (born 00.00.1937, died 00.00.1998)
◦ Elisabeth Margarete Prinzessin von Hessen (born 08.10.1940)
- Foreign language proficiency: English, French, and Italian.

Sources

Petropoulos, Jonathan: *Royals and the Reich. The Princes von Hessen in Nazi Germany*. Oxford University Press, 2006.
- *Gray Zones: Ambiguity and Compromise in the Holocaust and Its Aftermath* (with John K. Roth). Berghahn Books, 2007.
Speer, Albert: *Inside the Third Reich*. Macmillan, 1970.

Paul Friedrich Karl Hocheisen

Prof. Dr. med.
SA-Sanitäts-Obergruppenführer

Born:	27.05.1870 in Beilstein / Kreis Marbach / Württemberg.
Died:	22.12.1944 in Heidenheim an der Brenz.

NSDAP-Nr.: 145 058 (Joined 01.08.1929)

Promotions

00.00.1889	*Einjährig-Freiwilliger*
18.08.1894	*Assistenzarzt*
14.11.1895	*Oberarzt*
31.05.1899	*Stabsarzt*
22.03.1910	*Oberstabsarzt*
01.10.1920	*Generalarzt (= Oberstarzt)*
01.11.1926	*Generalstabsarzt (= Generalarzt)*
30.04.1929	*Charakter als Generaloberstabsarztes (= char. Generalstabsarzt)*
01.07.1930 (?)	*SA-Sanitäts-Oberführer*
01.08.1930	*SA-Sanitäts-Gruppenführer*
20.04.1933	*SA-Sanitäts-Obergruppenführer*

Career

00.00.18__-00.00.1888	Attended *Gymnasium* in Stuttgart (passed his *Abitur*, 00.00.1888).
00.00.1888-00.00.1892	Studied medicine at the *Kaiser-Wilhelm-Akademie für das militärärztliche Bildungswesen* in Berlin.
00.00.1888-00.00.1937	Member of the *Burschenschaft Suevo-Borussia*.
00.00.1889-00.00.1890	Military service as an *Einjährig-Freiwilliger* with *Kaiser-Alexander-Garde-Grenadier-Regiment 1* (Berlin).
00.00.1892	Passed his state medical examinations.
00.00.1892	Received his doctorate (Dr. med.) at the University of Berlin.
00.00.1892-00.00.19__	Active military service as *Sanitätsoffizier* with *Infanterie-Regimenter 125, 119, 123,* and *120,* and in the *Württembergische Kriegsministerium*.
00.00.1894	Passed his state physician's examinations.
00.00.1898-00.00.1938	Member of the students' association *Saxonia* at the *Forstakademie* in Hannoversch-Münden.
00.00.1899	Passed examinations for the *unteren ärztlichen Staatsdienst* (lower medical civil service) in Württemberg.
00.00.1900-00.00.1902	Training as a surgeon at the *Karl-Olga-Krankenhaus* in Stuttgart.
00.00.1903-00.00.1906	Gynecological assistant at the *Universitäts-Frauenklinik der Charité* in Berlin.
[00.03.1914]	*Regimentsarzt* of *Infanterie-Regiment "Kaiser Wilhelm, König von Preußen" (2. Württembergische) Nr. 120* (Ulm).

SA-Sanitäts-Gruppenführer **Dr. Paul Hocheisen, 1930.**

00.00.1914–00.00.1918	*Divisionsarzt* of *54.Reserve-Division* (severely wounded [shot in the neck and lungs] on 06.05.1916).
00.00.1919–31.10.1926	*Wehrkreisarzt V* and *Divisionsarzt* of *5.Division* (Stuttgart).
30.09.1919–00.00.1920(?)	Attached to the *Reichswehrministerium / Feldzeugmeister-Amt / Abt. Fz. 1 (Allgemeine Abteilung)*.
00.00.1920–31.10.1926	*Wehrkreisarzt V* and *Divisionsarzt* of the *Reichswehr's 5. Division* (Stuttgart).
01.11.1926–30.04.1929	*Gruppenarzt* of *Gruppenkommando 2* (Kassel).
30.04.1929	Retired from the *Reichswehr*.
01.08.1929	Joined the *NSDAP*.
01.07.1930	Joined the *SA*, assigned to the *Stab der Obersten SA-Führers*.
01.08.1930–13.04.1932	*Reichsarzt der SA* and *Leiter* of *Unterabteilung IVb (Sanitätswesen*, Medical Services) in the *Quartiermeisterstab der Obersten SA-Führung* (*SA* banned from 13.04.1932 to 01.07.1932).
01.07.1932–12.09.1932	*Reichsarzt der SA* and independent *Referent IVb (Sanitätswesen)* on the *Stab der Obersten SA-Führer*.
31.07.1932–12.09.1932	Member of the *Reichstag (Wahlkreis 28, Dresden-Bautzen)*.
06.11.1932–01.02.1933	Member of the *Reichstag (Wahlkreis 28, Dresden-Bautzen)*.
21.11.1932–31.10.1933	*Chef der Sanitätsamt der SA* (Chief of the *SA* Medical Office) and *Chef* of *Abteilung IVb (Sanitätswesen)* in the *Stab der Obersten SA-Führer*.
00.00.1933	Appointed as *besonderer Beauftragter des Reichsministers des Innern zur Regelung der Rot-Kreuz-Fragen* (Special Representative of the *Reich* Minister of the Interior for the Regulation of Red Cross Issues).
05.03.1933–14.10.1933	Member of the *Reichstag (Wahlkreis 29, Leipzig)*.

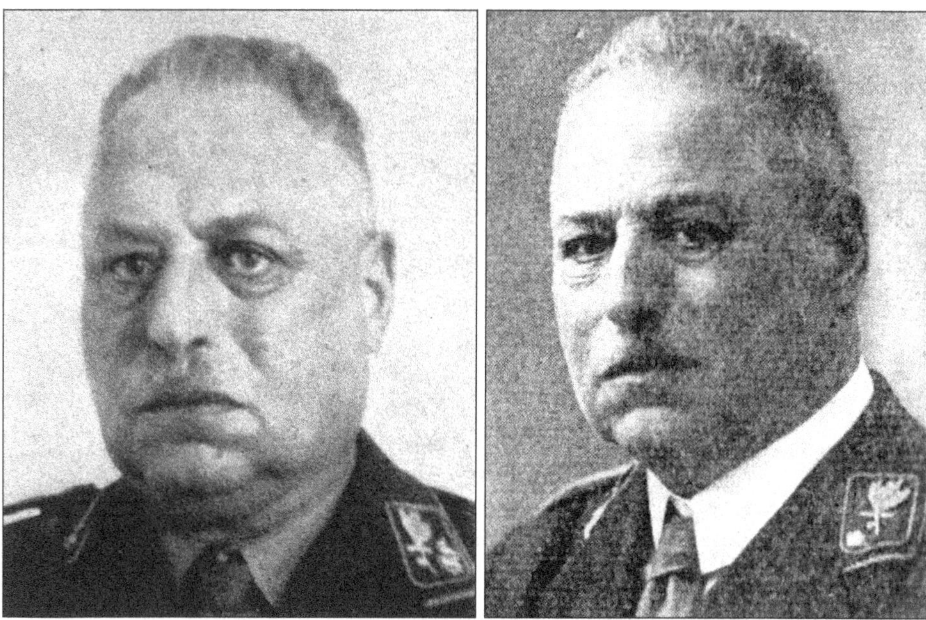

Prof. Dr. Hocheisen as an *SA-Sanitäts-Gruppenführer* (left) and *Obergruppenführer* (right).

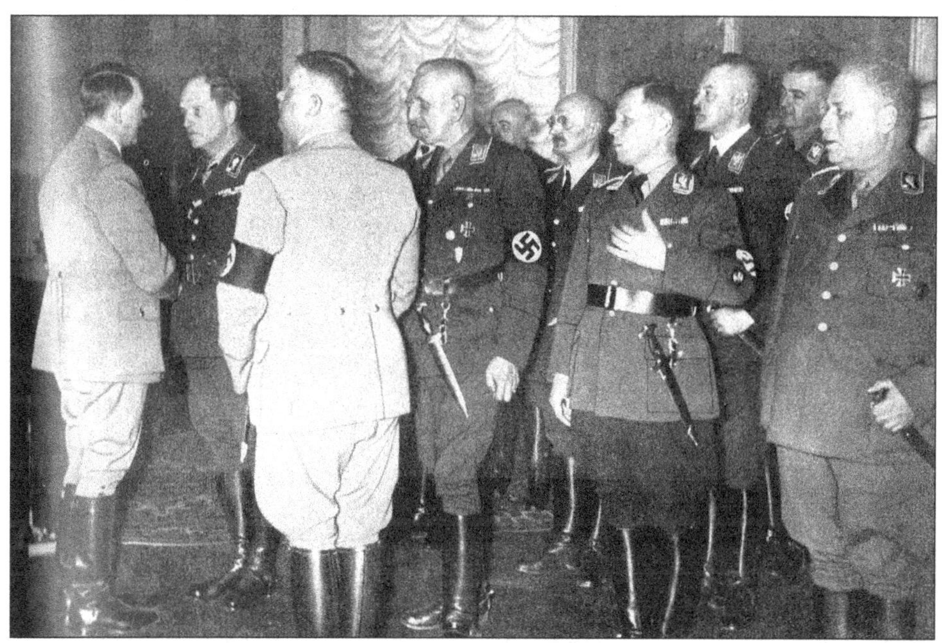

Berlin, 1934: *SA* leaders meet with the *Führer* after an *SA-Führertagung* (leaders' conference) in Friedrichsoda. From right to left: Dr. Hocheisen, unknown, Georg von Detten, Adolf Hühnlein, Hans Georg Hofmann, Curt von Ulrich, Ernst Röhm, Franz Ritter von Epp, and Hitler. (*Illustrierter Beobachter* photo)

SA-Sanitäts-Obergruppenführer Prof. Dr. Hocheisen, as Vice President of the *Deutschen Roten Kreuz* (*DRK*, German Red Cross) presides over a meeting of that organization in 1934. To his left are Carl Eduard Herzog von Sachsen-Coburg und Gotha, President of the DRK, and SA-Sanitäts-Gruppenführer Prof. Dr. Hermann Jensen. (NARA, Heinrich Hoffmann collection)

SA-Sanitäts-Obergruppenführer Prof. Dr. Hocheisen poses at his
desk in 1934 (NARA, Heinrich Hoffmann collection)

00.05.1933	Appointed as *Beauftragter des Reichsministers des Innern bei der Dienststelle des Kommissars der Freiwilligen Krankenpflege* (Representative of the *Reich* Minister of the Interior to the Office of the Commissioner of Voluntary Nursing).
01.11.1933-31.01.1934	*Generalinspekteur des Sanitätswesens der SA und SS* (Inspector General of *SA* and *SS* Medical Services).
12.11.1933-22.12.1944	Member of the *Reichstag (Wahlkreis 29, Leipzig*; after 29.03.1936, *Wahlkreis 15, Osthannover*).
01.12.1933-19.12.1936	*Stellvertretender Präsident* of the *Deutschen Roten Kreuzes (DRK)*. At the request of *Reichsminister* Dr. Frick, he submitted his resignation from this post on 11.12.1936.
00.01.1934-19.12.1936	*Erster Stellvertreter* (1st Deputy) to the *Kommissar für die Freiwillige Krankenpflege* in the *Reichsministerium des Innern*.
01.02.1934-00.00.19__	Assigned as *SA-Sanitätsführer z.b.V.* to the *Stab der Obersten SA-Führung*.
16.07.1934-01.01.1944	*Verbindungsführer* of the *Obersten SA-Führung* to the *DRK*.
01.01.1937-22.12.1944	*Ehrenpräsident der DRK* (Honorary President of the German Red Cross).

SA-Sanitäts-Obergruppenführer
Prof. Dr. Hocheisen inspects SA
medical units in 1934 (NARA,
Heinrich Hoffmann collection)

(NARA, Heinrich Hoffmann collection)

Published Works

Der Muskelsinn Blinder (dissertation, 1892)

Die intravenösen Kollargolinjektionen bei Puerperalfieber (The Intravenous Injection of Collargol in Puerperal Fever) (1906), summarized as follows in *The Medico-Pharmaceutical Critic and Guide, Vols. VI – VII*, 1906:

> Dr. Hocheisen details the histories of 52 cases which he had in the Charité during two years, dividing them into the following groups: 1. Sapremic and septic endometritis. II. Phlegmasia alba dolens. III. Pyemia. IV. Septicemia. V. Metritisdissecans. VI. Adnexal tumors and parametritis. VII. Peritonitis.

Die Gynäkologie des Truppenarztes (1925)

Decorations and Awards

16.09.1916	*1914 Eisernes Kreuz I. Klasse*
02.11.1914	*1914 Eisernes Kreuz II. Klasse*
16.10.1917	*Ritterkreuz des Ordens der Württembergischen Krone mit Schwertern*
05.10.1916	*Ritterkreuz des Württembergischen Militärverdienstorden*
00.00.191_	*Württembergischer Kronenorden mit Schwertern*
Spring 1914	*Württembergischer Fridrichsorden I. Klasse*
10.06.1915	*Sächsischer Albrechtsorden I. Klasse mit der Krone und Schwertern*
16.05.1918	*Verwundetenabzeichen, 1918 in Schwarz*
03.12.1915	*Dienstehrenzeichen*
04.03.1935	*Ehrenkreuz des Weltkrieges 1914-1918 mit Schwertern*

(NARA, Heinrich Hoffmann collection)

00.00.194_	*Dienstauszeichnung der NSDAP in Silber*
00.00.194_	*Dienstauszeichnung der NSDAP in Bronze*
12.03.1934	*Ehrenzeichen des Deutschen Roten Kreuzes I.-IV. Klasse*
28.11.1936	*Deutsches Olympia-Ehrenzeichen I. Klasse*
00.00.193_	*Ehrenwinkel für alte Kämpfer*

Notes

- Son of the physician *Dr. med.* Sigmund Wolfgang Hocheisen (born 26.12.1838 in Oberndorf/ Neckar) and his wife Marie Katherine, née Fuchs (born 02.12.1842 in Cannstatt bei Stuttgart).
- Married on 27.07.1909 to Alice Maag (born 10.0.1886 in Stuttgart). Three daughters (born 01.08.1910, 27.12.1911, and 03.03.1919).

Sources

Bundesarchiv, Berlin-Lichterfelde (former Berlin Document Center): *Personalunterlagen von SA-Angehörigen: SA-Personalakte* of Paul Hocheisen.

Lilla, Joachim; Döring, Martin; & Schulz Andreas: *Statisten in Uniform. Die Mitglieder des Reichstags 1933-1945*. Droste Verlag, 2004.

Wicke, Markus: *SS und DRK: Das Präsidium des Deutschen Roten Kreuzes im nationalsozialistischen Herrschaftssystem 1937-1945*. Books on Demand, 2002.

Franz Ritter von Hörauf

SA-Obergruppenführer

Born:	16.07.1878 in Landau / Regierungsbezirk Pfalz / Bayern.
Died:	08.12.1957 in München / Regierungsbezirk Oberbayern / Bayern.
NSDAP-Nr.:	374 771 (Joined 01.12.1930)

Promotions

14.07.1896	*Fahnenjunker*
23.01.1897	*Fähnrich*
06.03.1898	*Sekondelieutenant* (redesignated *Leutnant*, 01.01.1899)
26.10.1907	*Oberleutnant*
26.10.1912	*Hauptmann*
17.01.1917	*Major*
01.07.1921	*Oberstleutnant* (20.03.1922: Granted RDA vom 01.06.1919)
01.02.1926	*Oberst*
31.01.1928	*Charakter als Generalmajor*
01.03.1931	*SA-Gruppenführer*
01.12.1940	*Generalmajor z. V.*
30.01.1941	*SA-Obergruppenführer*

Career

00.00.189_-00.00.1896	Attended a humanistic *Gymnasium* in München.
14.07.1896-01.10.1908(?)	Entered the Royal Bavarian Army as a *Fahenjunker*, assigned to *10. Kgl. Bayerisches Infanterie-Regiment "König"* (Ingolstadt).
01.03.1897-01.02.1898	Attended the *Kriegsschule* in München.
01.10.1901-20.11.1901	Attending a *Reitkurs* (riding course) with *3. Bayerische Train-Bataillon*.
01.10.1901-00.00.19__	*Instruktionsoffizier für Einjährig-Freiwillige* (instructional officer for one-year volunteers) in *10. Kgl. Bayerisches Feldartillerie-Regiment*.
18.12.1905-01.10.1908(?)	*Regimentsadjutant* of *10. Kgl. Bayerisches Feldartillerie-Regiment*. Among his comrades during this period was his later *SA* superior, Ernst Röhm.
01.10.1908-30.09.1911	Attended the *bayerischen Kriegsakademie* in München.
25.09.1911-26.08.1912(?)	Attached to the *Zentralstelle* of the *Bayerisches Generalstab* (with effect from 01.10.1911).
26.08.1912-04.09.1912	Attached to the staff of *6. Kgl. Bayerisches Kavallerie-Brigade*.
05.09.1912-09.09.1912	Attached to the staff of *6. Kgl. Bayerisches Feldartillerie-Brigade*.
10.09.1912-18.09.1912	Attached to the staff of *6. Kgl. Bayerisches Infanterie-Division*.
01.10.1912-01.04.1913	Assigned to the *Zentralstelle* of the *Bayerisches Generalstab*.
01.04.1913-31.07.1914	Instructor at the *Kriegsakademie* in München.
01.08.1914-22.05.1915	*2.Generalstabsoffizier (Ib)* in the *Generalstab* of *III. Bayerische Armee-Korps*. While in this assignment he participated in the following engagements:

Oberstleutnant Franz Ritter von Hörauf, ca. 1923. (Laurens Hessels photo)

20.08.1914-21.08.1914	Battle of Lothringen (Lorraine).
22.08.1914-14.09.1914	Battle near Lunéville.
19.09.1914-22.05.1915	Fighting between the Meuse and Mosel.
22.05.1915-27.07.1916	Assigned as a *Hauptmann im Generalstab* to the *Alpenkorps*.
27.07.1916-01.12.1917(?)	Assigned as *1.Generalstabsoffizier (Ia)* to the *Generalstab* of *12. Bayerische Infanterie-Division*.
01.12.1917-31.12.1917	Attached to the *Heeresgruppe Kronprinz Rupprecht* for training on the Western Front.
30.01.1918-23.12.1918	Assigned as *Ia* to the *Generalstab* of *I. Bayerische Reserve-Korps*.
11.03.1919-10.05.1919(?)	*Major beim Stab* to the *bayerischen Freikorps* for the *Grenzschutz Ost (Freikorps Epp)*.
07.05.1919-30.09.1920	Assigned as *Ia* to the staff of *Reichswehr-Schützenbrigade 21*.
01.10.1920-01.06.1921	Assigned as *Ia* to the staff of *Infanterieführer VII* (München).
02.06.1921-30.09.1921	*Führer* (m.d.F.b.) of *I. Bataillon / Infanterie-Regiment 21*.
01.10.1921-30.09.1923	*Kommandeur* of *I. Bataillon / Infanterie-Regiment 21*.
01.10.1923-11.01.1924	Instructor at the *Infanterie-Schule* in München.
12.01.1924	*Zur Verfügung* (placed at the disposal of) *7.Division* (München).
01.04.1924-31.07.1925	Attached to staffs of *Gruppenkommando 2.* and *7.Divisions* as head of *(Führergehilfen-Ausbildung)* for the latter.
01.08.1925-31.01.1928	Assigned to the staff of *Infanterie-Regiment 21*.
31.01.1928	Retired from the *Reichswehr*.
00.00.1929-00.00.1930	Member of the *"Stahlhelm"- Bund*.

SA-Gruppenführer Franz Ritter von Hörauf, ca. 1933. (Left: Laurens Hessels
photo; Right: Baldur von Schirach, *Die Pioniere des Dritten Reiches*, 1933)

01.12.1930	Joined the *NSDAP / Ortsgruppe Bogenhausen*.
01.03.1931-00.00.193_	Joined the *SA* as a *hauptamtlicher SA-Führer*, assigned to the *Obersten SA-Führung*.
01.03.1931-00.00.1931	*Leiter der SA-Stabes* and *Ia* to the *Quartiermeisterstab* in the *Obersten SA-Führung*.
00.00.1931-13.04.1932	*Chef* of the *SA*-Führungsstab and *Leiter* of *Amt I* in the *Stab der Obersten SA-Führung*.
18.10.1931	Participated in the *SA-Aufmarsch in Braunschweig*.
01.07.1932-20.11.1932	Chef of the *Ausbildungsstab* (training staff) and *Inspekteur der Schulen der SA* (Inspector of *SA* Schools).
21.11.1932-10.03.1933	Inspekteur of the *Aufmarschinspektion der SA*. Released from active *SA* service, 10.03.1933.
10.03.1933-00.05.1945	Assigned as an *SA-Führer z. V.* to the *Stab der Obersten SA-Führung*.
00.05.1933-00.00.193_	Full-time *Referent für Militärangelegenheiten* (Advisor for Military Matters) in the *Bayerische Staatskanzlei* (Bavarian Government Chancellery).
00.00.193_-01.07.1937	*Leiter* of the *Bund Freikorps Epp* (absorbed by the *Deutschen Reichskriegerbund* on 01.07.1937).
22.12.1937-00.00.194_	*Ehrenamtlicher Richter* (honorary judge) of the *Volksgerichtshof* (5-year appointment; renewed in November 1941).
00.02.1939-00.00.19__	*Bezirksverbandsführer* of the *Bezirksverband Oberbayern der Volksbund Deutsche Kriegsgräberfürsorge e.V.* (District Association for Upper

Nazis and *Stahlhelm-Bund* members during the meeting of the "Harzburg Front"
in Bad Harzburg, 11.10.1931. From left to right, are: Wilhelm Friedrich Loeper,
Christian Mergenthaler, Dr. Bernhard Rust, unknown *NSKK-Oberführer*, Hermann
Göring, Adolf Hühnlein, Franz Ritter von Hörauf, Curt von Ullrich, Ernst Röhm,
Heinrich Himmler, and Gerret Korsemann. (*Bundesarchiv Bild* 102-02134)

	Bavaria of the National Union for the German Military Grave Registration Service).
26.08.1939	Mobilized as *zur Verfügung des Heeres gestellt* (at the disposal of the German Army).
10.11.1939-31.01.1943	*Kommandant* of Lodsch (Pol.: Łódź; renamed Litzmannstadt, 00.04.1940).
31.01.1943-31.03.1943	In *Führerreserve / OKH*.
31.03.1943	Mobilization order lifted.

Postwar Confinement

Interned by Allied authorities, 01.09.1945-30.04.1948.

Published Work

"Das Wesen der SA", in *Der SA-Mann, Nr.3* (19.01.1932)
"Das Führertum in der SA", in *Der SA- Mann, Nr. 3* (23.02.1932)
Pflichtenlehre des Sturm-Abteilungsmannes (ca. 1934)

Decorations and Awards

24.03.1919	*Ritterkreuz des Kgl. Bayerischen Militär-Max-Joseph-Ordens* (presented for an act of bravery performed on 11.11.1916)

Another image of Ritter von Hörauf and Wilhelm Friedrich
Loeper at Bad Harzburg, 11.10.1931.

22.02.1917	*Ritterkreuz des Kgl. Hausordens von Hohenzollern mit Schwertern*
08.09.1915	*1914 Eisernes Kreuz I. Klasse*
16.10.1914	*1914 Eisernes Kreuz II. Klasse*
04.01.1919	*Kgl. Bayerischer Militär -Verdienstorden III. Klasse mit Schwertern*
17.01.1917	*Kgl. Bayerischer Militär-Verdienstorden IV. Klasse mit der Krone und mit Schwertern*
05.11.1914	*Kgl. Bayerischer Militär-Verdienstorden IV. Klasse mit Schwertern*
26.03.1916	*Hessische Tapferkeitsmedaille*
13.05.1916	*Großherzoglich Mecklenburgisches Militär-Verdienstkreuz II. Klasse*
00.00.19__	*Kgl. Bayerisches Jubiläumsmedaille*
00.00.19__	*Kgl. Bayerisches Dienstauszeichnung II. Klasse*
12.05.1917	*Österreichischer Orden der Eisernen Krone III. Klasse*
00.00.19__	*Tiroler Landesdenkmünze*
ca. 1934	*Ehrenkreuz des Weltkrieges 1914-1918 mit Schwertern*
00.00.19__	*Wehrmacht-Dienstauszeichnungen*
00.00.194_	*Dienstauszeichnung der NSDAP in Silber*
00.00.194_	*Dienstauszeichnung der NSDAP in Bronze*
00.00.1931	*Abzeichen des SA-Treffens Braunschweig 1931*
00.00.193_	*Ehrendolch der SA*
00.02.1934	*Ehrenwinkel für alte Kämpfer*

Notes

- Son of the *Kasern-Inspekteur* (barracks inspector) Franz Hörauf (born 09.05.1846 in Schwabach, died 00.00.1921), and his wife Elisabeth ("Elise"), née Buhl (born 06.05.1849 in Hilpotstein, died 00.00.1923).
- Married to Carola Schwarzmeier (born 25.10.1874 in Prien). No children were born to the marriage.

Sources

Bayerisches Hauptstaatsarchiv, München, Abteilung IV Kriegsarchiv: Excerpts from various Kriegsranglisten containing data on the Royal Bavarian Army service of Franz Ritter von Hörauf.

Bundesarchiv, Berlin-Lichterfelde (former Berlin Document Center): *Personalunterlagen von SA-Angehörigen: SA-Personalakte* of Franz Ritter von Hörauf.

Schirach, Baldur von: *Die Pioniere des Dritten Reiches*. Zentralstelle fur der deutschen Freiheitskampf, 1933.

Hans Georg ("Trotski") Hofmann
SA-Obergruppenführer

Born: 26.09.1873 in Hof / Regierungsbezirk Mittelfranken / Bayern.

Died: 31.01.1942 in München (heart attack). A state funeral was held for him on 04.02.1942. In a private conversation on the night of 24./25.02.1942, Hitler commented:

> The death of *Staatssekretär* Hofmann has deeply grieved me. In 1919 I harangued his battalion at Passau. What a marvelous lot of men we had there! Blazing patriots. To start with, Hofmann trusted me—and yet at that time I stood for so little. Hofmann was already convinced that it was I who would save Germany. At the time of the Kapp-putsch, Hofmann sent a telegram: "Putting myself under Kapp's orders. What's regiment doing?" There were a lot of officers of that sort in Bavaria. Seeckt got rid of them all. The only ones who were kept were those who never wavered. (Henry Picker, ed, *Hitler's Table Talk 1941-1944*, p. 159)

NSDAP-Nr.:   550 075 (Joined 01.06.1931)

Promotions

23.11.1894	*Fahnenjunker-Unteroffizier*
23.11.1894	*Portepéefähnrich*
04.03.1895	*Sekondelieutenant* (redesignated *Leutnant*, 01.01.1899)
06.03.1905	*Oberleutnant*
07.03.1910	*Hauptmann*
17.01.1917	*Major* (01.02.1922: Granted RDA vom 28.12.1916)
01.04.1922	*Oberstleutnant*
31.01.1926	*Charakter als Oberst*
31.07.1931	*SA-Gruppenführer* (01.07.1932: Reconfirmed at that rank after lifting of the ban on the SA)
01.04.1933	*SA-Obergruppenführer*
04.04.1933-08.05.1934	*Regierungspräsident*
05.07.1934-31.01.1942	*Staatssekretär*
20.04.1937	*Charakter als Generalmajor*

Career

ca. 1879-ca. 1883	Attended Catholic *Volksschule*n in Hof and Steinwiesen (Frankenwald).
ca. 1883-00.00.1893	Attended a humanistic *Gymnasium* in Bamberg (passed his *Abitur*).
01.08.1893-01.10.1900(?)	Entered service as *Dreijährig-Freiwilliger* (3-year volunteer), assigned to *16. Kgl. Bayerisches Feldartillerie-Regiment* (Passau), then assigned to active duty as a *Fahnenjunker* in that unit.

SA-Gruppenführer Hofmann, ca. 1932. (Baldur von
Schirach, *Die Pioniere des Dritten Reiches*, 1933)

01.03.1894-22.01.1895	Attended the *Kriegsschule* in München.
01.10.1900-30.09.1903	Adjutant of the *Bezirkskommando* in Vilshofen.
01.10.1903-01.01.1904	On leave in France.
01.01.1904-19.08.1907	Assigned to *I. Bataillon / 16. Kgl. Bayerisches Feldartillerie-Regiment* (Landshut).
20.08.1907-00.00.19__	Assigned to *13. Kgl. Bayerisches Feldartillerie-Regiment* (Ingolstadt).
25.05.1911-09.12.1914	Chef of *2. Kompanie / 13. Kgl. Bayerisches Feldartillerie-Regiment* (frontline service in this post, 07.08.1914-09.12.1914).
09.12.1914-25.02.1915(?)	Hospitalized due to illness.
25.02.1915-01.03.1915(?)	Detached to the *Ersatz-Bataillon / 10. Kgl. Bayerisches Feldartillerie-Regiment* as *Führer* of *III.Rekrutendepots*.
01.03.1915-07.02.1916(?)	*Führer* of *II. Ersatz-Bataillon / 10. Kgl. Bayerisches Feldartillerie-Regiment*.
07.02.1916-25.03.1916(?)	*Führer* of *II. Ersatz-Bataillon / 21. Kgl. Bayerisches Feldartillerie-Regiment*.
25.03.1916-04.11.1916	Frontline service as *Führer* (until 31.03.1916, then full *Kommandeur* from 01.04.1916) of *III. Bataillon / 13. Kgl. Bayerisches Feldartillerie-Regiment*.
04.11.1916-10.11.1916	*Führer* of *I. Ersatz-Bataillon / 13. Kgl. Bayerisches Feldartillerie-Regiment*.

SA-Obergruppenführer Hofmann, ca. 1936.

10.11.1916-18.01.1917	Detached for service as an instructor at the *Unterführer-Schule* (NCO school) of *6. Armee* in Tournay, Belgium.
21.01.1917-00.00.1917	Detached for temporary duty as an instructor of a *Kompanieführer-Kursus* (company commander's course) in Nürnberg.
01.06.1917-26.09.1918	*Kommandeur* of II. *Bataillon* / 28. *Kgl. Bayerisches Feldartillerie-Regiment* (in Romania and on the Western Front).
03.01.1918-16.01.1918	*Stellvertretender Kommandeur* of 28. *Kgl. Bayerisches Feldartillerie-Regiment.*
22.01.1918-02.02.1918	*Führer* of 28. *Kgl. Bayerisches Feldartillerie-Regiment.*
15.02.1918-23.02.1918	*Führer* of 28. *Kgl. Bayerisches Feldartillerie-Regiment.*
14.03.1918-21.03.1918	Attended a training course in Bunzlau.
15.04.1918-17.05.1918	*Führer* of 28. *Kgl. Bayerisches Feldartillerie-Regiment.*
08.06.1918-19.06.1918	*Führer* of 28. *Kgl. Bayerisches Feldartillerie-Regiment* (until captured by British troops).
26.09.1918-00.12.1918	In British captivity.
00.12.1918-00.00.1919	Repatriated from British captivity and returned to service with 13. *Kgl. Bayerisches Feldartillerie-Regiment.*
00.00.1919-00.00.1919	*Kommandeur* of III. *Bataillon* / 1. *Bayerisches Schützen-Regiment,* a component of *Freikorps Epp.*
21.02.1919-00.00.1919	*Führer* of the *Bürgerwehr* in Ingolstadt.
00.05.1919	Served as judge of the *Kriegsgericht* (military tribunal) in München in the trial of the Russian-born Jewish Bolshevik and *KPD* leader Eugen

An illustration of *SA-Obergruppenführer* Hofmann from the cigarette card album
Männer des Dritten Reich. (Orientalischen Cigaretten-Compagnie "Rosma", 1934)

	Leviné-Nissen, a cofounder of the so-called "*Zweite Räterrepublik*" (the short-lived communist government in München). Leviné-Nissen was sentenced to death and executed on 06.05.1919.
00.06.1919-00.08.1919	Participated in the suppression of a communist revolt in Hamburg.
00.00.1920-31.03.1923	*Kommandeur* of *III. Bataillon / Reichswehr-Infanterie-Regiment Nr. 20* (Passau).
00.02.1920	First meeting with Adolf Hitler.
00.00.192_-00.01.1923	*Führer* of the *Verband "Bayern und Reich"* in *Kreis Niederbayern*.
00.01.1923-00.00.19__	*Führer* of the *Organisation Niederbayern* (later renamed *Organisation Unterland*, and, in April 1923, *Bund Vaterland*).
01.04.1923-31.01.1926	*Kommandant* of *Festung Ingolstadt*.
09.11.1923	Participated in the *München-Putsch* as intermediary between the *Reichswehr* and the Putschists.
31.01.1926	Retired from active *Reichswehr* service.
01.02.1926-31.05.1931	Civilian employee of the *Reichswehr*, assigned as organizer of the *Feldjägerkorps*.
31.05.1931	Joined the *SA*.
01.06.1931	Joined the *NSDAP / Ortsgruppe Ingolstadt*.
31.07.1931-14.11.1931	*SA-Oberführer Süd* and *Führer* of *SA-Gruppe "Bayern"* (Ingolstadt).
14.11.1931-13.04.1932	*Führer* of *SA*-Gruppe "Mittelland".

SA-Obergruppenführer Hofmann attends a *Hitler-Jugend* event, ca. 1937. (NARA photo)

18.10.1931	Participated in the *SA-Aufmarsch in Braunschweig*, during which he was severely injured- struck in the head with a stone- in an altercation with communists.
01.07.1932-14.09.1932	Assigned as *Inspekteur "Süd"* to the *Obersten SA-Führung* (responsible for *SA-Gruppen Hochland, Franken, Thüringen*, and *Österreich*).
31.07.1932-12.09.1932	Member of the *Reichstag* (representing *Wahlkreis 25, Niederbayern*).
15.09.1932-31.03.1933	*Führer* of *SA-Obergruppe IV* (consisting of *SA-Gruppen Bayerische Ostmark, Franken*, and *Hochland*) in Ingolstadt.
06.11.1932-01.02.1933	Member of the *Reichstag (Wahlkreis 25, Niederbayern)*.
05.03.1933-14.10.1933	Member of the *Reichstag (Wahlkreis 25, Niederbayern)*.
09.03.1933-04.04.1933	*(kommissarischer) Polizeidirektor* (acting Police Director) in Regensburg.
01.04.1933-31.01.1934	Assigned as an *SA-Führer z. V.* to the *Stab der Obersten SA-Führer*.
04.04.1933-08.05.1934	*Regierungspräsident* of *Oberfranken* and *Mittelfranken* (Seat: Ansbach). Fired by *Gauleiter* Julius Streicher.
12.11.1933-31.01.1942	Member of the *Reichstag (Wahlkreis 25, Niederbayern)*.
01.02.1934-30.06.1934	Assigned as *Inspekteur "Südost"* to the *Obersten SA-Führung*.
05.07.1934-31.01.1942	*Staatssekretär and politischer Referent* (political advisor) to the *Reichsstatthalter* of Bayern (Franz Xaver Ritter von Epp). After his death, *Ministerialrat* Dr. Franz Eduard Schachinger held acting control of the office. Hofmann had done most of the work of the *Reichsstatthalter*'s office, von Epp being content to serve predominantly as a figurehead.
13.07.1934-00.00.1939(?)	Member of the *Volksgerichtshof* (5-year appointment).
19.07.1934-31.01.1942	Assigned as an *SA-Führer z. V.* to the *Stab der Obersten SA-Führung*.
00.00.193_-00.00.19__	*Vorsitzender des Aufsichtsrates* of *Bayerische Elektricitäts-Lieferungs-Gesellschaft AG*, Bayreuth and *Ueberlandwerk Oberfranken AG*, Bamberg.
00.00.193_-00.00.19__	*Stellvertretender Vorsitzender des Aufsichtsrates* of *Frankische Licht- und Kraftversorgung AG*, Bamberg.

Decorations and Awards

00.00.191_	*1914 Eisernes Kreuz I. Klasse*
00.00.191_	*1914 Eisernes Kreuz II. Klasse*
00.00.191_	*Bayerischer Militärverdienstorden IV. Klasse mit Krone und Schwertern*
00.00.____	*Prinz-Regent-Luitpold-Medaille*
00.00.____	*Österreichisches Militär-Verdienstkreuz*
00.00.____	*Kaiser-Franz-Joseph-Medaille*
ca. 1934	*Ehrenkreuz des Weltkrieges 1914-1918 mit Schwertern*
30.01.1939	*Goldenes Ehrenzeichen der NSDAP*
20.04.1939	*Ehrenzeichen des 9. November 1923 (Blutorden)*
00.00.1931	*Abzeichen des SA-Treffens Braunschweig 1931*
00.00.194_	*Dienstauszeichnung der NSDAP in Silber*
00.00.194_	*Dienstauszeichnung der NSDAP in Bronze*
ca. 1936	*Deutsches Olympia-Ehrenzeichen I. Klasse*
00.00.1934	*Ehrendolch der SA* (mit Wirkung vom 03.02.1934)
00.02.1934	*Ehrenwinkel für alte Kämpfer*
23.05.1933	*Ehrenbürgerrecht der Stadt Bayreuth*

Notes

- Son of the businessman Franz Christoph Hofmann (born 25.03.1849 in Zeyern) and his wife Frieda, née Laupmann (born 00.00.1850 in Hof).
- Religion: Catholic.
- Married.
- In a posthumous de-Nazification hearing on 28.04.1948, the *Spruchkammer München II* placed the late Hofmann in "Gruppe II" ("Belastete", incriminated persons; the original indictment asked that he be placed in "Gruppe I" ["Hauptschuldige", major offenders). The court ordered the confiscation of 40% of his estate. In March 1949, the *Berufungskammer München (IV. Senat)* reduced the amount to 40% of his estate holdings located in Bavaria, and in 1951, granted a partial survivor's pension to his widow.

Sources

Bundesarchiv, Berlin-Lichterfelde (former Berlin Document Center): *Personalunterlagen von SA-Angehörigen: SA-Personalakte* of Hans Georg Hofmann.

Lilla, Joachim; Döring, Martin; & Schulz Andreas: *Statisten in Uniform. Die Mitglieder des Reichstags 1933-1945*. Droste Verlag, 2004.

Peterson, Edward N.: *The Limits of Hitler's Power*. Princeton University Press, 1969.

Schirach, Baldur von: *Die Pioniere des Dritten Reiches*. Zentralstelle fur der deutschen Freiheitskampf, 1933.

Adolf Andreas Hühnlein
SA-Obergruppenführer

Born:	12.09.1881 in Neustädtlein bei Kulmbach / Regierungsbezirk Oberfranken / Bayern.
Died:	18.06.1942 in München / Bayern.
NSDAP-Nr.:	375 705 (Joined 00.00.1930)

Promotions

01.10.1900	*Einjährig-Freiwilliger*
11.01.1901	*Fahnenjunker*
12.01.1901	*Unteroffizier*
03.05.1901	*Fähnrich*
09.03.1902	*Leutnant*
12.08.1908	*Oberleutnant*
01.10.1913	*Hauptmann* (ohne Patent; 16.12.1914: Granted Patent Nr. 5)
01.07.1921	*Major*
01.05.1931	*SA-Gruppenführer*
01.01.1933	*SA-Obergruppenführer*
30.04.1933-23.08.1934	*Korpsführer der NSKK*
02.09.1934-18.06.1942	*Korpsführer der NSKK*
14.05.1936	*Charakter als Generalmajor a. D.*
08.09.1938-18.06.1942	*Reichsleiter der NSDAP*

Career

00.00.1888-ca. 1892	Attended *Volksschule* in Bayreuth.
ca. 1892-00.00.1900	Attended *Gymnasium* in Bayreuth.
01.10.1900-00.00.1901	Entered service as an *Einjährig-Freiwilliger*, assigned to 7. *Kgl. Bayerisches Feldartillerie-Regiment.*
11.01.1901- 01.03.1901(?)	Transferred to active service as a *Fahnenjunker*, assigned to *1. Kgl. Bayerisches Feldartillerie-Regiment.*
01.03.1901-09.02.1902	Attended the *Kgl. Bayerisches Kriegsschule* in München.
15.02.1903-14.03.1903	Temporarily assigned to the *Gewehrfabrik* in Amberg.
01.10.1904-01.07.1906	Attended the *Kgl. Bayerisches Artillerie- und Ingenieurschule.*
08.03.1907-30.09.1912	Assigned to the *Kgl. Bayerisches Telegraphen-Detachement.*
01.10.1909-30.09.1912	Attended the *Kgl. Bayerisches Kriegsakademie* in München.
01.10.1912-30.09.1913	Adjutant of the *Kgl. Bayerisches Pionierinspektion* (Bavarian Engineering Inspectorate).
01.10.1913-25.04.1914(?)	Assigned to the *Fortifikation Ingolstadt*, with attachment to the *Inspektion der Kgl. Bayerisches Ingenieurkorps* (Inspectorate of the Bavarian Engineering Corps).
25.04.1914-31.07.1914	*Adjutant* of the *Inspektion der Kgl. Bayerisches Ingenieurkorps.*
01.08.1914-31.03.1916	*Hauptmann beim Stab* to the *General der Pioniere beim AOK 6.*

A formal portrait of *SA-Oberführer* Hühnlein in 1932. (Baldur
von Schirach, *Die Pioniere des Dritten Reiches*, 1933)

01.04.1916-10.01.1917	*Führer* of *Reserve-Pionier-Kompanie Nr. 5* in *8. Kgl. Bayerisches Reserve-Division*.
11.01.1917-21.01.1917	Attached as *Hauptmann der Pioniere* to the staff of a *Generalkommando* (Army Corps HQ).
21.01.1917-04.02.1917	On convalescent leave in Bayreuth.
22.01.1917-09.02.1917	*Hauptmann beim Stab* of *1. Kgl. Bayerisches Pionier-Ersatz-Bataillon*.
02.02.1917-00.02.1917	Attached to the *Kgl. Bayerisches Minenwerfer-Ersatz-Bataillon* as a participant in the *Februar-Kurs* at the *Minenwerferschule*.
10.02.1917-02.11.1917	*Kommandeur* of *Kgl. Bayerisches Pionier-Bataillon Nr. 15*, with which he deployed to the front on 01.03.1917.
20.10.1917-	Assigned, with effect from 03.11.1917, as a *Generalstabsanwärter* (general staff candidate) to the staff of *15. Kgl. Bayerisches Infanterie-Division*.
21.11.1917	Appointed, with effect from 23.11.1917, as *2. Generalstabsoffizier (Ib)* of *15. Kgl. Bayerisches Infanterie-Division*.
23.11.1917-18.05.1918	*2. Generalstabsoffizier (Ib)* of *15. Kgl. Bayerisches Infanterie-Division*.
29.11.1917-02.12.1917	Illness (febrile enteritis).
21.01.1918-31.01.1918	Participated in the *26. Führerkurs* (26th leadership course) in Sedan.
09.02.1918-26.02.1918	On sick leave due to *allgemeiner Abspannung* (general exhaustion).
26.02.1918-18.05.1918(?)	Returned to *15. Kgl. Bayerisches Infanterie-Division*.
18.05.1918-00.11.1918(?)	Assigned to the *Generalstab* of *Generalkommando z.b.V. 57* as *Leiter großer Nachschubaufgaben* (head of *Major* replenishment tasks).
06.01.1919	Granted 3 weeks' leave to Bayreuth.
07.01.1919	Transferred to the *Inspektion der Ingenieurkorps*.

SA-Obergruppenführer Hühnlein at a Party function in 1933.

SA-Oberführer Hühnlein, ca. 1932.

SA-Obergruppenführer Hühnlein, 1933.

Official portrait of *SA-Obergruppenführer* Hühnlein, 1933. (Max Williams photo)

SA-Obergruppenführer Hühnlein and Franz Seldte in 1933.
(NARA, Heinrich Hoffmann collection)

An autographed official portrait of *NSKK-Korpsführer* Hühnlein, dedicated to a loyal assistant and dated 12.09.1937. (Hermann-Historica, Auctioneers, München)

NSKK-Korpsführer Hühnlein, ca. 1936.

An autographed and framed presentation photo of *SA-Obergruppenführer* Hühnlein. (Hermann-Historica, Auctioneers, München)

10.01.1919	Transferred to *1. Kgl. Bayerisches Pionier-Bataillon*.
15.02.1919-26.02.1919	*Hauptmann beim Stab* (m.d.W.d.G.b.) of *1. Kgl. Bayerisches Pionier-Bataillon*.
26.02.1919	Placed on indefinite leave.
00.00.1919-00.00.1920	Assigned as a *Kompanieführer* to *Freikorps Epp*.
00.00.1919	First heard Adolf Hitler speak, later describing the event as follows:

> In 1919, in the barracks of the Bavarian Leib-Garde- Regiment when I first heard Adolf Hitler and the train of his thoughts of never submitting, never surrendering, I was possessed by his philosophical outlook on the world, it drew me to him, held me fast, and inspired that sort of discipleship which only ends with death. (excerpted from the *Völkischer Beobachter*, 19.06.1942)

00.00.1919-00.00.1920	Chef of *1. Kompanie / Reichswehr-Pionier-Bataillon 7* (München).
00.00.1920-01.10.1920	Assigned to *Reichswehr-Pionier-Bataillon 21*.
01.10.1920-00.00.1922	Assigned to *Pionier-Bataillon 7*.
00.00.1922-00.00.1923	*1.Generalstabsoffizier (Ia)* to *Infanterie-Führer VII* (München).
00.00.192_-00.00.1923	Member of the Ernst Röhm's *Wehrverband "Reichskriegsflagge"*.
00.03.1923-09.11.1923	*Chef des Truppenstabes (Ia)* of the *SA*.
09.11.1923	Participated in the *München Putsch*.
00.00.1923-00.00.1923	Assigned to the staff of *Gruppenkommando 1*.
17.09.1923-00.00.1923	Attended the *Artillerieschule Jüterbog* as a *Taktiklehrer* (tactics instructor).
00.11.1923-00.03.1924	Held in investigative detention at Stadelheim, Neudeck, and Landsberg.
16.04.1924	Discharged from the *Reichswehr*.
00.04.1924	Sentenced to 6 months' fortress arrest for his role in the *München Putsch*.
00.00.1925-00.00.1930	Employed as a business man in the tire industry.
00.00.1930	Joined the *NSDAP*.
00.00.1930-31.08.1934	Member of the *Stab der Obersten SA-Führung*.
01.04.1930	Joined the *NS-Automobilkorps (NSAK)*.
00.12.1930-00.00.193_	*Kraftfahrinspekteur der Motorstürme der SA und SS*.
00.12.1930-30.04.1933	*Stellvertretender Korpsführer der NSAK* (Redesignated *NS-Kraftfahrkorps* [*NSKK*], 20.04.1931).
01.05.1931-13.04.1932	*Leiter* of *Unterabteilung K* in the *Quartiermeisterstab der Obersten SA-Führung*.
01.07.1932-30.04.1933	*Quartiermeister der SA* and *Stellvertreter des Stabschefs der SA* (deputy to Ernst Röhm).
01.07.1932-31.08.1934	*Chef des Kraftfahrwesens der SA* and *Sachberater* of *Abteilung I* in the *Stab der Obersten SA-Führung*.
05.03.1933-18.06.1942	Member of the *Reichsta*g (Wahlkreis 31, Württemberg; from 29.03.1936, *Wahlkreis 25, Niederbayern*).
07.03.1933	Appointed as a member of the *Beirat für das Kraftfahrwesen* (Advisory Council for Motor Transport Matters).

Official portrait photos of *Korpsführer* Hühnlein, inscribed to
Reichsminister SS-Obergruppenführer Walther Darré, 08.06.1937 and
17.02.1941. (Hermann-Historica, Auctioneers, München)

NSKK-Korpsführer Hühnlein attends the *Sportveranstaltung der HJ*
(Hitler Youth sports event) in March 1935. (NARA photo)

Walter Miehe's 1939 oil painting of *NSKK-Korpsführer* Hühnlein. (U.S. Army photo)

01.04.1933	Appointed as a member of *Gauleiter* Julius Streicher's *Zentralkomitees zur Abwehr der jüdischen Greuel- und Boykotthetze* (Central Committee for Defense Against Jewish Atrocity and Boycott Agitation).
00.04.1933-00.00.193_	Member of the *Arbeitsausschuß der Beirat für das Kraftverkehrswesen* (Working Committee of the Advisory Council for Matters of Motor Transport).
30.04.1933-18.06.1942	*Korpsführer des NSKK* (originally under *SA* control). On 02.09.1934, the *NSKK*, having merged with the Motor-SA, became an independent organization, with Hühnlein being given the title of *Reichsführer des NSKK*. The title was soon changed, as it was apparently concluded that only the leader of the *SS*, Himmler, was entitled to be a *Reichsführer*. Under Hühnlein's leadership, the *NSKK*, which numbered 30,000 members as of 1933, grew rapidly; in September 1933, when it absorbed all of Germany's private motor clubs, its membership rose to 350,000, and by 1939, there were 500,000 *NSKK* men. It's most important role in the years leading up to World War II was in providing motorized transport training, and numerous members of the *Panzerwaffe* had previously served in Hühnlein's organization. In his memoirs, Germany's armored warfare innovator, *Generaloberst* Heinz Guderian, wrote:

In the summer of 1933 the head of the National-Socialist Motor Corps, Adolf Hühnlein, invited me to a festive meeting of the

Official portrait photo of *Korpsführer* Hühnlein, inscribed to
Reichsminister NSKK-Obergruppenführer Wilhelm Ohnesorge and dated
08.06.1942. (Hermann-Historica, Auctioneers, München)

leaders of his organization at which Adolf Hitler had promised
to be present. I thought that it might beinteresting to see Hitler
among his faithful followers. Since Hühnlein was also a decent,
upright man with whom it was easy to work, I accepted ...
(Guderian, *Panzer Leader*, p. 30)

Through his relationship with Hühnlein, Guderian was able
to ensure the training of Germany's future tank and truck drivers
(approximately 187,000 of them in the years 1933 to 1939) in the 24
NSKK-Reichsmotorsportschulen.

00.06.1933	Appointed as *Führer* of the *Deutsche Kraftfahrverband* (German Motor Transport Association) and as a member of the *Reichsführerringes des Deutschen Sports.*
27.09.1933	Cofounder of *Der Deutsche Automobilclub* (*DDAC*, The German Automobile Club).
27.09.1933-18.06.1942	*Präsident* of the *Obersten Nationalen Sportbehörde für die Deutsche Kraftfahrt* (ONS) and *Führer des Deutschen Kraftfahrsports* (Leader of German Motor Sports).

Visit of *NSKK-Korpsführer* Hühnlein with Hitler at *Führer* HQ "Wolfsschanze", Rastenburg / East Prussia, mid-September 1941. (NARA photos)

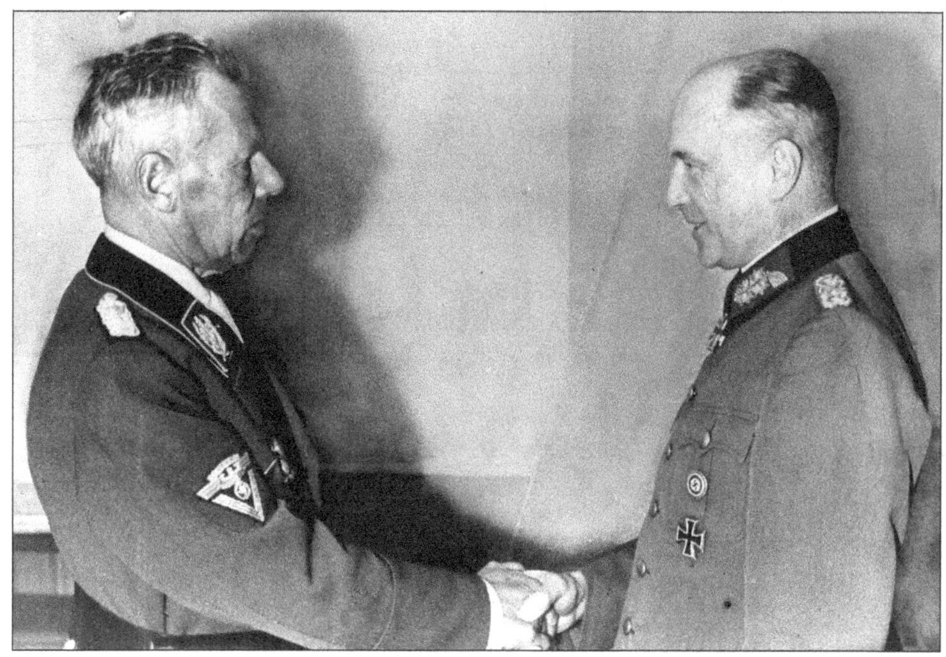

Angerburg, 04.10.1941: Hühnlein congratulates *Generalfeldmarschall* von Brauchitsch on his birthday. (NARA photo)

00.04.1935	Appointed as a member of the *Verwaltungsausschußes des Deutschen Museums* (Administrative Committee of the German Museum) in München.
21.12.1935-18.06.1942	Member of the *Reichsverkehrsrat* (*Reich* Transportation Council).
00.00.1936(?)-18.06.1942	Permanent member of the *Reichsarbeitskammer*.
00.00.1936(?)-18.06.1942	Member of the *Verwaltungsausschuß der Deutsche Museum* (Administrative Council of the German Museum).
00.00.1936(?)-18.06.1942	Member of *Akdademie für Deutsches Recht*, München.
11.07.1936	Appointed as *Ehrenführer der Motor-HJ*.
27.09.1936-04.10.1936	*Führer* of a *Deutsch-ungarischen Versuchsfahrt* (Berchtesgaden-Graz-Plattensee-Baja-Debrecin-Budapest-Wien-Linz-Braunau-München).
30.12.1936-00.00.19__	Member of the *Verwaltungsrat der Gesellschaft "Reichsautobahnen"*.
08.09.1938-18.06.1942	*Reichsleiter der NSDAP*.
22.02.1940-18.06.1942	*Beauftragter für den motorisierten Transport der Kriegswirtschaft* (Representative for Motorized Transport of the Wartime Economy), appointed by *Reichsmarschall* Göring.

Published Works

Das nationalsozialistische Kraftfahr-Korps (1938)
Die Verbundenheit des Nationalsozialistischen Kraftfahr-Korps mit den Werken der Motorisierung (1937)

Decorations and Awards

21.06.1942	*Goldenes Kreuz des Deutschen Ordens* (posthumous award)
18.11.1915	*1914 Eisernes Kreuz I. Klasse*
13.09.1914	*1914 Eisernes Kreuz II. Klasse*
00.00.194_	*Kriegsverdienstkreuz I. Klasse mit Schwertern*
00.00.194_	*Kriegsverdienstkreuz II. Klasse mit Schwertern*
03.03.1919	*Kgl. Bayerischer Militär-Verdienstorden IV. Klasse mit der Krone und mit Schwertern*
27.03.1915	*Kgl. Bayerischer Militär-Verdienstorden IV. Klasse mit Schwertern*
16.09.1908	*Kgl. Preußischer Kronen-Orden IV. Klasse*
12.03.1905	*Kgl. Bayerisches Jubiläumsmedaille*
15.09.1941	*Ehrendolch des Heeres*
ca. 1939	*Medaille zur Erinnerung an den 1. Oktober 1938*
ca. 1938	*Medaille zur Erinnerung an den 13. März 1938*
ca. 1934	*Ehrenkreuz des Weltkrieges 1914-1918 mit Schwertern*
09.11.1936	*Goldenes Ehrenzeichen der NSDAP*
00.05.1939	*Goldenes Hitler-Jugend Ehrenzeichen mit Eichenlaub*
00.00.1934	*Ehrenzeichen des 9. November 1923 (Blutorden)* (Nr. 8; with effect from 09.11.1933)
00.00.194_	*Dienstauszeichnung der NSDAP in Silber*
00.00.194_	*Dienstauszeichnung der NSDAP in Bronze*
00.00.1936	*Deutsches Olympia-Ehrenzeichen I.Klasse*
00.00.193_	*Großkreuz des Ehrenzeichens des Deutschen Roten Kreuzes*
00.00.1939	*Deutsches Schutzwall-Ehrenzeichen*
00.00.19__	Badge of the *Ring der nationalen Kraftfahrt- und Luftfahrtbewegung* (Association of the National Motoring- and Aviation Movement" [grade unknown])
12.09.1941	Received an *Ehrengabe* (gift of honor) of RM 250.000 from the *Führer* (to mark his 60th birthday)
14.12.1938	*Ehrenbürgerbrief der Stadt Bad Gandersheim*
12.06.1937	*Ehrenbürgerrecht der Stadt Ahrweiler*
00.00.1937	*Ehrenbürgerrecht der Stadt Bayreuth*
04.09.1938	Order of Saints Maurice and Lazarus, Grand Cross (*Gran Croce dell'Ordine dei Santi Maurizio e Lazzaro*)
00.00.193_	Star to the Grand Cross of the Order of Military Merit (Bulgaria)

Notes

- Son of the teacher Friedrich Hühnlein (died 00.00.1896) and his wife Kunigunde, née Opel.
- Married in Bayreuth, 00.00.1906, to Paula Däumling, daughter of the salesman Adolarius Däumling. Three daughters were born to this marriage, one of whom died in infancy.

Sources

Bayerisches Hauptstaatsarchiv, München, Abteilung IV Kriegsarchiv: Excerpts from various Kriegsranglisten containing data on the Royal Bavarian Army service of Adolf Hühnlein.

Hochstetter, Dorthee: *Motorisierung und "Volksgemeinschaft": Das Nationalsozialistische Kraftfahrkorps (NSKK), 1931-1945.* Oldenbourg Wissenschaftsverlag, 2005.

Lilla, Joachim; Döring, Martin; & Schulz Andreas: *Statisten in Uniform. Die Mitglieder des Reichstags 1933-1945*. Droste Verlag, 2004.

Macksey, Kenneth: *Guderian. Creator of Blitzkrieg*. Stein and Day, 1976.

Schirach, Baldur von: *Die Pioniere des Dritten Reiches*. Zentralstelle fur der deutschen Freiheitskampf, 1933.

Dietrich Wilhelm Bernhard von Jagow
SA-Obergruppenführer

Born:	29.02.1892 in Frankfurt an der Oder / Provinz Brandenburg.
Suicide:	26.04.1945 in the residence of Rudolf Rahn, the German Ambassador to the fascist *Repubblica Sociale Italiana* (*RSI*), in Meran (*It.*: Merano) / Südtirol (shot himself).
NSDAP-Nr.:	110 438 (First joined in Autumn 1920; Left, 00.00.1923; Reenrolled 01.01.1929)

Promotions

01.04.1912	*Seekadett*
12.04.1913	*Fähnrich zur See*
00.08.1914	*Leutnant zur See*
00.12.1917	*Oberleutnant zur See*
14.10.1931	*SA-Gruppenführer*
27.06.1933	*SA-Obergruppenführer*
28.04.1936	*Kapitänleutnant d. R.* (mit RDA vom 01.05.1936)
20.04.1938	*Korvettenkapitän d. R.*

Career

00.00.1909-00.00.1910	Attended *Gymnasium* in Ballenstedt and Blankenburg (Harz) through Untersekunda (Primareife).
01.04.1912-00.00.1914	Joined the *Kaiserliche Marine*, initially attending the *Marineschule* in Mürwik.
31.07.1914-20.11.1915	Assigned to the light cruiser *S.M.S. Wettin* (eventually as *Wachoffizier*).
21.11.1915-21.05.1917	Assigned as *Wachoffizier* to the light cruiser *S.M.S. Arkona*.
04.06.1917-15.07.1917	Assigned [as *Wachoffizier?*] to *S.M.S. Vulkan*.
16.07.1917-08.08.1917	Assigned as *Wachoffizier* to the U-Boat *S.M.S. UB 11*.
09.08.1917-04.09.1917	Assigned to *S.M.S. Kaiserin Augusta*.
05.09.1917-28.09.1917	Assigned to the *U-Boat S.M.S. UB 4*.
12.10.1917-11.12.1918	Assigned to the *U-Boat S.M.S. UB 77*.
01.12.1918-01.03.1919	Assigned to the cruiser *S.M.S. Blitz*.
01.03.1919-04.08.1919	*Kommandant* of *Torpedoboot 28 (T 28)*.
04.08.1919-14.11.1919	*Kommandant* of *Minensuchboot M 100*.
14.11.1919-31.05.1920	Assigned to *4.Regiment / 2.Marine-Brigade (Ehrhardt)*.
00.03.1920	Participated in the "Kapp-Putsch".
01.06.1920-31.05.1920	Assigned to *Schiffs-Stammdetachement "Nordsee"*.
01.10.1920	Officially discharged from the *Reichsmarine*, due to his refusal to swear an oath to the newly-established Weimar Republic.
00.09.1920	Joined the *Organisation Consul (OC)*.
00.00.1920-00.00.1921	Employed as a forestry and agricultural worker withe the *Bayerische Holzverarbeitungsgesellschaft* (a firm which served as a cover for the *OC*).

A postcard of Dietrich von Jagow, ca. 1931. (Laurens Hessels photo)

Autumn 1920	Joined the *NSDAP*.
Winter 1920	Joined the *SA*.
00.05.1921-00.06.1921	*Führer* of *1. Sturmzug / Sturmkompanie von Killinger* in Oberschlesien.
Summer/Autumn 1921	*Stellvertreter Führer der Turn- und Sportabteilung* ("Deputy Leader of the Gymnastics and Sports Detachment", a covername for the SA) in München.
Autumn 1921	*Führer* (under the nom-de-guerre "*Oberleutnant* Dietrich") of the *1. SA-Sturmabteilung* (München).
00.00.1921-00.00.19__	Employed as a traveling salesman (soap, washing powder, and toiletries) and *Volontär* (unsalaried commercial employee) with the *Osiandersche Buchhandlung* in Tübingen.
00.01.1922-00.00.192_	*Ausbilder* (training supervisor) of the *Tübinger Studentenbataillon* (Student Battalion in Tübingen).
00.01.1922-00.11.1923	*Inspekteur der württembergischen SA* (Inspector of the *SA* in Württemberg).
00.04.1922	Associate student of the *Staatswissenschaftlichen Fakultät* (political science faculty) at the University of Tübingen.
00.00.1922-00.00.1928	*Bezirksführer* (*Landesführer* per *SA-Personalfragebogen* dated 08.01.1937) of the *Bund Wiking* in Württemberg and Baden.
00.00.1923	Left the *NSDAP*.
00.00.1927-01.05.1928	Member of the "*Stahlhelm*"-*Bund*.
00.00.192_-00.00.19__	Member of the *Württembergische Heimatschutz*.
01.01.1929	Reenrolled in the *NSDAP / Ortsgruppe Esslingen a. Neckar*.

Dietrich von Jagow, ca. 1932.
(Laurens Hessels photo)

SA-Obergruppenführers von Jagow
(center) and Lutze (left) in 1933.

01.01.1929	Reentered the *SA* (*Brigade 3* [Esslingen]).
00.00.1929-00.04.1930	*Gau-SA-Führer Württemberg.*
00.00.1929-00.04.1930	*Ortsgruppenleiter* of the *Ortsgruppe Esslingen der NSDAP*. He also served as a *Parteiredner* (party speaker) in Esslingen am Neckar during this period.
01.01.1930-00.00.1931	Assigned as *Gaugeschäftsführer* to the *Gauleitung Württemberg-Hohenzollern der NSDAP* (per *SA-Personalfragebogen* [personnel questionnaire] dated 08.01.1937, began this assignment in 1929).
01.04.1930-31.10.1930	*Führer* of *SA-Brigade III* in *SA-Untergruppe Württemberg.*
01.11.1930-01.04.1931	*Führer* of *SA-Untergruppe Württemberg.*
02.04.1931-09.09.1931	*Führer* (m.d.F.b.) of *SA-Gruppe Südwest* (Stuttgart)(?- Assignment not mentioned in his *SA-Personalfragebogen* of 08.01.1937).
10.09.1931-19.03.1933	*Führer* of *SA-Gruppe Südwest* (*SA* banned by the German government from 13.04.1932 to 01.07.1932).
18.10.1931	Attended the *SA Aufmarsch in Braunschweig.*
00.12.1931-18.12.1931	Attended the *7.Lehrgang* of the *Reichsführerschule der SA*, München.
09.05.1932-04.06.1932	Member of the *Reichstag* (*Wahlkreis 31, Württemberg*).
31.07.1932-12.09.1932	Member of the *Reichstag* (*Wahlkreis 31, Württemberg*).
06.11.1932-01.02.1933	Member of the *Reichstag* (*Wahlkreis 31, Württemberg*).
05.03.1933-14.10.1933	Member of the *Reichstag* (*Wahlkreis 31, Württemberg*).
08.03.1933-31.03.1933	*Reichskommissar für die Polizei in Land Württemberg* (redesignated *Polizeikommissar für das Land Württemberg* on 18.03.1933). In this capacity, he initiated a wave of arrests beginning the day of his

SA-Obergruppenführer von Jagow in 1933. (Laurens Hessels photos)

Dietrich von Jagow about to board a *Lufthansa* flight in 1934
(NARA, Heinrich Hoffmann collection)

SA-Obergruppenführer von Jagow, ca. 1936.

appointment to the post. KPD politicians and functionaries were the main targets of this action, initially being detained in the state prison at Rottenburg and later, from 20.03.1933 on, to the newly established *KL-Heuberg*. In March 1933 alone, some 500 KPD members were arrested by von Jagow's *Hilfspolizei* (auxiliary police, composed of SS, *SA,* and *"Stahlhelm"-Bund* members).

21.03.1933-30.06.1933	*Führer* of *SA-Obergruppe III* (*SA-Gruppen West, Thüringen,* and *Südwest*) in Koblenz.
01.07.1933-19.07.1934	*Führer* of *SA-Obergruppe V* (*SA-Gruppen Thüringen, Hessen, Südwest, Westmark,* and *Kurpfalz*) in Stuttgart. In May 1934, *SA-Gruppen Westmark* and *Kurpfalz* were transferred to the jurisdiction of *SA-Obergruppe IX.*
14.09.1933-00.00.1945	*Preußischer Staatsrat.*
12.11.1933-26.04.1945	Member of the *Reichstag* (*Wahlkreis 31, Württemberg*; after 29.03.1936, representing *Wahlkreis 3, Berlin Ost*).
00.00.1934-00.00.1934	*Preußischer Provinzialrat* of *Provinz Hessen-Nassau.*
13.07.1934-00.00.1939(?)	Member of the *Volksgerichtshof* (5-year appointment).
20.07.1934-31.01.1942	*Führer* (m.d.F.b. until 14.09.1935, then permanent from 15.09.1935) of *SA-Gruppe Berlin-Brandenburg.* Succeeded Kurt Daluege, who had held acting command following the murder of Karl Ernst on 30.06.1934. Succeeded by Willi Künemund.
00.00.1934-00.00.1945	*Preußischer Provinzialrat* for *Provinz Brandenburg.*

A formal portrait of *SA-Obergruppenführer* von Jagow, ca. 1936.

The *Handwerker-Kongresse-Festsitzung* at the Kroll Opera House, 30.05.1938. From left to right: *Reich* Minister of Economics Funk, *Reichsleiter* Dr. Ley, deputy *Gauleiter* of Berlin Artur Görlitzer, unknown Army general, and Dietrich von Jagow. (Atlantic-Photo)

Berlin, 23.02.1940: Von Jagow (in naval uniform), *SA-Stabschef* Lutze, and *NSKK-Brigadeführer* Paul Heinsius salute the grave of Horst Wessel. (Roger Bender photo)

20.03.1935-00.00.193_	Member of the *Ehrenausschuß* (honorary committee) of the *2. Reichsberufswettkampfes der deutschen Jugend für den Gau Berlin* (2nd *Reich* Professional Competition of the German Youth for Gau Berlin).
00.00.1935-00.00.1945	*Ratsherr der Reichshauptstadt Berlin* (Councilman of the *Reich* Capital City Berlin).
00.00.1935-00.00.1935	*Leiter / Beisitzer* of the *Prüfungskommission für Brigade- bezw. Standartenführer* (Testing Commission for [SA]-*Brigadeführer* through *Standartenführer*).
00.00.193_-00.00.19__	Member of the *Vorstand* of the *Deutscher Adelsgenossenschaften* (Association of German Members of the Nobility).
15.07.1935-03.08.1935	Participated in reserve training exercises with *1.Minensuchhalbflotte* (1st minesweeper half-flotilla).
18.11.1935-30.11.1935	Reserve training exercises at the *Sperrschule* (mine warfare school) in *Kiel-Wyck*.
00.04.1936-00.00.19__	Assigned as an *ehrenamtlicher Richter* (honorary judge) to the *Obersten Ehren- und Disziplinarhof der DAF*.
03.09.1939-00.05.1940	Entered war service as *Kommandant* of the minelayer *Tannenberg*, tasked with laying mines in the North Sea. Commissioned in 1935, the 5,500-ton vessel could carry up to 460 mines.
00.05.1940-00.06.1940	*z. V. Marineoberkommando Ost* (at disposal of the Naval High Command East).
00.06.1940-00.09.1940	*Chef* of the *Hafenflottille Brest*.
00.09.1940-00.10.1940	*z. V. Führer der Vorpostenboote West*.

Dietrich von Jagow as German Minister to Hungary observes a parade of
Hungarian Volksdeutsche volunteers of the *Waffen-SS*, 1943. (Photo by *SS-
Kriegsberichter* [war correspondent] Ferdinand Fritsch; NARA photos)

00.10.1940-30.04.1941	*Flottillenchef* of *18.Vorpostenflottille*, then placed at disposal of the *Oberkommando der Marine (OKM)*. Succeeded *Korvettenkapitän d. R.* Karl Wolters. Succeeded by *Korvettenkapitän d. R.* Albrecht Boit.
01.05.1941	Discharged from the *Kriegsmarine*.
29.06.1941-19.03.1944	*Gesandter I. Klasse* in Budapest. He assumed this post on 24.07.1941, succeeding Otto von Erdmannsdorff. He was recalled and succeeded by Dr. Edmund Veesenmayer on 19.03.1944.
01.02.1942-26.04.1945	Assigned as *SA-Führer z. V.* to the *Stab der Obersten SA-Führung*.
08.05.1944-26.04.1945(?)	Resumed employment with the *Auswärtige Amt*, official reentering its service on 01.06.1944.
00.09.1944-20.01.1945	*Führer* of *Volkssturmbataillon 35* in Schlesien.
20.01.1945	Sustained a severe head injury, losing an eye.
20.01.1945-00.03.1945	Hospitalized in Leipzig.
00.04.1945-26.04.1945	Following convalescence, traveled as an *NS-Kurier* (Nazi Party courier) to Meran / Südtirol, where he remained until taking his own life.

Decorations and Awards

00.00.194_	*1939 Spange zum 1914 Eisernen Kreuz I. Klasse*
00.00.194_	*1939 Spange zum 1914 Eisernen Kreuz II. Klasse*
00.00.191_	*1914 Eisernes Kreuz I. Klasse*
00.00.191_	*1914 Eisernes Kreuz II. Klasse*
00.00.19__	*[Württembergischer?] Kriegsverdienstkreuz*
00.00.1935	*Karl Eduard Medaille*
00.00.1934	*Stern zum Sächsischen Ernestinischer Hausorden*

00.00.1926	*Komturkreuz mit Schwertern der Sächsischen Ernestinischer Hausorden*
ca. 1918	*Unterseebootkriegsabzeichen*
00.00.194_	*Kriegsabzeichen für Minensuch- und Vorpostenverbände*
21.02.1945	*Verwundetenabzeichen, 1939 in Schwarz*
ca. 1921	*Schlesischer Adler 1. Stufe*
ca. 1921	*Schlesischer Adler 2. Stufe*
ca. 1934	*Ehrenkreuz des Weltkrieges 1914-1918 mit Schwertern*
30.01.1940	*Goldenes Ehrenzeichen der NSDAP*
00.00.194_	*Dienstauszeichnung der NSDAP in Silber*
00.00.194_	*Dienstauszeichnung der NSDAP in Bronze*
ca. 1931	*Abzeichen des SA-Treffens Braunschweig 1931*
00.00.1929	*Nürnberger Parteitagsabzeichen 1929*
ca. 1936	*Deutsche Olympia-Ehrenzeichen I. Klasse*
00.00.193_	*SA-Sportabzeichen in Silber* (Nr. 33 178)
00.00.1934	*Ehrendolch der SA* (mit Wirkung vom 03.02.1934)
00.02.1934	*Ehrenwinkel für alte Kämpfer*
18.12.1931	*Tyr-Rune*
00.00.193_	Portrait photo of the *Führer*, personally inscribed by Hitler

Notes

- Parents (married, 14.02.1880):
  - ◊ Father: Kön. preuß. *Oberst Eduard* Ludwig von Jagow (born 08.09.1850 in Calberwisch [Altmark], died 00.00.1917.
  - ◊ Mother: *Elisabeth* Pauline Thusnelda, née von Kleist (born 25.09.1859 in Potsdam, died 00.00.1938), daughter of the kön. preuß. *Generalleutnant* Christian *Ewald* Leopold von Kleist [born 25.03.1824, died 29.12.1910; *Kommandeur* of *1. Garde-Infanterie-Division* and holder of the Pour le mérite, and his wife Ottilie Wilhelmine *Betty*, née von Knoblauch [born 12.08.1834]). At the time of their marriage, Eduard von Jagow was a *Premierlieutenant* in *1. hannoverschen Ulanen-Regiment Nr. 13* (Hannover).
- Religion: Lutheran until 00.00.1937, then left the church.
- Married in Esslingen on 21.09.1926 to Hedwig Sinner (born 28.10.1903 in Schwäbisch Hall). Three sons (one of them born 28.05.1934) and Four daughters (born 03.02.1928, 16.08.1930, 00.00.19__, and 21.07.1936) were born to this marriage.
- Posthumously classified by the *Spruchkammer* (de-Nazification court) of Freiburg as a "Minderbelaster" (minor offender), 13.02.1950.

Sources

Bundesarchiv, Berlin-Lichterfelde (former Berlin Document Center): *Personalunterlagen von SA-Angehörigen: SA-Personalakte* of Dietrich von Jagow.

Keiper, Gerhard & Kröger, Martin: *Biographisches Handbuch des deutschen Auswärtigen Dienstes 1871-1945. Bd. 2: G-K.* Paderborn, 2005.

Lilla, Joachim; Döring, Martin; & Schulz Andreas: *Statisten in Uniform. Die Mitglieder des Reichstags 1933-1945.* Droste Verlag, 2004.

Maier, Franz: *Biographisches Organisationshandbuch der NSDAP und ihrer Gliederungen im Gebiet des heutigen Landes Rheinland-Pfalz.* Verlag v. Hase & Koehler, 2007.

Schirach, Baldur von: *Die Pioniere des Dritten Reiches*. Zentralstelle fur der deutschen Freiheitskampf, 1933.

Wilhelm Jahn
SA-Obergruppenführer

Born:	02.02.1891 in Soest / Kreis Soest / Westfalen.
Died:	21.09.1952 in Weende (Göttingen) / Niedersachsen.

NSDAP-Nr.:	42 535 (First joined in 07.1922 with a Nr. under 2000; Party banned following the *München-Putsch* of 09.11.1923; Reenrolled with Nr. 42 535 on 16.08.1926)

Promotions

00.05.1916	*Leutnant d. R.*
01.01.1931	*SA-Standartenführer*
18.12.1931	*SA-Oberführer*
15.10.1932	*SA-Gruppenführer*
09.11.1937	*SA-Obergruppenführer*

Career

00.00.1897-00.00.1901	Attended *Volksschule*.
00.00.1901-00.00.1908	Attended *Gymnasium* (graduated Oberprima).
00.00.1908-00.00.1911	Served a 3-year apprenticeship in the banking field.
00.00.1911-00.00.1914	Employed as a bank clerk.
03.08.1914-00.12.1914	Entered service as a *Kriegsfreiwilliger*, assigned to *Fußartillerie-Bataillon 18*. Deployed to the Front, 00.11.1914.
00.12.1914-00.05.1916	Assigned to *Fußartillerie-Bataillon 39*.
00.06.1916-00.12.1916	Assigned to *Landwehr-Fußartillerie-Bataillon 11*.
00.01.1917-00.03.1917	*Stellvertretender Batterieführer* in *Fußartillerie-Bataillon 59*.
00.03.1917	Transferred to the *Fliegertruppe*.
00.04.1917-00.11.1917	Underwent pilot training with *Flieger-Ersatz-Bataillon 7*.
00.01.1918-00.03.1918	Assigned as a *Flugzeugführer* (pilot) and *technischer Offizier* (technical officer) to *Flieger-Abteilung 224*.
00.03.1918-15.12.1918	Assigned as a *Flugzeugführer* and *technischer Offizier* with *Flieger-Abteilung 40*.
15.12.1918	Discharged from military service.
00.00.1919-00.00.1922	Returned to work as a bank clerk.
Spring 1920-00.12.1920	Member of the *DNVP*.
01.01.1921-00.07.1922	*Geschäftsführer* of the *Organisation 'Escherich* in Osnabrück.
00.03.1921-00.07.1921	Member of the *"Stahlhelm"-Bund*.
00.07.1922	Joined the *NSDAP / Ortsgruppe Osnabrück* (left after the *München-Putsch* of 09.11.1923).
00.07.1922	Joined the *Bund Friedericus Rex* in Osnabrück.
00.00.1922-00.00.1924	Self employed as an electrical equipment salesman in Osnabrück.
Spring 1923	Founded the *SA* in Osnabrück.
00.00.1923	Joined the *Verband "Hindenburg"* in Osnabrück.

SA-Gruppenführer Jahn, ca. 1934.

00.00.1924	Joined the *Frontbann* in Osnabrück.
00.00.1925-00.00.1927	Employed as an automobile salesman in Osnabrück.
16.08.1926	Reenrolled in the *NSDAP / Ortsgruppe Osnabrück*.
00.00.1928-00.00.1930	Independent automotive sales manager.
00.12.1928	Joined the *SA-Reserve* in Osnabrück.
01.01.1931	Returned to the regular *SA* as a *hauptamtlicher SA-Führer*.
01.01.1931-14.04.1931	*Adjutant* to the *OSAF Nord* (Hannover).
15.04.1931-30.06.1932	*Stabsleiter* (m.d.W.d.G.b. until 18.12.1931, then permanent) of *SA-Gruppe Nord* (Hannover).
18.10.1931	Participated in the *SA-Aufmarsch in Braunschweig*.
01.07.1932-14.10.1932	*Führer* (m.d.F.b.) of *SA-Gruppe Nordsee* (Bremen). Succeeded by Viktor Lutze.
15.10.1932-30.06.1933	*Stabsführer* of *SA-Obergruppe II* (*SA-Gruppen Niedersachsen, Westfalen, Niederrhein*, and *Nordsee*) in Hannover.
24.01.1933-20.02.1933	*Referent Fl. (Fliegerreferent* [Aviation Advisor]) of *SA-Gruppe Nordsee*.
01.07.1933-14.07.1933	*Stabsführer* of *SA-Obergruppe VI* (*SA-Gruppen Niedersachsen, Nordsee, Niederrhein*, and *Westfalen*) in Hannover.
15.07.1933-24.09.1934	Reappointed as *Referent Fl. (Fliegerreferent)* to the staff of *SA-Obergruppe VI*. He was removed from this assignment on 24.09.1934 after he'd assumed a post with the *Deutsche Luftsportverband* (German Air-Sports Association; see next entry), as this was contrary to the official segregation between the *Deutsche Luftsportverband*.

Hitler visits Hannover in 1931. Beside him is Viktor Lutze (as *Obersten SA-Führer Nord*). Lutze's Adjutant, then *SA-Standartenführer* Jahn, stands at far left. (Roger Bender photo)

00.07.1934-31.03.1935	*Hauptamtlicher Führer der Flieger-Landesgruppe IV* of the *Deutsche Luftsportverband* in Niedersachsen and *Führer* of the *Flieger-Ortsgruppe Osnabrück.*
01.10.1935-31.08.1936	Reentered *SA* service at his original rank of *SA-Gruppenführer*, assigned as an *SA-Führer z. V.* to *SA-Gruppe Niedersachsen.*
12.02.1936-11.01.1939	*Polizeipräsident* (m.d.W.d.G.b. until 28.10.1936, then permanent with effect from 01.10.1936) of Halle an der Saale. Confirmed in this post by *Regierungspräsident* Dr. Sommer, 12.02.1936. Succeeded *Oberst der Schutzpolizei a. D.* Berend Roosen. Succeeded by *SA-Brigadeführer* Friedrich Habenicht. In addition to officially assuming this post, he was also assigned as *örtlicher Luftschutzleiter* (local air-raid protection coordinator) and *Leiter* of the *Kriminalpolizeileitstelle* in Halle an der Saale
01.09.1936-31.12.1938	Assigned as an *SA-Führer z. V.* to *SA-Gruppe Mitte.*
00.06.1937-00.06.1937	Participated in a 10-day *Luftschutzlehrgangs für Polizeiverwalter* (Air-raid defense instructional course for police administrators) at the *Reichsanstalt für Luftschutz* in Berlin.
20.07.1937-00.00.1942(?)	Honorary member of the *Volksgerichtshof* (a 5-year appointment).
10.04.1938	Unsuccessfully proposed for the *Liste des Führers zur Wahl des Großdeutschen Reichstages am 10. April 1938.* (List of the *Führer* for election to the Greater German *Reichstag* on 10. April 1938).
01.01.1939-24.02.1942	Assigned as an *SA-Führer z. V.* to *SA-Gruppe Pommern.*

11.01.1939-00.00.1943	*Polizeipräsident* in Stettin, simultaneously assigned as *örtlicher Luftschutzleiter* (local air-raid protection coordinator) in the city.
25.02.1942-00.05.1945	Assigned as an *SA-Führer z. V.* to several *SA-Gruppen*, as follows:

        25.02.1942-31.07.1943   *SA-Gruppe Tannenberg.*
        01.08.1943-30.11.1943   *SA-Gruppe Pommern.*
        01.12.1943-00.05.1945   *SA-Gruppe Elbe.*

Decorations and Awards

00.00.1918	*1914 Eisernes Kreuz I. Klasse*
00.00.1915	*1914 Eisernes Kreuz II. Klasse*
00.00.1918	*Flugzeugführerabzeichen*
00.00.1918	*Verwundetenabzeichen, 1918 in Schwarz*
00.00.193_	*Goldenes Ehrenzeichen der NSDAP*
00.00.193_	*Traditions-Gau-Abzeichen*
20.04.1940	*Dienstauszeichnung der NSDAP in Silber*
20.04.1940	*Dienstauszeichnung der NSDAP in Bronze*
ca. 1931	*Abzeichen des SA-Treffens Braunschweig 1931*
00.00.1934	*Ehrendolch der SA* (mit Wirkung vom 03.02.1934)
00.02.1934	*Ehrenwinkel für alte Kämpfer*

Notes

- Son of the newspaper publisher Wilhelm Jahn (born 12.04.1860 in Nörenberg / Pommern) and his wife Charlotte, née Rochol (born 15.01.1867 in Soest / Westfalen).
- Religion: Lutheran until 193_, then left the church and declared himself "gottgläubig".
- First married, 19__. Divorced, 00.03.1925. Remarried on 25.05.1927 to Ingeborg Moritz (born 17.08.1908 in Erfde / Kreis Schleswig). One son (born 15.04.1920 [from his first marriage]) and two daughters (born 16.03.1928 and 01.09.1933 [both from his second marriage]).
- Foreign language proficiency: French.

Sources

Bundesarchiv, Berlin-Lichterfelde (former Berlin Document Center): *Personalunterlagen von SA-Angehörigen: SA-Personalakte* of Wilhelm Jahn.

Karl *Rudolf* Jordan
SA-Obergruppenführer

Born:	21.06.1902 in Grosslüder bei Fulda / Hessen.
Died:	27.10.1988 in München-Haar / Bayern.

NSDAP-Nr.:   4 871 (Joined 15.05.1925)

Promotions

00.00.1926	*SA-Mann*
00.00.1929-00.00.1931	*Ortsgruppenleiter* der *NSDAP*
19.01.1931-02.05.1945	*Gauleiter* der *NSDAP*
03.04.1934	*SA-Gruppenführer*
09.11.1937	*SA-Obergruppenführer*

Career

00.00.1908-00.00.1916	Attended *Volksschule*.
00.00.1916-00.00.1918	Worked voluntarily as a laborer in a munitions factory.
00.00.1918-00.00.1924	Attended the *Präparandenanstalt* and *Lehrerseminar* (teachers college) in Fulda. He was a member of the Catholic *Neudeutschland* movement during the early part of this period.
00.00.1919-00.00.1920	*Reichswehr* service as a *Zeitfreiwilliger* (part-time volunteer).
00.00.1921	Joined the *Bund Oberland*.
00.00.1924	Passed his examinations as a *Volksschule* teacher.
00.00.1924-00.00.1926	Employed as an assistant in publishing advertising.
00.00.1924-00.00.1924	Active as a speaker for the *Völkisch-Sozialer Block* and the *Deutsch-Völkischen Reichspartei*, however he did not hold membership in these parties.
15.05.1925	Joined the *NSDAP*.
00.00.1925-00.00.19__	Founder and editor of the völkisch monthly *Notung*.
00.00.1926-00.00.192_	Assigned as an *SA-Mann* to *SA-Sturm 5* in Fulda.
00.00.1926	Emigrated to Australia.
00.00.1927	Returned to Germany.
00.00.1927-00.00.1928	*Gaugeschäftsführer* of the *Gauleitung Hessen-Nassau der NSDAP* (Seat: Kassel).
00.00.1927-00.00.1928	Worked in an assisting capacity for the *Finanzamt* (finance office) in Kassel
00.00.1928-22.12.1929	Employed as a substitute teacher at a *Volksschule*, a *Berufsschule* (trade school), and at the *Heeresfachschule für Wirtschaft und Verwaltung* (Army Vocation School for Economics and Administration) in Fulda. He was ultimately forced out of his position at the Heeresfachschule due to his National Socialist political activities.
00.00.1929-00.00.1931	*Ortsgruppenleiter* of the *Ortsgruppe Fulda der NSDAP*.

Gauleiter Jordan in 1934.

17.11.1929-00.00.1931   Member of the *Kommunallandtag* (local parliament) in Kassel and the *Provinziallandtag* (provincial parliament) of Hessen-Nassau.

00.11.1929-00.00.19__   Founder and editor of the newspaper *Der Fuldaer Beobachter.*

00.12.1929-00.00.1931   *Stadtrat* in Fulda.

22.12.1929   Stricken from the list of education officer applicants and from the school service due to his National-Socialist political activities. In a letter from Dr. Kuchen, head of the *Abteilung für Kirchen und Schulwesen* (Department for Churches and Public Education) in the office of the *Regierungspräsident* in Kassel, dated 13.12.1929, *Schulamtsbewerber* (education office applicant) Jordan was informed of that office's "conviction that [he was] unsuitable for the occupation of a public school teacher ... The reproaches, insults, and slanders against the State and its responsible representatives which you hurled as a speaker at political meetings far exceed the measure that can be accepted in [these] excited political times ... " He later wrote in his memoirs of his belief that he had been expelled solely for his political opinions, bitterly asserting that "the State had made [him] unemployed, unemployed and without rights." At a conference held at the *Domschule* in Fulda on 02.06.1933, arranged by the school's rector Karl Josef Hagemann at the instruction of *Schulrat* Dr. Wilhelm Hamacher (almost surely initiated by the *NSDAP Kreisleitung* in Fulda), Jordan was deemed "rehabilitated" in the school service. The *Gauleiter* arrived at the school together with Hitler and *Reich* President von Hindenburg. The minutes of the conference read: "After greetings

Gauleiter Jordan, ca. 1938.

Gauleiter Jordan in civilian attire, ca. 1938.
(*Bundesarchiv Bild* 146-2005-0027)

by the rector, the Herr *Schulrat* [Dr. Hamacher] spoke on behalf of the government that Herr Jordan [had been] unjustly removed from his office. Retroactively, complete acknowledgement was granted to him [Jordan] as a teacher. In a longer speech, Herr Jordan spoke about the present state of educators ... In the meantime a detachment of the *SA* and *SS* arranged itself in the courtyard and ... greeted Herr Jordan ... "

00.00.1930-00.01.1931	*Aussenpolitischer Schriftleiter* (foreign policy editor) of the *NSDAP Gau* newspaper *Sturm*, Kassel.
19.01.1931-20.04.1937	*Gauleiter* of *Gau Halle-Merseburg der NSDAP* (Seat: Halle an der Saale). Succeeded Paul Hinkler, who only reluctantly gave up the post. Jordan was summoned to München, his first visit to that city, by Gregor Straßer, and was personally greeted by Hitler on 19.01.1931. The *Führer* welcomed him as a capable and selfless National-Socialist idealist, then personally appointed him as the new *Gauleiter* of *Halle-Merseburg*. Succeeded by Joachim Albrecht Eggeling, he then assumed Eggeling's post as *Gauleiter* of *Magdeburg-Anhalt*.
00.00.19__-00.00.19__	Editor of the *Mitteldeutschen Nationalzeitung* (Halle) and the weekly newspaper *Der Kampf*.
24.04.1932-14.10.1933	Member of the *Preußische Landtag (Wahlkreis 11, Merseburg)*.
00.04.1932	Appointed as a member of the *Provinziallandtag der Provinz Sachsen* (Provincial Parliament of the Province of Saxony).
00.03.1933-14.10.1933	Member of the *Vorstand* of the *Preußische Landtag*.

Gauleiter Jordan on 23.04.1937. (NARA photo)

12.03.1933-00.00.193_	*Fraktionsführer der NSDAP* (leader of the *NSDAP* faction) in the *Provinziallandtag der Provinz Sachsen* and the *Stadtparlament der Stadt Halle/Saale* (Municipal Parliament of the City of Halle an der Saale).
12.03.1933-14.02.1934	*Bevollmächtigter* (Plenipotentiary) for *Provinz Sachsen* in the *Reichsrat*.
12.02.1933	Together with the *Landrat* of Eisleben, Ludolf-Hermann von Alvensleben (who later earned infamy as a particularly murderous *SS* and Police leader), and the local *SS* leader Hermann Florstedt, played a key role in the "Eisleber Blutsonntag" (Eisleben Bloody Sunday), which resulted in the murder of several communists.
00.07.1933-08.05.1945(?)	Member of the *Preußische Staatsrat*.
14.09.1933-08.05.1945	*Preußischer Staatsrat*.
12.11.1933-08.05.1945	Member of the *Reichstag* (representing *Wahlkreis 11, Merseburg*; after 10.04.1938, *Wahlkreis 10, Magdeburg*).
03.04.1934-20.04.1937(?)	Appointed as an *Ehrenführer der SA* with honorary assignment to *SA-Standarte 36*.
00.00.1934-00.05.1945	*Preußischer Provinzialrat* of *Provinz Sachsen*.
04.09.1935-00.00.1944(?)	Member of the *Akademie für Deutsches Recht*, München.
27.04.1936	Attended a ceremony marking the establishment of the *Obersten Ehren- und Disziplinarhofs der DAF* (Supreme Honor and Disciplinary Court of the German Labor Front) in the Berlin Ratshaus.
02.07.1936	Participated in the ceremony marking the 1000th anniversary of the death of Heinrich I in Quedlinburg, together with Heinrich Himmler, Reinhard Heydrich, and Karl Wolff.

Gauleiter Jordan dines with the *Reichsfrauenführerin*, Gertrud Scholtz-Klink, ca. 1937.

20.04.1937-02.05.1945	*Gauleiter* of *Gau Magdeburg-Anhalt der NSDAP* (Seat: Dessau). Succeeded Joachim Albrecht Eggeling, who then assumed Jordan's previous post as *Gauleiter* of *Gau Halle-Merseburg der NSDAP*.
20.04.1937-02.05.1945	"*Reichsstatthalter* in Braunschweig und Anhalt" (Seat: Dessau).
20.04.1937(?)-02.05.1945	Assigned as *SA-Führer z. V.* to *SA-Gruppe Mitte* (Dessau).
01./02.07.1939	Participated in the "König-Heinrich-Feier der *Reichsführer-SS*" in Quedlinburg.
08.01.1940-02.05.1945	*Führer der Landesregierung* (Leader of the State Government) of *Land Anhalt* (with effect from 01.01.1940). In this capacity, he was involved in the "euthanasia" killings- as part of the "T4" program- at the Anstalt Bernburg located in Land Anhalt, and was a convinced proponent of this program. In 1934, he had declared: "Sterilization is the best way to liberate humanity from its misbegotten specimens, degenerates, the mentally ill, and hereditary criminals."
01.09.1939-16.11.1942	*Reichsverteidigungskommissar für den Wehrkreis XI*, and as such *Vorsitzender* of the *Verteidigungsausschuß* of *Wehrkreis XI*.
15.11.1940-02.05.1945	*Gauwohnungskommissar* for *Gau Magdeburg-Anhalt*.
06.04.1942-02.05.1945	*Beauftragter des GBA für den Gau Magdeburg-Anhalt*.
16.11.1942-02.05.1945	*Reichsverteidigungskommissar für den Gau Magdeburg-Anhalt*.
07.02.1943	At *Führer* HQ "Wolffschanze" in Rastenburg / Ostpreußen, during which Hitler addressed most of the *Reichs-* and *Gauleiter* on the subject of the recent German defeat at Stalingrad.
01.07.1944-02.05.1945	*Oberpräsident* of Provinz Magdeburg ("mit der Verwaltung der Stelle beauftragt" [Charged with administration of the post] until 18.08.1944,

Heinrich Hoffmann's official portrait of *Gauleiter* Jordan, ca. 1941. (NARA photo)

	then officially assigned). Confirmed in office by Heinrich Himmler, 06.07.1944.
00.00.1944	Appointed as a *Gauarbeitsführer (ehrenhalber)*.
00.09.1944-02.05.1945	*Führer der Deutschen Volkssturm in Gau Magdeburg-Anhalt*. The *Gaustabsführer* of the *Volkssturm* in the *Gau* was *SA-Gruppenführer* Albert Heinz.
24.02.1945	Attended a conference of *Reichsleiter* and *Gauleiter* with Hitler in the *Reichskanzlei*, Berlin. This was his last meeting with the *Führer*.
02.05.1945	Dissolved the *Gaustab* of *Gau Magdeburg-Anhalt*. During this period he also disbanded the local *Volkssturm* and prohibited any *"Werwolf"* resistance activity.

Postwar Prosecution and Activities

Fled westward over the Elbe together with his family, 07.05.1945, then lived under the assumed name "Richard Gabriel". Arrested by British troops in Dannstedt / Ostharz, 30.07.1945. He was interned and repeatedly interrogated by British authorities, first in Wernigerode, then in Magdeburg, and reportedly suffered severe mistreatment at their hands. Transferred to a prison in Oschersleben, then to the internment camp at Westertimke / Niedersachsen, he was, in September 1945, handed over to U.S. custody and held in the Zuffenhausen camp near Stuttgart to undergo further interrogation by the American CIC. Briefly held at the internment camp in Dachau, he was handed over to Soviet authorities on 26.07.1946, and subsequently interned in Dresden, Potsdam, and Berlin-Hohenschönhausen. Relocated to Lubjanka Prison, Moscow on 30.11.1949. Sentenced by a Soviet court to 25 years' hard labor, 00.12.1950, although he was not present during his "trial" and no witnesses were called. He was thereafter confined to a total of seven Soviet prisons and prisoner of war camps. Erroneously declared dead by the Amtsgericht in München, 08.01.1951; alleged eyewitnesses had reported seeing him executed by shooting in Magdeburg in 1946. The declaration of death was corrected on 11.10.1954. Released from Soviet captivity, 13.10.1955 (thanks to the intervention of *Bundeskanzler* Konrad Adenauer), and repatriated to Germany. He subsequently lived with his family in München, receiving assistance from *Heimkehrerhilfe*, an organization providing aid to people returning to Germany from imprisonment in the Soviet Union. Jordan eventually found employment, working successively as a sales representative, insurance inspector, publishing representative, and administrator for an aircraft manufacturing firm.

Published Works
"Der wissenschaftliche Sozialismus" (article, 1925)
"Deutschland als Kolonie der Wallstreet" (article, 1925)
Das demaskierte Zentrum (1932)
Erlebt und erlitten. Weg eines Gauleiters von München bis Moskau (1971)
Im Zeugenstand der Geschichte. Antworten zum Thema Hitler (1974)
Der 30. Juni 1934. Die sog. 'Röhm-Revolte' und ihre Folgen aus der Sicht eines Erlebniszeugen (1984)

Decorations and Awards
00.00.194_	*Kriegsverdienstkreuz I. Klasse ohne Schwerter*
00.00.194_	*Kriegsverdienstkreuz II. Klasse ohne Schwerter*
00.00.193_	*Goldenes Ehrenzeichen der NSDAP*
00.00.19__	*Goldenes Hitler-Jugend Ehrenzeichen mit Eichenlaub*
00.00.1933	*Traditions-Gauabzeichen*
ca. 1941	*Dienstauszeichnung der NSDAP in Gold*
ca. 1940	*Dienstauszeichnung der NSDAP in Silber*
ca. 1940	*Dienstauszeichnung der NSDAP in Bronze*
00.05.1934	*Ehrendolch der SA*
00.00.193_	*Ehrenwinkel für alte Kämpfer*

Notes
- Fourth and youngest child of the farmer and textile merchant Anton Jordan and his wife Mathilde (died at the age of 84 in Fulda, 00.00.1946). He had two brothers and a sister. His brother, Ludwig Jordan, was an officer of the *SS*.
- Married, 00.00.192_. His first wife died ca. 1938, and he later remarried. One son (H.-J.) and two daughter (S. and A. [born 00.00.1945]).

Sources
Höffkes, Karl: *Hitlers politische Generale: Die Gauleiter des Dritten Reiches*. Grabert-Verlag-Tübingen, 1986.

Jordan, Rudolf: *Erlebt und erlitten. Weg eines Gauleiters von München bis Moskau*. Druffel Verlag, 1971.

Lilla, Joachim; Döring, Martin; & Schulz, Andreas: *Statisten in Uniform. Die Mitglieder des Reichstags 1933-1945*. Droste Verlag, 2004.

Schirach, Baldur von: *Die Pioniere des Dritten Reiches*. Zentralstelle fur der deutschen
   Freiheitskampf, 1933.

Appendix I

Comparative Table of Ranks

Sturmabeilung	Schutzstaffel	Reichsheer (German Army)	U.S. Army Equivalent
Oberster SA-Führer (OSAF) (Adolf Hitler)	No Equivalent	No Equivalent	No Equivalent
SA-Stabschef	*Reichsführer-SS*	No Equivalent	No Equivalent
No equivalent	No Equivalent	*Generalfeldmarschall*	General of the Army (5-star)
No equivalent	*SS-Oberst-Gruppenführer*	*Generaloberst*	General (4-star)
SA-Obergruppenführer	*SS-Obergruppenführer*	*General der Infanterie, Kavallerie, Artillerie, etc.*	Lieutenant-General (3-star)
SA-Gruppenführer	*SS-Gruppenführer*	*Generalleutnant*	Major-General (2-star)
SA-Brigadeführer	*SS-Brigadeführer*	*Generalmajor*	Brigadier-General (1-star)
SA-Oberführer	*SS-Oberführer*	No Equivalent	No Equivalent
SA-Standartenführer	*SS-Standartenführer*	*Oberst*	Colonel
SA-Obersturmbann-führer	*SS-Obersturmbann-führer*	*Oberstleutnant*	Lieutenant-Colonel
SA-Sturmbannführer	*SS-Sturmbann-führer*	*Major*	Major
SA-Sturmhauptführer	*SS-Hauptsturmführer*	*Hauptmann or Rittmeister*	Captain
SA-Obersturmführer	*SS-Obersturmführer*	*Oberleutnant*	First-Lieutenant
SA-Sturmführer	*SS-Untersurmführer*	*Leutnant*	Second-Lieutenant
SA-Haupttruppführer	*SS-Sturmscharführer*	*Stabsfeldwebel, Stabswachtmeister*	Sergeant Major
SA-Obertruppführer	*SS-Hauptscharführer, SS-Standarten-oberjunker*	*Oberfeldwebel, Oberwachtmeister (Hauptfeldwebel), Oberfähnrich*	Master Sergeant
SA-Truppführer	*SS-Oberscharführer*	*Feldwebel, Wachtmeister*	Technical Sergeant
SA-Oberscharführer	*SS-Scharführer*	*Unterfeldwebel, Unterwachtmeister*	Staff Sergeant
SA-Scharführer	*SS-Unterscharführer*	*Unteroffizier, Oberjäger*	Sergeant
SA-Rottenführer	*SS-Rottenführer*	*Stabsgefreiter, Obergefreiter*	Corporal
SA-Sturmmann (after 1938, *SA-Obersturmmann*)	*SS-Sturmmann*	*Oberschütze*	Private First Class
SA-Mann (after 1938, *SA-Sturmmann*)	*SS-Mann*	*Schütze*	Private

Glossary of German Military and Political Terms

Abitur (Abi.) (aka *Reifeprufung*): Most prestigious of several certificates issued after completion of secondary schooling in Germany. It typically allowed admission to university in any subject.

Abschnitt (Abs.; Abschn.): Sector; district. In the SS: Regional division of Germany and its headquarters; In the *SD*: Subordinate Regional HQ and its area.

Abteilung (Abt.; A.): Unit, battery, battalion, department.

Abwehr (Abw.): *Wehrmacht* intelligence service.

Abwehrpolizei: Counter-espionage police. A function of the *Grenzpolizei* under the jurisdiction of the *Gestapo*.

Abzeichen (Abz.): Badge of rank, appointment, or distinction.

Adjutantur: i. Staff department dealing with routing personnel and administrative matters; ii. The earlier title of the *Personlicher Stab RfSS*.

Agrarpolitischer Apparat (a.A.): The *NSDAP*'s Office for Agricultural Policy.

Ahnenerbe Forschungs- und Lehrgemeinschaft: Society for Research into and Teaching of Ancestral Heritage. Administered by the *Pers.Stab RfSS*, this office promoted research of family and national hereditary history and dissemination of racial theories.

Akademie fur Deutsches Recht: The Academy of German Law, founded by *Reichsleiter Dr. jur.* Hans Frank in 1933 and based in Munchen.

Allgemeines Heeresamt (A.H.A.): General Army Directorate.

Allgemeine-SS (Allg.SS): The general body of the SS, composed of part-time, full-time, and honorary members.

Allgemeines Wehrmachtsamt (A.W.A.): General Armed Forces Office in *OKW*, concerned with matters of personnel, training, equipment, etc.

Alte Kampfer: "Old Fighter". Designation applied to members of the *NSDAP* or its affiliated organizations prior to the Nazi assumption to power on 30.01.1933.

Amt: A main office, branch, or directorate of a Ministry, or an independent Ministry such as the *Auswartiges Amt* (Foreign Office).

Amtsgruppe: A subordinate branch of a *Hauptamt*.

Anhaltelager: Temporary detention camp.

Anschluss: The unification of Germany and Austria on 13.03.1938.

Anwarter (Anw.): Cadet or candidate.

Arbeitsgemeinschaft: Labor collective. Associations of former *Freikorps* members formed to facilitate finding employment and maintaining contact as a covert military force.

Armee-Korps (A.K.): Infantry corps.

Armee-Oberkommando (A.O.K.): Army Headquarters.

Artillerie-Kommandeur (Arko): Artillery commander.

Artillerie-Regiment (Art.Rgt.; A.R.): Artillery regiment.

Aufsichtsrat: Supervisory board; board of directors.

Aufklarung (Aufkl.): lit. enlightenment; military reconnaissance.

Aufstellungsstab: Formation staff.

Ausbildung (Ausb.): Training.

Ausbildungs-und Ersatz (Ausb.u.Ers.; A-u.E.): Training and replacement.

Ausland/Abwehr (Ausl./Abw.): *Amt* (previously an *Amtsgruppe*) of *OKW* responsible for Espionage, Counterespionage, Sabotage, and Foreign Information. Headed by *Admiral* Wilhelm Canaris until his arrest for anti-Nazi activities in 1944, whereupon it was absorbed by the *RSHA*.

Auslands-Organisation (AO): Foreign Organization. The *NSDAP* agency concerned with care and supervision of Germans in foreign countries. Ranked as the 43rd *Gau* of the *NSDAP* under *Gauleiter* and Foreign Office *Staatssekretar* Ernst Wilhelm Bohle.

Aussendienststelle or *Aussenstelle* (Aust.): Outstation or outpost of the *Sipo* and *SD*.

Ausser Dienst (a. D.): retired; on inactive list.

Auswartiges Amt (AA): Foreign Office; the *Reich* Ministry for Foreign Affairs.

Bahnschutzpolizei: Railway protection police. Transferred to *SS* control in 1942.

Bann: A subdivision of the *Hitler-Jugend Gebiet* (region), approximately comparable to a *Kreis* (district) of the *NSDAP*.

Bataillon (Btl.): Battalion.

Batterie (Bttr.): Battery (artillery troop).

Bayern: Bavaria. The southernmost Land (State) of Germany, comprising the following *Regierungsbezirke* (government districts; capital cities follow in parentheses): Unterfranken (Wurzburg); Oberfranken (Bayreuth); Mittelfranken (Ansbach); Oberpfalz (Regensburg);Niederbayern (Landshut); Oberbayern (Munchen); and Schwaben (Augsburg).

Bayerisch: Bavarian.

Bayerische Landtag: Bavarian State Parliament.

Beamter: An official or functionary.

Beauftragt (b.): authorized; commissioned; in charge.

Beauftragter (Beauftr.): A representative, commissioner, or administrator.

Befehl: An order or command.

Befehlshaber: Senior commander.

Befehlshaber des Ersatzheeres (B.d.E.): C.-in-C. Replacement Army.

Befehlshaber der Ordnungspolizei (BdO): Commander of the *Orpo* in a *Wehrkreis* or similar territory.

Befehlshaber der Sicherheitspolzei und des Sicherheitsdienstes (B.d.S.): Commander of the *Sipo* and *SD* in a *Wehrkreis* or similar territory.

Behorde (Beh.): An authority or administrative body.

Bei / beim (b.): on / near.

Bereitschaftspolizei: Mobile barracks police units of the *Landespolizei*, administered by the various *Länder* before the police were centralized in 1935/1936.

Bevollmachtigter (Bev.): Plenipotentiary.

Bezirk (Bez.): A district; sub-region; administrative unit.

Blockleiter: Lowest official of the *NSDAP*, responsible for the political supervision of 40 to 60 households. Subordinate to a *Zellenleiter*.

Bund: League, Federation.

Bund Deutscher Madel (BDM): League of German Girls. The *Hitler-Jugend* organization for girls aged 14 to 18.

Bund Oberland: A *Wehrverband* (paramilitary formation) of predominantly Bavarian membership formed from the *Freikorps* Oberland. Under the leadership of *Dr. med. vet.* Friedrich Weber, it was the primary ally of the Nazi Party during the Munich Putsch of.11.1923.

Bund Wiking: A *Wehrverband* formed by Hermann Ehrhardt, led by members of his former *Freikorps*.

Burgermeister: Mayor of a mid-size town or smaller community

Charakterisiert (Char.): Brevet rank (a temporary promotion, generally render as "Charakter als [*Oberst*, etc.]").

Chef (Ch.): chief.

Chef des Ausbildungswesens (Chef-AW): A covert military training organization based in part on the *SA* which instructed young men in military skills, 1933-1935.

Chef der deutschen Polizei (Ch.d.Dt.P.; ChdDtP; ChdDtPol.): Commander-in-Chief of the German Police (Heinrich Himmler).

Chef der Heeresrustung und Behfehlshaber des Ersatzheeres (Ch.H.Ru.u.B.d.E.): Chief of Army Equipment and Commander of the Replacement Army (*GeneralOberst* Friedrich Fromm until July 1944, then Heinrich Himmler).

Chef der Ordnungspolizei (CdO): Chief of the Order Police (Kurt Daluege, then Wünnenberg assumed duties due to Daluege's illness from 1943 onward).

Chef der Sicherheitspolizei und des SD (CSSD): Chief of the Security Police and *SD* (Reinhard Heydrich until 04.06.1942, then Himmler until 30.01.1943, and finally *Dr. jur.* Ernst Kaltenbrunner).

Chef der Zivilverwaltung (CdZ): Head of the Civilian Administration in an occupied territory.

Der Reserve (d. R.): of the Reserve.

Deutsch (dt.): German.

Deutsche Arbeitsfront (DAF): The German Labor Front. Largest of the *NSDAP*'s affiliated organizations comprising all the corporations, guilds, and professional associations. Led by *Reichsleiter* Dr. Robert Ley.

Deutsche Arbeiterpartei (DAP): German Workers' Party. Founded in München, January 1919. Later redesignated National-Socialist German Workers' Party.

Deutsche Ausrustungswerke (DAW): German Equipment Works, *SS* enterprise founded in 1939.

Deutsche Erdund Steinwerke GmbH (DEST): German Clay and Brickworks Ltd. *SS* enterprise established in 1938, primarily to construct brickworks and exploit quarries by the use of prisoner labor.

Deutsches Rotes Kreuz (DRK): The German Red Cross.

Deutschnationale Volkspartei (DNVP): The conservative German National People's Party.

Deutsch-Volkischer Schutz- und Trutzbund (DVSuTB): German-Folkish Protection and Defense League. Founded in 1919 with support of the Pan-German League. An anti-Semitic organization specializing in publication of propaganda and rallies among the folkish movement in Germany. Its branch leaders were designated as *Gauleiter*, and this system was duplicated by the *NSDAP*. This league was officially banned in 1922.

Deutsche Umsiedlungs-Treuhand GmbH (DUT): German Resettlement Trust Ltd., affiliated with the *RKFDV*.

Deutsche Wirtschaftsbetriebe GmbH (DWB): German Economic Enterprises Ltd., an enterprise controlled by the *SS-WVHA*.

Deutsches Jungvolk (DJV; DJ): German Young People. The *Hitler-Jugend* branch for boys aged 10 to 14.

Deutschnationaler Handlungsgehilfen Verband (DHV): German National Trade Assistants' Association. A rightist union of white-collar clerks.

Deutsch-Nationale Volkspartei (DNVP): German Nationalist People's Party. The leading German conservative party during the Weimar era.

Deutsch-Sozialististische Partei (DSP): German People's Party. An anti-Semitic political party that merged with the *NSDAP* in 1923.

Deutsche Volkspartei (DVP): German People's Party. A moderate political party active during the Weimar years.

Deutschevolkische Freiheitspartei (DVFP): German Volkisch Freedom Party. A splinter party of the *DNVP* with an anti-Semitic slant operating in northern Germany until ca. 1929.

Dienstgrad: Rank or grade.

Dienststelle: Headquarters, administrative office, station, or depot.

Dr. agr.: Doctor of Agricultural Sciences.

Dr. chem.: Doctor of Chemistry.

Dr. h.c.: doctor honoris causa-Honorary doctor's degree.

Dr. Ing.: *Doktor Ingenieur*-Doctor of Engineering.

Dr. jur.: doctor juris-doctor of Jurisprudence / Law.

Dr. med.: doctor medicinae-Doctor of Medicine (M. D.).

Dr. med. vet.: Doctor of Veterinary Medicine.

Dr. phil.: doctor philosophiae-Doctor of Philosophy (Ph. D.).

Dr. rer. nat.: doctor rerum naturalium-Doctor of Natural Science.

Dr. rer. pol.: doctor rerum politicarum-Doctor of Political Science.

Dr. theol.: doctor theologiae-Doctor of Theology or Divinity (D. T.).

Durchgangslager (Dulag): POW transit camp.

Ehrenburger: Honorary citizen of a town, city, or other community.

Ehrenburgerrecht: Honorary citizenship of town, city, or other community.

Ehrenfuhrer: Honorary leader. Distinction accorded by Himmler to leading *NSDAP* and State figures nominally attached to his staff.

Ehrenhalber (e.h.): honorary.

Eingetragener Verein (e.V.) : A registered society (e.g., Lebensborn e.V.).

Einheit (Einh.): Unit.

Einjahriger Freiwilliger (Einj.Frw.): One-year volunteer. A type of voluntary officer cadet in the Imperial German Army.

Einsatz: Action, operation, employment.

Einsatzgruppe (EGru.; EG): Operational group or task force of the *Sipo* and *SD* for special missions (particularly liquidations of Jews, communist officials, etc.) in occupied territory.

Einsatzkommando (Ekdo.; EK): Detachment of the *Sipo* and *SD*; subordinate element of an *Einsatzgruppe*.

Einwohnerwehr: Part-time citizens' militia formed by the German Government during the revolutionary period, 1918/1919.

Erganzungs(Erg.-): Reserve; supplementary.

Erlass: Edict, decree, or order.

Ernahrung (Ern.): Food; nutrition.

Ersatz (Ers.): Training; depot; reinforcement.

Europaisch (europ.): European.

Fachreferat: A specialist subsection or "desk" in an office or headquarters.

Fahnrich: Officer candidate (second stage).

Fahnenjunker: Officer candidate (first stage).

Feldgendarmerie (Feld.): Military police.

Feldjagerkorps (FJK): A shock formation of the SA, dissolved 1935 and incorporated into the Police.

Feldkommandantur (F.K.): Administrative headquarters in occupied countries (of regimental status).

Feldlazarett (Feldlaz.): Field hospital.

Feme: The following is Robert M. W. Kempner's definition of this term: "'*Feme*' refers to the criminal practice of the German secret para-military. Organizations, such as the Free Corps, of executing their members for disciplinary reasons, and for informing the State authorities about details of these secret organizations. These executions were made without trial during the period from 1918-33, at which time such procedure became the "legal" right of the [*SS*]. Democratic groups in the country and in the *Reichstag* which exposed the *Feme* were stigmatized as "*Feme agitators*" by the National Socialists. Furthermore, the *Feme* organizations murdered outstanding representatives of the new democratic regime, among them the *Reich* Finance Minister Matthias Erzberger (August 26, 1921) and the *Reich* Foreign Minister Walther Rathenau (June 24, 1922). The *Feme* organizations were financially supported by secret funds of the Army and certain industrialists." (Robert M. W. Kempner, "Blueprint of the Nazi Underground- Past and Future Subversive Activities", in *Research Studies of the State College of Washington, Volume XIII, Number 2, June 1945*)

Feuerschutzpolizei (FSchp.): Fire Protection Police; a branch of the *Orpo*.

Fordernde Mitglied der SS (F.M.): Sponsoring member of the *SS* paying monthly dues for that status; did not hold actual *SS* rank. In many cases the wives of *SS* officers held this title.

Freiherr (Frhr.): Baron (aristocratic title; e.g. Constantin Freiherr von Neurath).

Freikorps (F.K.): Temporary and voluntary paramilitary formations created by the German Army and Navy during the 1918/1919 revolutionary period in Germany.

Freiwilligen (freiw.): volunteer.

Freiwilliger Arbeitsdienst (FAD): Voluntary Labor Service. Formed under Government control as a remedy to unemployment during the Great Depression. Forerunner of the obligatory *Reichsarbeitsdienst.*

Frontbann: Front Union. An enormous *Wehrverband* established by Ernst Rohm and other nationalist leaders in 1924.

Fuhrer (Fhr.; F.): Leader; commander; chief. The title "*Der Fuhrer*" was only used in reference to Adolf Hitler.

Fuhrerhauptquartier (FHQ): Hitler's Field Headquarters.

Fur (f.): for.

Funk: Radio.

Gau: Main territorial division of the *NSDAP*. There were 42 *Gaue* in the *Reich* and annexed territories, with a further 43rd

Gau comprising the *Auslandsorganisation* (*AO*; Foreign Organization of the Nazi Party).

Gauleiter (GL): Highest-ranking *NSDAP* official in a *Gau*, with responsibility for all matters of politics, economics, labor mobilization, and civil defense.

Gauwohnungskommissar: Regional Housing Commissioner, title held by the majority of the *Gauleiter*. In this capacity, they were subordinated to *Reichsleiter* Dr. Robert Ley, who was appointed as *Reichskommissar für den sozialen Wohnungsbau* (*Reich* Commissioner for Social Housing Construction on 15.11.1940 (redesignated as *Reichswohnungskommissar* [Reich Housing Commissioner] on 23.10.1942).

Gebirgs-Armee: Mountain Army; a formation of the German Army.

Gebirgs-Division (Geb.Div.): Mountain Division of the *Heer* or *Waffen-SS*.

Gebirgs-Korps: Mountain Corps (a formation of the *Heer* or *Waffen-SS*).

geheim (geh.): secret (adj.).

Geheime Feldpolizei (GFP): Secret Field Police.

Geheimes Staatspolizei (*Gestapo*): Secret State Police; *Amt IV* of the *Reichssicherheitshauptamt* from 01.10.1939.

Gemeinde (Gem.): Municipality, community.

Gemeindepolizei (Gem.Pol.): Municipal Police.

Genannt (gen.): Called; named; known as.

Gendarmerie (Gend.): Rural police, including motorized traffic control units.

Generalbevollmachtigter: Plenipotentiary General.

Generalgouvernement (Gen.Gouv.; GG): Government General. German-occupied Poland and the administration thereof under *Dr. jur.* Hans Frank.

Generalkommando (Gen.Kdo.): Headquarters of an Army Corps.

Generalquartiermeister des Heeres: Quartermaster-General of the German Army.

Generalstab des Heeres (Gen.St.d.H.): The German Army General Staff.

Germanizche Leitstelle: Germanic Liaison Office of the *SS-Hauptamt*, responsible for supervision of Germanic *SS*.

Gesellschaft: Society; association; company.

Gesellschafter: Associate; partner; member of a society or company.

Gottgläubig: Lit.: "God-believing". Essentially a form of agnosticism, this designation was used in place of a religious denomination after one had left the church ("*Kirchenaustritt*"). This action was strongly encouraged and expected of *SS* members by Himmler, however it was not mandatory.

Graf: Count (aristocratic title; e.g. Georg-Henning Ernst Adolf Graf von Bassewitz-Behr).

Grenadier (Gren.): adj.: Infantry; n.: infantryman.

Grenzpolizei (Grepo; GP): Frontier Police (a branch of the *Gestapo*).

Grenzschutz: Border protection. Military formations tasked with the defense of Germany's, particularly in the east and northeast following the First World War.

Groß (gr.): great; large.

Großdeutsche Volksgemeinschaft (GDVG): Greater German People's Community. Organization composed of the *NSDAP* remnants in southern Germany, 1924-1926.

Großer Generalstab (Gr.Gen.St.): Great General Staff. The General Staff of the Imperial German Army.

Gruppe (Gr.): Group; section. From 1927 to 1931, term used to identify a group of 3 to 13 men (redesignated as *Schar* in 1931). From 1931 on it was applied to a large regional unit of the *SA* and *SS* (*SS-Gruppen* were redesignated as *SS-Oberabschnitte* in late-1933).

Gymnasium (Gymn.): Type of secondary school in German-speaking countries. Its primary purpose was preparation for university study and positions of higher administrative work. The following are names for the different grades in a *Gymnasium*:

5th grade = *Sexta*
6th grade = *Quinta*
7th grade = *Quarta*
8th grade = *Untertertia*
9th grade = *Obertertia*
10th grade = *Untersekunda*
11th grade = *Obersekunda*
12th grade = *Unterprima*
13th grade = *Oberprima*

Passing into the *Untersekunda* equals a *Hauptschulabschluss* (basic-level school graduation); successfully finishing the *Untersekunda* is equal to the *Realschulabschluss* (mid-level school graduation); successfully finishing the *Unterprima* means having *Fachholschulreife* (qualification for certain general higher education studies); and successful completion of the *Oberprima* means having *Abitur*, i.e. the *Hochschulreife*, necessary for university studies (e.g. in medicine or law). A *Gymnasium* provides a classical education including subjects such as history, geography, biology, and foreign languages (including Latin).

Handelshochschule (H.H.): University of Commerce.
Handelsschule: Secondary school emphasizing business skills.
Hauptamt (HA): Main office.
hauptamtlich: full time; fully paid; professional.
Hauptaussenstelle (HAust.; HASt.): Large branch office of the *SD*.
Heer: Army.
Heeresgruppe (Hgr.): Army Group.
Hilfsgemeinschaft auf Gegenseitigkeit (HIAG): A postwar Mutual Aid Society for veterans of the *Waffen-SS*.
Hilfspolizei (Hipo): Auxiliary Police formed in 1933 from *SA*, *SS*, and *Stahlhelm* members.
Hitler-Jugend (HJ): Hitler Youth. The Nazi youth organization.
Höhere SS- und Polizeifuhrer (HSSPF): Higher and Police Leader.
Hundertschaft: Century. Units of around 100-men, approximately equivalent to a company (often found in the *Polizei* and *SA*).

In; im (i.): in; in the.
Infanterie (Inf.; I.): Infantry.
Infanterie-Division (Inf.Div.; I.D.): Infantry division.
Infanterie-Regiment (Inf.Rgt.; I.R.): Infantry regiment.
Ingenieur (Ing.): Engineer.
Inspektor; Inspekteur (Insp.): Inspector.
Inspekteur der Ordnungspolizei (IdO): Original title for the BDO (see *Befehlshaber der Ordnungspolizei*).
Inspekteur der Sicherheitspolizei und des Sicherheitsdienstes (Insp.d.Sipo u.d.SD; IdS): Inspector of the Security Police and Security Service.
Inspektion (Insp.; In.): Inspectorate.

Jager: Hunter.

Kadettenanstalt: Designation for a military academy of the Imperial German or Austro-Hungarian Army.

Kaiserlich und kφniglich (k.u.k.): Of or pertaining to the Imperial and Royal armed forces of the Austro-Hungarian Empire.

Kampfbund fur Deutsche Kultur: Fighting League for German Culture. A Nazi organization established to combat Jewish influences in German cultural life.

Kaserne (Kas.): Barracks.

Kavallerie (Kav.; K.): Cavalry.

Kavallerie-Division (Kav.Div.; K.D.): Cavalry division.

Kavallerie-Regiment (Kav.Rgt.; K.R.): Cavalry regiment.

Kφniglich (kgl.): Royal.

Kommandant (Kdt.): Commandant.

Kommandantur (Kdtr.): Administrative HQ; Garrison HQ.

Kommandeur (Kdr.): Commander.

Kommandeur der Ordnungspolizei (KdO): Commander of the uniformed police in a general commissariat; subordinate to the *BdO*.

Kommandeur der Sicherheitspolizei und des Sicherheitsdienstes (Kdr.d.Sipo u.d.SD; KdS): Commander of the Security Police and Security Service.

Kommando (Kdo.): Command; detachment; detail.

kommissarisch (k): Acting; temporarily put in charge.

Kompanie (Komp.): Company.

Konzentrationslager (KL-[official abbr.]; KZ [unofficial abbr.]): Concentration Camp.

Kraft durch Freude (KdF): The Nazi "Strength through Joy" movement.

Kreis (Kr.): District (US: County).

Kreisleiter (KL): District Leader, Head of an *NSDAP Kreis*.

Kriegs(K): war-; wartime.

Kriminalpolizei (Kripo): Criminal Police; *Amt V* of the *Reichssicherheitshauptamt* from 1939.

Kriminalpolizei-Aussendienststelle (KPAuDSt.): Large branch office of the *Kripo*.

Kriminalpolizei-Aussenposten (KPAuP.): Small branch office of the *Kripo*.

Kriminalpolizei-Leitstelle (KPLSt.): A regional HQ of the *Kripo*.

Kriminalpolizeistelle (KPSt.): A sub-regional HQ of the *Kripo*.

Kyffhauserbund: A large German veterans' organization which evolved into the *NS-Reichskriegerbund*, the primary veterans' organization of the Third Reich.

Land: State or Province (Plural: Lander). One of the 15 territorial divisions of Republican Germany, each with its own government. Controlled by the central *Reich* government through *Reichsstatthalter* from 1933. [*Kasernierte*] *Landespolizei* (Lapo): [Barracks-housed] State Police. Militarized Barrack Police Forces of the *Lander*; taken over by *Reich* after 1935 and placed under *Wehrmacht* control.

Landwirtschaft (Landw.): Agriculture.

Landtag: The Parliament of a Land (State).

Lebensborn e.V.: The "Fountain of Life" society. Established by the *SS* in 1936. Attached to the Personal Staff of the *Reichsfuhrer-SS* and affiliated with the *SS-RuSHA*. Main functions of the society were adoption of "racially suitable" children for childless *SS* families, encouragement and facilitation of procreation between "Aryan" men and women, and the promotion of the racial policies of the *SS*.

Lehrerseminar: Training college for teachers.

Leitabschnitt: Regional HQ: of the *SD*, approximately coinciding with a *Wehrkreis*.

Legationsrat (Leg.Rat): Legation or embassy counselor in the Foreign Office.

leitend(er): governing; leading; executive.

Leitender Regierungsdirektor (LtRDir.): Title of an executive assistant in a high government office.

Leiter (Ltr.): chief; head; director; coordinator.

Leitstelle: Regional HQ: of the *Gestapo* or *Kripo*, established at the HQ of a *Wehrkreis* or capital of a Land.

Luftschutzpolizei (LSP): Air raid protection police established in 1942 from personnel of the *SHD* and *TeNo* as an element of the *Schutzpolizei*. Recruited primarily from reservists of the *Polizei*. *Marine Kusten Polizei* (MKP): Naval Coastal Police.

Mark (Mk.): Region; province; also type.

Maschinengewehr (MG.): Machine gun.

Militar; militarisch (Mil.; milit.): military.

Minister (Min.): Minister.

Ministerialdirektor (Min.Dir.): A senior civil service official. Department head in a ministry with the approximate equivalent rank of *Generalleutnant*.

Ministerialdirigent (Min.Dirig.): Executive official in a Ministry approximately equivalent to a *Generalmajor*.

Ministerialrat (Min.Rat): Senior ministerial counselor in civil service; usually head of a section within a ministry; approximately equivalent to an *Oberst*.

Ministerprasident (Min Pras.): Minister-President; Prime Minister of a Land government.

Mit (m.): with.

Mit der Fuhrung der Geschafte beauftragt (m.d.F.d.G.b.): Temporarily charged with the conduct of affairs.

Mit der Fuhrung beauftragt (m.d.F.b.): Temporarily charged with leadership or command.

Mit der vertretungsweisen Fuhrung beauftragt (m.d.vertretungsw.F.b.): Temporarily charged with the conduct of affairs.

Mit der Wahrnehmung der Geschafte beauftragt (m.d.W.d.G.b.): Temporarily charged with the conduct of affairs.

Mitglied (Mitgl.): Member.

Mitglied des Reichstages (M.d. R.): Member of the *Reichstag*.

Mit Patent vom (m.Pat.v.): A term relating to seniority for promotion (later known as *Rangdienstalter* [RDA]; see below). An example: If an officer was promoted *Oberstleutnant der Polizei* on 18.12.1934 mit RDA vom 01.10.1934, he would actually receive the promotion on 18.12.1934 but the date of calculation for salary, retirement pay, etc. would be 01.10.1934.

Mit Wirkung vom (m.W.v.): With effect from; effective date (e.g. of rank). For example, if an officer was promoted to *SS-Obergruppenfuhrer* on 08.04.1944 mit Wirkung vom 20.04.1944, he was not entitled to wear the uniform or receive the salary of that rank until 20.04.1944. Between 08.04.1944 and 20.04.1944, the officer in question would remain in the rank of *Gruppenfuhrer*. *motorisiert* (mot.): motorized.

Munchen: Munich. Capital city of the German state of Bayern (Bavaria).

Nachrichten (N; Na.; Nachr.): Signals.

Nationalsozialistische Betriebszellenorganisation (NSBO): National-Socialist Industrial Cell Organization. The Nazi factory workers' organization.

Nationalpolitische Erziehungsanstalten (NPEA; Napola): National Political Educational Institutes. Secondary schools organized along the lines of the *Hitler-Jugend* and under the jurisdiction of the *Dienststelle Heissmeyer*.

Nationalsozialistische Deutsche Arbeiter Partei (*NSDAP*): National-Socialist German Workers' Party (commonly known as the Nazi Party).

Nationalsozialistische Deutscher Arzten-Bund (NSDAB): National-Socialist German Association of Physicians.

Nationalsozialistische Deutscher Juristen-Bund (NSDJB): National-Socialist German Association of Jurists.

Nationalsozialistische-Fliegerkorps (NSFK): National-Socialist Flying Corps. The *NSDAP* flying enthusiasts' organization.

Nationalsozialistische Freiheitsbewegung (NSFB): National-Socialist Freedom Movement. Formed as a continuation of the *vφlkisch* movement during the ban on the *NSDAP* and other such organizations following the *Munchen-Putsch*.

Nationalsozialistische-Kraftfahrkorps (NSKK): National-Socialist Motor Corps. Played an important role in the military and paramilitary training. Initially known as the *Nationalsozialistische Automobilkorps* (*NSAK*).

Nationalsozialistische Volkswohlfahrt (NSV): National-Socialist People's Welfare Organization. Primarily responsible for the care of mothers and juveniles and headed by Erich Hilgenfeldt.

Nord (N): North.

Nordost (NO): Northeast.

Nordwest (NW): Northwest.

Nummer (Nr.): Number.

Ober (O.): higher; senior.

Oberbefehlshaber des Heeres (OB d.H.): Commander-in-Chief of the Army.

Oberburgermeister: Lord Mayor of a large town or city.

Oberkommando: Headquarters Staff.

Oberkommando des Heeres (OKH): High Command of the Army.

Oberkommando der Marine (OKM): German Navy High Command.

Oberkommando der Wehrmacht (OKW): High Command of the Armed Forces (Hitler was supreme commander, while *Generalfeldmarschall* Wilhelm Keitel headed *OKW*).

Oberprasident (OPras.; OPrs.): Chief administrator and government executive of a Prussian Province.

Oberrealschule: Higher modern school (modern languages, mathematics, science; nine years' course).

Oberregierungsrat (ORR): Senior government counselor of the civil service; approximately equivalent to a Lieutenant-Colonel.

Oberschlesien (O.S.; OS.): Prussian Province of Upper Silesia.

Oberste SA-Fuhrer (OSAF): Supreme *SA* Leader (title held by Adolf Hitler from 1930).

Oberstes Parteigericht: Supreme *NSDAP* Court under *Reichsleiter* Walter Buch.

Offizier (Offz.; Off.): Officer.

Offiziers-Aspirant: Probationary or aspirant officer.

Offizierstellvertreter: A German Army rank created during World War I, used to denote an NCO acting as an officer.

Ordnungspolizei (Orpo): Order Police. Regular uniformed police force of the Reich, consisting of the *Schutzpolizei, Gendarmerie,* and *Feuerschutzpolizei* as well as various technical and auxiliary services (such as the *Technische Nothilfe* [*TeNo*]).

Ordonnanz-Offizier (Ord.Offz.): Orderly officer.

Organisation Todt (Org. Todt; OT): Semi-military government construction force, established 1933 under *SA-Standartenführer* (later *SA-Obergruppenführer*) *Dr.-Ing.* Fritz Todt, responsible for building strategic highways and military installations/fortifications. Controlled by Todt's successor, Albert Speer, after Todt's death in February 1942.

Ortsgruppenleiter: Local Group Leader. *NSDAP* official in charge of one or more parts of a town; subordinate to a *Kreisleiter.*

Osten (O): East.

Ostindustrie GmbH (Osti): Eastern Industries Ltd. *SS* economic concern formed in March 1943 with the primary purpose of exploiting Jewish labor in Poland.

Ostpreußen (Ostpr.): Prussian Province of East Prussia.

Panzer (Pz.): Armor; tank.

Parteikanzlei (PK): Hitler's chancellery as leader of the *NSDAP*. Headed by *Reichsleiter* Martin Bormann.

Personalhauptamt (SS-PHA): Main personnel office of the SS, responsible for records of all *SS* officers.

Politische Bereitschaft: Political alarm squad. Forerunner units of the *SS-Verfugungstruppe.*

Politische Organisation (PO): The *NSDAP*'s political organization (as separate from its paramilitary branches, such as the *SS* and *SA*).

Polizeidirektor (Pol.Dir.): Police Director. Head of the regular police in a small city.

Polizeiprasident (Pol.Pras.): Police President. Head of the regular police in a large city.

Polizeiverwaltung (PV): Police administration.

Polizeiverwaltungsgesetz (PVG): Police Administration Law.

Praparandenschule: Training college for elementary school teachers.

Preu.en / preu.isch: Prussia / Prussian.

Preu.ischen Landtages: Prussian State Parliament.

Rangdienstalter (RDA): A term relating to seniority for promotion (previously known as "*Patent*" [see "mit Patent vom" above for definition]).

Rasse- und Siedlungshauptamt (RuSHA): *SS* Race & Settlement Main Office. Responsible for racial purity of the *SS* and settlement of *SS* colonists in occupied territories.

Realschule: A non-classical secondary school (6 years' course), less prestigious than the *Gymnasium*, with the purpose of preparing students for employment in practical or technical fields.

Redner: Speaker; orator.

Referat: A sub-section or "desk" within an *Amtsgruppe* .

Referent: Official in charge of a *Referat.*

Regierungsbezirk (Reg.Bez.): Sub-division of a Prussian province. Also, an administrative district of Bavaria.

Regierungsdirektor (Reg.Dir.): Rank of administrative official in a regional government, approximately equal with that of *Oberst.*

Regierungsprasident (Reg.Pras.): Senior government official in a *Regierungsbezirk.*

Regierungsrat (Reg.Rat; RR): Government counselor. Lowest rank in the higher civil service; approximately equivalent to the rank of *Major.*

Reichsarbeitsdienst (RAD): *Reich* Labor Service. A national compulsory labor service, organized along paramilitary lines (superseded the earlier *Freiwilligen-Arbeitsdienst* [*FAD*]).

Reichsarbeitsministerium (RAM): *Reich* Labor Ministry.

Reichsbevollmachtigter: *Reich* Plenipotentiary controlling the civil affairs of an occupied country (e.g. *SS-Obergruppenfuhrer Dr. jur.* Karl Werner Best in Denmark).

Reichsfinanzministerium (RFM): *Reich* Ministry of Finance.

Reichsfuhrer-SS (RF SS): *Reich* Leader of the *SS*. The following men held this title:
> Joseph Berchtold: 1926-1927
> Erhard Heiden: 1927-1929
> Heinrich Himmler: 1929-28.04.1945
> Karl Hanke: 30.04.1945-08.05.1945

Reichsfuhrer-SS und Chef der Deutschen Polizei (RF SS u. Chef d. Dtsch.Pol.): "Reich *SS* Leader and Chief of the German Police". Heinrich Himmler's title from 17.06.1936 to 28.04.1945. He was succeeded by Karl Hanke.

Reichsfuhrung-SS: Supreme command of the SS, comprising the *Persοnlicher Stab RfSS* and the *Hauptamter*.

Reichsgau: One of eleven regions formed from territories annexed from Austria, Czechoslovakia, and Poland in 1938 and 1939.

Reichsgesetzblatt (RGB): Official legal gazette issued in two parts by the *Reichsministerium des Innern*. Part 1 concerned current legislation; Part 2 dealt with international treaties, etc.

Reichskommissar fur die Festigung Deutschen Volkstums (RKFDV): *Reich* Commissioner for the Consolidation of Germandom. Title given to Heinrich Himmler in October 1939 when Hitler entrusted him with the repatriation of *Volksdeutsche* ("racial Germans") and the settlement of German colonies in the East. A *Stabshauptamt RKFDV* was established under Ulrich Greifelt to put Himmler's plans for Germanic mastery into effect.

Reichskommissariat (RK): Title for the German civil administration in various occupied territories, headed by a *Reichskommissar*. The occupied Eastern Territories were divided into *Reichskommissariat Ostland* (comprising the Baltic States [Latvia, Estonia, Lithuania] and White Russia), under *Reichskommissar* Hinrich Lohse while *Reichskommissariat Ukraine*, under *Reichskommissar* Erich Koch, controlled the Ukraine and portions of the Crimea and Caucasus. These *Reichskommissariate* were broken down into *Generalbezirke* [general districts] under the control of *Generalkommissare*. Subordinate to the *Generalbezirke* were *Kreisgebiete* [district regions] headed by *Gebietskommissare*).

Reichskriminalpolizeiamt (RKPA): The headquarters of the *Kriminalpolizei* (Amt V of the *RSHA*).

Reichsleiter (RL): The highest rank in the *NSDAP* hierarchy.

Reichsluftschutzbund (RLB): *Reich* Air-Raid Protection League.

Reichsministerium fur die besetzten Ostgebiete (RMBO; Ostministerium; Ostmin.): *Reich* Ministry for the Occupied Eastern Territories, run by Reichsminister Alfred Rosenberg and his deputy, *Gauleiter* Dr. Alfred Meyer from 1941 to 1945.

Reichsministerium des Innern (RmdI): *Reich* Ministry of the Interior (Dr. Wilhelm Frick, 1933-1943; Heinrich Himmler, 1943-1945).

Reichsnahrstand (RNS): *Reich* Food Estate. Established in 1933 by Richard Walter Darré to control agricultural production.

Reichsparteitag der NSDAP (RPT): *Reich* (National) Party Day of the *NSDAP*. The annual Nazi Party Congress. On 30.08.1933, Hitler designated the city of *Nürnberg* as the "*Stadt der Reichsparteitage*". A total of ten RPT's were held as follows (the 11th, very ironically named the *Reichsparteitag des Friedens (Reich* Party Day of Peace), was cancelled due to the outbreak of war on 01.09.1939).

> Dates / City / Title
> 27.-29.01.1923 / Munchen
> 03.-04.07.1926 / Weimar
> 19.-21.08.1927 / Nürnberg
> 01.-04.08.1929 / Nürnberg
> 30.08-03.09.1933 / Nürnberg / *Reichsparteitag des Sieges* [Victory]
> 04.-10.09.1934 / Nürnberg / *Reichsparteitag der Einheit und Starke* [Unity and Strength]
> 10.-16.09.1935 / Nürnberg *Reichsparteitag der Freiheit* [Freedom]
> 08.-14.09.1936 / Nürnberg / *Reichsparteitag der Ehre* [Honor]
> 06.-13.09.1937 / Nürnberg / *Reichsparteitag der Arbeit* [Labor]
> 05.-12.09.1938/ Nürnberg / *Reichsparteitag Großdeutschland* [Greater Germany]
> [To start 02.09.1939] Nürnberg / *Reichsparteitag des Friedens* [Peace]

Reichsrat: One of two legislative organs under the Weimar Republic in Germany. The *Reichsrat* was the representative body for the various German *Lander* (states), superseding the former *Bundesrat* in 1919, while the Reichstag dealt with the nation as a whole.

Reichssicherheitshauptamt (RSHA): *Reich* Security Main Office.

Reichsstrafgesetzbuch (RStGB): *Reich* Penal Code.

Reichstag: The German National Parliament. Largely a figurehead body after its legislative powers were taken away and granted to the *Reich* Government by way of Hitler's "Enabling Act" of 24.03.1933.

Reichsstatthalter (Rsth.): *Reich* Governor; Hitler's representative in a German Land or *Reichsgau*.

Reichsverteidigungskommissar (RVK): *Reich* Defense Commissioner. Official in charge of a *Reichsverteidigungsbezirk (Reich* Defense Region). These regions originally corresponded to the *Wehrkreise*, but following a decree of 16.11.1942 they were made identical to the 42 *NSDAP Gaue* in Germany. From that date onward, each *Gauleiter* also held the post of *Reichsverteidigungskommissar*.

Reichswehr (RW): *Reich* Defense. The armed forces of Germany from 06.03.1919 until the enactment of the *Wehrgesetz* (Defense Law) of 21.05.1935; on that date, its title was changed to *Wehrmacht*.

Reichswirtschaftsministerium (RWM): *Reich* Ministry of Economics.

Reifeprufung: School-leaving examination (also known as *Abitur*).

SA-Aufmarsch in Braunschweig ("*SA-Treffens* Braunschweig 1931"): A rally of over 100,000 *SA* and *SS* members held in Braunschweig on 17./18. October 1931. It was hosted by *SA-Gruppe Nord* (under then-*SA-Gruppenfuhrer* Viktor Lutze). John R. Angolia writes of the rally in *For Fuhrer and Fatherland, Volume II*: "It was at this assembly, which followed closely on the heels of the "Stennes Putsch", that Hitler gained the assurance of the *SA* rank and file and at which Lutze gained a reputation as a totally loyal Party member... It was at this meeting also that Hitler authorized the creation of 24 new *Standarten*, thus expanding the *SA*, and recognized the *Motor-SA* and *NSKK*. All Party members who had officially attended the rally were authorized to wear the [Abzeichen des SA-Treffens Braunschweig 1931 {see Appendix III – Awards Glossary}] on their left breast."

Sachbearbeiter: Officer or official responsible for a specific matter.

Sanitatsdienst: Medical service.

Schar: An *SS* or *SA* unit consisting of 3 to 13 men.

Schutzmannschaften (Schuma): Auxiliary police units composed of foreign elements and Volksdeutsche; the first *Schuma* unit was set up in the Ukraine in August 1941

Schutzpolizei (Schupo; Schp.): Protection Police. The regular uniformed municipal and country police forces, comprising most of the membership of the *Ordnungspolizei*.

Schutzstaffel (SS): lit. Protection or Guard Detachment. Officially established in 1925, the *SS* became, under Heinrich Himmler, the most powerful organization in the Third Reich.

Selbstschutz: i. A German nationalist self-protection organization formed in Silesia in 1920; ii. A self protection militia recruited from the *Volksdeutsche* in Poland by the *SS*; iii. Self-Protection Service, an element of the *Luftschutzdienst* composed of air raid wardens and other nonmilitary air raid protection personnel.

Sicherheitsdienst des RfSS (SD): The *SS* Security Service established by Reinhard Heydrich in 1931 as the intelligence organization of the Nazi Party.

Sicherheitspolizei (Sipo): Security Police, composed of the *Gestapo* and *Kripo*.

Sicherheits- und Hilfsdienst (SHD): Security and Assistance Service. An auxiliary police unit responsible for air raid-related tasks. Superseded by the *Luftschutzpolizei*, 1942.

Sigrunen: The runic double "S" insignia of the *Schutzstaffel*.

Sonderkommando (Skdo.): Special commando of the *Sipo* or *SD*.

SS- und Polizeifuhrer (SS-u.Pol.F.; SSPF): *SS* and Police Commander in the occupied territories, subordinate to an HSSPF.

Staat: State, country.

Staatssekretar (Sta.Sek.): State Secretary in a *Reich* or Land Government Ministry.

Stab (St.): Staff.

Stabschef: Chief of Staff of the *SA*.

Stabsfuhrer (Stabsf.): Chief of Staff (e.g., to the *Fuhrer* of an *SS-Oberabschnitt*).

Stadthauptmann: Senior administrative official in a *Stadthauptmannschaft*, a subdivision of a district in the *Generalgouvernement* (occupied Poland).

Stadtrat: Town-council; town-councillor, alderman.

Stadtverordneter: Town-councillor.

Stahlhelm: "Steel Helmet", also known as the *Bund der Frontkampfer* (League of Front Soldiers). The most prominent German veterans' organization, established by Franz Seldte on 25.12.1918.

Standortfuhrer: Garrison commander.

Stellvertreter (Stellv.): Deputy.

Sturmabteilungen (SA): Storm Troops. The original defense formations of the *NSDAP*, founded in 1921. Purged in June/July 1934 when it became too radical and unwieldy for Hitler's tastes, prompting him to wipe out numerous members of its leadership, including *SA-Stabschef* Ernst Rohm. The purge of the *SA* was carried out by Himmler's SS, initially a subunit of the *SA*, and in its wake the *SS* gained considerable power.

Sturm: A company-sized unit of the *SS, SA*, or *NSKK*, consisting of several *Truppen*.

Sturmbann: A battalion-sized unit of the *SS, SA*, or *NSKK* consisting of several *Stürme*.

Technische Nothilfe (TeNo; TN): Technical Emergency Corps, established in 1919. Auxiliary force of the *Ordnungspolizei* consisting of engineers, technicians, and specialists involved with construction work, communications, salvage, public utilities, etc.

Teilkommando: A sub-unit; Smallest element of an *Einsatzgruppe* of the *Sipo* and *SD*.

Totenkopfverbande (SS-TV): Death's Head units, employed in the concentration camps as guards. Formed the nucleus of the *SS-Totenkopf-Division* when it was formed in October 1939.

Trupp (Tr.): Squad or detail. An *SA* (and in its early years, SS) unit equivalent to a platoon, composed of several *Scharen* .

Truppenubungsplatz: Troop training area.

Unterfuhrer: Non-commissioned officer of the *SS, SA,* or *NSKK.*

Unternehmen: Operation; undertaking; enterprise.

Verband (Verb.): A formation or unit.

Verbindungsoffizier (Verb.Offz.) or *Verbindungsfuhrer* (Verb.Fhr.): Liaison officer.

Vertrauensmann (V-Mann): Intelligence agent or informer.

Verwaltung (Verw.): Administration.

Verwaltungspolizei: Administrative branch of the *Ordnungspolizei* and *Sicherheitspolizei.*

Volkisch: "An adjective denoting an ultranationalist, antidemocratic populism that claimed to represent a kind of integral Germanness. Xenophobic in general, the *völkisch* movement was particularly identified with a virulent anti-Semitism. It was strongly influenced by Social Darwinism." (definition from Prof. Bruce Campbell's *The SA Generals and the Rise of Nazism*).

Verein [later redesignated Volksbund] fur das Deutschtum im Ausland (VDA): The League for Germans Abroad. Pre-Nazi organization concerned with activities of the *Volksdeutsche* (ethnic Germans). Taken over by the *NSDAP*, 1930 and eventually absorbed by the *Volksdeutsche Mittelstelle* under *SS-Obergruppenfuhrer* Werner Lorenz.

Volksdeutsche: Ethnic German.

Volksdeutsche Mittelstelle (VoMi): Ethnic German Assistance Office. Established as the *Büro von Kursell*, 1936 and renamed in 1937. Led by *SS-Obergruppenführer* Werner Lorenz, it obtained the status of a *Hauptamt* of the *SS* in 1941.

Volksgruppe: Ethnic group.

Volkstum: Nationality.

Vorsitzender: Chairman.

Vorsitzender des Aufsichtsrates: The Chairman of a Supervisory Board (in business and industry).

Vorstand: Board (or directors/management); Governing body.

Waffen-SS (W-SS): The fully militarized combat formations of the *SS.*

Wehrkreis (Wkr.): Military District.

Wehrmacht (WH; Wehrm.): The German armed forces, consisting of the German Army (*Heer*), Navy (*Kriegsmarine*), and Air Force (*Luftwaffe*).

Wehrmachtbefelshaber: Senior Armed Forces commander in an occupied territory.

Wehrmachtfuhrungsstab: Armed Forces Operations Staff, headed by *Generaloberst* Alfred Jodl.

Wehrwirtschaft: Military or war economy.

Wehrwirtschaftsamt: War Economics Directorate of *OKW.*

Wien: Vienna. The capital city of Austria.

Zur besonderer Verwendung (z.b.V.): For special employment.

Zellenleiter: Cell leader. An *NSDAP* official responsible for four to five blocks of households; subordinate to an *Ortsgruppenleiter.*

Zollgrenzschutz: Border Customs Protection Service. Controlled by the *Sicherheitspolizei*, with personnel from the Customs Service.

Zug: Platoon.

Zugfuhrer (Zugfhr.): Platoon leader.

Glossary of German Military, Political, and Civil Decorations and Awards.

Abzeichen des SA-Treffens Braunschweig 1931: Badge of the *SA* Meeting at Braunschweig 1931. Instituted 1931 and officially recognized as a Party honor award on 06.11.1936. Issued to commemorate the rally of 100,000+ members of the *SA* and *SS* at Braunschweig on 17/18.10.1931.

Ärmelband "Afrika": "Afrika" cuff title. Instituted by order of Army General Staff, 15.01.1943. Equivalent of a campaign medal. .

Allgemeines-Sturmabzeichen (Allg.St.Abz.): General Assault Badge.

Bandenkampfabzeichen (BKA): Anti-Partisan War Badge. Instituted by *Reichsführer-SS* Heinrich Himmler on 30.01.1944. Issued in three classes (Bronze, Silver, and Gold).

Coburger Abzeichen 1922: Coburg Badge. Instituted by Hitler on 14.10.1932 to commemorate participation in the patriotic rally known as the "3.Deutscher Tag in Coburg" of 14/15.10.1922 (to which Hitler and other Nazis, including 700 or 800 *SA* men, had been invited); the rally was occasioned by many street skirmishes between Nazis and communists, Hitler's followers emerging victorious (436 men were registered as recipients of this extremely rare award, which was declared an official *NSDAP* and Reich decoration by *Führer* decree on 06.11.1936).

Danzig Kreuz (aka ***Kreuz von Danzig***): Danzig Cross. Instituted 31.08.1939 to recognize meritorious service to the *NSDAP* in Danzig. Awarded in two classes.

Deutsche Olympia-Ehrenzeichen: German Olympic Games Decoration. Instituted by Hitler on 04.02.1936 in two classes (*I.* and *II. Klasse*) to recognize contributions toward the preparation and execution of the XIth Summer and IVth Winter Olympic Games (in Berlin and Garmisch-Partenkirchen, respectively).

Deutscher Orden: German Order. Instituted by Hitler on 11.02.1942 as the highest decoration of the Nazi Party. The first award was presented posthumously to *Reichsminister / SA-Obergruppenführer Dr.-Ing.* Fritz Todt after his death in an air crash on 06.02.1942. .

Deutsches Kreuz (DK): Shortened form of ***Kriegsorden des Deutschen Kreuzes*** (War Order of the German Cross). Instituted by Adolf Hitler on 28.09.1941. Issued in Gold (DKiG; for combat bravery or leadership) and Silver (*DKiS*) for meritorious service in furtherance of the war effort. .

Deutsches Reichssportabzeichen: German National Sport Badge. Instituted in 1913 by the *Deutscher Reichsbund für Leibesübungen* (German National Physical Training Union). Initially available only in Bronze and Gold, a silver edition was introduced in 1920. Award was based on various tests of physical skill, endurance, and completion time of athletic events.

Deutsches Reitersportabzeichen: German Horseman's Sports Badge. Instituted by the German National Federation for the Breeding and Testing of Thoroughbreds, 09.04.1930. Awarded in three classes: 3rd class (bronze) for achievement in horse racing; 2nd class (silver)awarded for achievement in equestrian shows and tournaments; 3rd class (gold) for outstanding accomplishments in equestrian sports.

Deutsches Schutzwall-Ehrenzeichen: German Defense Wall Honor Award. Instituted 02.08.1939 to recognize planning and labor (between 15.06.1938 and 31.03.1939) which led to completion of the "Siegfried Line" defenses.

Dienstauszeichnung der NSDAP (***D.A. D. NSDAP***): Long Service Awards of the *NSDAP*. Instituted by Hitler on 2.04.1939 in three classes: Bronze (for ten years service in the *NSDAP*); Silver (for 15 years service in the *NSDAP*); and Gold (for 25 years service in the *NSDAP*). The *Kampfzeit* ("Time of Struggle") years 1925 to 1933 counted as double toward service time accumulation).

Ehrenblatt-Spange des Heeres (***EBSdH***): Honor Roll Clasp of the Army. Instituted by order of Adolf Hitler on 30.01.1944. Awarded to those who already held the Iron Cross I. and II. Class and again distinguished themselves in action.

Ehrendegen des Reichsführers-SS: Sword of Honor of the *Reichsführer-SS*.

Ehrendolch der SS: Honor Dagger of the SS.

Ehrenkreuz des Weltkrieges 1914-1918: Honor Cross of the World War 1914-1918. Instituted by Reich President *Generalfeldmarschall* Paul von Hindenburg on 13.07.1934. Issued in three forms:.

  1) für Frontkämpfer (for Combatants); with Swords.

  2) für andere Kriegsteilnehmer (for non-combatants).

  3) für Witwen und Eltern (for widows and parents).

Ehrenkreuz des Weltkrieges 1914-1918 mit Schwertern: (see *Ehrenkreuz 1914-1918* above).

Ehrenplakette für die Mitglieder des Reichs-Kultur-Senats: Honor Badge for Members of the National Senate of Culture. First awarded on 28.11.1936 (to all 125 members of the *Reichskultursenat*).

Ehrenwinkel für alte Kämpfer: Honor Chevron for Old Fighters. Issued for wear by all persons who had entered the *NSDAP*, the SS, *SA, NSKK*, or any other Party affiliated organizations prior to Hitler's assumption of power on 30.01.1933. Another version, the *Ehrenwinkel mit Stern* (Honor Chevron with Star), was instituted on 25.07.1935 to recognize *SS* members who had formerly served in the *Wehrmacht* and *Polizei*.

Ehrenzeichen des 9. November 1923 (***Blutorden***): Honor Decoration of 9. November 1923 (also known as the Blood Order). Instituted 15.03.1934 by Hitler to honor participants in the Munich Putsch of 09.11.1923. The conditions were expanded on 30.05.1938, with the inclusion of non-Putsch participants who had rendered outstanding service to the Party during the "Time of Struggle." These included:.

  1. Receipt of a death sentence (later commuted to life imprisonment);.

  2. Imprisonment for one or more years for Nazi political activities;.

  3. Suffering severe wounds in the service of the *NSDAP*.

Ehrenzeichen des Deutschen Roten Kreuzes: German Red Cross Decoration. Originally instituted in 1922. Redesigned with Third Reich motifs in 1934. Issued in a number of classes. Superseded by *Führer* order of 01.05.1939 with the introduction of the *Ehrenzeichen für Deutsche Volkspflege* (German Social Welfare Decoration), as Hitler believed awards should cover all facets of social welfare, rather than the limited area of the Red Cross) .

Ehrenzeichen für Deutsche Volkspflege: German Social Welfare Decoration. Instituted 01.05.1939 to supersede the *Ehrenzeichen des Deutschen Roten Kreuzes* (see above).

Ehrenzeichen "Pionier der Arbeit": Pioneer of Labor Decoration. Instituted by Hitler, 07.08.1940 to recognize exceptional achievement in industry and society.

Eisernen Kreuzes (***EK***): Iron Cross. The basic German award for bravery. Originally instituted in 1813, reinstituted again in 1870, 1914, and finally on 01.09.1939 by Adolf Hitler. Awarded in two classes (*Eisernes Kreuz II. Klasse / EK II* and *Eisernes Kreuz I. Klasse / EK I*) for bravery in combat. Open to all members of the *Wehrmacht, Waffen-SS*, personnel in organizations providing direct support to the German military, and personnel of Germany's Axis allies and foreign volunteer units. In cases where an individual had received one or both classes of the Iron Cross in the First World War, he would receive the *1939 Spange zum 1914 Eisernes Kreuz I. or II. Klasse* (1939 Clasp to the 1914 Iron Cross). (See also *des Eisernes Kreuzes*.)

Flugzeugführerabzeichen: *Luftwaffe* Pilot's Badge. Instituted by Hermann Göring on 12.08.1935. Awarded to all active military personnel who had qualified for a military pilot's license.

Frontbannabzeichen: Frontbann Badge. Instituted 1932 by *SA-Gruppe Berlin-Brandenburg* (under Kurt Daluege). Officially recognized as an honor badge of the *NSDAP* in 1933. Authorization for wear of this badge was discontinued in 1934.

Frontflug-Spange: Operational Flight Clasp (*Luftwaffe*). Instituted by Hermann Göring on 30.01.1941, initially in three design pattes (for Fighters, Bombers, and Reconnaissance), later expanded and broadened to include *Tagjäger* (Day Fighters), *Kampf- und Sturzkampfflieger* (Heavy, Medium, and Dive Bombers), and a number of other categories of air units. Issued in Bronze for 20 operational missions, silver for 60, and gold for 110.

Gau-Ehrenzeichen des Gaues Sudetenland der NSDAP: *Gau* Sudetenland Commemorative Badge. Instituted by *Gauleiter* Konrad Henlein on 25.12.1943.

Gau-Traditionsabzeichen Berlin: *Gau* Berlin Commemorative Badge. Instituted by *Gauleiter* Dr. Goebbels on 29.10.1936, awarded in silver and gold classes.

Gau-Traditionsabzeichen des Gaues Danzig-Westpreußen: *Gau* Danzig-West Prussia Commemorative Badge. Instituted ca. 05.1939 by *Gauleiter* Albert Forster.

Gau Wartheland-Traditionsabzeichen: *Gau* Wartheland Commemorative Badge. Instituted by *Gauleiter* Arthur Greiser, probably early in 1940, to commemorate the establishment of the *Warthegau* on 26.10.1939.

Gau-Traditionsabzeichen Essen: *Gau* Essen Commemorative Badge. Instituted by *Gauleiter* Josef Terboven, 1935 in commemoration of the 10th anniversary of the establishment of *NSDAP Gau* Essen.

Goldenes Ehrenzeichen der NSDAP (aka ***Goldenes Parteiabzeichen***): Golden Badge of Honor of the *NSDAP*. A *Führer* decree of 13.10.1933 stated that all *NSDAP* members with uninterrupted service since 27.02.1925, and with *NSDAP*-Nr.'s 1 to 100,000, were to receive the badge on 09.11.1933. In later years, the award was presented to certain individuals who, though not meeting the original criteria as to *NSDAP* membership number or entrance date, had made significant contributions to Party and State. Such honorary awards were bestowed upon *Gauleiter / SS-Obergruppenführer* Konrad Henlein (*NSDAP*-Nr. 6 600 001), *Generalfeldmarschall* Werner von Blomberg (a non-*NSDAP* member honored for his contributions to Germany's military rebirth), and numerous others. 1,795 women received the award.

Goldenes Hitler-Jugend Ehrenzeichen: Golden Hitler Youth Honor Badge. Instituted by *Reichsjugendführer* Baldur von Schirach on 23.06.1934.

Goldenes Hitler-Jugend Ehrenzeichen mit Eichenlaub: Golden Hitler Youth Honor Badge with Oakleaves. Instituted by Reichsjugendführer Baldur von Schirach in 1935. Awarded to members of the Hitler Youth, as well as various non-members (such as *Reichsführer-SS* Heinrich Himmler) who had contributed to the growth and advancement of the Nazi youth movement.

Infanterie-Sturmabzeichen (*Inf.St.Abz.*): Infantry Assault Badge. Instituted on 20.12.1939. Initially awarded only in Silver (for infantry and mountain infantry), a bronze edition was made available on 01.06.1940 to recognize troops of the motorized infantry.

Julleuchter der SS: Yule candleholder of the *SS*. Earthenware candleholders used to adorn the Yuletide tables of *SS* families, manufactured by the Allach Porcelain company at Dachau.

Kriegsorden des Deutschen Kreuzes: Official title of the Deutsches Kreuz (see above).

Kriegsverdienstkreuz (*KVK*): War Merit Cross. Instituted by *Führer* order on 18.10.1939 to recognize meritorious service in furtherance of the war effort of a non-combatant nature. Awarded "mit Schwertern" (with swords) for bravery and "ohne Schwerter" (without swords) for service. Issued in the following classes:

Ritterkreuz des Kriegsverdienstkreuzes (Knight's Cross of the War Merit Cross [*mit Schwertern* or *ohne Schwerter]*) (118 awarded with swords and 137 without swords): Two awards in gold (both without swords) were presented on 20.04.1945.

Kriegsverdienstkreuz I. Klasse (*mit Schwertern* or *ohne Schwerter*).

Kriegsverdienstkreuz II. Klasse (*mit Schwertern* or *ohne Schwerter*).

Luftschutz-Ehrenzeichen: Air Raid Protection Honor Badge (*1.* and *2. Stufe*). Instituted 30.01.1938 for honorable service in the *Reichsluftschutzbund* (RLB / Reich Air Raid Protection League), the *Werkschutz* (Factory Police), the *Ordnungspolizei*, the *Feuerschutzpolizei*, and the *Technische Nothilfe* (Technical Emergency Corps); issued in two classes (*I.* and *II. Stufe*).

Medaille zur Erinnerung an den 13. März 1938: Commemorative Medal of 13 March 1938. Instituted by Hitler on 01.05.1938 to recognize all military, political, and civil service personnel involved in the annexation of Austria.

Medaille zur Erinnerung an den 1. Oktober 1938: Commemorative Medal of 1 October 1938. Instituted by Hitler, 18.10.1938 to recognize persons who had participated in the occupation of the Sudetenland region of Czechoslovakia (criteria for award expanded to include those involved with the creation of the Reich Protectorate of Bohemia & Moravia, 01.05.1939. For those involved in the occupation of the remainder of Czechoslovakia, Hitler instituted the Spange "Prager Burg" (Prague Castle Bar), a small metal device affixed to the ribbon of the medal.

Medaille zur Erinnerung an die Heimkehr des Memellandes: Commemorative Medal of the Return of the Memel District. Instituted 01.05.1939 by Hitler to recognize participation of military, political, and civil service personnel in the annexation of the Memel (Klaipeda) region of Lithuania.

Medaille "Winterschlacht im Osten 1941/42" (aka *Ostmedaille*): Medal for the Winter Campaign in Russia 1941/1942. Instituted by Hitler, 26.05.1942 to recognize members of the *Wehrmacht*, *Waffen-SS*, and *Polizei* who had endured the first winter campaign against the Soviet Union.

Nahkampfspange (*NKS*): Close Combat Bar. Instituted by Hitler on 25.11.1942 to recognize participation in hand-to-hand combat unsupported by armor. Issued in three grades – *I. Stufe* (Bronze); *II. Stufe* (Silver); and *III. Stufe* (Gold).

Nürnberger Parteitagsabzeichen 1929: Nürnberg Party Day Badge of 1929. Instituted 15.08.1929 to commemorate the 4th Party Rally which convened at Nürnberg from 01.08.1929-04.08.1929. Officially classified a Party honor badge by *Führer* decree of 06.11.1936.

Panzerkampfabzeichen (*Pz.K.Abz.*): Tank Battle Badge. Instituted on 20.12.1939 for award to all members of German Army tank crews. Initially awarded only in silver, but from 01.06.1940 a bronze edition was available for *Panzer-Grenadiers*, armored car crewmen, and crewmembers of self-propelled assault guns.

Polizei-Dienstauszeichnungen: Police Long Service Awards. Instituted by *Führer* order of 30.01.1938, initially in 3 classes (*3. Stufe*: 8 years' loyal service; *2. Stufe*: 18 years'; *1. Stufe*: 25 years' service). A higher class was authorized on 12.08.1944 to recognize 40 years' loyal service; this was indicated by a gold metal device with the number 40 in an oakleaf pattern affixed to the first class awards ribbon.

[Orden] Pour le Mérite (***PlM***): Order for Merit (popularly known as the "Blue Max"). Highest military award of Imperial Germany before and during the First World War.

Ritterkreuz des Eisernes Kreuzes (***RK d. EK***): Knight's Cross of the Iron Cross. Instituted 01.09.1939 by Adolf Hitler for bravery or leadership in combat (approximately 7,300 awards of the RK d. E.K. were rendered between 1939 and 1945). A further award, the *Ritterkreuz des Eisernes Kreuzes mit Eichenlaub*, was instituted on 03.06.1940. The *Ritterkreuz des Eisernes Kreuzes mit Eichenlaub und Schwertern* (Oakleaves & Swords) was instituted on 21.06.1941. To recognize those who performed further outstanding acts of heroism or decisive leadership, the *Ritterkreuz des Eisernes Kreuz mit Eichenlaub, Schwertern und Brillanten* (Oakleaves, Swords, & Diamonds), was instituted on 15.07.1941. Finally, a special award was created specifically for *Luftwaffe* Stuka ace *Oberst* Hans Ulrich Rudel- this was the *Ritterkreuz des Eisernes Kreuz mit Goldenem Eichenlaub mit Schwertern und Brillanten* (Golden Oakleaves, Swords, and Diamonds).

Ritterkreuz des Kriegsverdienstkreuzes (***RK d. KVK***): Knight's Cross of the War Merit Cross (see *Kriegsverdienstkreuz* above for details).

SA-Sportabzeichen (***SA-Sp.Abz.***): *SA*-Sport Badge. Instituted by *SA-Stabschef* Ernst Röhm on 28.11.1933. Initially issued only in bronze; *Führer* decree of 15.02.1935 introduced a silver and a gold version of the award.

SS-Dienstauszeichnungen (***SS-D.A.***): SS Long Service Awards. Instituted 30.01.1938 in four classes (*4. Stufe*: 4 years' loyal service; *3. Stufe*: 8 years' loyal service; *2. Stufe*: 12 years' loyal service; *1. Stufe*: 25 years' loyal service (the *Kampfzeit* ["time of struggle"] years 1925 to 1933 counted as double toward service time accumulation).

SA or SS-Zivilabzeichen (***SA/SS-Z.A.***): *SS* Civil Badge. A small circular badge bearing the *SS* runes worn on civilian clothes by members of the organization.

Totenkopfring der SS: Death's Head Ring of the *SS*. Instituted by Heinrich Himmler on 10.04.1934 to recognize members of the *SS* who had exhibited noteworthy achievement, devotion to duty, and loyalty to the *SS* and Third Reich. Approximately 14,000 had been awarded by 17.10.1944, when Himmler halted further production and presentation of the rings.

Traditions-Gauabzeichen für Thüringen: *Gau* Thüringen Commemorative Badge. Instituted by *Gauleiter* Fritz Sauckel, 00.06.1933.

Traditions-Gau-Abzeichen: *Gau* Commemorative Badges. Instituted 1933 for loyalty and meritorious service during the "Kampfzeit" (Time of Struggle of the Nazi movement).

Treudienst Ehrenzeichen: Faithful Service Decorations. Instituted by Hitler on 30.01.1938 to recognize loyal civilian service (issued in 3 classes: *2. Stufe* in Silver (for 25 years' service); *1. Stufe* in Gold for 40 years' service; and a *Sonderstufe* (special grade) for 50 years' service.

Verwundetenabzeichen (***Verw.Abz.***): Wound Badge. First instituted on 03.03.1918 by Kaiser Wilhelm II; reintroduced by Adolf Hitler on 01.09.1939; issued in three classes: *Schwarz* (black)1 to 2 wounds; *Silber* (silver): 3 to 4 wounds; Gold: 5 or more wounds. In event of loss of eyesight or limb, the silver grade was awarded automatically. The gold award was granted in cases of death or total physical disability.

Wehrmacht-Dienstauszeichnungen (***WH-D.A.***): Armed Forces Long Service Awards. Instituted by Adolf Hitler on 16.03.1936 in four classes (for 4, 12, 18, and 25 years' service. A 40-year award was introduced on 10.03.1939).